Laser Interaction

and Related Plasma Phenomena

Volume 3B

A Continuation Order Plan is available for this series. A continuation order will bring delivery of each new volume immediately upon publication. Volumes are billed only upon actual shipment. For further information please contact the publisher.

Laser Interaction
and Related Plasma Phenomena

Volume 3B

Edited by
Helmut J. Schwarz
Professor of Physics
Rensselaer Polytechnic Institute

and
Heinrich Hora
Adjunct Associate Professor of Physics
Rensselaer Polytechnic Institute
and
Max-Planck-Institut für Plasmaphysik
Garching, Germany

PLENUM PRESS • NEW YORK AND LONDON

Library of Congress Cataloging in Publication Data

Main entry under title:

Laser interaction and related plasma phenomena.

Proceedings of the 1st- workshops held at Rensselaer Polytechnic Institute, Hartford Graduate Center, Hartford, Conn., 1969-

Includes bibliographies.

1. High temperature plasmas—Congresses. 2. Controlled fusion—Congresses. 3. Lasers—Congresses. I. Schwarz, Helmut J., 1915- ed. II. Hora, Heinrich, ed. III. Rensselaer Polytechnic Institute, Troy, N.Y.

QC718.5.H5L37 530.4'4 79-135851

ISBN 978-1-4684-8418-2 ISBN 978-1-4684-8416-8 (eBook)

DOI 10.1007/978-1-4684-8416-8

Second half of the Proceedings of the Third Workshop on Laser Interaction and Related Plasma Phenomena held at Rensselaer Polytechnic Institute, Troy, New York, August 13-17, 1973

A Division of Plenum Publishing Corporation
227 West 17th Street, New York, N.Y. 10011

United Kingdom edition published by Plenum Press, London
A Division of Plenum Publishing Company, Ltd.
4a Lower John Street, London W1R 3PD, England

PREFACE

As was the case in the two preceding workshops of 1969 and 1971, the Third Workshop on "Laser Interaction and Related Plasma Phenomena" held in 1973 was of international character. The main purpose was to review the advanced status of this particular and turbulent field of physics as it had developed vigorously in all major laboratories of the world since 1971. Due to recently accelerated advancements, it was hardly possible to present a complete tutorial review; the subject is still in its premature stages and changing rapidly. A topical conference would have been too specific for a group of physicists with broad backgrounds working in the field or for those just about to enter it.

It was the aim of the workshop and it is the aim of these proceedings to help this large group of scientists find their way within the highly complex and sometimes confusing results of a new field. We optimized the task of the workshop with extensive reviews on several topics and at the same time included more detailed information for specialists. The differences in their conclusions were not a matter of contention but rather served to complement the advanced results.

As in the preceding workshops, we directed our attention toward critical realism in respect to the complexity of the field. What is meant here is exemplified in the contribution by R. Sigel (p.667). R. Sigel illustrated how different the real intensity distribution in the focus of a neodymium glass laser is from the theoretical assumption of an ideally distributed figure-eight plane-wave intensity. A pulse of a few nanoseconds duration resolved by a picosecond image converter revealed this striking fact.

Not only does the curiosity of the scientists working in this fascinating field provide motivation for solving these particular problems, but beyond this the exigencies of alleviating future energy crises is of major concern.

The Janus-like nature of these proceedings was manifested in the extensive treatment of laser produced nuclear fusion and, for the first time, the discussion of experimental and theoretical results achieved in the application of lasers to high energy physics. Measurement of X-ray emission at 1 to 10 MeV energy was reported for the first time by J. W. Shearer et al. (see p.803) and production of anti-particles by lasers was discussed by several authors. Again the laser may open up a very unique method in physics. Compared with the "classical" way of producing anti-matter in accelerators, the laser allows an interaction at 10^{12} times higher densities and therefore results in substantially higher efficiencies.

Another historical aspect of these workshops is evidenced by the kickoff of the proceedings as introduced by the forwards of renowned scientists: the first workshop by one of the discoverers of the laser, Nobel Laureate Nicolai Basov; the second by the co-discoverer of thermonuclear fusion reactions, Paul Harteck; and this time, Edward Teller who attained the first exotherm nuclear fusion reaction on earth.

Other highlights included the first detailed presentations concerning two-dimensional codes of laser compression in fusion plasmas. They showed a very favorable stabilization of the dynamics, spatial and temporal irregularities and hot spots produced by the incident radiation (pp.399,427,449). A favorable decrease of the reflectivity at intensities which are desirable for nuclear fusion was observed by several authors (pp. 267, 291). The first direct measurement of the compression process was reported (p. 547). The development of high power lasers exceeding 10^{10} watts (the most important basis for nuclear fusion) was discussed. These lasers are: the CO_2 laser (pp.11,39,109), the iodine laser (p. 133) and the neodymium glass laser (pp.149, 427).

In order to understand the present situation of neutron generation from laser produced plasmas, one should realize that a 200 Joule laser irradiating a spherical deuterized polyethylene plasma produces no more than 10^7 neutrons (see "Laser Interaction and Related Plasma Phenomena" Vol. II, p. 389). Even under optimized inertial confinement conditions one can theoretically expect from deuterium at solid state density and under the same experimental set-up between 10^8 and 10^9 fusion neutrons (see "Laser Interaction and Related Plasma Phenomena" Vol. II, p. 520, Eq. 7). A considerable increase of neutron production can be expected only under high compression conditions. The contribution of K. Brueckner (p. 427) made it understandable that the experiments within the next months might result in the essential breakthrough.

Finally, we would like to mention what we somewhat neglected to treat in this volume: gas breakdown due to high power lasers.

Except for the contribution of G. H. Canavan (p. 177), we thought that the progress gained in gas breakdown since the report of the Second Workshop ("Laser Interaction and Related Plasma Phenomena" Vol. II) was of such small scale compared to the progress of the other topics that a report on the state-of-the-art could be postponed. The same holds true for the progress on interaction of laser produced plasma and on laser radiation in plasmas with external magnetic fields. However, the topic of spontaneous generation of magnetic fields in laser plasmas was represented by several contributions (pp. 213, 237), showing interesting advances.

As in the previous laser workshops, the success of the Third Workshop was not only due to the high caliber lecturers, but also due to the very active discussions stimulated and carried out by the participants. The participants also carried the financial load except for what was allocated to them by their respective organizations since this time the Workshop itself did not receive any outside support. Therefore, we would like to express our appreciation to the enthusiastic laser plasma groups (see list of participants p. 893) converging to the campus of Rensselaer Polytechnic Institute, Troy, New York, from 12 different foreign countries.

We would also like to thank the following Advisors of the Third Workshop for their counsel: N. G. Basov (P. N. Lebedev Institute, USSR), K. Boyer (Los Alamos Laboratories, New Mexico), J. M. Dawson (Princeton University), A. J. DeMaria (United Aircraft Research Labs), A. H. Guenther (Kirtland Air Force Base), P. Harteck (Rensselaer Polytechnic Institute), R. E. Kidder (Lawrence Livermore Laboratory), A. M. Prokhorov (P. N. Lebedev Institute, USSR) and W. Witkowski (Max-Planck-Institut, Germany).

We are indebted to our University, Rensselaer Polytechnic Institute and to its President, Dr. R. J. Grosh for hosting our guest of honor, Dr. Edward Teller, and for providing us with a competent staff.

We are grateful to Dr. Stephen E. Wiberley, Acting Provost, Vice Provost and Dean of the Graduate School of RPI who took time off to welcome the participants in his opening address at the first session which was chaired with stimulating vigor by Professor Paul Harteck.

The Office of Continuing Studies under the direction of Richard J. Teich was invaluable for the organization of this International Conference. Particularly, we wish to acknowledge the outstanding service rendered by the Director of Special Programs, Marvin G. Clark and his assistants.

Thanks are also due to Dr. W. C. Stoker, President of Rensselaer Polytechnic Institute, Hartford Graduate Center, who encouraged and supported us in organizing the Third Workshop on the Troy campus of Rensselaer Polytechnic Institute.

Last but not least, we want to thank our colleague, Professor George C. Baldwin, who so effectively co-operated as an Associate Director of the Third Laser Workshop.

We would also like to acknowledge the excellent assistance provided by Ms. Jean Jakoboski of Rensselaer Polytechnic Institute, Hartford Graduate Center, in preparing these proceedings.

March 1974 Helmut Schwarz

Heinrich Hora

CONTENTS

VOLUME 3B

V. EXPERIMENTS FOR NUCLEAR FUSION

VOLUME 3A

IV

THEORY OF COMPRESSION AND ENERGY GAIN FROM LASER PLASMAS

LASER-INDUCED IMPLOSION AND THERMONUCLEAR BURN†*

John H. Nuckolls

University of California Lawrence Livermore Laboratory

Livermore, California 94550

ABSTRACT, INTRODUCTION

In high density laser induced fusion, the key idea is laser implosion of hydrogen isotope microspheres to approximately 10,000 times liquid density in order to initiate efficient thermonuclear burning[1]. Fusion yields 50-100 times larger than the laser energy for laser energies of 10^5-10^6 joules have been achieved in sophisticated computer simulation calculations. Most of the dense pellet is isentropically compressed to a high density Fermi-degenerate state, while thermonuclear burn is initiated in the central region. A thermonuclear burn front propagates radially outward from the central region heating and igniting the dense fuel.

The laser fusion implosion system consists of a tiny spherical pellet of deuterium-tritium surrounded by a low density atmosphere extending to several pellet radii, located in a large vacuum chamber, and a laser capable of generating an optimally shaped pulse of light energy. Figure 1. The atmosphere may be produced by ablating the pellet surface with a laser prepulse.

---

†Presented at the Third Workshop on "Laser Interaction and Related Plasma Phenomena" at Rensselaer Polytechnic Institute, Troy, New York, August 13-18, 1973.

*Work Performed Under the Auspices of the U.S. Atomic Energy Commission.

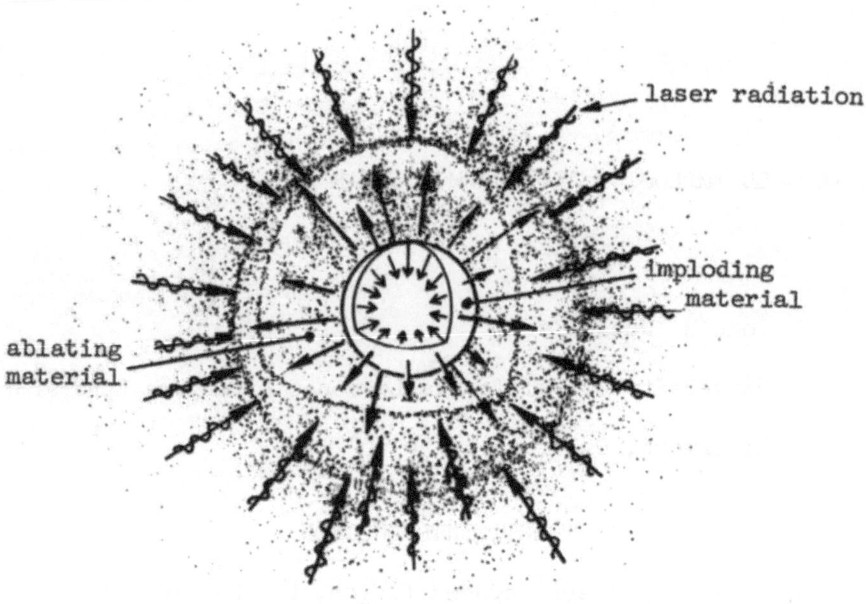

Fig. 1 - Multiple laser beams are focussed on the atmosphere ablated from the surface of the imploding pellet. Laser heated electrons generated in the outer atmosphere transport energy inward through the atmosphere to heat and ablate the pellet surface. Ablation reaction forces implode the pellet.

Mirrors (or lenses) are used to focus the laser light on the atmosphere more or less uniformly from all sides. Absorption of the laser light in the outer atmosphere generates hot electrons. The atmosphere and the pellet surface are heated by electron diffusion and transport. As the electrons move inward through the atmosphere, scattering and solid angle effects greatly increase the spherical symmetry. Violent ablation and blowoff of the pellet surface generates the pressures which implode the pellet. This is essentially a spherical rocket. The laser pulse is shaped (in time) to achieve ultra-high compression of the pellet while fully exploiting electron degeneracy and thermonuclear propagation to maximize the fusion yield.

THERMONUCLEAR BURN

The feasibility of thermonuclear explosions was demonstrated by Los Alamos scientists in the early 1950's[2].

Thermonuclear micro-explosions scale as the density-radius product ρR. The rates of burn, energy deposition by charged reaction products, and electron-ion heating are proportional to the density, and the inertial confinement time is proportional to the radius. Consequently the burn efficiency, self-heating, and feasibility of thermonuclear propagation are determined by ρR. In spherical compression, ρR increases (because $\rho \propto R^{-3}$) and is proportional to $(M\rho^2)^{1/3}$, where M is the mass. Compression by 10^4 reduces the mass--and laser energy--required to initiate an efficient thermonuclear micro-explosion by up to 10^8-fold, depending on the efficiency of the compression process.

The burn efficiency, ϕ, is proportional to the product of the burn rate, $\rho\overline{\sigma v}$, and the inertial confinement time, $R/4C$, where $\overline{\sigma v}$ is the Maxwell velocity-averaged reaction cross section and C is the sound speed. The factor of 4 arises because in a sphere half the mass is beyond 80% of the radius. Both $\overline{\sigma v}$ and C depend on temperature, but their ratio is approximately constant in the 20-50 keV temperature range characteristic of efficient deuterium-tritium micro-explosions. Using $\overline{\sigma v}/C$ evaluated at 20 keV, and correcting for depletion, it follows that[3]

$$\phi \approx \frac{R\rho}{6+R\rho} \quad .$$

At $\rho R = 3$ g/cm^2, $\phi \approx 1/3$, corresponding to a fusion energy release of $\sim 10^{11}$ joules/g.

The average specific heat energy required for ignition is

$$C_P \, \theta_{ign} \, \beta$$

where C_P is the heat capacity, θ_{ign} is the ignition temperature, and β is a correction for the self-heating by the 3.6 MeV DT alpha particles, and for thermonuclear propagation. If $\rho R >> 0.3$ g/cm^2, then only about 0.3 g/cm^2 in the central region need be heated to approximately 10 keV in order to initiate a radially propagating burn front which ignites the entire pellet. In this case 1.6×10^{10} joules/g of fusion energy will be released from the central region; one fifth of the energy is in alpha particles, sufficient to heat 3 times more DT to 10 keV. The alpha particles will deposit their energy in approximately this mass since their range in 10 keV DT is about 0.3 g/cm^2.

Due to the effects of shock convergence and pulse shaping during implosion, the DT temperature just prior to ignition may be made to vary approximately R^{-2}. Then β is proportional to $(\rho R)^{-2}$. Because of practical limitations on implosion symmetry, a minimum of ~ 0.03 is imposed on β, which occurs for $\rho R \gtrsim 3$ g/cm^2. Then the average ignition energy is 3×10^7 joules/g. This is also the minimum compressional energy of DT at a density of 1000 g/cm^3, and this minimum occurs if the DT electrons are Fermi-degenerate[4]:

$$\varepsilon = \varepsilon_F\left[\frac{3}{5} + \frac{\pi^2}{4}\left(\frac{\theta_e}{\varepsilon_F}\right)^2 + \ldots\right]$$

where ε_F, the Fermi-energy, is $\frac{h^2}{8m}\left(\frac{3}{\pi} n_e\right)^{2/3}$, h is Planck's constant, m is the electron mass, θ_e is the electron temperature, and n_e is the electron density. For $\rho = 1000$ g/cm^3 ($n_e \approx 2.5 \times 10^{26}$/cm^3), $\varepsilon_F \approx 1$ keV, and the DT electrons are degenerate if the implosion is carried out so that $\theta_e \ll 1$ keV (except in the central region where ignition occurs).

The minimum average energy of ignition and compression with $\rho R \sim 3$ g/cm^2, assuming propagation and degeneracy, is roughly the sum of the compression and ignition terms, or $\sim 6 \times 10^7$ joules/g. Since the fusion energy produced at this ρR is 10^{11} joules/g, the gain is ~ 1500. Approximately 95% of the laser energy absorbed by the pellet during implosion is lost to kinetic and internal energy of the blowoff. Consequently the energy gain relative to the laser light employed is about 75 fold. This is sufficient for CTR applications with a 10% efficient laser, a 40% thermal-to-electric efficiency, and about 30% of the electrical energy circulated internally.

Figure 2 shows the variation of gain (relative to laser light energy) with compression and laser light energy[1,16]. The curves have been slightly normalized to computer calculations of the implosion and burn. Gains approaching 100 are predicted for laser energies of 10^6 joules. The calculations indicate that less than 1 kJ of laser light may be sufficient for breakeven (gain ≈ 1) and 10^5 joules may be sufficient to generate net electrical energy with a 10% efficient laser. These predicted gains are probably upper limits to what can be achieved. Unforeseen difficulties may cause significant performance degradations when the predictions are experimentally tested.

Similar gain curves may be generated for D_2, DHe^3, and $B^{11}p_x$ pellets seeded with a small percentage of tritium to facilitate ignition. Since these reactions have smaller $\overline{\sigma v}$'s than DT, higher ρR's are required for efficient burn. This may be

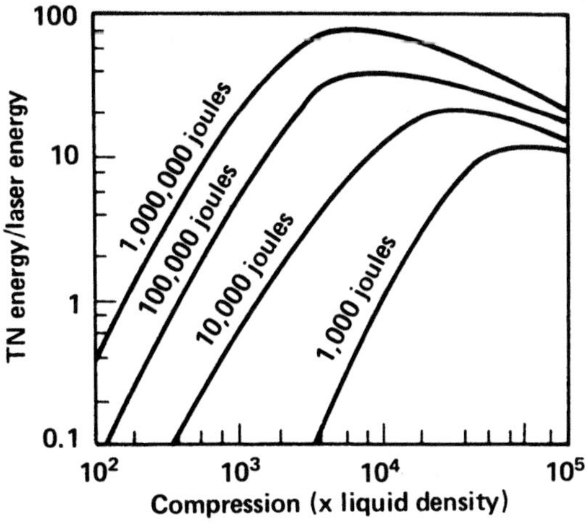

Fig. 2 - Calculated gain (ratio of fusion energy output to laser light energy input) as a function of fuel compression for several different laser light energy inputs. At compressions less than about 10^3, the gain increases strongly with increasing compression because of increasing burn efficiency and propagation. The gain decreases with compressions much greater than 10^4 because the energy of compression (against degeneracy pressures) becomes larger than the energy of ignition.

achieved either by use of larger pellets and higher energy lasers, or by compressing the pellet to higher densities ($\sim 10^4$ g/cm^3). The D_2 and DHe^3 fuels produce fewer neutrons than DT, and the $B^{11}p_x$ fuel produces essentially no neutrons[5].

IMPLOSION

Conditions involving pressure, symmetry, and stability must be satisfied in order to implode a DT sphere to a state at 10^4 times liquid density, in which both Fermi-degeneracy and thermonuclear propagation can be exploited to achieve maximum gain[1]. A pressure of at least 10^{12} atmospheres must be generated. Much higher pressures are required if the electrons in the high density DT are not Fermi-degenerate, i.e. if the implosion is not essentially isentropic. The pressures applied to implode the pellet must be uniform spatially and temporally to less than one part in twenty in order to preserve effective spherical symmetry. During compression the radius decreases by about 20-fold ($10{,}000^{1/3}$), and it is necessary to maintain sufficient symmetry within the central region in order to utilize central ignition and propagation. The hydrodynamic Rayleigh-Taylor instability must be controlled. Otherwise the pellet surface cannot be relatively gradually accelerated during the implosion as required by the optimum pulse shape. These conditions may be satisfied by the characteristics of the laser pulse shape and of the laser implosion scheme.

Pressure

The minimum pressure of DT at 10^4 times liquid density is 10^{12} atmospheres, and occurs if the electrons are Fermi-degenerate. If $\theta = 5$ keV, the ideal ignition temperature for DT, the electrons are not degenerate, and the pressure is 10^{13} atmospheres. The required 10^{12} atmosphere pressure is generated by focussing the high power laser light to high intensities on the outer atmosphere, multiplying the absorbed energy flux by electron transport in the spherically convergent density gradient from the outer atmosphere to the pellet surface, ablating the pellet surface, and multiplying the resulting ablation pressures by the concentration of energy density which occurs in the implosion process.

Absorption. Laser light is focussed to intensities as high as 10^{15} watts/cm^2 on the outer atmosphere. Absorption occurs at densities less than the critical density (where the plasma frequency equals the optical frequency) by inverse bremsstrahlung and by plasma instabilities if the intensity exceeds the instability thresholds. Figure 3 from Shearer[6] shows the inverse bremsstrahlung opacity in matter at one keV temperature for Nd laser light ($\lambda \gtrsim 10^{-4}$cm). The light slows down as it approaches the critical density (10^{21} electrons/cm^3 for Nd light), the electromagnetic field increases, and at high intensities the quivering velocity of the electrons may exceed their thermal velocity. The inverse bremsstrahlung opacity scales as follows:

$$K_{\text{inverse brem}} \sim \frac{Z^3}{A^2} \frac{n^2}{(1-n/n_c)^{1/2}} \lambda^2 \theta_e^{-3/2} \text{ cm}^{-1}$$

where θ_e is the electron temperature, λ is the laser wavelength, n is the electron density, n_c is the critical electron density ($\sim \lambda^{-2}$), A is the atomic weight, and Z is the effective charge. Since in typical laser fusion pellets the scale height of density is comparable to one mm, efficient absorption via inverse bremsstrahlung is not possible with Nd at electron temperatures $\gtrsim 10$ keV, or with CO_2, at electron temperatures $\gtrsim 1$ keV. Plasma simulation calculations show that the instability absorption length near the critical density is approximately ten wavelengths[7], or 10^{-3}cm for Nd light. The threshold for the modified two-stream

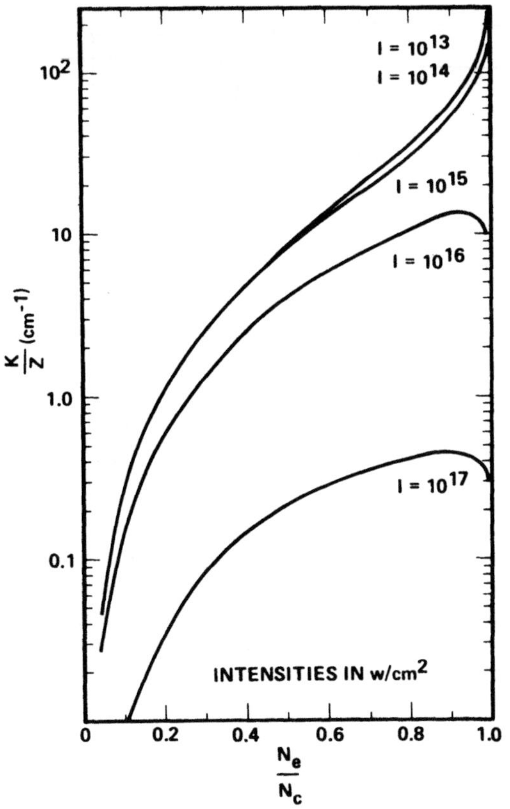

Fig. 3 - Inverse bremsstrahlung opacity for laser light absorption, 1 μm wavelength, one keV electron temperature[6].

instability is approximately[7]

$$\text{Instability threshold} \sim \left(\frac{n_c^3}{\theta_e}\right)^{1/2} Z \times S$$

where the swelling factor, S, corrects for the increasing electromagnetic field as the light approaches the critical density. For 1 keV DT, this threshold is approximately 10^{12} watts/cm^2.

Electron Spectrum. Absorption of intense laser light generates hot electrons. If absorption is by inverse bremsstrahlung the electron spectrum is near Maxwellian

$$f(v_e)_{\text{inverse brem}} \sim v_e \exp\left(-\frac{mv_e^2}{2\theta_e}\right)$$

where v_e is the electron velocity. If absorption is by plasma instabilities the electron spectrum may be non-Maxwellian[8]. Figure 4 shows an electron spectrum calculated by a plasma

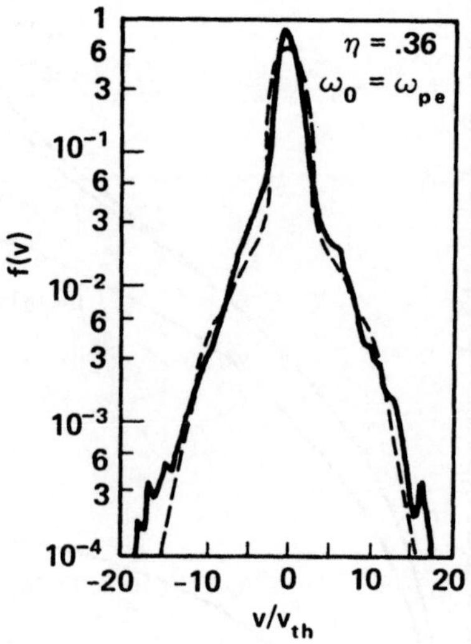

Fig. 4 - Calculated electron spectra generated by absorption of intense laser light via plasma instabilities[8]; v is the electron velocity, and v_{th} the thermal electron velocity. The central core $(v/v_{th} \approx 0)$ is near Maxwellian and the wings are non-Maxwellian. (Dashed is theory).

simulation computer program[8]. An exponential-like tail extends to 100 times the thermal electron energy.

The laser heated electrons heat the pellet atmosphere by electron-electron collisions. The mean free path, λ_{e-e}, is

$$\lambda_{e-e} \sim E_e^2\, n^{-1}$$

$$\sim 1 \text{ cm},\ E_e \sim 60 \text{ keV},\ n \sim 10^{21} \text{ cm}^{-3}\ .$$

where E_e is the electron energy. After a tiny fraction of the electrons escape into the vacuum chamber, electrostatic potentials comparable to θ_e are generated and the remaining electrons are trapped within the atmosphere which expands at ion velocities. Sufficiently high energy electrons will reflect back and forth across the atmosphere many times before losing a significant fraction of their energy.

Ablation Pressure. Typically the density falls off inversely as radius cubed from the ablating radius to the absorption radius. So long as electron-electron collisions are sufficiently rapid energy is strongly coupled by electron transport in this spherically convergent gradient from the absorbing surface to the much smaller ablating surface. Then

$$\dot{E} \approx 2\, P_A C A_p + \text{smaller terms}$$

where P_A is the ablation pressure, A_p is the ablating surface area, the factor 2 corrects for heating of the blowoff--through which energy is transported--and $\dot{E}$ is the absorbed laser power. The blow-off velocity, C, is approximately sound speed in the hot atmosphere

$$C \approx \sqrt{\frac{P}{\rho}} \sim \theta_e^{1/2} \qquad (\theta_e >> \theta_i)$$

where θ_e is the electron temperature. This temperature is determined largely by the equation for flux limited electron transport

$$n_c \, v_e \, \Delta E \, A_c \approx \dot{E}$$

where $n_c \, v_e$ is the incident electron flux, ΔE is the net energy carried by each electron from the absorption region, and A_c is the absorption area. Then it follows that the ablation pressure is approximately

$$P_A \approx \frac{A_c}{A_p} \left(\frac{\dot{E}}{A_c}\right)^{2/3} \rho_c^{1/3}$$

where ρ_c is the critical density in g/cm^3. If $A_c = A_p$ (plane geometry) this is the equation derived by Kidder[9]. The pressure is multiplied by a factor A_c/A_p by electron transport in the convergent density gradient.

The ablation coupling efficiency is

$$\frac{P_A A_p}{\dot{E}} \approx \frac{v}{2C}$$

where v is the implosion velocity. Typically v varies from 10^6 to 3×10^7 cm/s while C varies from 2×10^7 to 10^8 cm/s, so that the coupling efficiency varies from 2 to 15 percent. Most of the energy is coupled near the end of the implosion, so that the average coupling efficiency is 5-10 percent.

Decoupling. If the hot electron range becomes comparable to the absorption radius, the electron may cross the pellet, re-enter the region in which laser light is absorbed and be further heated before its energy is lost to colder electrons. Then the electrons are heated to even higher energies, have a larger range, and lose energy even more slowly to the colder electrons[10]. The threshold for onset of this decoupling determines a maximum hot electron energy and laser intensity, and thereby a maximum ablation pressure. If self-generated electric fields and anomalous effects are neglected, then the maximum ablation pressure is

$$P_{max} \approx 10^2 \frac{A_c}{A_p} A_c^{1/4} \rho_c^{3/2} .$$

However if the density gradient is steeper than R^{-3}, (as in implosion of hollow pellets), higher pressures may be generated. Decoupling may be a serious problem for CO_2 laser implosions (ρ_c is smaller), and weakly affects Nd laser implosions. For larger pellets and laser energies decoupling becomes less severe because the hydrodynamic time is larger. Decoupling can be compensated for if the initial volume is increased by making the pellet hollow. Then the required implosion velocity may be achieved with limited implosion pressures (acting over a larger volume).

Preheat. Electrons with energies of 50 keV have a range of 100 μ in liquid density DT, comparable to the thicknesses of typical laser fusion pellets. Electrons of this energy are generated by long wavelength lasers even if the electron spectrum is assumed to be Maxwellian because of decoupling effects and by lasers of any wavelength via instabilities if non-Maxwellian electron spectra are assumed. If the flux of these electrons is too high the resulting preheating may make compression to a Fermi-degenerate state impossible, and a high compression extremely difficult[1]. The tolerable fraction of electron energy in hot electrons with range comparable to the pellet thickness is exceeded when the preheat energy flux is comparable to the implosion energy flux:

$$f \approx \frac{P_A v}{P_e v_e}$$

where P_A is the ablation pressure, v is the implosion velocity, P_e is the electron pressure, and v_e the hot electron velocity. In the initial stages of a CO_2 implosion P_A is 1 Mb and v is 1 cm/μs, while P_e is 10^{-2} Mb ($\theta_e \sim 1/2$ keV) and v_e is 10^4 cm/μs, so that f is only one percent. For Nd f is initially 10^{-2}%. Because of the preheat problem, absorption of the laser light by plasma instabilities may not be feasible for some laser fusion pellet designs. Since high intensity CO_2 light must be absorbed by instabilities, it is possible that CO_2 is not suitable for high density laser fusion. In order to increase the instability threshold and the inverse bremsstrahlung opacity, seeding of the pellet with higher Z material, and use of short wavelength lasers is advantageous. The instability threshold may also be increased by frequency modulation of the laser[20]. [However recent studies at Livermore by J. Lindl show that with 1 μ light if $\lesssim$ 20% of the

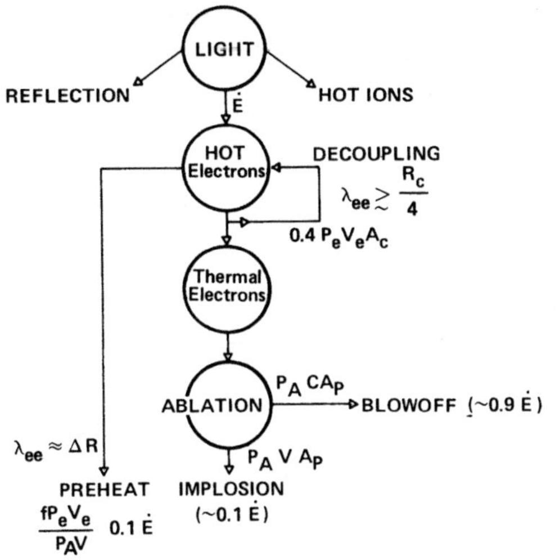

Fig. 5 - Modes of energy transfer and coupling in laser implosion. In the preferred mode ablation occurs via hot and thermal electrons. The implosion may be perturbed or destroyed by the other channels, i.e. reflection, anomalous ion heating, decoupling, and preheat.

electron spectra is non-Maxwellian then the laser implosion may not be strongly degraded.]

Magnetic Field. Spontaneous magnetic fields are generated due to the non-uniform irradiation of the atmosphere with the laser[11]. Gradients in the electron pressure and in the electron density are generated which have a non-zero cross product (via the electron temperature). Then[12]

$$\dot{B} \approx \nabla x\ [\vec{v} \times \vec{B} + \frac{c}{en_e} \nabla P_e + \cdots] \quad .$$

Implosion calculations with long wavelength lasers (e.g. CO_2) show magnetic fields sufficiently large to inhibit electron transport. However the magnetic energy density is much smaller than the electron energy density. The effect of the magnetic field may be reduced by increasing the collision frequency (via short wavelength lasers and seeding with small amounts of high Z material).

Implosion. In the implosion the applied ablation pressure first generates kinetic energy

$$\bar{P}_A\ (V_o - V_f) \approx \frac{1}{2} Mv^2$$

where $\overline{P}_A$ is the average ablation pressure, V_o is the initial volume, and V_f is the final volume. At maximum compression most of this kinetic energy is internal, so that the pressure multiplication ratio is proportional to the compression

$$P_{max} \approx (\gamma-1)\,\frac{\frac{1}{2}Mv^2}{V_f} \approx (\gamma-1)\,\frac{\overline{P}_A(V_o-V_f)}{V_f}$$

$$\therefore \quad \frac{P_{max}}{\overline{P}_A} \approx (\gamma-1)\,\frac{V_o}{V_f}\,, \quad V_o \gg V_f \quad .$$

Convergence transforms the kinetic to internal energy nearly isentropically--except for a small region near the implosion center. If the matter is compressed to a Fermi-degenerate state (the initial implosion velocity is near sonic and the subsequent ablation pressure increases sufficiently gradually) then

$$P \sim \rho^\gamma, \; \gamma = \frac{5}{3}$$

and

$$\left(\frac{P_{max}}{\overline{P}_A}\right) \approx \left(\frac{\overline{P}_A}{P_{Ao}}\right)^{3/2} \left(\frac{2}{3}\, V_o\, \frac{\rho_o}{M}\right)^{5/2}$$

where P_{Ao} is the initial ablation pressure. Large pressure multiplications may be achieved even if $\overline{P}_A$ is limited (e.g. by decoupling) by use of hollow pellets (V_o term). However V_o cannot be made indefinitely large because of symmetry and stability constraints. The multiplication ratio is also increased by making the average ablation pressure large compared to the initial pressure (via pulse shaping). Because of the magnitude of the Fermi-energy in matter under ordinary conditions, the entropy is not significantly decreased by making P_{Ao} much less than one half megabar.

Pulse Shape. The optimum laser pulse shape--which maximizes the fusion yield for a given laser energy--satisfies five conditions.

(1) Densities greater than 1000 g/cm^3 are achieved.

(2) The implosion occurs in less than one sonic transit time so that long wavelength Taylor unstable growth is tolerable.

(3) In most of the compressed pellet the electrons are Fermi-degenerate. This minimizes the laser energy required for compression.

(4) The central core of the compressed pellet--having a radius comparable to the DT alpha range at ignition temperatures--reaches temperatures of $\sim$ 10 keV, initiating rapid thermonuclear burn and propagation. This minimizes the laser energy required for ignition.

(5) The implosion and blowoff velocities are increased together so as to maximize the ablation efficiency.

These five conditions are satisfied by a pulse shape having the following two properties.

(1) The initial implosion velocity is $\sim$ 1 cm/μs, slightly larger than sound speed. This shock is sufficiently weak so that most of the pellet can be compressed to a Fermi-degenerate state, but--with convergence--is strong enough so that the center is significantly shock heated. The peak implosion velocity is sufficiently large ($\gtrsim 3 \times 10^7$ cm/s) so that densities of 1000 g/cm^3 are reached, and central ignition occurs.

(2) The implosion pressure is increased with time so that the hydrodynamic characteristics coalesce within most of the compressed mass, but beyond one alpha particle range from the center. This insures isentropic Fermi-degenerate compression of most of the pellet while igniting the central region.

By means of hundreds of implosion/burn computer calculations, the optimum pulse shape (laser power history) has been determined. In these calculations the pulse shape was represented by an eight element histogram. The amplitudes and durations of the elements were varied to optimize the ratio of fusion energy to laser light energy. For solid DT pellets in which preheat, decoupling, and other effects are not important, the optimum histogram can be approximated by the following equation:

$$\dot{E} = \dot{E}_o \, \tau^{-s}, \text{ where } \tau = 1 - \frac{t}{\tau'}, \; t < t'$$

$$s \approx \frac{3\gamma}{\gamma+1} \sim 2 \; (\text{if } \gamma = \frac{5}{3}), \text{ and } \gamma = \frac{PV}{\varepsilon} + 1.$$

where $\dot{E}_o$ is the initial laser power (which generates an initial implosion velocity of $\sim$ 1 cm/μs), and t' is the implosion time.

If the pellet is hollow or seeded, or if preheat, decoupling, or other effects are important, the optimum pulse shape is significantly modified. Symmetry requirements also affect the optimum pulse shape.

Symmetry

In compression of a sphere by 10^4-fold, the radius decreases somewhat more than 20-fold. If after compression spherical symmetry is required to within 1/2 the compressed radius (1/40 the initial radius) then the implosion velocity (and time) must be spatially uniform (and synchronized) to ∿ one part in 40, or a few percent[1]. In general, for a spherical implosion in which the ratio of initial to final volumes is η, and in which the tolerable error in the final radius, R, is wR, the tolerable fractional error in the integral of velocity and time is approximately

$$\frac{\Delta \int v dt}{\int v dt} \sim \frac{w}{\eta^{1/3}}, \eta \gg 1.$$

Implosion errors may be reduced to about 10% by a many-sided irradiation system, consisting of a laser, beam splitters, mirrors, lenses, and other optical elements. This error can then be reduced to less than 1% by means of electron transport in the atmosphere surrounding the pellet.

Multiple Beam Irradiation. If a sphere is irradiated from all sides with many laser beams focussed to the diameter of the sphere, intensity variations over the spherical surface as small as 10% may readily be achieved. However light which is not incident perpendicular to the spherical surface is refracted in the density gradient[6]. If the incident angle is more than 30° from normal, the light trajectory in the density gradient does not approach the critical density surface. Then absorption by plasma instabilities is not possible. Also absorption by inverse bremsstrahlung is reduced and occurs at low densities, so that the electron decoupling problem is more severe.

Using f/1 optics the entire surface of a sphere cannot be illuminated by as few as six circular laser beams focussed on the center of the sphere. However if each of six beams is focussed to a point one spherical radius beyond the center (along the axis of each beam) then the entire surface is illuminated, and the maximum angular deviation from normal incidence is less than twenty degrees. If the overlapping edges of the beam are blocked (so that each beam is four sided instead of circular) then the maximum error in intensity of illumination is 10-20%. This error may be reduced to less than 10% by using more laser beams, e.g. 12 beams with pentagonal cross section.

Complex lenses and mirrors may also be used to increase the symmetry of irradiation.

Electron Transport in Atmosphere. The region in the atmosphere where the laser light is absorbed has a radius several times that at which ablation occurs. Consequently each point on the ablating surface is heated by nearly 2π steradians of the absorbing region. If many laser beams are used, e.g. 12 pentagons, then each beam occupies a small fraction of 2π steradians, and each point on the ablating surface is effectively heated by many beams. These effects reduce the irradiation error.

The range of the laser heated electrons in the low density region where the laser light is absorbed is comparable to the diameter of one of the multiple focussed laser beams. An initial prepulse may be used to satisfy this condition at the beginning of the implosion (after the atmosphere is formed). Hence non-uniformities which occur over small areas are strongly smoothed. In addition there are many electron scattering mean free paths through the density gradient between the absorbing and ablating surfaces. The smoothing due to scattering during electron transport may be estimated by use of spherical harmonic analysis if a steady state, a short electron mean free path, and a uniform density are assumed[13]. This analysis shows that scattering in the atmosphere reduces the heating error from 10-20% at the absorbing surface to less than 1% at the ablating surface with heating by eight laser beams. Solution of this problem with the real (non-uniform) density gradient would reduce the heating errors even further.

Stability

The unstable growth of small surface perturbations during acceleration of a fluid interface is described by the Rayleigh-Taylor theory[14].

$$A = A_o e^{\int \sigma dt}, \quad \sigma^2 = aK$$

where A_o is the initial amplitude of a perturbation with wave-number K, and a is the acceleration. Unless some stabilization mechanism can be utilized, a fluid shell can be accelerated through only about five thicknesses before it is essentially destroyed by this instability. However in the laser fusion implosion scheme described here the optimum pulse shape may accelerate shells through many tens of thicknesses. Stabilization is achieved by several means.

(1) Growth of perturbations with wavelength less than ∿ 2π times the shell thickness is prevented because the implosion pressure is generated by surface ablation driven by diffusing electrons. Under these conditions

$$\sigma^2 = aK - \frac{P_A K^2}{\rho}$$

where P_A is the ablation pressure[15]. This stabilizing effect occurs largely because the temperature gradient near the peak of the perturbation is steeper than the gradient near the valley. Consequently the rate of transport of energy is higher to the peak than to the valley, and higher ablation rates and ablation pressures are generated. This reduces the amplitude of the perturbation.

(2) Growth of long wavelength perturbations--which have a slower growth rate--is sufficiently limited by imploding the pellet as rapidly as possible consistent with compression to a Fermi-degenerate state. Consequently the initial implosion velocity is somewhat supersonic.

(3) Pellets are used which have an initial ratio of shell thickness to radius of more than 2%. If thinner shells are used, perturbations with wavelengths greater than 2π times the shell thickness (which are not stabilized by ablation) grow so large that the shell is destroyed.

Because the implosion stability is affected by the symmetry of the pellet irradiation, and because of other complex factors (convergence, compressible DT, finite sound speed, long range electrons, etc.) the theory is uncertain and stability has been demonstrated with two-dimensional computer calculations.

COMPUTER CALCULATIONS

The compression and burn processes which have been described are illustrated in results of a typical computer simulation calculation of the implosion of a fusion pellet to 10,000 times liquid density, and of the resulting thermonuclear micro-explosion. This calculation was carried out at the Livermore Laboratory by Albert Thiessen on the CDC 7600 computer with the LASNEX program developed by George Zimmerman[16]. LASNEX is a two-dimensional (axially symmetric) finite difference code which includes coupled non-linear partial differential equations for the following physical processes (Figure 6)

Hydrodynamics--Lagrangian; real and generalized Von Neumann artificial viscosities; pondermotive, electron ion, photon, magnetic, and alpha particle pressures.

Laser light--absorption via inverse bremsstrahlung and plasma instabilities; reflection at critical density.

Coulomb coupling of charged particle species.

Suprathermal electrons--Multigroup flux-limited diffusion with self-consistent electric fields; non-Maxwellian electron spectra determined by results of plasma simula-

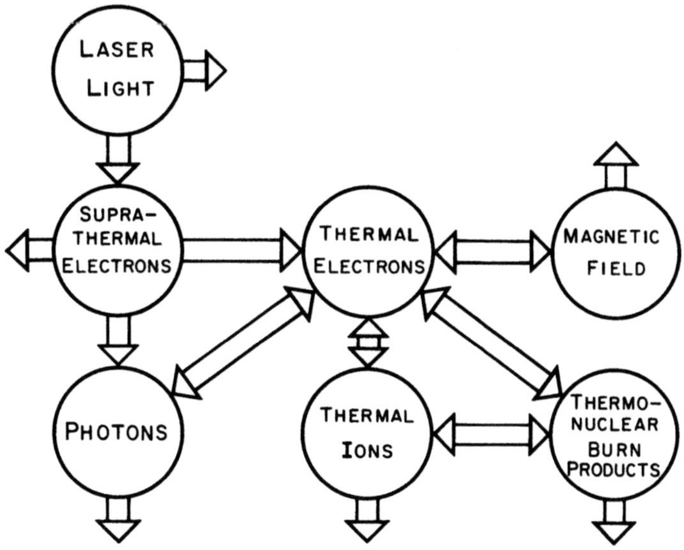

Fig. 6 - Physical processes included in the LASNEX laser implosion /fusion computer program. Arrows represent coupling and possibilities for diagnostics.

tion calculations for laser light absorption by plasma instabilities; inverse bremsstrahlung electron spectrum for classical absorption.

Thermal electrons and ions--flux-limited diffusion.

Magnetic field--includes modification of all charged particle transport coefficients, as well as most of the equilibrium MHD effects described by Braginskii[12].

Photonics--Multigroup flux-limited diffusion[17]; LTE non-LTE average-atom opacities for free-free, bound-free and bound-bound processes[18]; Fokker-Planck treatment of Compton scattering[19].

Fusion--Maxwell velocity-averaged reaction rates; the DT alpha particle is transported by a one group flux-limited diffusion model with appropriate energy deposition into the electron and ion fields; one group transport of the 14 MeV neutron.

Material properties--opacities, pressures, specific heats, and other properties of matter are used which take into account nuclear Coulomb, degeneracy, partial ionization, and other significant effects.

In this implosion calculation, a 54 KJ pulse of laser light is focussed symmetrically onto a low density atmosphere (generated by a laser prepulse) surrounding a 400 μm radius spherical pellet of liquid deuterium-tritium. The applied laser power is optimally increased from $\sim 10^{10}$ to $\sim 10^{15}$ watts in 24 nanoseconds, while the laser wavelength is varied from 4 µm initially to 1/2 micron at the end of the implosion. Figure 7. The initial spike in the pulse shape increases the implosion symmetry by preheating the atmosphere. Long wavelength light also minimizes asymmetries in the early stages of the implosion and short wavelength light avoids decoupling in the late stages. (Recent calculations show that much less frequency variation is required than was used in this calculation.) The laser light is absorbed via inverse bremsstrahlung and plasma instabilities near the critical density in the atmosphere, generating hot electrons with an assumed near-Maxwellian spectra. Coupling between the hot and cold electron distributions due to self-generated electric fields is assumed to be strongly degraded by runaway and other effects, and coupling due to anomalous effects is assumed to be negligible.

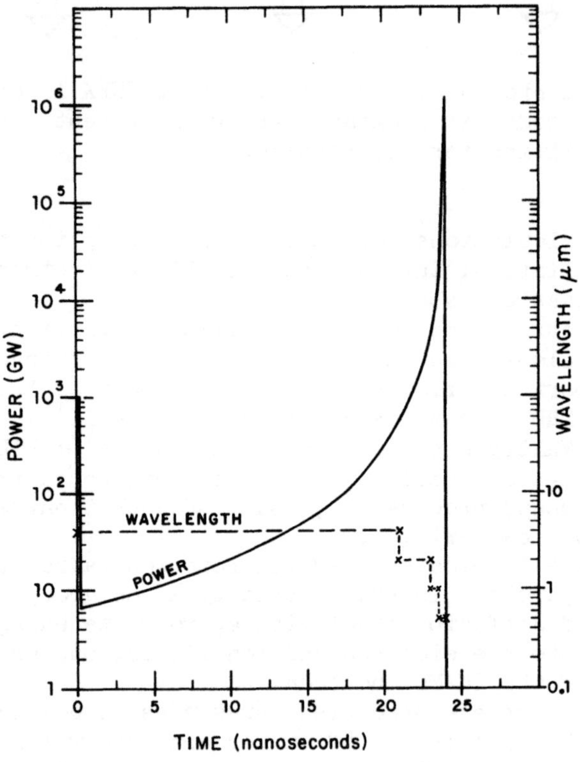

Fig. 7 - Pulse shape: power and wavelength histories. This pulse shape achieves a symmetric implosion in which degeneracy and thermonuclear propagation are fully exploited to maximize the gain.

If these pessimistic assumptions are relaxed, up to 20% of the hot electrons may be generated with a non-Maxwellian (i.e. exponential) distribution without significantly degrading the implosion. If the electron spectrum is hotter than this, then seeding, hollow pellets, frequency modulation, and short wavelength light may be utilized[1]. The atmosphere is heated by the hot electrons to electron temperatures which increase in time from $\sim 10^6$ to 10^8 ^oK at the absorption radius. The surface of the pellet is heated and ablated by electron thermal conduction through the hot atmosphere, generating implosion pressures which optimally increase from $\sim 10^6$ to $\sim 10^{11}$ atmospheres. Figure 8. This five order of magnitude increase in implosion pressure occurs at an optimal rate during transit of the initial shock to the center. Consequently the outer part of the pellet is isentropically compressed into a high

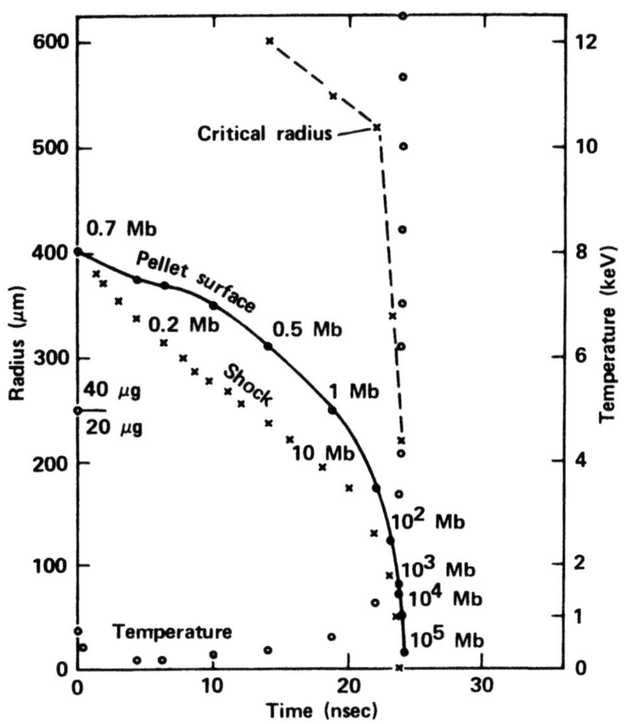

Fig. 8 - Implosion history. The solid line is the surface of the dense pellet and the "x's" inside of this line show the trajectory of the initial shock. Numbers along the solid line indicate the ablation pressures. Circles show the electron temperatures at the radius where the critical density occurs. The dashed line indicates the variation in the critical radius during the implosion.

density spherical shell ($\rho > 100$ g/cm^3) while at the same time this shell is inwardly accelerated to velocities which increase in time from 10^6 to 3.5×10^7 cm/s. As the internal pressure becomes larger than the ablation pressure the rapidly converging shell slows down and is compressed near isentropically, at sub-Fermi-temperatures, to densities greater than 1000 g/cm^3. Figure 9. The inner region is compressed by the outer shell to densities approaching 1000 g/cm^3, and heated to ion temperatures greater than 10^8 ^{0}K, initiating thermonuclear burn. A thermonuclear burn front then propagates outward. Figures 10, 11. About 1800 KJ of fusion energy is produced in less than 10^{-11} seconds. The energy gain is about 35 fold.

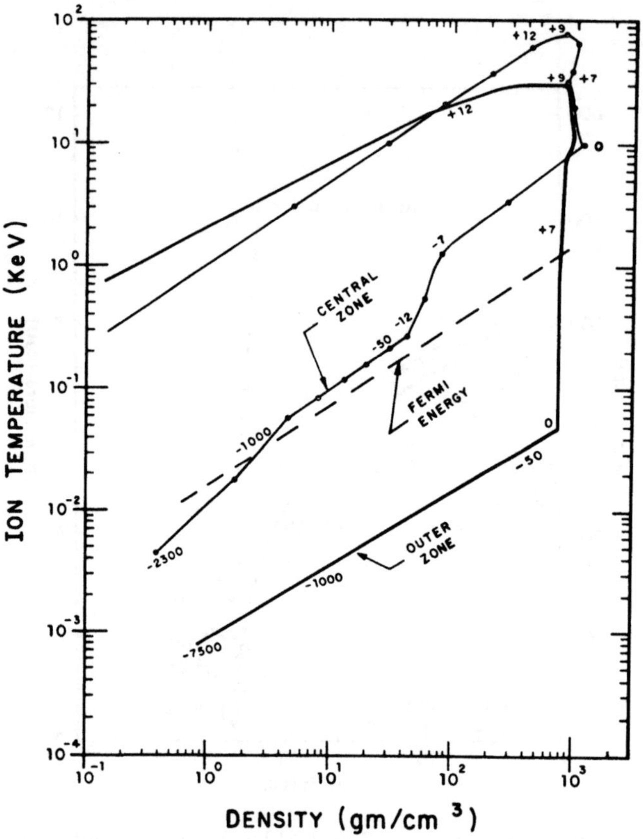

Fig. 9 - Thermodynamic histories of the central and outer regions of the dense pellet. Dashed line shows the Fermi energy. The numbers along curves are times relative to ignition (t=0 ps). The outer material is isentropically compressed to a highly degenerate state while the central material is compressed to thermonuclear ignition conditions.

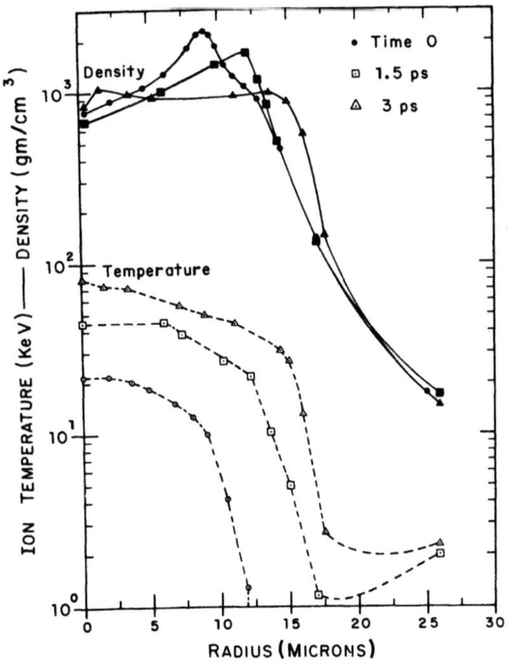

Fig. 10 - Propagation. Density and ion temperature profiles at three times spanning 3 ps during thermonuclear burn.

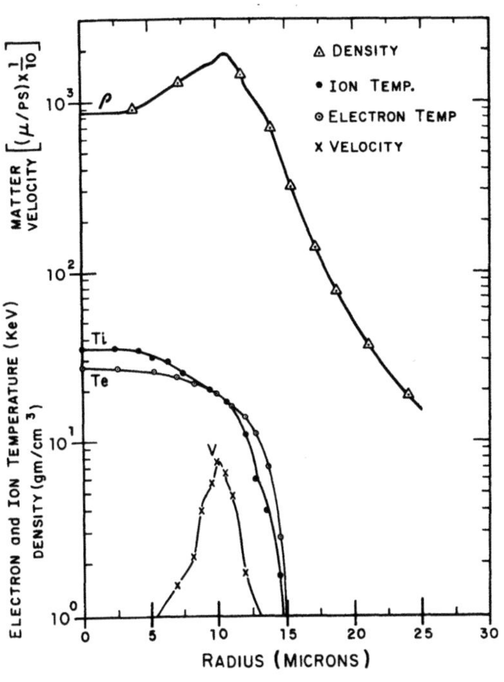

Fig. 11 - Structure of thermonuclear burn front. An electron temperature precursor is caused by deposition of most of the DT alpha particle energy in the electrons at temperatures around 10 keV as well as by thermal conduction.

A two-dimensional calculation has been made of this same pellet configuration to study symmetry and Taylor instability. The pellet was irradiated with ten laser beams consisting of adjacent equiangular (18°) conical shells symmetric about a common axis. The laser intensity varied sinusoidally ± 10% across each beam. Selected frames from a computer movie of the resulting implosion and explosion are reproduced in Figures 12 and 13. A cross section of one quadrant is shown (the horizontal axis is a line of symmetry and the vertical axis is a plane of symmetry). The lines are the Lagrangian mesh and are attached to the matter. The counters show the maximum electron and ion temperatures (TE,TI) and density (D), the inputted laser energy (EL), fusion energy (EP), time, and scale size. The thermonuclear yield produced in this two-dimensional calculation was 1750 KJ, only a few percent less than that in the corresponding one-dimensional calculation. For reasons discussed earlier, the magnetic field physics was not used in this calculation. If the calculation is run with magnetic fields, decoupling effects are observed but the symmetry is not strongly perturbed.

A two-dimensional calculation was made of the same pellet in which the thickness of the initial atmosphere was reduced by a factor of two, the laser wavelength was 1 μ (and was not varied), and the pulse shape did not contain the initial spike. The central region of the pellet is destroyed by a severe axial jet when the radius is 100 μm and the pellet surface is strongly perturbed. Figure 14. Negligible thermonuclear yield is produced.

CTR APPLICATION

Cheap fuel pellets are required for commercial power production. Ten million joules of electrical energy is currently worth roughly one cent, and most of this cent must be used to pay capital and operating costs of the power plant. In the scheme described here the pellet may be a droplet of DT (or deuterium, using multi-megajoule lasers) ejected from a sophisticated eye dropper and spherized by surface tension and viscosity effects, while freely falling in an evacuated drop tower. The pellet may also be a hollow sphere of fusion fuel. The high cost of tritium is obviated by regenerating the consumed tritium by means of neutron reactions in lithium blankets.

Thermonuclear microexplosions producing of the order of 10^7-10^8 joules (5-50 pounds TNT equivalent) may be suitable for commercial power production. Then multigigawatt-electric average power levels could be achieved by burning of the order of 100 pellets per second, perhaps 10 per second in each of 10 explosion

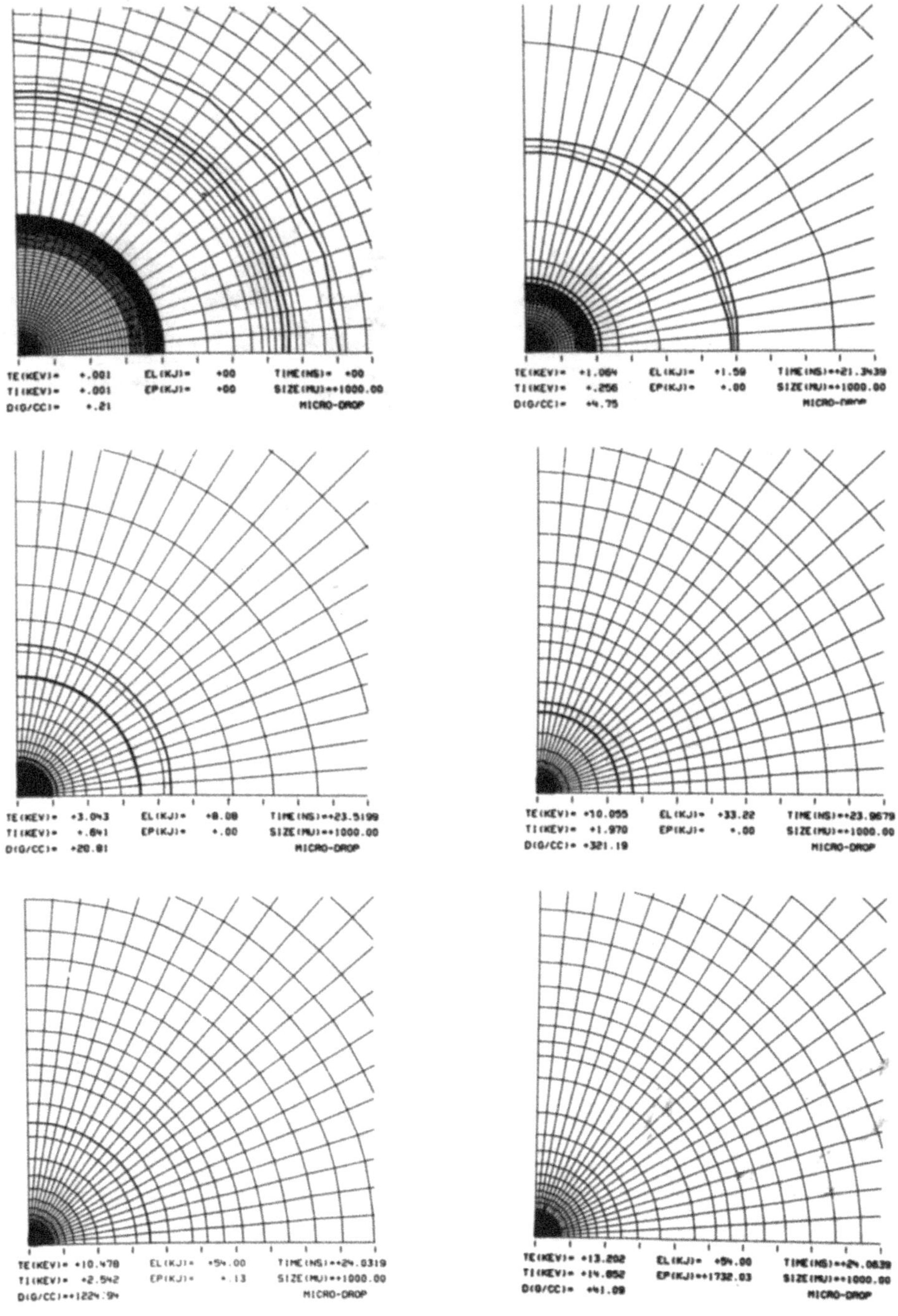

Fig. 12 - Frames from computer movie of two-dimensional implosion of a laser fusion pellet with initial radius 400 μm surrounded by an atmosphere extending beyond 1000 μm. Lagrangian mesh and electron temperature isotherms (dark arcs) are shown (0.1, 0.3, 1, 3, and 10 keV).

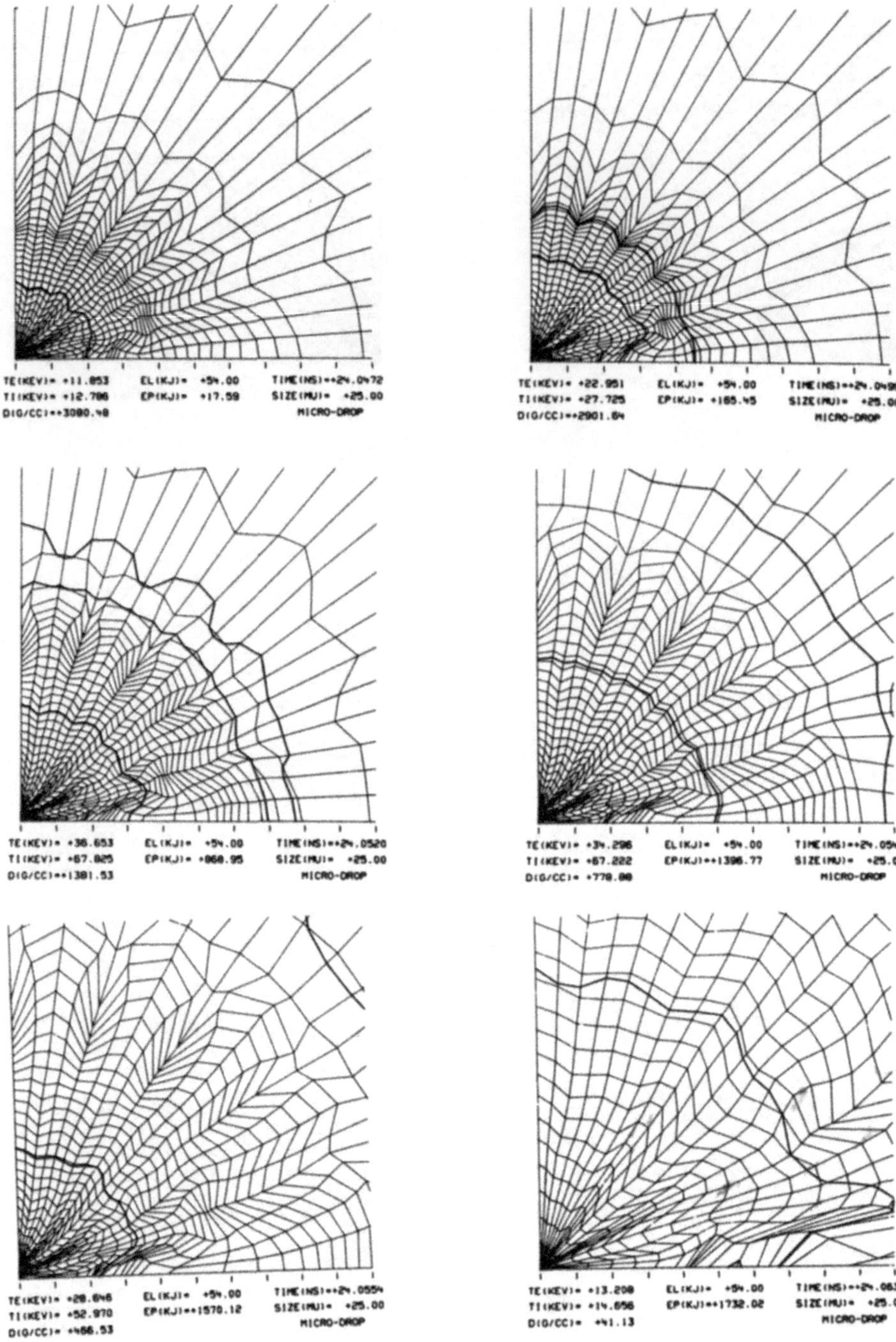

Fig. 13 - Frames from computer movie of two-dimensional calculation of thermonuclear explosion of pellet imploded in Figure 12. Dark arcs are 10, 20, and 50 keV ion temperature isotherms.

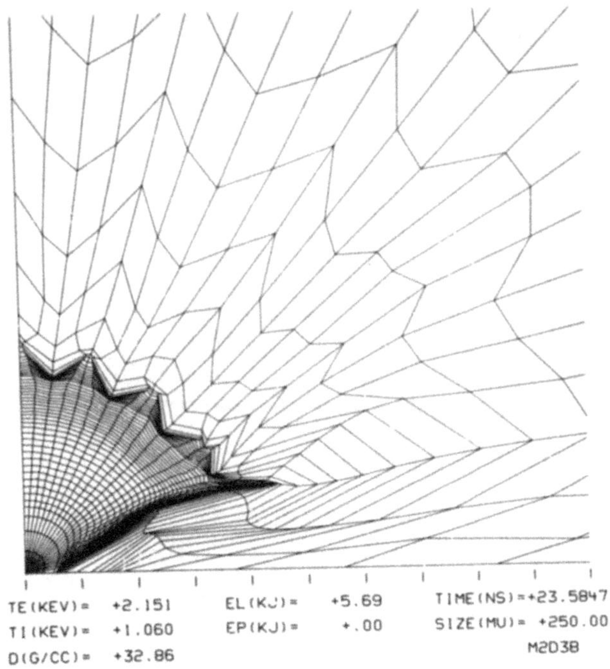

Fig. 14 - Two-dimensional failure of implosion calculation in which atmosphere was too thin and was not preheated, and in which initial laser wavelength was too short.

chambers.

The electrical power, P, is related (via energy conservation) to the laser energy, L, explosion frequency, N, gain, G_L, and thermal-electric efficiency ϕ_L by

$$P = LN\ (G_L\ \phi_t - \phi_L^{-1})\ .$$

where it is assumed the laser is electrically pumped and the hot gas exhausted from the laser is not used to generate electricity. Less than one third of the electrical energy circulates internally if $G_L\ \phi_t \geq 3\ \phi_L^{-1}$. Then if $\phi_t = 0.4$, $\phi_L = 0.1$, G_L must be 75 or more, which occurs for $L \sim 10^5$-10^6 joules--see Figure one. Consequently if $P = 10^9$ watts, $L = 1/2 \times 10^6$ joules and $N = 100\ sec^{-1}$.

The explosive impulse from a fusion energy pulse is much smaller than that from a chemical explosion of the same energy, because the fusion pellet weight is $< 10^{-6}$ that of the chemical, and the impulse generated is proportional to the square root of the mass. However, a specially designed combustion chamber is required to withstand the neutrons, x-rays, and hot plasma. A chamber several meters in diameter can be designed to withstand the neutron flux. The walls may be protected from x-rays and plasma by a thin film of Li[21].

Conventional thermal cycles may be used to generate electricity if the fusion neutrons are absorbed in lithium blankets. If relatively large DT pellets or essentially pure deuterium pellets are burned (requiring a multi-megajoule laser energy), the fusion neutrons will deposit much of their energy in the fuel plasma, and this plasma may be expanded against a magnetic field, and much of the fusion energy directly converted to electricity[22]. Then higher electrical generating efficiencies may be achieved. With deuterium pellet burning, lithium utilization and tritium storage and cycling may be greatly reduced.

Hybrid reactors--in which the 14 MeV DT neutrons which escape from the combustion chamber are used to fission natural uranium (or thorium)--may be capable of generating more energy than used to pump the laser even with low efficiency, low energy lasers (e.g. 1% efficient, 10 KJ lasers).

A high energy, high efficiency, possibly short wavelength laser which is sufficiently cheap and capable of firing up to 100 times per second must be developed for CTR applications. This laser, the associated electrical power source, and the explosion chamber must be designed to function up to 10^{10} times without replacement. The overall capital cost of this equipment must be less than about $100 per installed kilowatt (electrical). Although these are formidable technological problems, it is believed that they are soluble.

Acknowledgments

I would like to acknowledge important and useful discussions with S. Bodner, J. DeGroot, J. Katz, and W. Kruer. The calculations were made by A. Thiessen and the LASNEX computer program was developed by G. Zimmerman. This work was done in collaboration with my colleague L. Wood who has made many key contributions.

REFERENCES

1. J. Nuckolls, L. Wood, A. Thiessen, G. Zimmerman, Nature 239, 139, (1972).
2. E. Teller, Science 121, 267 (1955).

3. J. Nuckolls, et al., Livermore report UCRL-74116 (1972).
4. J. Mayer, M. Mayer, Statistical Mechanics, Wiley (1940), p. 385.
5. T. Weaver, G. Zimmerman, L. Wood, Livermore report UCRL-74191/ UCRL-74352 (1972).
6. J. Shearer, UCRL-72400 (1970).
7. W. Kruer, private communication.
8. J. Katz, J. Weinstock, W. Kruer, J. DeGroot, R. Faehl, Livermore report UCRL-74334 (1972).
9. R. Kidder, Physics of High Energy Density, (1969), Academic Press, p. 315.
10. R. Kidder, J. Zink, Nucl. Fusion 12, 325 (1972).
11. J. Stampen, et al., Phys. Rev. Lett. 26, 1012 (1971).
12. S. Braginskii, Reviews of Plasma Physics 1, 2-5 (1965).
13. L. Wood, et al., Livermore report UCRL-74115 (1972).
14. G. Taylor, Proc. Royal Society 201A, 192 (1950).
15. C. Leith, LLL internal report (1962).
16. G. Zimmerman, Livermore report UCRL 50021-72-1, 107 (1972).
17. D. Post, J. Wilson, LLL internal document (1972).
18. W. Lokke, J. Ramus, LLL internal document (1972).
19. J. Chang, G. Cooper, Jour. of Comp. Phys. 6, No. 1, 1 (1970).
20. S. Bodner, Livermore report UCRL-74074 (1972).
21. L. Booth, et al., Los Alamos report LA-4858-MS, Vol. 1 (1972).
22. B. Freeman, L. Wood, J. Nuckolls, Livermore report UCRL-74486 (1971).

THEORY AND EXPERIMENT IN LASER DRIVEN FUSION *

Keith A. Brueckner

KMS Fusion, Inc.

Ann Arbor, Michigan

The production of fusion energy from a pellet of thermonuclear fuel can be achieved on a level useful for power production only if the pellet is highly compressed with efficient energy transfer from the external energy source into the pellet. The simple model of a uniformly compressed DT sphere can be used to determine the fusion energy production. Figure (1) gives the ratio of fusion energy output to initial thermal energy for a uniform initial temperature of 5 kev. The energy multiplication, for an initial thermal energy of one kilojoule, is 5 at a density of 300 gm/cm^3, 16 at 600, 40 at 1000, and 80 at 2000. For high energy input on high compression, the energy multiplication levels off at about 200 corresponding to about 35% burnup of the DT. The energy multiplication can be increased if the fuel is only centrally heated to the ignition point of 5 kev, with the rest of the fuel ignited by an expanding supersonic burning front propagating outward from the fuel center. Figure (2) shows a typical example of the propagation of a supersonic burning front. Figure (3) shows the energy multiplication with the fuel center heated to 5 kev over a few micron radius and the rest of the fuel at 500 ev. With an initial thermal energy of one kilojoule, the energy multiplication is 130 at $\rho = 600$ gm/cm^3, 400 at $\rho = 1000$ gm/cm^3, and 700 at $\rho = 2000$ gm/cm^3. The energy multiplication reaches a maximum of about 1200 for initial thermal energy of 5-10 kilojoules, independent of initial density, corresponding to about 35% fuel burnup. The effect of the centrally-initiated burning wave increases the energy multiplication by about a factor of ten over the uniformly heated case.

* "Presented at the Third Workshop on 'Laser Interaction and Related Plasma Phenomena' held at Rensselaer Polytechnic Institute, Troy, New York, August 13-17, 1973".

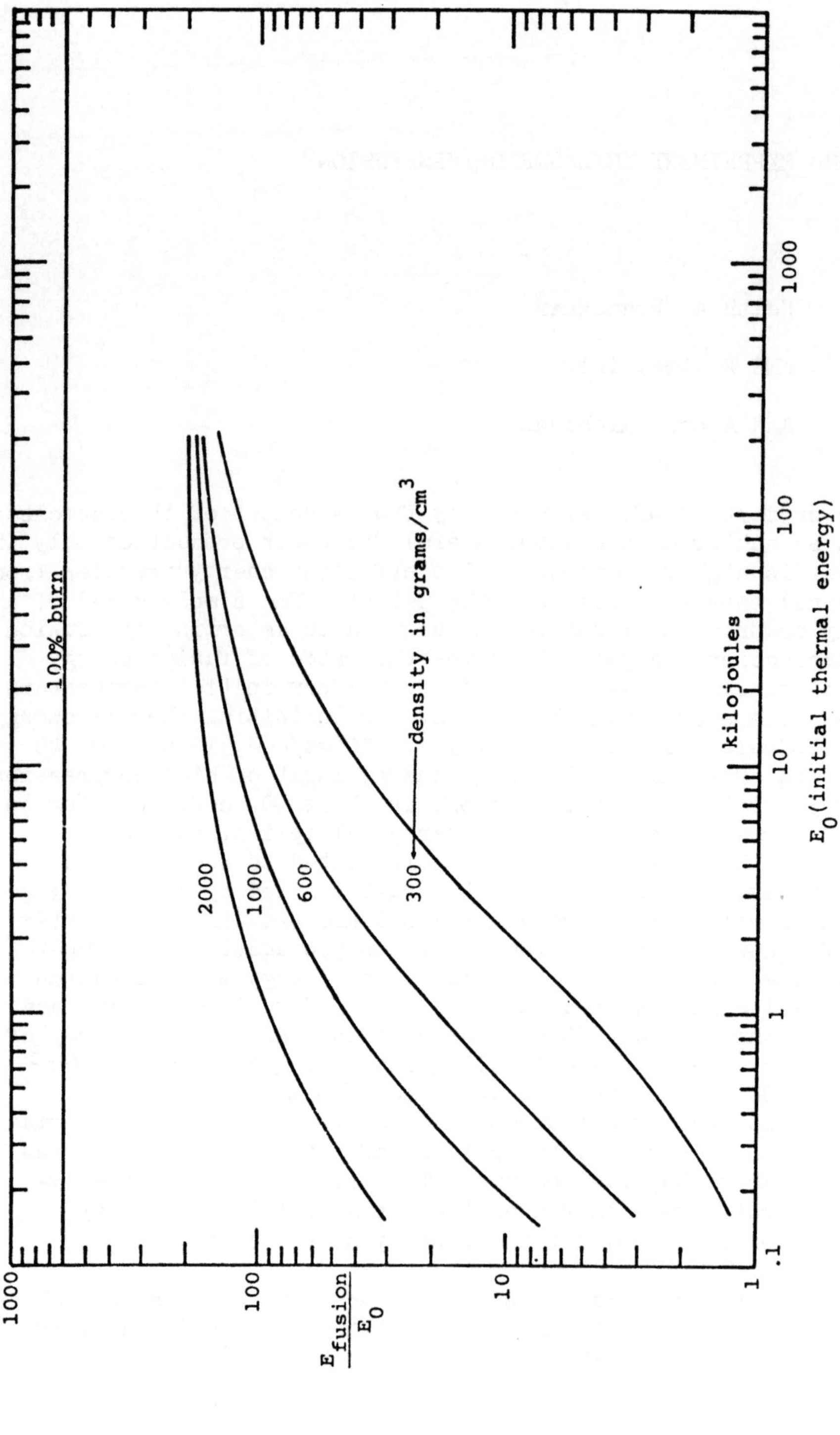

FIGURE 1 - Ratio of fusion energy to initial thermal energy for a uniformly compressed DT sphere initially heated to 5 kev.

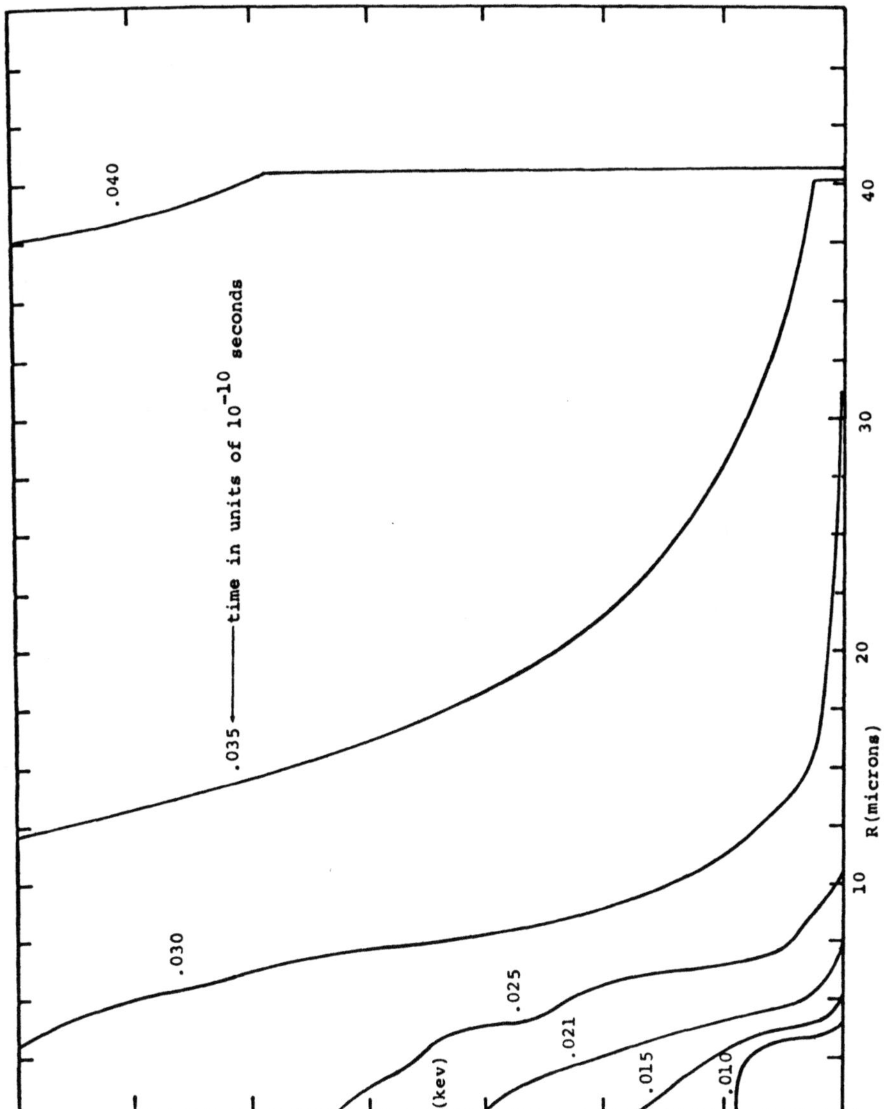

FIGURE 2 - Thermonuclear burning front in DT at an initial density of 2000 gm/cm^3.

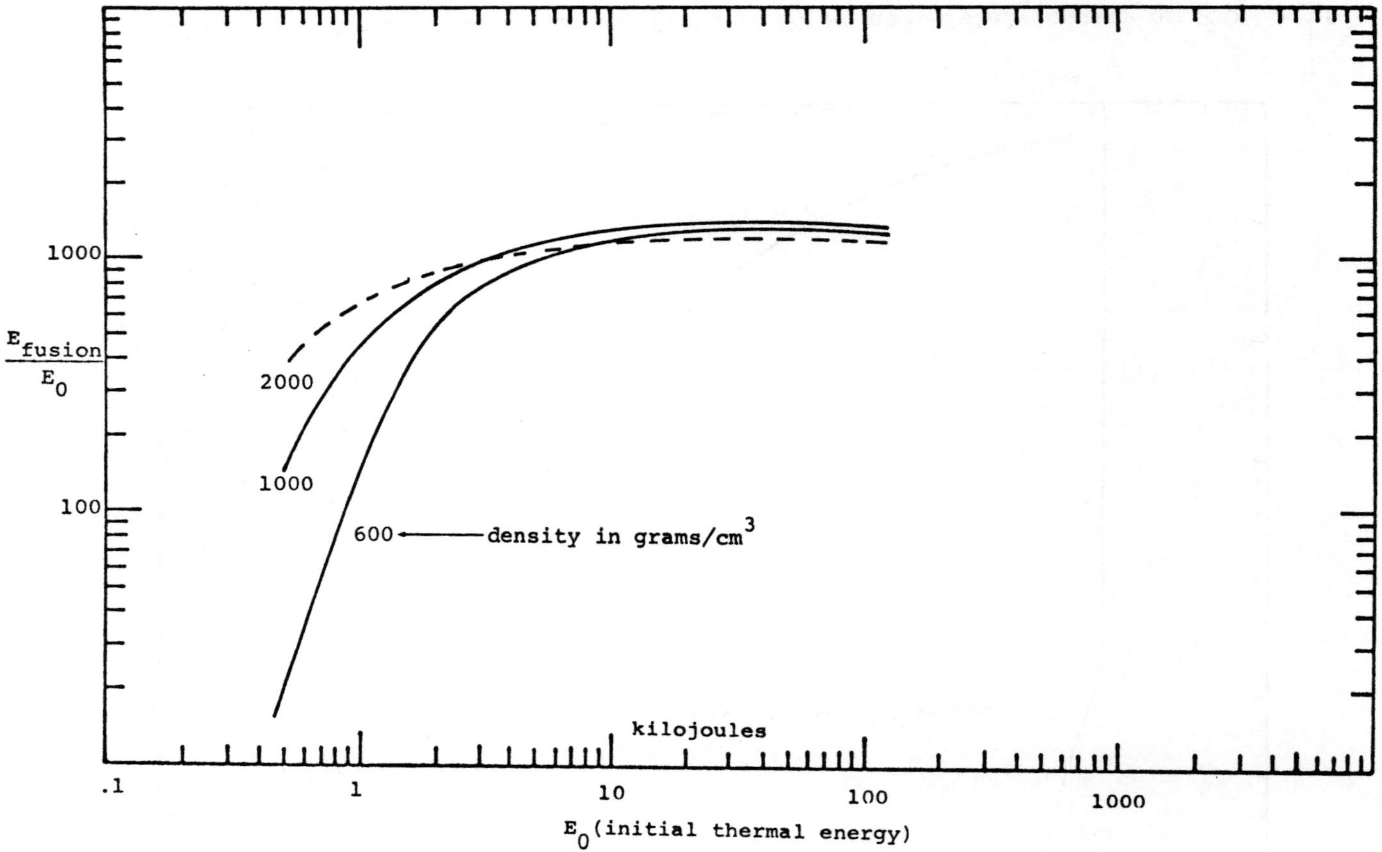

FIGURE 3 - Ratio of fusion energy to initial thermal energy for a uniformly compressed DT sphere initially heated to 5 kev over a few micron radius, with the rest of the DT at 500 ev temperature.

Calculations of the implosion of a DT sphere in spherical symmetry show that high compression can be produced by a laser pulse with proper time variation. The pressure driving the implosion is produced by the penetration of energy from the underdense laser deposition region into the dense plasma which results in ablation of the dense pellet surface. The efficiency of this process depends on the efficiency of energy absorption from the laser beam into the underdense plasma and on the subsequent energy partition between the compression of the dense pellet core and the energy removed in the high temperature expanding plasma produced by the ablation process. The latter partition can be estimated from a hydrodynamic model or determined by computer simulation of the energy deposition, energy transfer, and hydrodynamic processes. The result is that 6 to 10% of the absorbed energy is transferred into compression and heating of the dense pellet core.

Detailed computer simulations of the full process of laser coupling for 1.06 micron wavelength, thermal energy flow, hydrodynamics, nuclear reactions, and of the energy transport in the nuclear reaction products and radiation, give results in approximate agreement with the numbers just given. Of particular interest for the experiments now being undertaken by several groups in the USA and other countries is the prediction that the "breakeven" condition, with fusion energy equal to laser energy, can be reached with laser energy of about one kilojoule. This prediction holds, however, only if several conditions are satisfied. These are:

A. Configuration requirements
   1) laser pulse time variation properly matched to pellet configuration
   2) spherical symmetry of pellet illumination
   3) spherical symmetry of pellet configuration

B. Physics requirements
   1) stable hydrodynamic motion
   2) adequate laser-plasma coupling
   3) absence of appreciable pellet preheat.

If these requirements are not met, the breakeven energy can be very markedly affected. Particularly striking is the effect of a poorly matched laser pulse. The breakeven energy for a square laser pulse and a DT sphere is several hundred megajoules. A drop in compression of a factor of ten as a result of imperfect convergence can increase the breakeven energy by a factor of ten to one hundred.

The configuration requirements on the laser energy variation in space and time, although difficult to meet, can be satisfied with properly designed illumination systems and laser oscillators giving controllable sequences of stacked pulses. The pellet symmetry can also be provided by careful pellet fabrication and selection methods. We have studied the effect of variation in the laser and pellet parameters, using 2-dimensional computer simulation, and determined the allowable departures from complete symmetry. These conditions are imposed on our laser system and pellets.

The problems of the physics of the laser-driven process are too difficult to resolve without experimental results, although very important analyses and calculations have been made.

The stability of the pellet implosion has been studied analytically and by computer simulation, using a 2-dimensional code. The results show stable motion, with initial disturbances not being amplified during the implosion. The laser-plasma coupling presents difficult problems which are of a complexity very familiar for the past two decades in the controlled-fusion programs. Closely associated with the coupling problem is the effect of anomalous laser-coupling on the energy flow into the pellet. Present theories estimate that a wide range of anomalous phenomena can occur which may seriously alter the predictions of the laser-driven process. Experiments are intended to resolve these uncertainties.

We have carried out a number of experiments using a neodymium-glass laser brought to full operation during July of this year. The laser configuration is shown in Figure (4). The laser driving the main amplifier train is a VK800 laser built by CGE, with some modifications and with Owen-Illinois ED-2 glass replacing the original French laser glass. This laser operates reliably on a six minute cycle with an energy output from the 80 millimeter output amplifier of 250 to 350 joules. A considerably higher output is possible, but has not been used because of possible glass damage from self-focusing. The output from the 80 mm amplifier is expanded to 100 mm diameter and further amplified in seven amplifier modules built by GE. Each module contains three disks of glass at Brewster's angle. The path length of the laser beam in each disk is 3 centimeters and the effective aperture is 122 cm^2. The glass stores 0.32 joules/cm^3 with 8 kev pumplamp voltage and 0.36 joules/cm^3 with 9 kev pumplamp voltage. With 200 joules input with 3 nsec pulse width (FWHM) to the GE system, the measured output from the first six modules is approximately 840 joules at 8 kev flashlamp voltage. The predicted output of seven modules at 8 kev is 990 joules and at 9 kev approximately 1400 joules. The measured gain is in good agreement with the design predictions.

The laser is protected against damage from reflected laser energy by Pockel cell isolators at the entrance to the 16 mm rod of the main amplifier chain, between the 23 and 32 mm rods, by a Faraday rotator between the 45 and 64 mm rods, and by a Faraday rotator at the exit of the GE amplifiers. The protection is adequate against the full output energy reentering the exit end of the rotator.

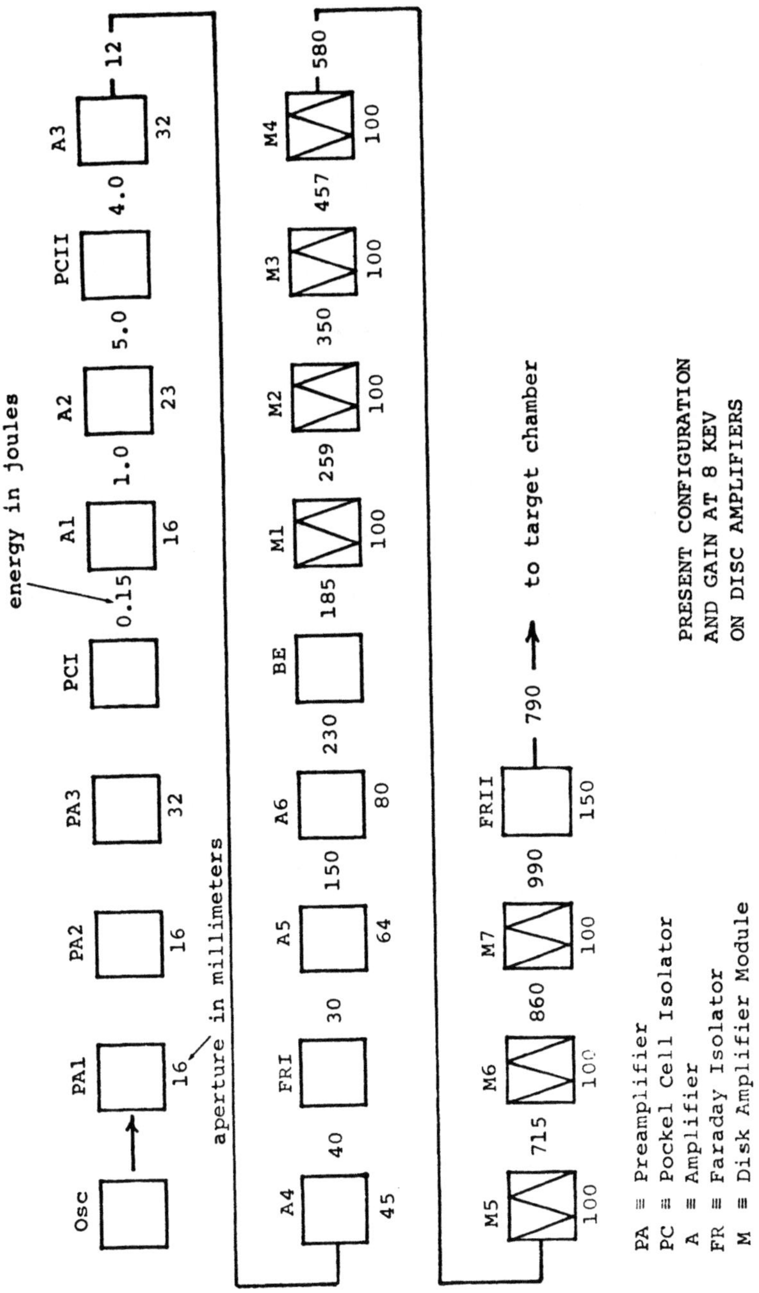

FIGURE 4 - Configuration of KMSF laser.

The laser pulse beam from the CGE oscillator is approximately Gaussian with a pulse width (FWHM) of 1.3 or 3 nanoseconds. The pulse is strongly distorted through the CGE and GE amplifiers due to partial saturation of the amplifiers. A pulse-stacking oscillator has been built to replace the CGE oscillator to give controllable pulse beams. The arrangement of the oscillator is shown schematically in Figure (5). This oscillator will be installed by mid-August.

The target area is shielded to allow breakeven experiments at the kilojoule level. The arrangement of the target area is shown in Figure (6). The chamber configuration allows the measurement of X-ray spectra by spectrometers, fast diodes, thermoluminescent diodes, and photographic plates. The neutron production is measured with several calibrated scintillators with large aperture and with time resolution of a few nanoseconds. In addition, the integrated neutron production is measured through silver foil activation. Provision has been made for time-of-flight measurement of the neutron energy spectrum. For measurement of the spectrum from a single neutron pulse, a DT neutron yield of about 10^8 is required.

The illumination system of the present laser configuration is shown in Figure (7). The illumination of spherical targets is uniform to 5-10%, the indicated correction plates perturbing the laser flux sufficiently to give uniform absorbed intensity on the target, after correction for the non-normal incidence. Other lens and mirror arrangements are being completed which will give further improvement in the illumination pattern.

Two-dimensional, cylindrical geometry computer simulations have been made of the response of CD_2 shells to non-uniform illumination by the KMS Fusion laser. Two configurations are shown here. The first is two-sided illumination with two f/2.6 lenses. The second is two-sided illumination with two f/1.0 lenses. The calculated energy absorptions versus angle from the axis of symmetry are plotted in Figure (8) for the two cases. The calculated plasma distribution at the time of collapse of a spherical 2 μm thick shell of 200 μm diameter is plotted in Figures (9) and (10) for the f/2.6 and f/1.0 lenses, respectively. The grossly non-spherical convergence resulting from the f/2.6 lenses is in striking contrast to the much improved convergence resulting from using f/1.0 lenses.

The reflectivity of targets under intense laser illumination has been the subject of intensive theoretical and computational analysis and of more limited experimental study. The theory of laser deposition predicts two principal classes of instabilities resulting from coupling of ion density waves, electron plasma oscillation, and the incident and reflected laser waves. One class of instabilities resulting from the excitation of transverse

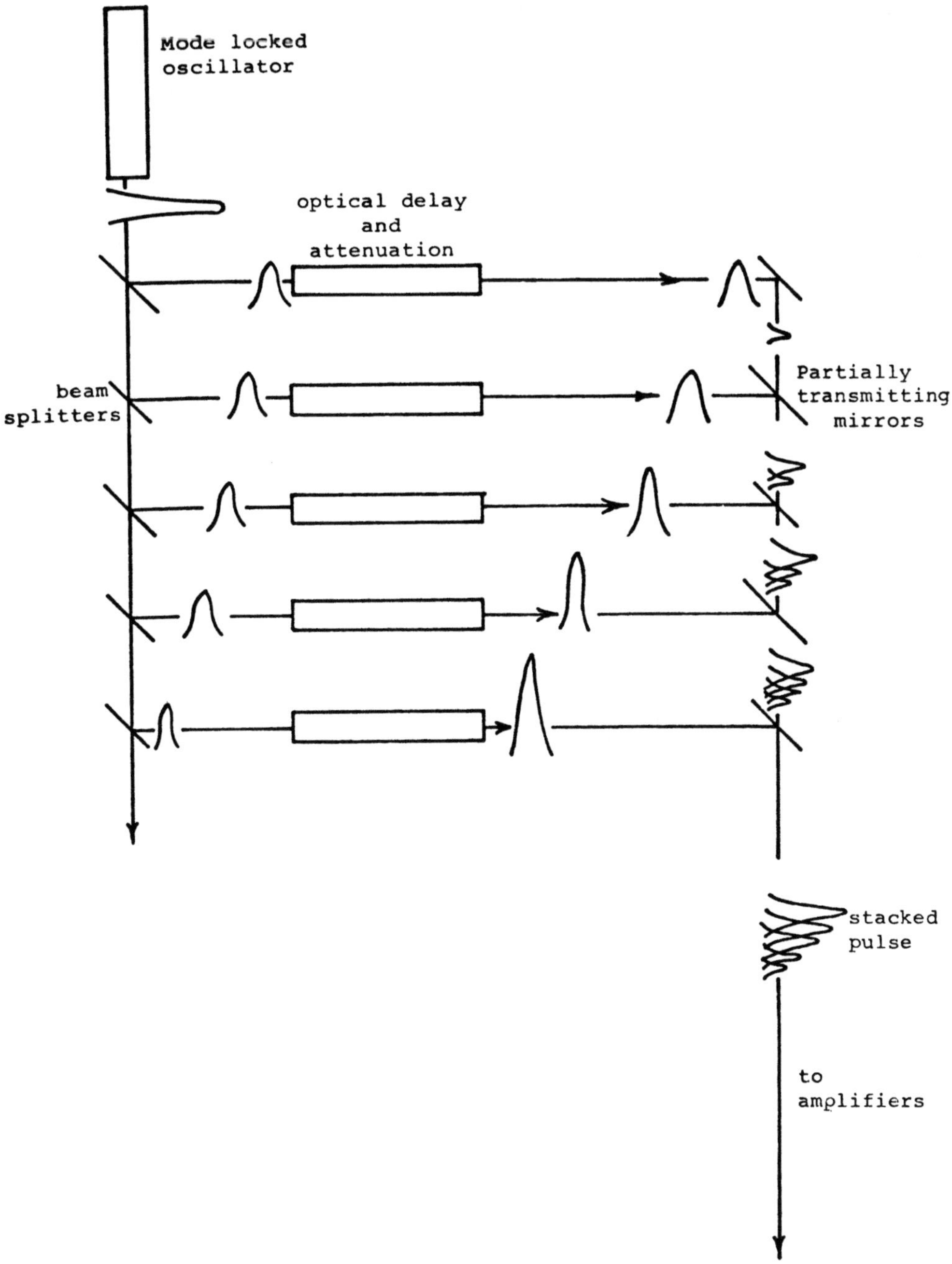

FIGURE 5 - Schematic layout of pulse stacker.

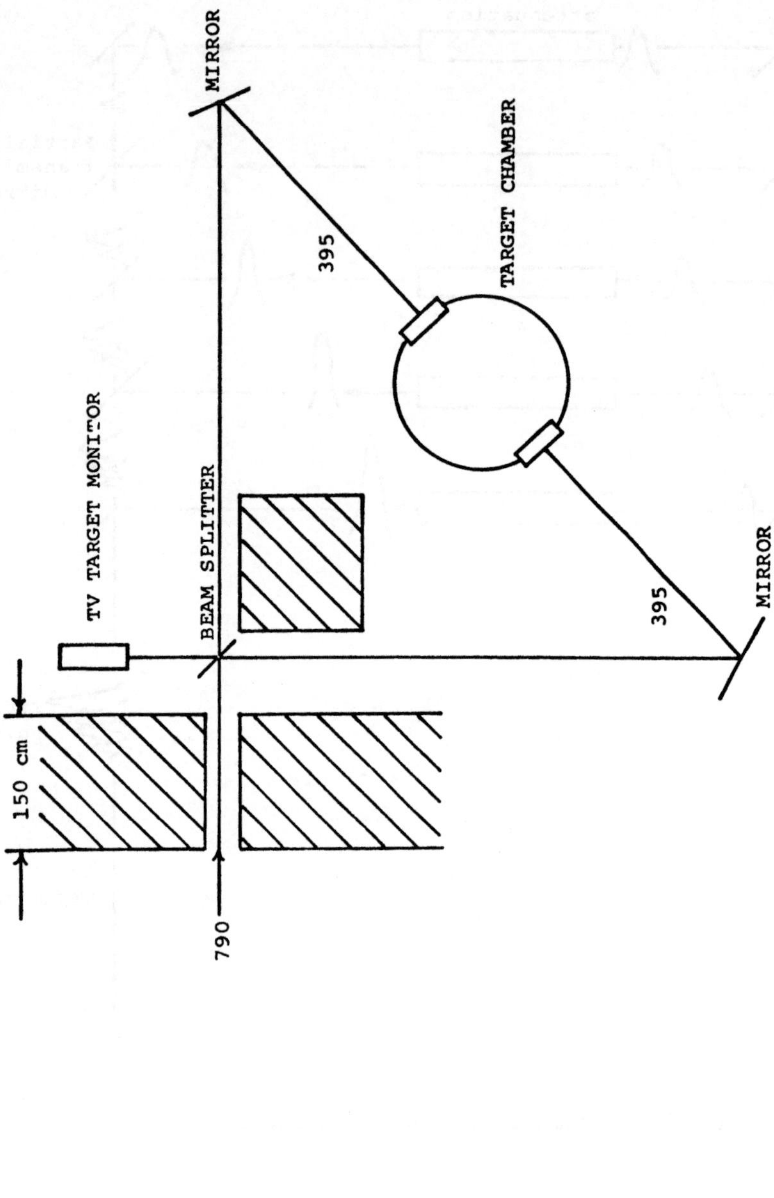

FIGURE 6 - Arrangement of target area.

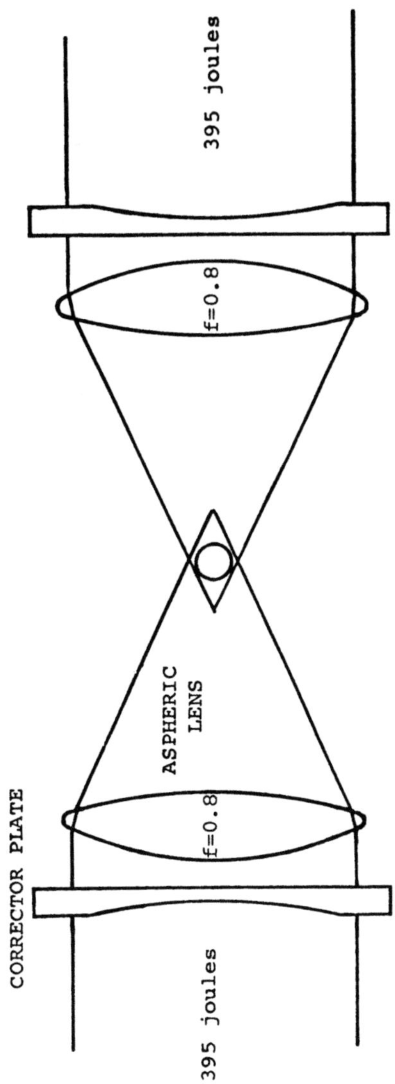

FIGURE 7 - Illumination system.

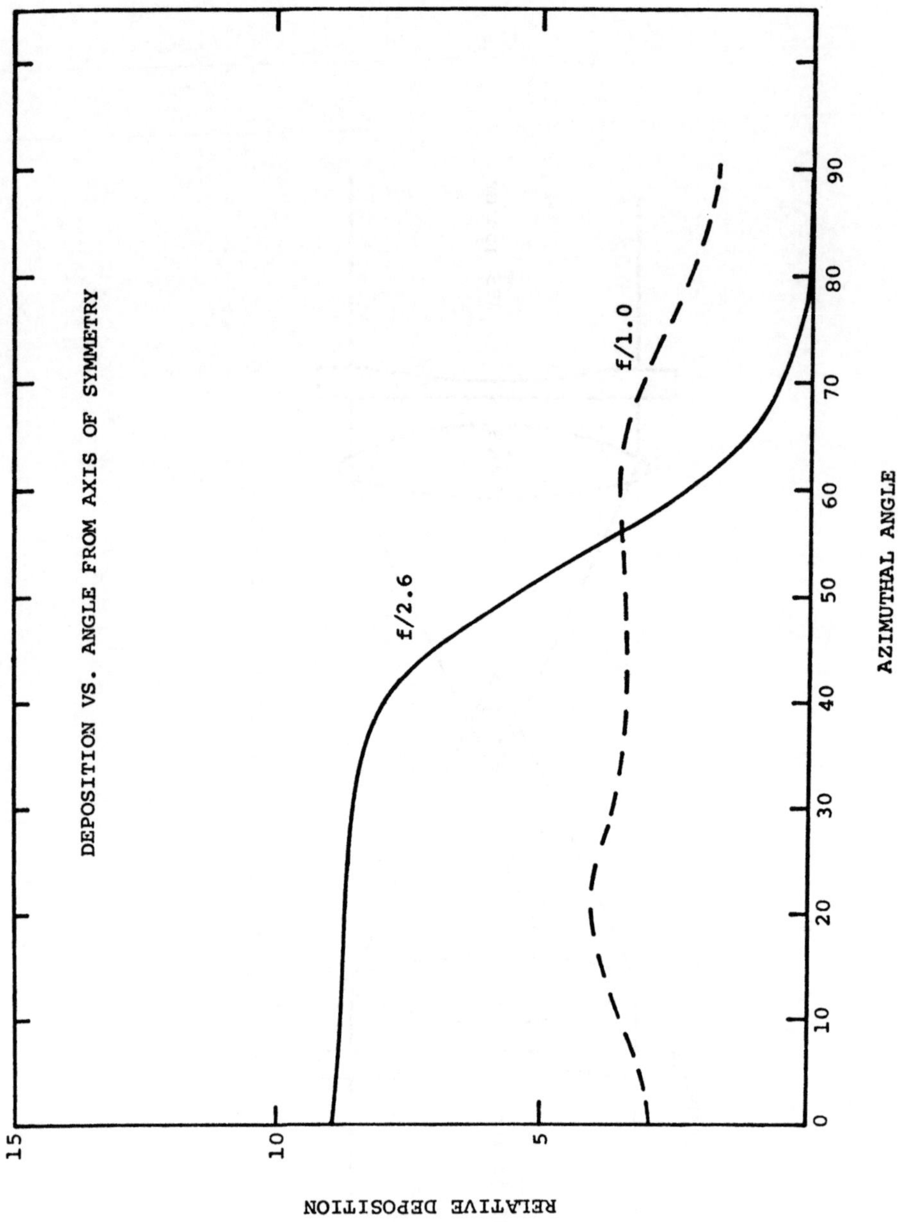

FIGURE 8 - Calculated energy absorption for two lens configurations.

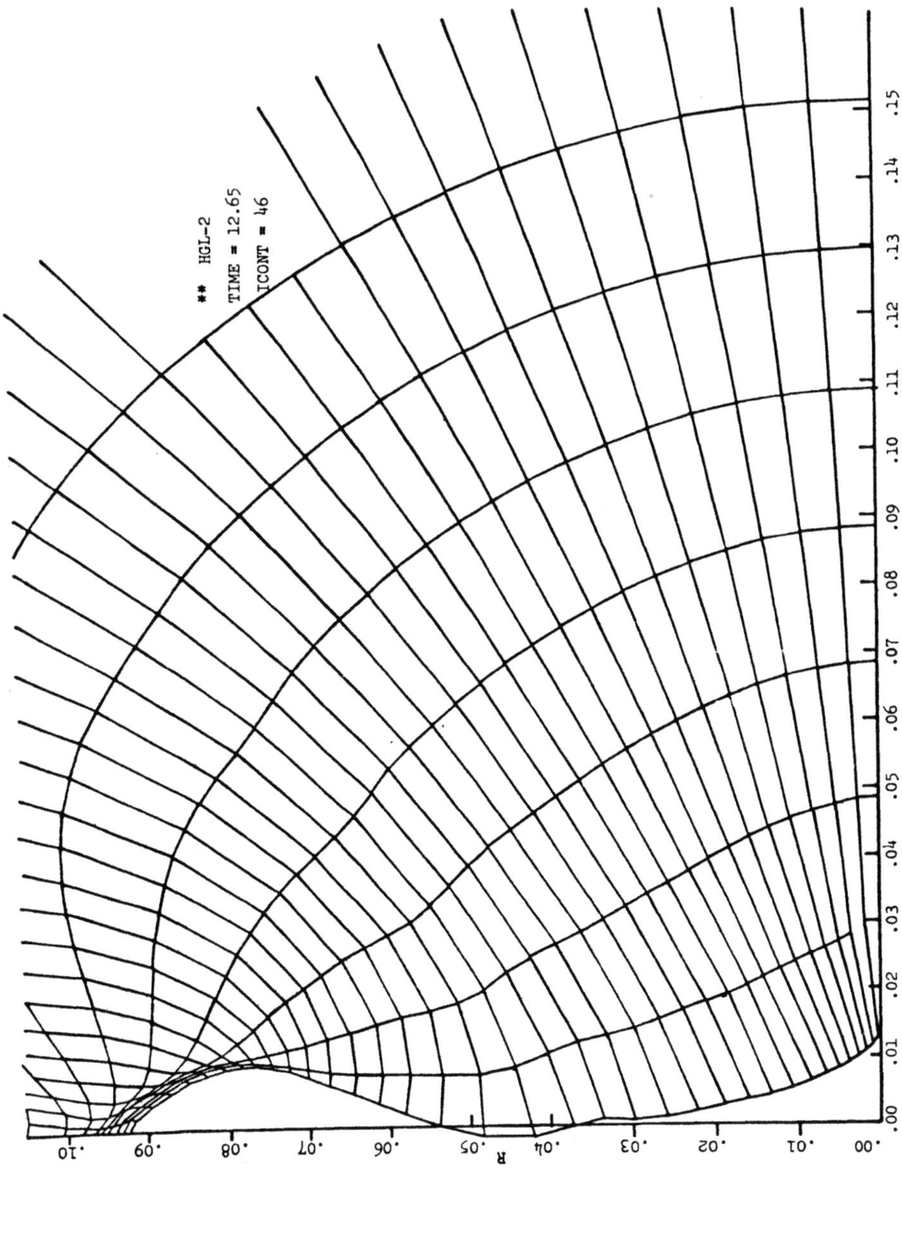

FIGURE 9 - Two-dimensional calculations of shell implosion. Dimensions are given in millimeters. The time unit is 10-10 seconds.

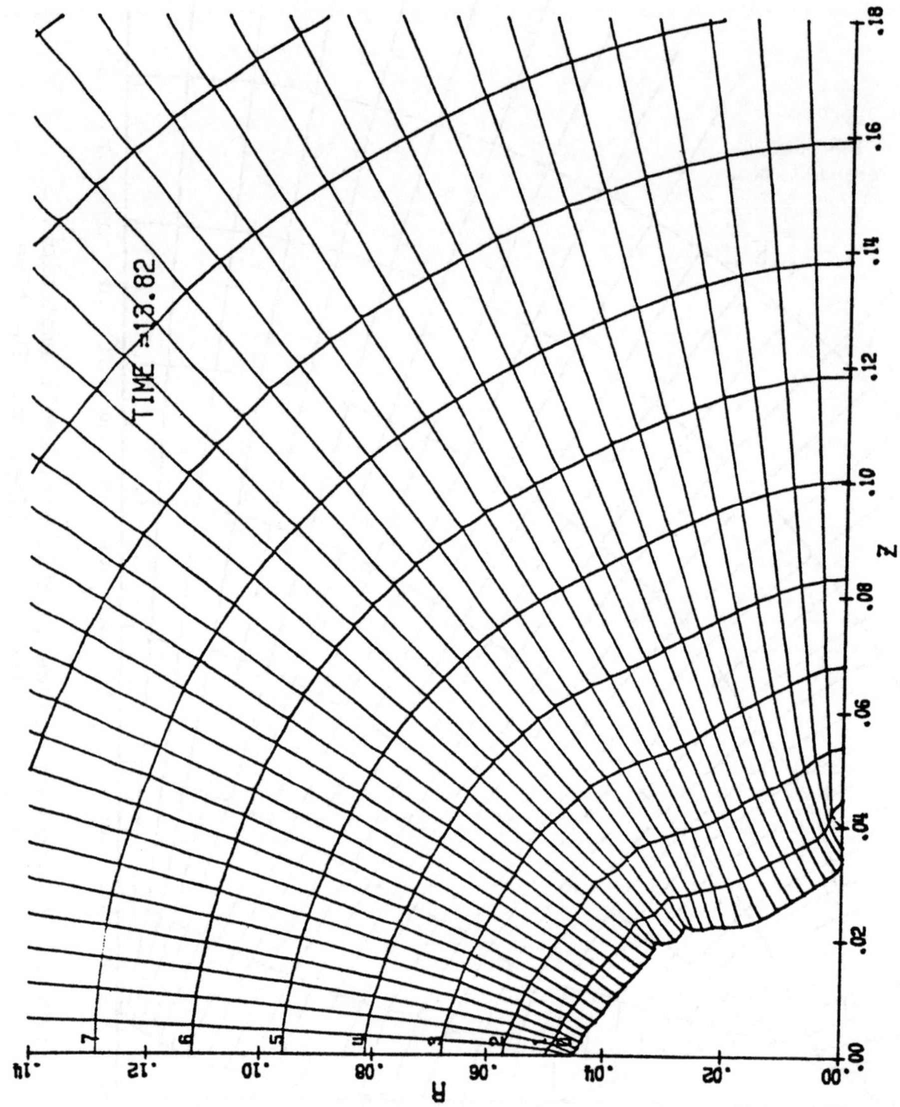

FIGURE 10 - Two-dimensional calculations of shell implosion. Dimensions are given in millimeters. The time unit is 10-10 seconds.

ion and plasma waves with wavelength much less than the laser wavelength is expected to increase the laser energy absorption and to produce a marked increase in electron energy together with a strong departure of the energetic electron distribution from Maxwellian. Another class resulting from longitudinal ion or electron density fluctuation, i.e., stimulated Brillouin or Raman scattering, is predicted to lead to a marked increase in reflectivity, possibly saturating at high laser power with reflectivity coefficient approaching unity. The threshold for the instability is expected to depend on target material and the illuminated area and pulse length, the latter since they determine the density gradients in the region of laser deposition. The thresholds are, however, expected to be reached for one-micron laser wavelength for power in the range of 10^{12} to 10^{13} watts/cm^2, for illuminated areas with characteristic dimensions of the order of 100 microns.

We have measured target reflectivity and electron temperature using an energy output up to 300 joules in a pulse with 3 nanosecond duration (full width at half maximum), delivered on plane CD_2 and CH_2 targets through a f/1.5 aspheric lens giving a measured vacuum focal spot of 60-80 microns diameter. The laser energy incident on the target chamber was measured calorimetrically; the energy on target was obtained by correction for window and focusing lens losses. The reflected energy was measured with fast photodiodes which directly compared a reference signal reflected from the incident laser pulse with the reflected energy from the target. An independent measurement was made calorimetrically of the energy from the target reflected from the 80 mm output face of the output amplifier of the laser. Relative measurements of the target reflections were also made calorimetrically at stations in the amplifier train of the laser.

The apparent electron temperature for photon energies in the 8-12 kev range was measured by a pair of fast diodes with aluminum foil attenuators. More complete measurements over a wide range of photo energies were also made on selected laser pulses using a film detector with graduated aluminum attenuators.

The reflectivity was found to vary rapidly with target position, with the maximum reflectivity associated with the vacuum focus 50 to 75 microns above the target surface. This position also corresponded with the point of maximum hard X-ray yield and with the maximum soft X-ray flux.

The measured reflected energy was that collected by the 8 cm diameter f/1.5 illuminating lens. This target reflection was also monitored at 45° and appeared to be very low, in agreement with previous measurements of the angular distribution of the reflected energy. We cannot at the present, however, exclude the possibility that some diffuse scattering occurs, increasing the true target reflectivity over our measurements. Improved reflectivity measurements will be carried out in the near future.

The variation of CH_2 and CD_2 reflectivity with target position for laser energy on target in the range of 100 to 110 joules is given in Figure (11). Figure (12) gives the measured peak reflectivity in CH_2 and CD_2 as a function of laser power on target, for a 80 micron vacuum focal diameter. The reflectivity peaks in the range of 40 to 80 joules on target or 3 to 4×10^{14} watts/cm^2, and drops by a factor of two to three at the maximum energy of 160 joules on target or 10^{15} watts/cm^2. The high peak reflectivity at 3×10^{14} watts/cm^2 may be associated with the onset of stimulated Brillouin scattering and the subsequent drop to saturation of the stimulated Brillouin scattering and the onset of anomalous absorption.

The variation of reflectivity during the laser pulse also clearly shows the onset of markedly increased absorption at high laser power. For energy less than approximately 40 joules, the reflected signal has the same time variation as the laser pulse. As the power increases, however, the reflected signal shows increased absorption, the signal becoming markedly distorted and the reflected power appearing to saturate. The details of the reflected pulse show some pulse-to-pulse variation, probably associated with small displacements of the target position relative to the laser focus. A characteristic pulse form at high power shows a strong late reflected signal following a saturated reflectivity plateau. This is an indication of the rising reflectivity with decreasing laser power for the strongly heated plasma produced by the maximum laser power.

We have also observed neutron production associated with a laser pulse directed on a solid CD_2 target. The neutron production has been observed with laser energy delivered to the CD_2 target in the range of 65 to 175 joules, corresponding to a peak power at the target in the range of 4×10^{14} to 11×10^{14} watts/cm^2. The plane target was positioned at the point of maximum reflectivity which also corresponded with the point of maximum production of hard X-rays. The neutron yield was measured with two plastic scintillators heavily shielded against X-rays, placed at 25 to 30 centimeters from the target. The time resolution of the scintillators was calibrated against a plutonium-beryllium source of known strength and against a beta source giving calibration pulses.

The characteristic signals detected consisted of an initial pulse of hard X-rays (several hundred kev) followed by a neutron pulse at 12-15 nanoseconds after the hard X-ray pulse. Delayed neutrons were also observed following the X-ray pulse by 60 to 70 nanoseconds. In several cases the delayed neutron signal was much

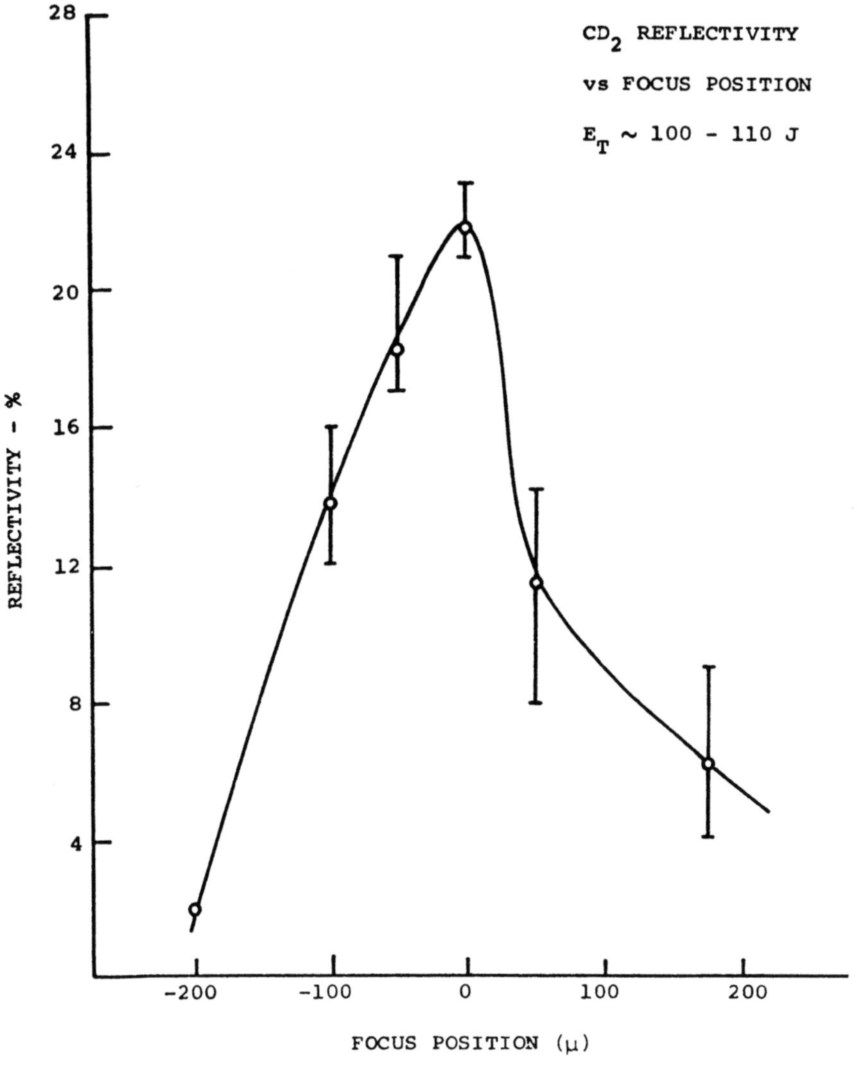

FIGURE 11 - Variation of target reflectivity with position.

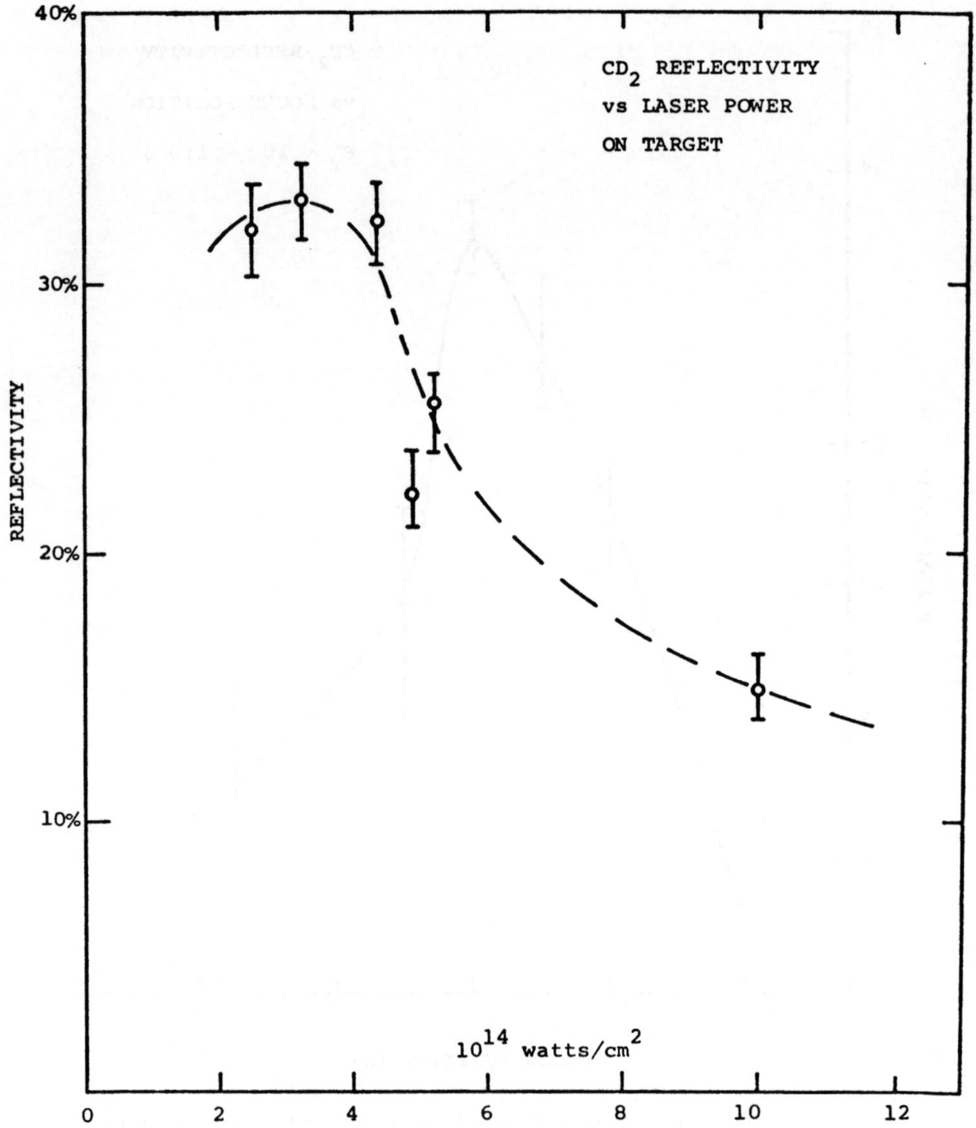

FIGURE 12 - Variation of target reflectivity with laser power.

stronger than the prompt neutron signal and occurred with a time spread of tens of nanoseconds.

The prompt neutron signals were not observed with laser energy under approximately 50 joules on target and were not always seen at the maximum energies of 175 joules on target. The strongest prompt pulses observed corresponded to several neutrons on the detector. From the detector sensitivity and geometry we infer a total neutron production of 4000 to 8000, with a weak dependence on laser energy. The time delay of 12-15 nanoseconds from the hard X-ray signal is in agreement with the transit time of the neutron from the target to the detector; we conclude that these neutrons are produced by DD reactions in the target.

The delayed neutron pulse, which is often much more intense than the prompt pulse, might arise from primary neutrons from the target scattered in the chamber walls or from other more localized scattering sources distributed around the chamber. This explanation can, however, be ruled out since 1) the secondary scattered signal can be estimated and is easily seen to be considerably weaker than the prompt signal, and 2) the scattered neutrons should show time delays continuously distributed from the onset of the prompt neutron pulse to many tens of nanoseconds, in contradiction with the clearly defined delay of 50-70 nanoseconds observed. The delayed neutron pulse is consistent with fast deuterons from the target impacting on CD_2 target debris from previous laser pulses distributed as a contaminant on the exposed inner surfaces of the chamber and on the illuminating lens.

The response of thick deuterated polyethylene foils to the KMS Fusion CGE laser has been computed in two-dimensional cylindrical geometry. The single fluid, two-temperature plasma model is used. Thermal conduction and electron-ion energy exchange are included. Shocks are treated by introducing a von Neumann-Richtmyer artificial viscosity.

The computed peak electron temperatures, θ_e, and neutron production, N, for laser energy outputs, E_L, of 100j, 200j, and 300j are presented in Table I for various spot sizes, R_0. (The spatial distribution of the radiation is taken to be proportional to $\exp\left[-(r/R_0)^2\right]$.)

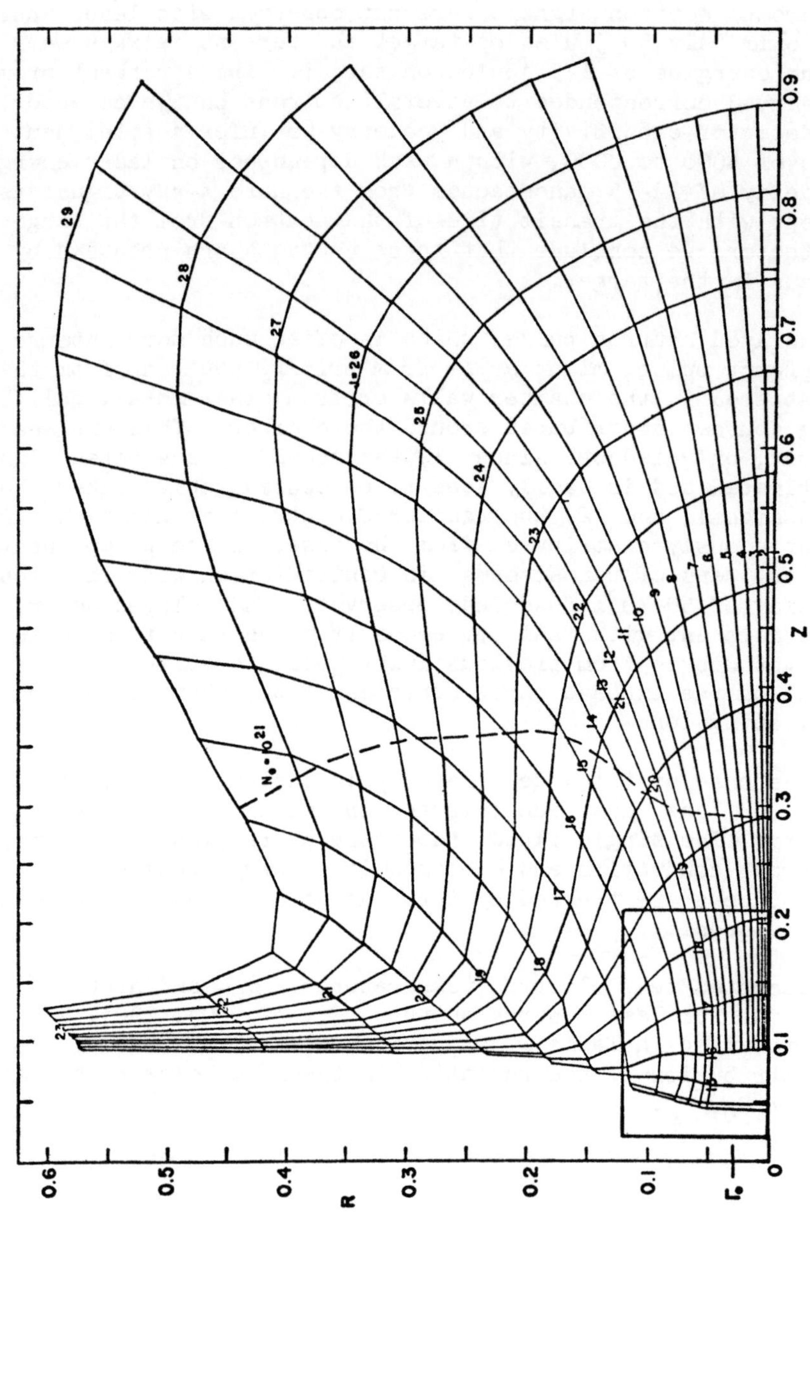

FIGURE 13 - Two-dimensional calculation of CD_2 slab illuminated by a Gaussian laser beam. The dimensions are given in millimeters. The time unit is 10-10 seconds.

TABLE I - CD_2 FOIL RESPONSE

Pulse Width (ns)	E_L (j)	R_0 (μm)	θ_e (kev)	N
3	100	50	1.2	$8 \cdot 10^2$
3	200	40	1.55	$6.5 \cdot 10^3$
1.2	100	30	2.0	$4.5 \cdot 10^3$
1.2	200	30	2.4	$3 \cdot 10^4$
1.2	300	30	2.56	$9 \cdot 10^4$

Although the computed peak electron temperatures range from 1.2 to 2.56 kev, the peak ion temperatures range from 0.4 to 0.7 kev. Peak ion temperatures are typically obtained in a region, between the critical density surface and the thermal conduction front, where the plasma density is of order 0.02 gm/cm^3 .

The computed neutron ouptuts of several hundred to a few thousand neutrons for 3 nanosecond pulses containing 100 to 200 joules are in agreement with the prompt neutron measurements. The data does not indicate any need to invoke any anomalous ion heating effects.

The calculated plasma distribution at peak conditions for the 200 joule, 1.2 nanosecond, 30 μm case is plotted in Figure (13). The relatively planar shape of the calculated critical density surface over the spot size can be seen in Figure (13).

THE THEORY OF HOMOGENEOUS ISENTROPIC COMPRESSION AND ITS APPLICATION TO LASER FUSION*

Ray E. Kidder

University of California, Lawrence Livermore Laboratory

Livermore, California 94550

ABSTRACT

The hydrodynamic theory and properties of spherical homogeneous isentropic compression are discussed. Computer results are described showing that a close approximation to compressions of this type may be accomplished by light-induced pellet ablation if the light absorbed by the pellet is properly programmed with time, and is spatially uniform over the pellet surface. Pellet compressions in excess of ten thousand-fold are computed.

1. INTRODUCTION

In order that a laser-heated pellet of thermonuclear fuel represent a practical source of fusion power, it must satisfy two criteria: It must be brought to a high temperature, and its specific surface area (= $3/\rho R$) must be reduced to a low value. A high temperature is required to provide an adequate rate of thermonuclear reaction. A low specific surface area is required to reduce energy lost from the surface of the pellet (principally hydrodynamic expansion losses) to an acceptable value as compared with energy gained throughout the mass of the pellet by thermonuclear fusion. This latter requirement may be immediately translated into a requirement for high pellet compression, since the specific surface area S of a pellet of any given mass M decreases with increasing pellet density ρ ($S \propto 1/\rho^{2/3}$).

It is well-known that extremely high pressures can be achieved by focusing an intense laser pulse on a solid surface in vacuum (1). The high pressure is derived from the violent

*Presented at the Third Workshop on "Laser Interaction and Related Plasma Phenomena" held at Rensselaer Polytechnic Institute, Troy, New York, August 13-17, 1973.

blowoff of the strongly-heated surface matter into the vacuum, in the same manner that a rocket derives thrust from its exhaust. In the application of lasers to the fusion of pellets, therefore, we are led to enquire how the high pressure derived from laser-induced blowoff can be effectively employed to compress fusionable pellets to high density. The required high temperature will be obtained during the compression by a combination of shock heating and compressive work.

There are two basic ways in which laser-induced blowoff pressure can be employed to produce high compression, which represent opposite limits. They are the strong spherically convergent shock (S) (2), and the shockless or isentropic compression (I). The compression of a pellet may in general be accomplished by either of these processes individually, or by some appropriately timed sequence of them such as:

(S,S), (S,I), (I,S), (S,S,S), (S,I,S), ···.

We shall confine our attention to the simple case of isentropic compression alone, and indeed to the special case of homogeneous isentropic compression. That is, an isentropic compression in which every volume element is compressed to the same degree. An important property of an isentropic compression is that maximum compression is achieved with a given applied pressure. The focused light intensity needed to achieve a given compression is therefore minimized; and the quality, or brightness, required of the laser beam is reduced.

It might appear that the limitation to homogeneous compression is overly restrictive. However, it has the virtue of permitting an exact solution of the hydrodynamic equations to be obtained in simple terms, and leads to results which have previously been obtained only through the inspection of (hundreds of) computer calculations. (3)

We shall first present the analytic theory of homogeneous isentropic compression and some of its consequences, including the pressure-time profile that is required to accomplish the compression (4). Results of computer calculations will then be described which determine the sensitivity of the compression process to variations in the initial state of the pellet. Finally, computer results will be presented in which a close approximation to homogeneous isentropic pellet compression is accomplished by properly programming the laser light absorbed by the pellet.

2. THEORY AND PROPERTIES OF HOMOGENEOUS ISENTROPIC COMPRESSION

The analysis of the process of homogeneous isentropic compression is similar to that given in reference (1) describing the homogeneous expansion of an isothermal sphere into a vacuum, and is as follows.

A homogeneous compression is one in which the element of volume d^3x in the Lagrangian description, in which $\underline{x}(t)$ denotes the instantaneous position vector of a mass point, decreases everywhere at the same rate. That is,

$$d^3x = h^3(t)d^3x_o, \tag{2.1}$$

where $\underline{x}_o = \underline{x}(t=0)$, and $h(t)$ is a dimensionless scale factor. Writing Eq. (2.1) in spherical co-ordinates we obtain immediately,

$$r(r_o,t) = r_o h(t), \tag{2.2}$$

$$u(r_o,t) = \dot{r} = r_o \dot{h}(t). \tag{2.3}$$

From the law of mass conversion we may write

$$dm = \rho r^2 dr = dm_o = \rho_o r_o^2 dr_o, \tag{2.4}$$

which together with Eq. (2.2) implies that

$$\rho(r_o,t) = \rho_o(r_o)/h^3(t). \tag{2.5}$$

The equation of motion is,

$$\dot{u} = - r^2(dp/dm) = - (r^2/\rho_o r_o^2)\ (dp/dr_o). \tag{2.6}$$

We wish to write dp/dr_o in terms of $d\rho_o/dr_o$. For that purpose we invoke the relation between pressure p and density ρ along an isentrope of an ideal gas,

$$\begin{aligned} p(r_o,t) &= a(r_o)\rho^\gamma(r_o,t) \\ &= p_o(r_o)/h^{3\gamma}(t). \end{aligned} \tag{2.7}$$

Making use of these relations, the equation of motion can be written,

$$\begin{aligned} h^{3\gamma-2}\ \ddot{h} &= - (1/\rho_o r_o)dp_o/dr_o \\ &= - 1/t_o^2 = \text{constant}, \end{aligned} \tag{2.8}$$

where the r.h.s. of Eq. (2.8) is a function of r_o alone and the l.h.s. is a function of t alone. Since r_o and t are independent Lagrangian variables, each side of the equation must be constant.

2.1 Time-Dependence of the Scale Factor and Boundary Pressure

Assuming that $\gamma = 5/3$, which describes either a fully degenerate Fermi electron gas or a non-degenerate fully-ionized plasma, the general solution of the temporal part of Eq. (2.8) is found to be

$$h^2 = (1+b\tau)(1-\tau), \tag{2.9}$$

where

$$b = (1-q)/(1+q), \tag{2.10}$$

$$\tau = t/t_c, \; t_c = t_o/(1+q), \tag{2.11}$$

$$q = -\, t_o\dot{h}(0) \geqslant 0. \tag{2.12}$$

We note that the time t_c is the time of "total collapse", i.e., the time at which the pellet radius shrinks to zero; and that the scale factor h vanishes in proportion to $\sqrt{1-\tau}$ as τ tends to unity.

There are two special cases of particular interest and simplicity. The first is the case in which the fluid is initially at rest (b=1). The second is the case in which the Mach number of the flow is independent of the time (b=0). In these two cases the scale factor h is given by,

$$h = \sqrt{1-\tau^2}\;, \quad (b=1), \tag{2.13}$$

$$h = \sqrt{1-\tau}\;, \quad (b=0), \tag{2.14}$$

respectively.

For the case (b=0) of constant Mach number, the required boundary pressure follows from Eqs. (2.7) and (2.14),

$$p(R,t) = p(R,0)/(1-\tau)^{5/2}\;, \tag{2.15}$$

where $R(=hR_o)$ denotes the pellet radius. If, as in reference (3) we define the quantity H according to

$$H = {}_0\!\int^R \rho dr, \tag{2.16}$$

we find that Eq. (2.15) can be written:

$$p(R,t) = p(R,0)/[(H_o/H)(1-\tau)]^{5/4} \tag{2.17}$$

since

$$H = {}_{o}\int^{R(0)} \rho_o (r_o/r)^2 dr_o = H_o/(1-\tau). \qquad (2.18)$$

The boundary pressure given by Eq. (2.17) is identical with that presented in reference (3).

For the case (b=1) in which the fluid is initially at rest, we obtain instead the result that

$$p(R,t) = p(R,0)/(1-\tau^2)^{5/2}. \qquad (2.19)$$

Henceforth, unless otherwise stated, we shall confine our attention to the case in which the fluid is initially at rest. The fluid particle paths are then elliptical as shown in Fig. 1.

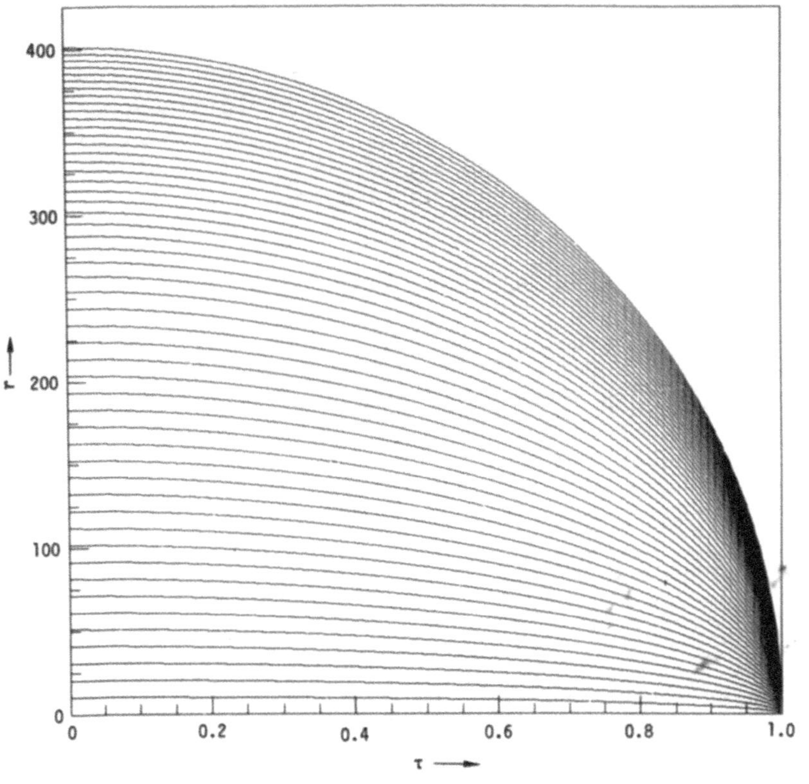

Homogeneous Isentropic Compression. Figure 1. Radius r (microns) of fluid particles vs. dimensionless time τ.

2.2 Radial-Dependence of the Fluid Pressure, Density, and Temperature

Assuming for simplicity that the initial fluid state is one of uniform entropy ($da/dr_o = 0$), the solution of the spatial part of Eq. (2.8) becomes

$$p_o(r_o)/p_o(0) = (1+\beta x_o^2)^{5/2} \tag{2.20}$$

$$\beta = (R_o/c_a t_c)^2/3 \tag{2.21}$$

$$x_o = r_o/R_o, \ (0<x_o<1), \tag{2.22}$$

$$c_a^2 = \gamma p_o(0)/\rho_o(0). \tag{2.23}$$

The radial dependence of the density and temperature of the fluid is given by

$$\rho_o(r_o)/\rho_o(0) = (1+\beta x_o^2)^{3/2} \tag{2.24}$$

$$T_o(r_o)/T_o(0) = 1+\beta x_o^2 \ . \tag{2.25}$$

If the compression takes place slowly ($\beta<<1$) we see that the compressed pellet is spatially uniform. Rapid homogeneous compression ($\beta>>1$), however, requires that the density increase strongly with radius according to Eq. (2.24) so that the collapsing pellet somewhat resembles a hollow shell.

2.3 Mach Number and Mach Lines

The adiabatic sound speed c is given by the relation

$$c = \sqrt{\gamma p/\rho} = (c_a/h)\sqrt{1+\beta x_o^2} \ , \tag{2.26}$$

and the Mach number is found to be

$$M = |u|/c = M^* \, [q + (1-q)\,\tau] \ , \tag{2.27}$$

where $M^* = \sqrt{3\beta x_o^2/(1+\beta x_o^2)}$

$$< \sqrt{3\beta/(1+\beta)} < \sqrt{3} \ . \tag{2.28}$$

We observe that the Mach number approaches a constant value M^* ($<\sqrt{3}$) as τ tends to unity, and that the Mach number is independent of time for the case ($q=1$), or equivalently ($b=0$), as stated earlier. We also note that if β is less than (1/2) the flow is always subsonic ($M<1$), and that for arbitrarily rapid compression the Mach number remains less than $\sqrt{3}$.

The equation defining the Mach lines, or trajectories of an infinitely small disturbance, is

$$dr_{\pm}/dt = u \pm c, \tag{2.29}$$

where the (+) and (-) signs denote outgoing and ingoing Mach lines, respectively. Employing the results of Eqs. (2.3), (2.13), and (2.26), we find, after some manipulation, the solution to Eq. (2.29) to be

$$r_{\pm} = R_o \sqrt{(1-\tau^2)/\beta}\ f_{\pm}(\tau), \tag{2.30}$$

where

$$f_{\pm}(\tau) = (F_{\pm} - F_{\pm}^{-1})/2, \tag{2.31}$$

$$F_{\pm} = \left[f_{\pm}(0) + \sqrt{f_{\pm}^2(0)+1}\right]\left[(1+\tau)/(1-\tau)\right]^{\pm 1/2\sqrt{3}} \tag{2.32}$$

Setting $r_-(0) = R_o$, $f_-(0) = \sqrt{\beta}$, and $f_-(\tau^*) = 0$, we find from Eqs. (2.30)-(2.32) the earliest (dimensionless) time τ^* of

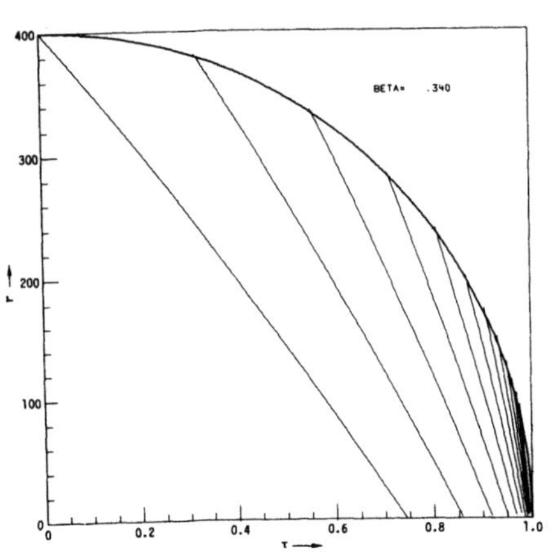

Figure 2

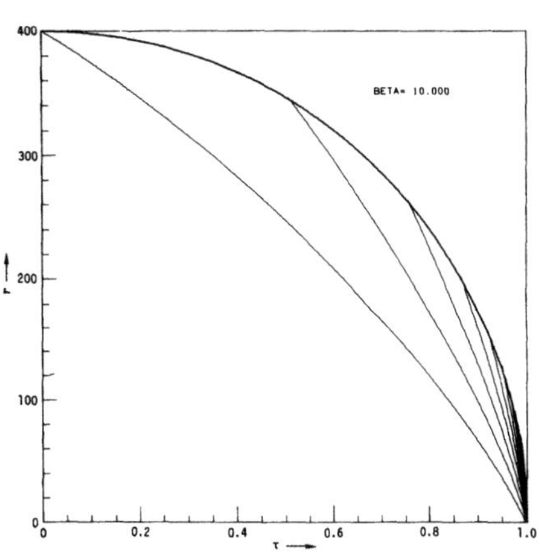

Figure 3

Homogeneous Isentropic Compression. Figure 2. Radius r (microns) of pellet surface and ingoing Mach lines vs. dimensionless time τ. (Moderate rate of compression: β = 0.340) Figure 3. Radius r (microns) of pellet surface and ingoing Mach lines vs. dimensionless time τ. (Rapid compression: β = 10)

arrival at the origin of an ingoing Mach line to be

$$\tau^* = (1-g)/(1+g) , \tag{2.33}$$

$$g = [\sqrt{\beta+1} - \sqrt{\beta}\,]^{2\sqrt{3}} . \tag{2.34}$$

In the case of rapid compression we may write

$$\tau^* \cong 1 - 2/(4\beta)^{\sqrt{3}}, \quad (\beta>>1) \tag{2.35}$$

from which it is clear that τ^* tends to unity as β becomes large. It follows immediately that all ingoing Mach lines tend to arrive simultaneously at the origin in the limit of very rapid compression ($\beta>>1$).

The trajectory of the pellet surface, together with representative ingoing Mach lines, are shown in Figs. 2 and 3 for moderately rapid ($\beta = 0.34$) and very rapid ($\beta = 10$) rates of compression, respectively. The simultaneous arrival of the Mach lines in the case of rapid compression is evident in Fig. 3.

2.4 Kinetic and Internal Energy

The work W done on the pellet is transformed into kinetic energy W_k and an increase ΔW_i in the pellet internal energy, that is,

$$W = W_k + \Delta W_i , \tag{2.36}$$

where

$$W = - \int p dv = A(1+\beta)^{5/2} , \tag{2.37}$$

$$W_k = A\beta I(\beta), \tag{2.38}$$

$$A = 2\pi R_o{}^3 p(0)\ [\tau^2/(1-\tau^2)] , \tag{2.39}$$

$$I(\beta) = 5 \int_0^1 (1+\beta x^2)^{3/2}\, x^4\, dx$$

$$= 1 + (15/14)\,\beta + (5/24)\,\beta^2 + \cdots . \tag{2.40}$$

The ratio of kinetic energy to internal energy increase satisfies the inequalities

$$0 < W_k/\Delta W_i < 5/3 \tag{2.41}$$

as the parameter β ranges from zero to infinity. The fact that the kinetic energy never exceeds five-thirds the increase in internal energy, however rapid the compression, is consistent with the fact that the Mach number of the flow, according to Eqs. (2.27) and (2.28), does not exceed the value $\sqrt{3}$.

2.5 Energy Addition Time

We define the energy addition time t_w to be the final doubling time of the internal energy of the pellet in reaching an η-fold compression. In the limit of large compression ($\eta >> 1$), this time can easily be shown to be

$$t_w = t_c / 2\eta^{2/3}. \tag{2.42}$$

To achieve pellet compressions η on the order of 10^4, therefore, the last doubling of the pellet's internal energy must be accomplished during the last thousandth of the total compression time t_c.

3. APPLICATION TO PELLETS WITH UNIFORM INITIAL STATE

The theory of homogeneous isentropic compression that has been presented strictly applies only to the somewhat artificial

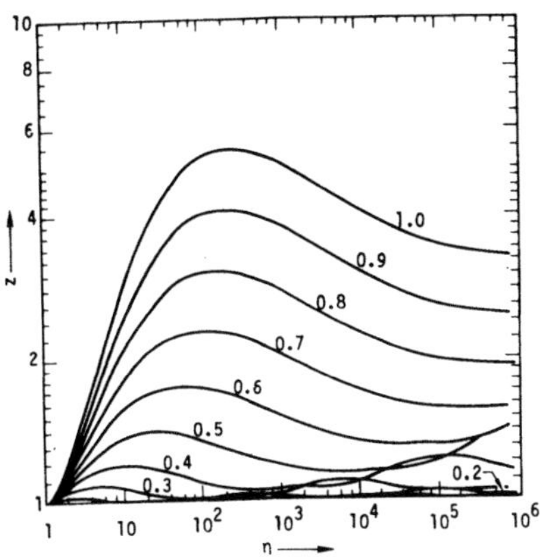

Figure 4

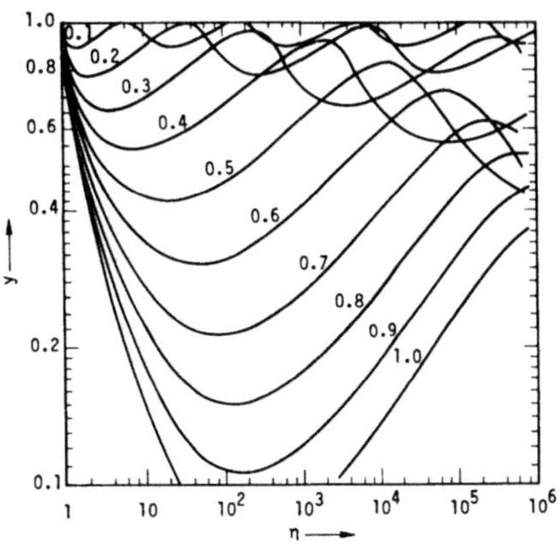

Figure 5

Compression from Uniform Initial State. Figure 4. Ratio z of pellet internal energy to corresponding internal energy for isentropic compression vs. compression η. (Compression rate parameter $\beta = 0.1, 0.2, \cdots, 1.0$.) Figure 5. Ratio y of pellet compression η to corresponding isentropic value η_{isent} vs. compression η. (Compression rate parameter $\beta = 0.1, 0.2, \cdots, 1.0$.)

situation in which the initial pressure, density, and temperature of the fluid increase with radius according to Eqs. (2.20), (2.24), and (2.25), the magnitude of this required increase being greater the more rapid the compression. A natural question is then, what happens if the boundary pressure profile of Eq. (2.19) is employed to compress a pellet in an initially uniform state? One expects that the compression will then no longer be isentropic, the entropy increase being generally greater, the more rapid is the compression.

To answer this question, the hydrodynamic code WAZER (5) was used to calculate the compression of an initially uniform pellet subjected to the pressure-time profile of Eq. (2.19). The results are shown in Figs. 4 and 5 for values of β ranging from (0.1) to (1.0) in increments of (0.1). Inspection of these results shows that if the compression is strictly subsonic ($\beta < 1/2$), then the internal energy is at most 50% higher (Fig. 4) and the compression achieved is at least 40% as great (Fig. 5) as would be achieved isentropically.

4. COMPRESSION BY LIGHT-INDUCED PELLET ABLATION

So far, we have been concerned with the compression of a pellet by means of a specified time-dependent boundary pressure, and have determined the time-dependence required to produce homogeneous isentropic compression. We now ask how this pressure profile can be produced by light-induced blowoff of the pellet surface.

If we assume that the pressure $p(R,t)$ exerted against the pellet is proportional to the two-thirds power of the absorbed light intensity I (1), then

$$\Psi/\Psi_o = (R/R_o)^2 \, I/I_o$$
$$\sim 1/h^{11/2} = 1/(1-\tau^2)^{11/4}, \qquad (4.1)$$

where Ψ is total laser power that is absorbed by the pellet. In reality, however, the situation is more complicated than this due to the effects of electron heat conduction and pellet ablation. Computer calculations which take these effects into account have been performed (3), with the result that the exponent on the r.h.s. of Eq. (4.1) should be approximately equal to 2 (instead of 2.75), i.e.,

$$\Psi/\Psi_o = (1-\tau^2)^{-s}, \quad s \cong 2. \qquad (4.2)$$

The WAZER code was used to calculate the implosion of a uniform spherical pellet of DT subjected to an ultraviolet light pulse with a power-time dependence of the form given by Eq. (4.2). The pellet was treated as an ideal gas with an adiabatic exponent γ equal to 5/3. Other pellet properties, together with implosion parameters and results, are listed in Appendix I.

The laser output power Ψ used in the WAZER calculations was as follows,

$$\Psi/\Psi_o = (1-\tau^2)^{-s} + f(\tau), \qquad (4.3)$$

$$f(\tau) = (1.4)\tau^{1/4} - 1, \qquad (4.4)$$

with $s = 1.9$, $t_c = 35$ nsec, and $\Psi_o = 7.5 \times 10^9$ watts. The added term $f(\tau)$ compensates for the non-negligible initial pellet pressure, but quickly becomes insignificant. All of the laser power Ψ was assumed to be absorbed by the pellet; at the critical plasma density ρ_c ($=1/v_c$), if not at lower density by inverse bremsstrahlung absorption.

The results of these implosion calculations are shown in Figs. 6-12. Figures 6 and 7 show the pellet particle paths during the implosion. (Fig. 7 is a magnification of the last 200

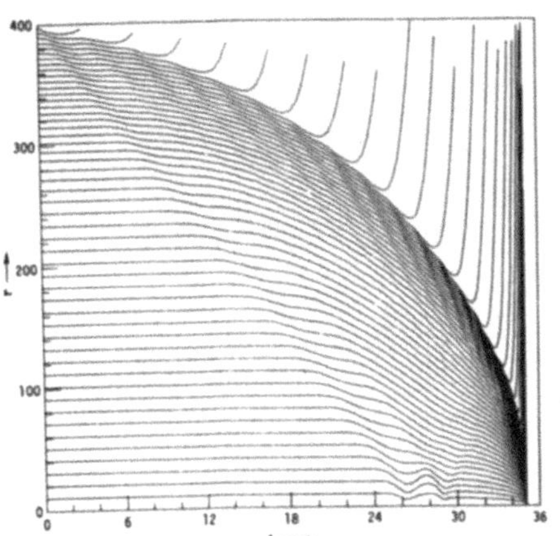

Figure 6

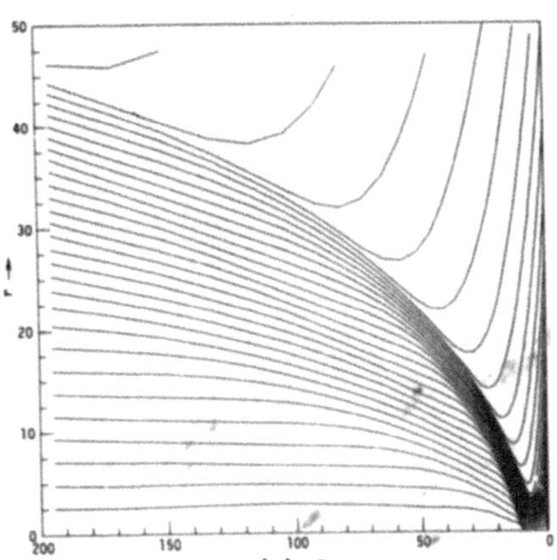

Figure 7

Light-Induced Pellet Compression. Figure 6. Radius r (microns) of fluid particles vs. time t (nanoseconds). Figure 7. Radius r (microns) of fluid particles vs. time until collapse t_c-t (picoseconds).

picosecond time interval of the pellet compression.) The particle paths shown in Fig. 6, with the exception of a weak initial disturbance, are seen to be similar to those shown in Fig. 1. During the last four nanoseconds of the implosion, which accounts for all but the first factor of ten in compression, the incident light is absorbed at a radius R_c (where $\rho = \rho_c$) that remains approximately constant with time.

Figures 8, 9, and 10 show the radial dependence of the electron T_e and ion T_i temperatures, the pressure p, and the density ρ, respectively, near the end of the implosion when the compression of the pellet core (15% of the original pellet mass) had attained a value of approximately ten thousand-fold.

The mean temperature <T> of the compressed core at this time is 820 eV, only 56% greater than would have been obtained in an ideal isentropic compression, showing that a good approximation to isentropic core compression may in principle be achieved by means of laser-induced pellet ablation.

Figure 11 shows the core compression η as a function of the laser energy W_L absorbed by the pellet during the course of the

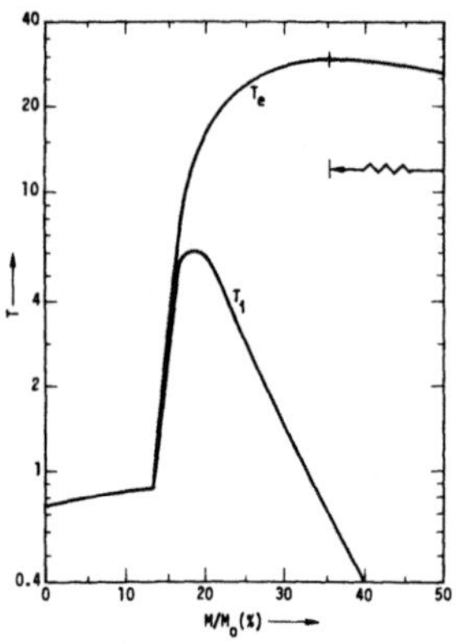

Figure 8

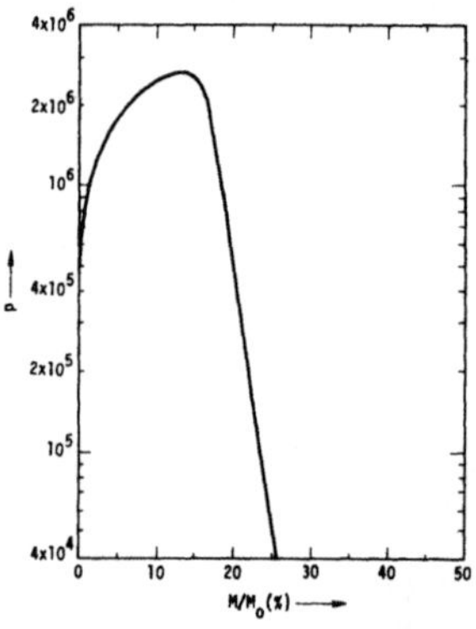

Figure 9

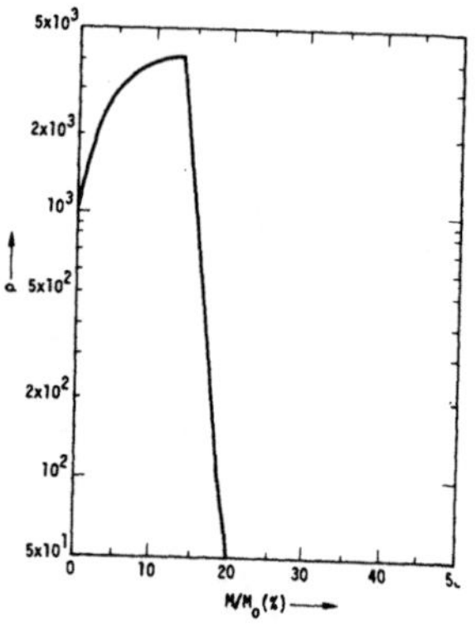

Figure 10

Light-Induced Pellet Compression. Figure 8. Ion and electron temperatures T_i, T_e (keV) vs. pellet mass M/M_o (percentage, measured from pellet center). Wiggly arrow indicates depth to which the light pulse penetrates the pellet corona. Figure 9. Pressure p (megabars) vs. pellet mass M/M_o (percentage, measured from pellet center). Figure 10. Density ρ (gm/cm^3) vs. pellet mass M/M_o (percentage, measured from pellet center).

implosion. The efficiency with which the core is compressed is quite low, amounting to only a few percent in terms of the light energy expended. The poor efficiency is the result of the fact that the compressive work is done by a spherical (as opposed to planar) rocket whose exhaust is too hot and in which there is incomplete conversion of heat into kinetic energy. (The latter source of inefficiency may be partly avoided by turning off the laser before peak compression, i.e., as soon as additional laser energy can no longer influence the compression.)

Figure 12 shows the focused light intensity I_c and the electron temperature T_{ec} at the critical radius R_c as a function of the pellet core compression η. The focused intensity I_c becomes very large during the later stages of compression, with the result that the light pressure (I_c/c) ranges from 10% to nearly 40% of the electron pressure ($N_{ec}kT_{ec}$) as the core compression increases from 10^3 to 10^4.

Since our purpose has been to illustrate the isentropic compression phase of the pellet implosion, we have not included results showing the subsequent ignition and burn of the thermonuclear fuel (DT). Ignition did occur at the center of the pellet, however, by the time the mean compression of the core had increased by an additional factor of four (i.e., to 40,000-fold), and rapidly spread throughout the pellet core.

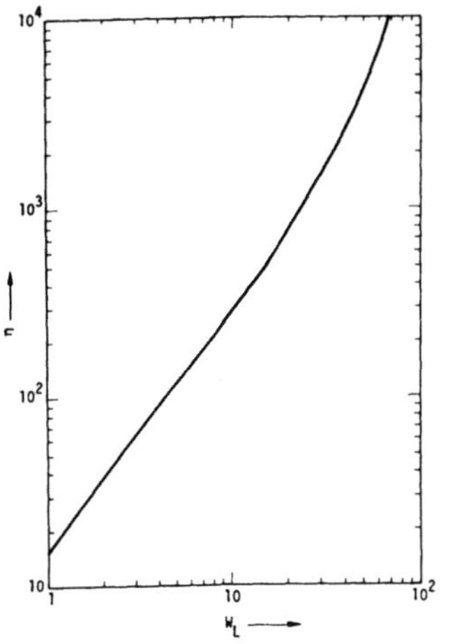

Figure 11

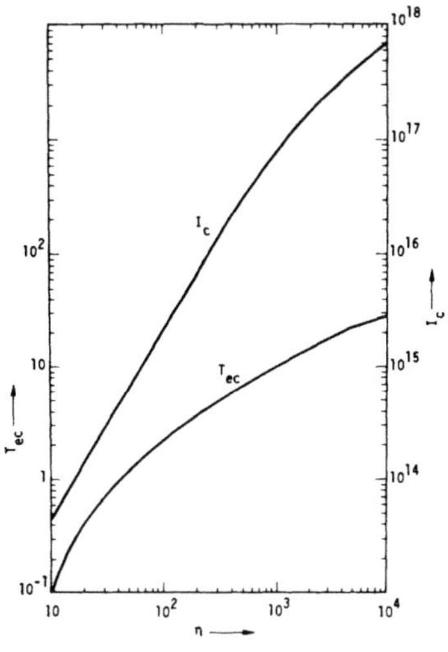

Figure 12

Light-Induced Pellet Compression. Figure 11. Core compression η vs. absorbed light energy W_L (kilojoules). Figure 12. Electron temperature T_{ec} (keV) and light intensity I_c (w/cm^2) at the critical density surface within the pellet corona vs. core compression η.

5. SUMMARY AND CONCLUSIONS

We have provided a resonably thorough account of the analytic properties of spherical homogeneous isentropic compression, and shown that a close approximation to compressions of this type may be accomplished by light-induced pellet ablation if the light absorbed by the pellet is properly programmed with time, and is spatially uniform over the pellet surface. It appears that roughly one-sixth of the original pellet mass may be compressed ten thousand-fold by this means. The compression efficiency is rather low, but the advantage of such extremely high compression to laser pellet fusion (6) would nonetheless repay the cost in energy by a large margin.

A sufficiently uniform deposition of light energy over the pellet surface may be difficult to achieve in practice, however, in view of polarization, interference, and propagation effects coupled with the inevitable and substantial nonuniformity of illumination produced by a limited number of confocal beams. In addition, it may be difficult to obtain a desired time-dependence of the absorbed light energy due to uncertainties in the relation between incident and absorbed laser light. This relation is complicated by such effects as corona-core decoupling (7) and the onset and growth of predicted absorbtive and reflective light-plasma instabilities (8), for example, which we have attempted to suppress by employing short-wavelength light (0.265μ) in our calculations.

Focused laser pulses are known to be effective in producing very hot (multi-kilovolt) thermonuclear plasmas of moderate density. Whether they will also be effective in producing very dense plasmas (many thousand times normal solid density) remains to be seen, but there is hope that this may be so.

ACKNOWLEDGMENTS

The author is indebted to David Bailey and James Shearer for many helpful suggestions, and to Alex Cecil and Maria DeGraaf for their able assistance in carrying out the computer calculations.

APPENDIX I

1. Initial Pellet Properties:	
Temperature (T_o)	1 eV
Specific volume (v_o)	5 cm^3/gm
Pressure (p_o)	150 kbar
Radius (R_o)	400 μ
Mass (M_o)	54 μgm (DT)
2. Optical Parameters:	
Wavelength (λ_L)	0.265 μ
Laser-pulse parameter (Ψ_o)	7.5 GW
Critical Sp. Vol. (v_c)	15 cm^3/gm
Laser pulse exponent (s)	1.9
3. Compression Parameters:	
Collapse time (t_c)	35 nsec
Rate parameter (β)	0.34
4. Implosion Results:	
Mass ablated ($\Delta M/M_o$)	85%
Laser energy delivered (W_L)	70 kJ
Energy addition time (t_w)	33 psec
Time before collapse (t_c-t)	19 psec
4.1 Core Properties:	
Core compression (η)	10^4
Core temperature ($\langle T\rangle$)	820 eV
Core radius (r_c)	9 μ
Specific surface area (S)	1.3 cm^2/gm
$\langle T\rangle/T_{isentropic}$	1.6
4.2 Corona Properties:	
Critical temperature (T_{ec})	29 keV
Critical radius (R_c)	180 μ
Peak light intensity (I_c)	7 x 10^{17} W/cm^2
Light pressure/Electron pressure	0.38

REFERENCES

1. R.E. Kidder, "Interaction of Intense Photon and Electron Beams with Plasmas", in Physics of High Energy Density, P. Caldirola and H. Knoepfel, Eds. (Academic Press Inc., New York, 1971) pp. 306-352.

2. G. Guderley, Luftfahrforsch, 19, 302 (1942).

3. J. Nuckolls, et al, Nature 239, 139 (1972). See also J. Nuckolls, et al, "Laser Implosion of DT to Densities > 1000 gm/cm^3: Optimum Pulse Shape; Fusion Yield vs. Laser Energy", UCRL-74116, October 26, 1972.

4. R.E. Kidder, "Interaction of Intense Photon Beams with Plasmas (II)", Proceedings of the Japan-U.S. Seminar on Laser Interaction with Matter, C. Yamanaka, Ed. (Tokyo International Book Co. Ltd., Tokyo, 1973),pp. 331-353.

5. R.E. Kidder and W.S. Barnes,"'WAZER' A One-Dimensional Two-Temperature Hydrodynamic Code", UCRL-50583, Jan. 31, 1969.

6. See reference 4 above, pp. 343-347.

7. R.E. Kidder and J.W. Zink, Nuclear Fusion, 12, 325 (1972).

8. J.M. Dawson, "Anomalous Absorption of Radiation by Plasmas I"; K. Nishikawa, "Anomalous Absorption of Radiation by Plasmas II"; reference 4 above, pp. 235-282.

NUCLEAR FUSION REACTIONS IN FRONTS PROPAGATING IN SOLID DT*

J. L. Bobin

Commissariat à l'Energie Atomique, Centre d'Etudes
de Limeil
BP n° 27, 94190-Villeneuve-Saint-Georges, France

I. INTRODUCTION

The use of strong shocks in controlled thermonuclear devices was thought about long ago[1]. However, it turns out that a so-called "thermonuclear shock" should have a tremendously high velocity of propagation[2]. Many means are known which drive strong shocks through matter: high explosives, magnetic fields, electron beams, laser beams. The shock may or may not be followed by some kind of rarefaction.

In the former case, hot regions can be found even when the shock itself propagates comparatively slowly. Furthermore the amount of thermonuclear energy released in the hottest part may be sufficient to sustain the propagating front by absorption of the reaction products (charged particles) in the nearby unperturbed medium.

Thus, from the viewpoint of C.T.R. two questions relative to propagating fronts are of interest:

1) Can such a front be self sustained by thermonuclear reaction (a process analogous to the detonation of a chemical explosive);

2) if and when the preceding regime does not exist, can externally driven shocks yield a sufficient amount of energy from thermonuclear reaction in order to get a positive energy balance.

---

*Presented at the Third Workshop on "Laser Interaction and Related Plasma Phenomena" held at Rensselaer Polytechnic Institute, Troy, New York, August 13-17, 1973.

These topics are to be reviewed in the following: the former rather briefly, and since many approaches have been developed, the latter with some more details.

II. SELF SUSTAINED FRONTS

Consider in an infinite medium, a high temperature zone in which thermonuclear reactions occur, separated from an unperturbed medium by some interface. The tendency of such a localized high energy density is to spread over the surrounding matter. Since the pressure difference between the two zones is exceedingly high, a shock front is expected to build up and propagate. The condition for a self sustained propagation is easily set up on a qualitative basis for one dimensional geometries.

Let $\nu = 3$, 2 or 1, characterize the spherical, cylindrical or plane geometry respectively. Initially a volume:

$$V_o(\nu) = \frac{\mathrm{Sup}[2(\nu - 1)\pi, 1]}{\nu} R_o^\nu$$

is at thermonuclear temperature T_o. The ion density is $n(Z = 1)$. The energy released by thermonuclear reactions is:

$$V_o(\nu) \ \langle\sigma v\rangle \ \frac{n^2}{4} Q \tag{1}$$

in which Q is the energy yield per individual reaction and $\langle\sigma v\rangle$ is the usual cross section averaged over the ion thermal distribution. Now, assume a fraction α of tiis energy is used to heat a layer of thickness dR up to T_o. The energy balance reads:

$$\alpha \frac{\mathrm{Sup}[2(\nu - 1)\pi, 1]}{\nu} R_o^\nu \ \langle\sigma v\rangle \ \frac{n^2}{4} Q = 3nkT_o \mathrm{Sup}[2(\nu - 1)\pi, 1] R_o^{\nu-1} \frac{dR}{dt} \tag{2}$$

If a shock front starts:

$$\frac{dR}{dt} = \beta \sqrt{\frac{kT_o}{m_i}} \tag{3}$$

β being a numerical factor and m_i the mass of the ions. Equation 2 is then a relationship between n, R_o and T_o viz.:

$$R_o = \frac{12\nu\beta(kT_o)^{3/2}}{\alpha n m_i^{1/2} \ \langle\sigma v\rangle \ Q} \tag{4}$$

Now the main point is the invested energy i.e., the thermal energy W_o one should transfer to the volume V_o in order to start the self sustained process:

$$W_o \geq 3nkT_o \frac{\mathrm{Sup}[2(\nu - 1)\pi, 1]}{\nu} R_o^\nu \tag{5}$$

Putting Eq. 4 into Eq. 5 yields:

$$W_o n^{\nu-1} \geq f(\nu, T_o) \tag{4}$$

with:

$$f(\nu, T_o) = \frac{3\ \mathrm{Sup}[2(\nu - 1)\pi, 1]}{\nu} \left[\frac{12\nu\beta}{\alpha m_i^{1/2} <\sigma v> Q}\right]^\nu (kT_o)^{\frac{3\nu}{2}+1} \tag{5}$$

Due to the presence of $<\sigma v>$ which is a rapidly varying function of T_o, $f(V, T_o)$ has a minimum value for a temperature around 10 keV in the case of a DT mixture.

Equation 4 shows the importance of the density for cylindrical and most of all, spherical geometry. In the latter case, an initial compression strongly diminishes the minimum energy to be invested. Cumulation of shocks[3] and laser driven adiabatic compression[4] have been proposed to achieve densities significantly higher than solid state. On the contrary, for plane one dimensional geometries there is no dependence on the density for the energy to be invested by unit surface of cross section.

$f(V, T_o)_{min}$ and the corresponding value of T_o can be evaluated accurately for each geometry and various initial conditions. Assuming first a uniform high temperature zone with components, deuterons, tritons and electrons at the same temperature T_o, one finds in spherical geometry[5]:

$$W_o \left(\frac{n}{n_s}\right)^2 \geq 7 \times 10^{15}_{\mathrm{ergs}} = 700\ \mathrm{MJ}$$

where n_s is the solid state particle density. Since losses are not taken into account, this evaluation is considered optimistic. If heat losses through electron thermal conduction which generates a heat precursor[6], and nonlocal deposition of helions which are the charged reaction products, are taken into account the result is:

$$W_o \left(\frac{n}{n_s}\right)^2 \geq 4000\ \mathrm{MJ}$$

However, the electron thermal conduction coefficient used in this derivation viz.:

$$\chi = a\, T^{5/2} \sim 0.4 \times 10^{-5} T^{5/2} \qquad (6)$$

applies only if λ_e being the electron mean free path,

$$\lambda_e \frac{\partial T_e}{\partial r} << T_e$$

This is not the case in such violent events as considered here. One rather has the equality and the value of χ in Eq. 6 turns out to be too large. Furthermore, the range of helions was calculated disregarding collective effects. Thus the corresponding value of f_{min} is to a certain extent over estimated.

More refined values come from numerical simulations[7]. Instead of a simple minded energy balance, one was the dimensional fluid equations with different temperatures for ions and electrons. Thermal conductivities, energy exchange between species and radiation losses (bremsstrahlung) are taken into account. The case of instantaneous energy release in an infinitely small volume, electrons only being initially hot, provides especially interesting boundary conditions. The ions are progressively heated as the energy spreads over a larger volume. Two characteristic values of f_{min} are accurately defined: f_{crit} critical value for which the amount of thermonuclear energy is equal to W_o, f_{ign} ignition value for which the temperature of the hot zone undergoes an irreversible* temperature increase due to helion deposition. Numerical results are given in Table I for DT mixtures and the three geometries. The value corresponding to self sustained thermonuclear waves lies somewhere in between.

Table I

geometry	f_{crit}	f_{ign}	units
plane	1.3×10^2	4×10^2	MJ/cm^2
cylindrical	2.6×10^2	10^3	MJ/cm
spherical	7×10^2	1.5×10^4	MJ

*At least as far as the fuel is not exhausted.

The ratio f_{ign}/f_{crit} increases with ν. Indeed the expansion in spherical geometry obviously consumes more energy.

The numerical simulation also shows that the hot zone in which thermonuclear reactions take place is definitely distinct from the shock front proper. This property was also inferred thinking about the thermonuclear detonation according to the Zeldovich, von Neuman, Doring model of the ordinary detonation[8]. However, more detailed studies[9] show that this model is not always valid. Very high temperature steady thermonuclear waves exhibit somewhat different density and temperature profiles after ignition has taken place. A criterion for ignition is given in Ref. 9 as a function of the energy per nuclear reaction which is coupled back to the plasma. The result is reported in Fig. 1.

III. SCHEME OF A LASER DRIVEN FLAME

Contrary to the above mentioned model, the flow driven by a laser beam impinging onto a solid target has destructive features. These come from the well known cut-off properties of plasmas with respect to electromagnetic waves. Indeed, a wave of frequency ω_k and wave vector k cannot propagate if ω is smaller than the plasma frequency ω_p. The plasma dispersion relation is

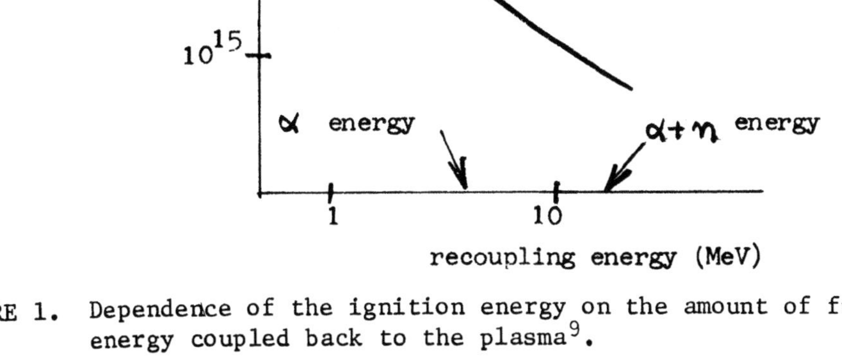

FIGURE 1. Dependence of the ignition energy on the amount of fusion energy coupled back to the plasma[9].

$$\omega^2 = \omega_p^2 + k^2c^2 \quad (7)$$

with:

$$\omega_p^2 = \frac{4\pi e^2}{m} n_{ec} \quad (8)$$

The corresponding cut-off density is:

$$n_{ec} = \frac{m}{4\pi e^2}\omega^2$$

For lasers in the visible or infrared range n_{ec} is by more than one order of magnitude smaller than the electron density in a solid.

Furthermore from Eq. 7 the group velocity is:

$$\frac{d\omega}{dk} = c\left(1 - \frac{\omega_p^2}{\omega^2}\right)^{1/2} \quad (9)$$

It goes to zero as $\omega \to \omega_p$. Thus the energy does not propagate at cut-off. The absorption is therefore enhances. All calculated absorption coefficients whether linear[10] or nonlinear (anomalous A.C. resistivity) exhibit a:

$$\left(1 - \frac{\omega_p^2}{\omega^2}\right)^{-1/2}$$

divergence. Also note that those nonlinear effects which lead to anomalous A.C. resistivity are resonant for $\omega = \omega_p$.

Thus laser light is strongly absorbed at densities close to cut-off. So the density of the hottest part of the flow structure is determined by the frequency of the light. It is considerably smaller than the density of the unperturbed solid. There is also a one to one correspondence between the absorbed radiative flux Φ_A and the thermodynamical state of the fluid in the region of absorption. Setting:

$$V_c = 1/n_{ec}$$

one readily derives from dimensional analysis the dependence of the pressure:

$$p \propto \Phi_A^{2/3}/V_c^{1/3} \quad (10)$$

and of the temperature :

$$T \propto (\Phi_A V_c)^{2/3} \tag{11}$$

For a DT mixture expressing Φ_A in watt/cm^2, V_c in cm^3 and T in eV, the proportionality coefficient is about 1.6×10^8 [11].

Basically as shown on Fig. 2 the laser driven flow is made of a shock wave followed by a rarefaction. The absorption sheet divides the rarefaction into two regions. On the vacuum side the plasma expands freely. It is convenient to assume for computational purposes that the profile is isothermal. Since most of the thermonuclear reaction due to the n^2 term in Eq. 1 occurs in dense zones, such an assumption does no harm. Accordingly the ion density in such a flow follows a relation of the type:

$$n = n_{ec} \exp\left(- \frac{x}{c_T t}\right)$$

where x is the distance from the plane where the cut-off density is reached and c_T is the isothermal velocity of sound.

On the solid state side electron thermal conduction strongly heats an over dense region. It has been shown[12] that for a constant

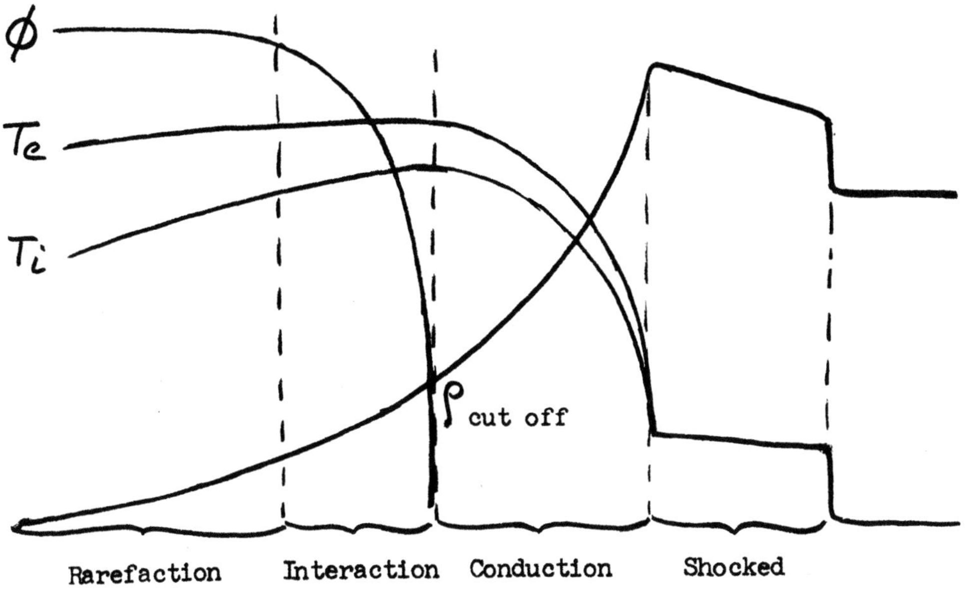

FIGURE 2. Scheme of a laser driven flame.

incoming flux steady solutions exist for this zone. The time independent equations of conservation of mass, momentum and energy read:

$$\rho u = J$$

$$Ju + p = C \tag{12}$$

$$J\left(\frac{u^2}{2} + \varepsilon + pV\right) - k\frac{dT}{dx} = E$$

where u is the flow velocity, ε the specific internal energy, V the specific volume (local variables). The electron thermal conductivity is given by Eq. 6 and J, C and E are constant of the motion.

Computing J, C and E at the cut-off plane it is possible to derive numerically, accurate spatial profiles for the temperature, the density and the velocity by combinations of the above equations. It turns out that the pressure does not vary very much along the structure. Then assuming a constant pressure, an analytical solution has been derived and yields a conduction length (measured from the plane where the cut-off density is reached) equal to (Z = 1):

$$x_o = \frac{4}{5}\,\frac{\gamma - 1}{5\gamma - 1}\left(\frac{m_i}{k}\right)^{3/2} \propto \frac{T_c^2}{n_{ec}} \tag{13}$$

Since the pressure is almost constant and putting numbers in Eq. 10 turns out to be very large (megabars), the rarefaction cannot have the unperturbed solid as a boundary condition. A shock wave followed by a high density region should propagate ahead of the rarefaction.

Such structures can be calculated with different ion and electron temperatures. Since ions are caught by the wave and travel along the structure as it propagates, they stay in the over dense region during a time:

$$\frac{x_o}{c_T} \propto \frac{T^{3/2}}{n_{ec}} \tag{14}$$

This time has the same dependence on temperature and density as the relaxation time between ions and electrons. Thus, the ratio between ion and electron temperatures does not depend on the flux (for DT it is about 0.7).

For obvious reasons Eq. 14 also gives the temperature and density dependence of the time t_D necessary to build up the structure.

Table II

Boundary condition	$\partial X_o/\partial t$	$T^{1/2}$	Dominant regime for longer times $t \to \infty$
Constant T_o at surface	$t^{-1/2}$	$T_o^{1/2}$	Flow
Exponential T at surface	$e^{n\alpha t}$	$e^{\alpha t}$	Heatwave if $n > 1$
Constant Φ at surface	$t^{-\frac{1}{n+2}}$	$t^{\frac{1}{2(n+2)}}$	Flow if $n > -4/3$
Exponential Φ at surface	$e^{\frac{n\alpha t}{n+2}}$	$e^{\frac{\alpha L}{(n+1)(n+2)}}$	Heatwave if $n > 0.366$
Instantaneous Φ (pico pulses)	$t^{-\frac{2n+1}{2(n+1)}}$	$t^{-\frac{1}{2(n+1)}}$	Flow if $n > 0$
$T \propto t^{\alpha}$	$t^{(n\alpha-1)/2}$	$t^{\alpha/2}$	Flow if $\alpha < 1/(n-1)$
$\Phi \propto t^{\alpha}$	$t^{\frac{n\alpha-1}{n+2}}$	$t^{\frac{2\alpha+1}{2(n+1)}}$	Heatwave if $n > 1$ $\alpha > 3/2(n-1)$

Numerically one finds for DT:

$$t_D \sim 3.10^{-20} T_e^{3/2} \lambda_L^2 \quad (\text{sec}) \tag{15}$$

where T_e is in degrees Kelvin and λ_L is the laser wavelength in μm. At thermonuclear temperature (eg. $T_e \sim 10^8$ °K) one finds:

$t_D \sim 3 \times 10^{-8}$ sec for Nd glass laser light ($\lambda \sim 1\mu m$)

$t_D \sim 3 \times 10^{-6}$ sec for CO_2 laser light ($\lambda \sim 10\mu m$)

Comparison of these times with needed laser pulse duration is a way of checking the validity of the partially steady flow model.

The validity should also be checked according to another criterion. Indeed, electron thermal conductivity should not propagate heat in the over dense region beyond the boundary with the compressed medium. Since the conductibility coefficient (Eq. 6) is strongly nonlinear, heat waves with a definite boundary may take place. Propagation of such fronts is known in a lot of relevant cases[6,13]. From one dimensional theories one derives the position of the boundary X_o and the temperature T as a function of time which depends on the way energy is delivered at the surface of a semi-infinite medium. Since flows have characteristic velocities of order $T^{\frac{1}{2}}$, the time dependence of this quantity is to be compared with $\partial X_o/\partial t$. The result is given on Table II. For the sake of generality the conductivity coefficient is written aT^n.

For plasmas n = 5/2, the heat wave is dominant only in rapidly increasing boundary conditions.

IV. EXPERIMENTS ON LASER DRIVEN PLASMAS

Interaction of laser beam with solid targets has been extens extensively studied for the last ten years. Obviously high fluxes are required to reach fusion conditions. So far they were obtained by focusing the beam with an aberration free lens. As a consequence energy is concentrated in a time volume and the geometry of a typical experiment on a deuterium ice looks definitely two dimensional: Figure 3. Cryogenic targets were used at Frascati[14], Garching[15], Limeil[16] and Nagoya[17]. Other labs favor deuteriated polyethylene in the form of massive targets[18] or pellets[19,20].

Obviously the model developed in Section III cannot accurately describe such experiments. However, in the case of long (3×10^{-8} sec) 1.06 μ pulses, a good agreement was obtained between experimental results and predictions of the model[11,16]. The situation is quite different in the case of shorter pulses (a few nanoseconds).

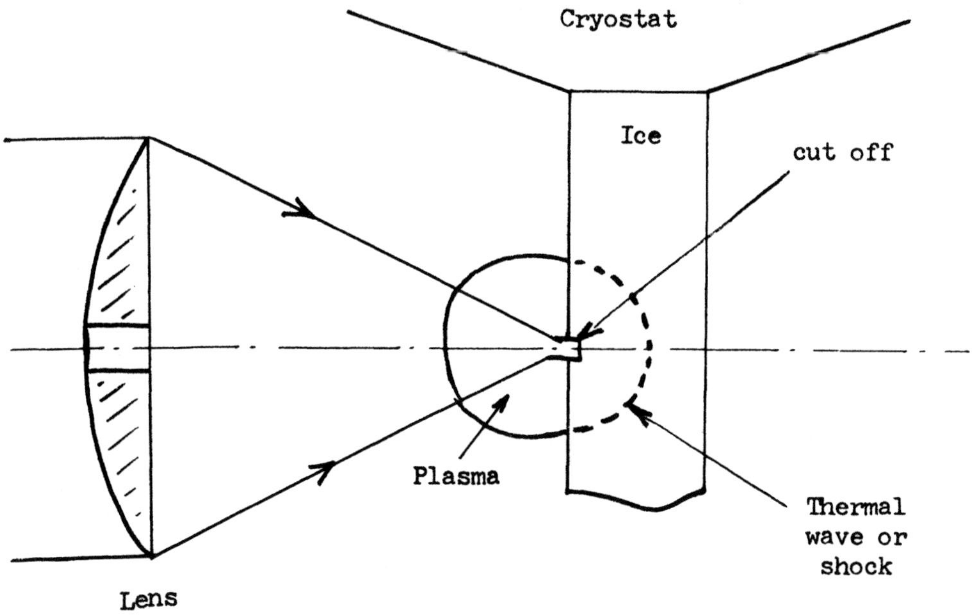

FIGURE 3. Geometry of the single beam interaction with solid D_2.

Then, numerical simulations are required[21]. Before we look more closely at such cases, it is worthwhile to mention a low density experiment which shows the importance of the ratio of the cut-off density[22].

In the setup, gas at any density and temperature (up to 10 atmospheres) expands supersonically through an orifice, into an evacuated chamber. The laser which is aligned with the jet has its output focused into the jet across the orifice. The subsequent breakdown creates a luminous region, the expansion of which from the orifice into the high density reservoir is recorded by a streak camera. The cut-off density for the incoming ruby laser light is 2×10^{21} cm^{-3}. Experiments were run in hydrogen or air. For hydrogen the propagation of the luminous front in the same direction as does the laser beam exhibits different behaviors according to the filling pressure as shown on Table III. In air, which can be ionized more than once the velocity varies as $(\Phi/\rho)^{1/3}$ for pressures ranging from 1 to 9 atmospheres.

These results show that the density in the shocked region should be higher than the cut-off in order that the regime sets up. Furthermore since the ratio of these densities is small, a completely steady structure (shock and rarefaction propagating at the same velocity) is likely to hold.

Table III

filling pressure	initial electron density	electron density in the shock x 6	velocity of the luminous front
4 atm	2.4×10^{10}	1.4×10^{21}	erratic
6 atm	3.6×10^{20}	2.6×10^{21}	$\alpha\ (\Phi/\rho)^{1/3}$
7 atm	4.2×20^{20}	2.5×10^{21}	$\alpha\ (\Phi/\rho)^{1/3}$

Such a situation would require a laser beam with frequency in the UV range in the case of solid targets. On the contrary the most recent experiments show a strongly unsteady behavior.

The neutron yield and the plasma interferometry can bring elements to determine the dominant regime. Assume first that the heat wave is dominant. The evaluation is simpler in the case of short pulses depositing instantaneously an energy W_A in the focal spot. According to Ref. 23, the volume heated at maximum ion temperature ($\alpha\ W^{11/7}$) is proportional to $W_A{}^{6/7}$ and the characteristic time to $W_A{}^{3/14}$. Since in the keV range $\langle\sigma v\rangle$ varies roughly as $T_i{}^{5.5}$, the number of expected thermonuclear neutrons is proportional to $W_A{}^{13/7} \sim W_A{}^2$.

On the contrary in the laser driven flow, we have seen that there is a definite relationship (Eq. 11) between the temperature and the incoming flux. Moreover most thermonuclear reactions come from the over dense region the width of which is given by Eq. 13. Hence the number of neutrons is proportional to:

$$T^{7.5} \propto \Phi_A{}^5 \propto W_A{}^5$$

Experimental results yield for the number of neutrons versus the absorbed laser energy the plot of Figure 4. These results were obtained with short (2-4 nsec) pulse with a steep leading edge (a factor 5×10^5 in 1.5 nsec)[24]. It can be seen that if the neutrons were of thermonuclear origin, they would be generated in a radiation driven flame at the lower fluxes in a thermal wave at the higher fluxes associated with fast rising pulses and sharper focusing.

The occurence of a thermal wave is indirectly confirmed by interferometric measurements of the electron density[25]. It was

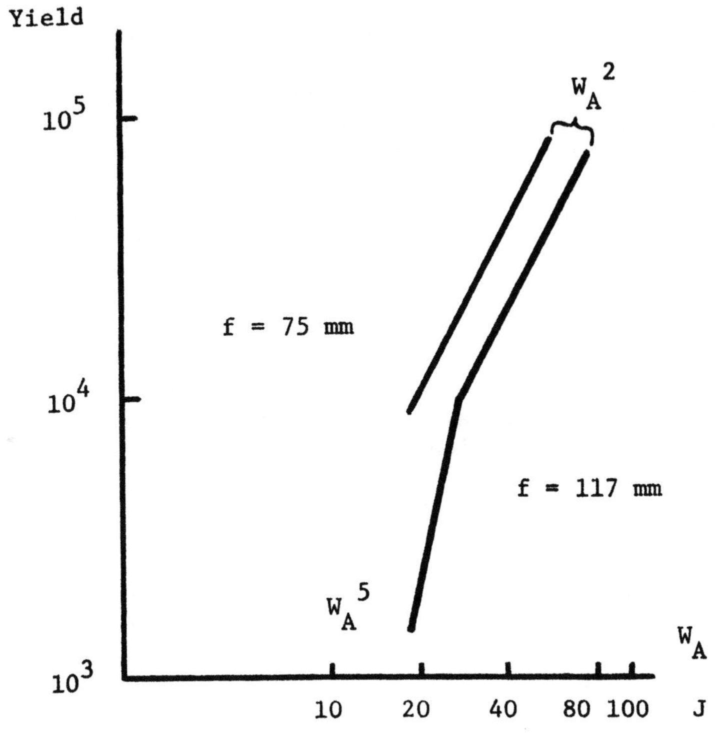

FIGURE 4. Neutron yield versus absorbed energy in a single beam laser interaction with solid D_2.

found that the density of the plasma expanding outside the target is well below the cut-off. Furthermore the plasma comes from an area significantly larger than the focal spot. The cut-off density is inside the target according to such measurements whereas it would be 100 μm outside if a radiative flame had time to build up.

The origin of the observed neutrons is still unknown. Indeed the numbers are too small to perform any spectroscopy. If the neutrons were not of thermonuclear origin, they would be due to some instability associated with nonlinear absorption of laser light by the plasma. Evidence of nonlinear parametric decay was obtained at Limeil through the occurence of harmonics $3/2\ \omega_L$ and $2\ \omega_L$ with an asymmetric spectrum in the reflected light[25]. In order to correlate the neutron emission with such an instability, an experimental study was made varying the location of the focal spot with respect to the surface. A typical result is shown on Fig. 5. The maximum neutron yield corresponds to a definitely different position from the maximum of reflectivity and of the harmonic $3/2\ \omega_L$ which indicates the onset of the instability[26].

V. FUSION BY LASER DRIVEN FLAMES

The efficiency of a given thermonuclear device is usually evaluated in terms of the Lawson criterion[27]. An intermediate criterion on the way to fusion, includnng no efficiency coefficients, is the breakeven condition. These criteria are now going to be defined for laser driven flames and the conditions under which they can be fulfilled will then be computed.

Most generally let us consider a pulsed system in which the reaction products escape from the hot plasma. This constant density plasma is heated up to a temperature T which is supposed to be maintained during the time τ; afterwards, it is allowed to cool down and the cycle can be repeated. Let R be the ratio of the energy released in the hot gas to the incident absorbed energy. Now, the energy released by nuclear reactions is converted into heat within the walls of the reactor and has to be converted into any kind of directly usable energy which is then fed back into the plasma. Let η be the overall efficiency of such an operation (η is hereafter called the cycle efficiency). Then the condition for a net power gain is

$$\eta(R + 1) > 1 \qquad (16)$$

Evaluation of R for a plasma where all the energy is thermal leads to the well known conditions for a DT mixture:

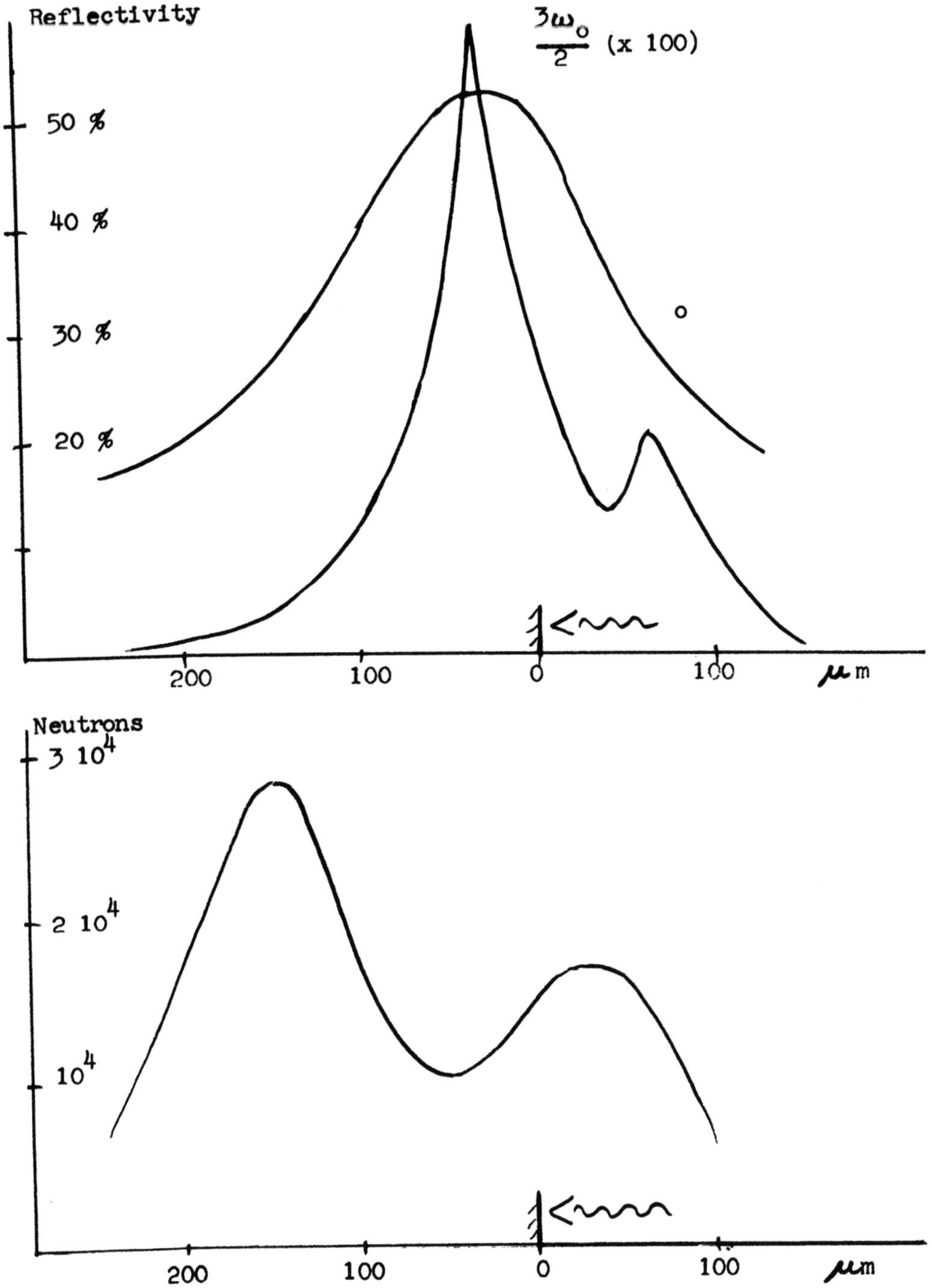

FIGURE 5. Neutron yield, reflectivity and $3\omega_0/2$ harmonic generation as a function of the focal spot location with respect to the surface.

$$n\tau > 6 \times 10^{13} \text{ cm}^{-3} \text{ sec}$$

at T = 10 keV.

Now the condition for breakeven is simply given by

$$\frac{\text{energy released}}{\text{incident absorbed energy}} > 1$$

In terms of R, it reads:

$$R > 1$$

Obviously the above well known numerical results do not apply directly to a nonstationary situation such as a flame propagation. R now has the specific meaning:

$$R = \frac{\text{energy released}}{\text{radiation losses + (thermal+kinetic)energy in the plasma}} \tag{17}$$

As in the Lawson derivation, the radiation term takes into account bremsstrahlung emission only. Reabsorption of this radiation will not be examined since at thermonuclear temperatures (e.g. 10 keV) the plasma can be considered transparent to its own radiation. Even at solid state density the Planck mean free path is much longer than the eventual dimension of the plasma.

Now, the various terms coming into the Lawson criterion are to be computed separately for the steady state deflagration structure and for the rarefaction wave. As was mentioned previously computations for the characteristics of the deflagration have been performed numerically and so will be the corresponding terms in Eq. 17.

On the other hand, integration of the terms in the expansion region can be performed analytically by using the isothermal approximation.

Let Q be the energy released in a single DT reaction. Then in the isothermal rarefaction an energy

$$W_R = \int_0^\infty\int_0^\tau \frac{n_c^2}{4} e^{-2x/c_T t} <\sigma v>_{DT} Q \, dx \, dt \tag{18}$$

is released as reaction products whereas

$$W_B = \int_0^\infty\int_0^\tau n_c^2 \, e^{-2x/c_T t} f(T) \, dx \, dt \tag{19}$$

with

$$f(T) = 1.4 \times 10^{-27} T^{1/2}$$

is radiated. Let us assume a constant incoming laser flux ϕ. Then according to Eqs. 8 and 11 T depends on the flux and the light frequency only. Thus f(T) and g(T) = $\langle\sigma v\rangle$ Q/4 do not depend on position and time. Therefore the integrals can readily be performed:

$$n_c^2 \int_0^\infty \int_0^\tau e^{-2x/c_T t}\, dx\, dt = \frac{n_c^2 C_T \tau^2}{4} = \frac{N_R n_c \tau}{4}$$

Physically $N_R = n_c c_T$ is the total number of ions per unit surface within the rarefaction zone at time τ and $n_c\tau$ is a characteristic parameter of the Lawson criterion.

The energy of the plasma (kinetic + thermal) is then calculated for the rarefaction at any time

$$W_{plasma} = \frac{11}{2} N_R kT \qquad (20)$$

Contributions from the two parts of the plasma flow are finally added and inserted into Eq. 17. R is thus obtained as a function of temperature and reduced interaction time τ^* and plotted in Fig. 6. The reduced time can be expressed as $\tau^* = \tau\lambda_L^2$ where τ is in seconds and λ in microns. From Eq. 8 we see that the product $n_c\tau^*$ does not depend on the incoming radiation. The reduced time amounts to the real interaction time for Nd light and is to be multiplied by 100 to obtain the real interaction time in the case of CO_2 laser light.

In the figure it can also be seen that for a cycle efficiency of 0.3 and at a temperature of 10^8 °K, pulses of 2×10^{-7} s for Nd glass light and 2×10^{-5} s for CO_2 laser are needed in order to get a power producing system. Minimum required interaction times are about 3×10^{-8} s and 3×10^{-6} s respectively at a temperature T = 5×10^8 °K.

Now, in addition to determining the temperature and interaction time (the density being automatically given by the choice of the laser radiation in our scheme) required for a power producing device, it is also important to know what the laser flux should be. This can readily be obtained from the Lawson criterion. In the case of breakeven condition, if W_{TR} is the total energy released during time τ, it is also equal to the energy density W_o/S_o provided by the laser pulse (W_o is the total energy delivered by the laser pulse at time τ and S_o is the target impact area). The laser flux is then equal to

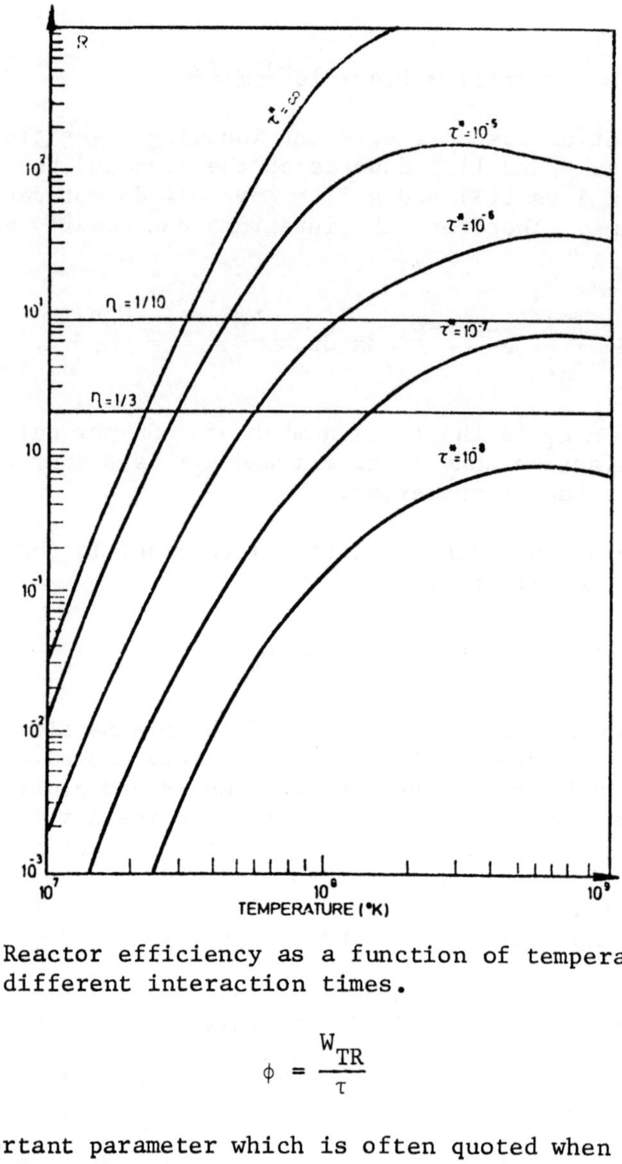

FIGURE 6. Reactor efficiency as a function of temperature for different interaction times.

$$\phi = \frac{W_{TR}}{\tau} \tag{21}$$

An important parameter which is often quoted when fusion devices are evaluated is the minimum energy required to operate these devices. It provides an adequate measure for the gap between the present state of the art and the fusion requirements. This minimum energy will obviously depend on the smallest area on which the laser flux is to be applied. If no limiting plasma effects such as instabilities are taken into account, the surface will be bounded from below by the optical properties of the incoming laser beam.

In most actual or conceivable devices the laser light is to be focused on the target by a property designed lens. The cross section of the focal zone should be approximately constant along a length equal to the displacement of the absorbing layer (derived from Eq. 41 of Ref. 12) plus the length of the over dense structure. Hence, for a given temperature and time we get the minimum value of the longitudinal spherical aberration L.

According to well known optical formulae[28] the corresponding nearly constant radius r is given by

$$r \sim \frac{1}{2}(L\lambda)^{1/2} \qquad (22)$$

provided the beam is diffraction-limited.

As a consequence of considerations to be mentioned later we restrict ourselves to CO_2 laser light for these breakeven computations. From the numerical results shown previously, an approximate expression can be derived for the energy W_p released in the deflagra tion structure by the 14 MeV neutrons:

$$W_D \sim 4.5 \times 10^6 T^6_{(keV)}\tau, \quad J\ cm^{-2}\ [2 < T < 15]$$

Then using Eq. 18 and the following equation for the thermal speed of sound:

$$c_T = 2.2 \times 10^7 T^{1/2}_{(keV)} \quad cm\ sec^{-1} \qquad (23)$$

we find

$$W_{TR} = 4.5 \times 10^6 T^6 \tau + 3.12 \times 10^{-6} T^{1/2} n^2_{ec} <\sigma v> \tau^2, \ J\ cm^{-2} \qquad (24)$$

If the breakeven temperature is given, e.g. by T = 10 keV, the required characteristics for a CO_2 laser become:

pulse duration $\tau = 10^{-5}$ sec

energy density $W_o/S_o = W_{TR} = 5.5 \times 10^7$ J cm^{-2}

laser flux $\phi = 5.5 \times 10^{12}$ W cm^{-2}

Then using Eqs. 21, 24 and 22 we obtain the laser energy for breakeven conditions $W_o = \phi\tau\pi r^2$ equal to

$$W_{o_J} = (7 \times 10^6 T^8 + 4.9 \times 10^{-6} T^{5/2} n^2_{ec} <\sigma v> \tau)\lambda\tau \qquad (25)$$

where T is in keV and $<\sigma v>$ in cm^3/s. This expression depends on laser light parameters through r^2 only and thus scales as λ. For the case T = 10 keV the minimum required energy to obtain breakeven conditions with a CO_2 laser is approximately equal to 10^7 J. The corresponding interaction time and focal zone radius are respectively 10^{-5} s and 0.25 cm. The laser energy required for a power producing reactor would be of the order of 2×10^7 J.

The corresponding invested energy per unit surface is of the order of 3×10^8 J/cm^2. It is of the same order of magnitude as the requirements for self sustained fronts listed in Table I. A similar value was also found when investigating the long wavelength heating of a confined low density plasma[29].

A plane one dimensional geometry has been assumed all along and the energy yields depend on it in a crucial way. The solution to the lateral confinement problem could be the use of a magnetic field parallel to the laser beam. The field magnitude required to confine a $\beta = 1$ plasma is:

$$B_{Gauss} = 9 \times 10^{-6}\sqrt{nT} \qquad (26)$$

where n is in cm^{-3} and T in eV. For T = 10 keV and n equal to the cut-off value for CO_2 light, B is equal to 3 MG. For an Nd glass laser the required value of the confining field becomes 30 MG. This latter figure is beyond present day technology and has been approached only in a transient way by imploding high magnetic fields with explosives[30]. So this process for Nd laser created plasmas does not seem to be practical. As for the CO_2 laser light, we note that the field should be applied all along the shocked medium, the over dense structure and a fraction of the rarefaction region, i.e. all along the focal volume of the lens. Now for fusion conditions, the characteristic length of the rarefaction region is about 15 meters.

Although fields of several megagauss have been observed in the laboratory[31], obtaining them in such a large volume looks like a real challenge. At these field values the ion Larmor radius is of the order of 100 μm< radius imposed by optical considerations.

An experiment with combination of inertial and magnetic lateral confinement has been in operation at the Kurchatov Institute in Moscow[33].

VI. FUSION BY ELECTROMAGNETICALLY DRIVEN SHOCK WAVES

In the preceding section an evaluation of the magnetic field necessary to confine hot plasma was given. Electromagnetic forces can also be used to move conductive fluid layers. Obviously, according to the numbers, such a mechanism would need a field far beyond present day technology if the fluid had a density comparable to solid state. So electromagnetically driven shock tubes use low density gases. However, it was shown in Section II that the requirements on invested energy per unit surface do not depend on the density.

In the earliest devices[34] an electrical discharge locally heats the gas. A subsequent shock wave expands. If the lead carrying the return current is located close to the discharge, electromagnetic forces contribute to the acceleration of the shock. Such a principle is used in the T tube geometry[35] or in the magnetic annular shock tube[36] shown on Fig. 7. This last system is very efficient in accelerating hydrogen plasmas up to velocities of interest for C.T.R. (~ 10^8 cm/sec) a significant neutron yield in such an experiment was reported[37].

The main problem concerning these shocks is the structure. For low density gas dynamic shocks the thickness is a few mean free paths in the post shock state. Detailed theoretical analysis shows that for the conditions of actual and predictable experimental devices, the thickness is several tens of meters[38]. Fortunately it can be reduced down to values compatible with the laboratory scale by using a strong magnetic field whose lines of force are perpendicular to the direction of shock propagation.

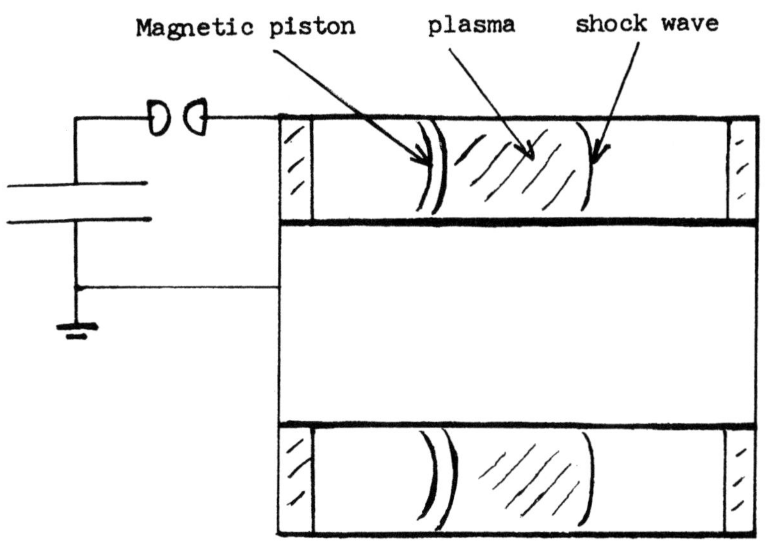

FIGURE 7. Scheme of an annular magnetic shock tube.

The efficiency of the electromagnetically driven shock tube as a C.T.R. device can be greatly enhanced by considering the reflexion on the end wall. Indeed, the macroscopic kinetic energy of the incident plasma is largely converted into internal energy of a medium which is at rest with respect to the laboratory frame. The subsequent increase of temperature and density may yield effective thermonuclear conditions. Such a fusion cycle is investigated in Ref. 39. When the plasma has reached the end of the tube an additional magnetic field is applied in order to prevent the onset of a rarefaction wave. The plasma is thus confined by this "magnetic dam" on one side and by the walls of the tube. The cooling of the plasma is due to the contact with the walls. The transverse magnetic field trapped in the plasma reduces the thermal conductivity coefficient. For instance, it turns out that the time required to cool a plasma with $n = 10^{16}$ cm^{-3} and $B \sim 95$ kGauss from 4×10^8 °K to 2×10^8 °K is 1.4 sec. Although a better understanding of the energy transfer rate from the plasma to the walls is needed, wall evaporation may not be as catastrophic as often feared.

Calculations were made for a DT plasma surrounded by a lithium blanket. Energy is released both in the plasma: fusion energy E_F, and in the blanket by exothermal absorption of neutrons, E_b. Moreover, the blanket provides tritium breeding. Assuming a thermal to electrical energy conversion efficiency of 1/3 and denoting E_{IN} the invested electrical energy, the ratio:

$$R = \frac{E_F + E_b}{E_{IN}}$$

should be greater than 2 for a system producing usable electric power. For a given shock geometry R has a sharp maximum when plotted versus shock velocities and transverse field. As shown on Fig. 8, R varies smoothly with the filling pressure: L is the shock tube length; the major diameter is equal to the length and the minor diameter is smaller by about 20% in the investigated scheme.

Since this concept actually involves a motionless plasma and a very long confinement time which permits almost complete fuel burning, the invested energy per unit surface is by several orders of magnitude below the values given on Table I.

VII. FUSION BY INTERACTION OF ELECTRON BEAMS WITH SOLID TARGETS

The use of relativistic electron beams for C.T.R. was proposed independently in 1968 by USSR and US scientists[40]. Recent improve-

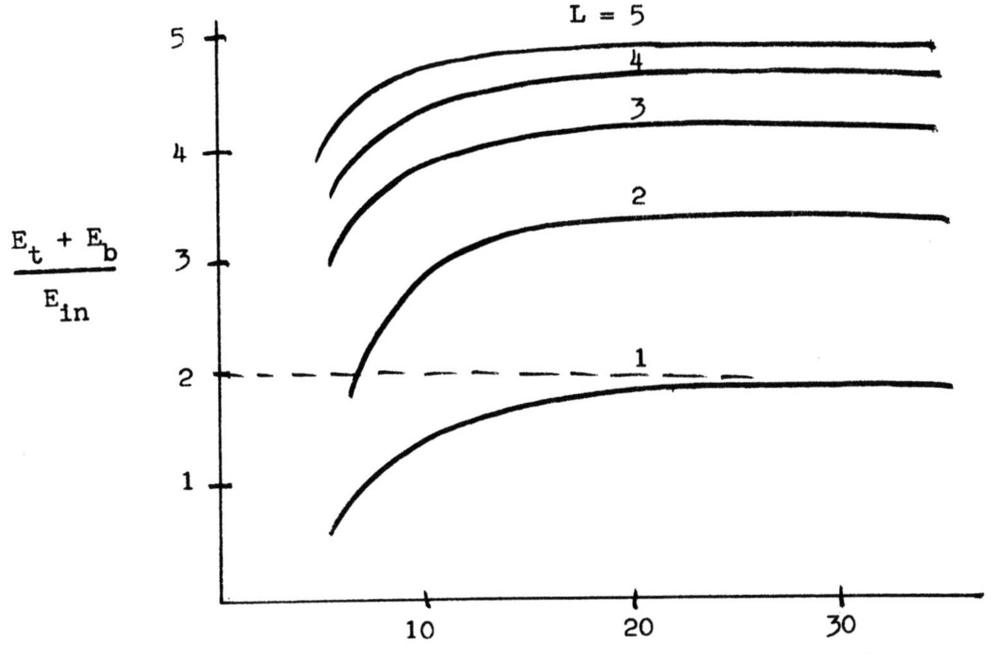

FIGURE 8. Efficiency of shock tube for fusion after Ref. 36.

ments[41,42] in tight focusing of electrons lead to fluxes comparable with those obtained by focusing laser beams. Accordingly fairly high neutron yields were observed when using deuteriated targets[43,44].

Basically the tight focusing of the beam in a high intensity high voltage diode is associated with the presence of a thin plasma filament along the axis. The physics of the mechanism is not yet completely understood. However, numerical simulations predict current densities of 10^8 A/cm^2 for a 3 MeV 3 Amp. beam. Experimentally current densities above 10^7 A/cm^2 have been achieved with voltages ranging from 0.2 to 2 MeV. The current is about 100 kA. The beam cross section is around 1 mm^2. The pulse duration is less than 100 nsec.

Although such data yield flux densities of the same order of magnitude as for Nd laser beams, the interaction physics is quite different on two main points: first, the way energy is transferred to the target, then, the flow which is subsequently driven.

In laser interaction with solid targets light is absorbed in a comparatively thin layer. The radiation energy density:

$$\Phi/c$$

is transferred to a material which is set into motion with a velocity:

$$u \ll c$$

The energy density in the matter is:

$$\phi/u \gg \phi/c$$

Due to the smallness of the cut-off density with respect to the solid state density, light is efficiently absorbed in the rarefaction at cut-off density.

There is no cut-off in the case of electron beams. Fast electrons are absorbed over a range which depends on their energy and on the density of the medium. If only classical scattering is taken into account the stopping power for solid deuteriated targets is 0.01 g/cm^2 at 100 keV and 1.0 g/cm^2 at 10 MeV, thus yielding in the latter case comparatively low energy densities inside the material.

Fortunately collective effects significantly reduce the range of electrons. The impact of an intense electron beam induces a two stream instability. For nonrelativistic beams a growth rate is derived[45] as

$$\gamma_I = 0.7\left(\frac{n'}{n}\right)^{1/3} \omega_p \tag{27}$$

where n' is the particle density inside the electron beam impinging onto the plasma of density n. Denoting by v the velocity of the electrons the range over which they transfer their energy to the instability is:

$$\lambda \sim \frac{2v}{\gamma_I} = 1.4\left(\frac{n}{n'}\right)^{1/3} \frac{v}{\omega_p} \tag{28}$$

In the case of relativistic electrons, denoting by λ_c the Coulomb scattering range, by $<\theta^2>$ the mean thermal angular scatter of the beam particle velocities, the following stopping distance was found[46]:

$$\lambda_R \sim \lambda_c <\theta^2> \left(\frac{\varepsilon_k}{m_o c^2}\right)\left(\frac{T_e}{m_o c^2}\right) \tag{29}$$

for electrons of kinetic energy ε_k; m_o is the rest mass. Since high azimuthal magnetic fields are associated with such beams λ_R can also be expressed as a function of the ratio of the magnetic energy ε_m to the kinetic energy ε_k, which should be greater than 1 in order to ensure the stability of the focused beam, thus:

$$\lambda_R \propto \frac{\varepsilon_k}{\varepsilon_m} \propto \frac{\gamma}{\nu} \tag{30}$$

where

$$\gamma = (1 - v^2/c^2)^{-1/2} = \varepsilon_k/m_o c^2$$

$$\nu = \pi a^2 n' \frac{e^2}{m_o c^2} = \varepsilon_m/n' m_o c^2 \tag{31}$$

(a = radius of the beam)

a high ν/γ characterizes a tightly focused relativistic electron beam.

The driven flow is generally investigated theoretically through computer codes. Due to the difficulties of plasma diagnostic comparison with experiment can be made only about the energy deposition into the target and neutron production. Since photo-production of neutrons by bremsstrahlung radiation is rather unlikely, a thermonuclear process is assumed in the calculations. In order to find out a production consistent with experiments[40], it is necessary that energy be deposited in a thin layer significantly beneath the surface. The outer layers thus provide an extra inertial confinement[47].

The application to fusion is thought about according to the approach of Section II[48] indeed due to the different way energy is deposited. It seems unlikely that a deflagration wave can be driven as in the laser case. Finally it should be noted that the high magnetic field associated with the beam is an efficient lateral confinement.

VIII. DISCUSSION

A positive energy balance from nuclear reaction in driven fronts requires a large invested energy unless the medium is highly compressed. Recently it was proposed to use laser radiation in order to obtain densities higher than solid state by several orders of magnitude[4]. The scheme of Fig. 2 is used in spherical geometry and the induced shock wave is expected to to the job. Such a shock could also result from the interaction of specially designed electron beams with spherical pellets.

The case of electromagnetically driven shocks is somewhat different. Indeed, spherical boundary conditions are incompatible with D.C. solutions of the Maxwell equations. However, it should be noted that the use of the shock reflexion on the end wall belongs to the same approach. Moreover, contrary to high density pellet explosion in which only inertial confinement is available, a very efficient containment comes from the walls and a supplementary magnetic field.

Not all lasers are convenient for pellet compression or for one dimensional flame propagation. Since a lateral confinement is needed in the latter case, long wavelength ($\lambda >> 10\ \mu m$) lasers seem better. On the contrary, radiation in the visible or near UV range would be more suitable for compression[49].

On Table IV the way each one of the energy sources reviewed in this paper fits with fusion by driven shocks or pellet compression is given. Some of the main problems which should be solved in energy special case are also listed.

A more detailed comparison can be made between lasers and relativistic electron beams. The present state of the art is given on Table V.

Table IV

	Driven shock or flame	Pellet compression
Low frequency laser (infrared)	Long pulses - Lateral confinement - Stability -	
High frequency laser (visible, near UV)		High flux absorption
Electron beams	Ignition	Avoiding core heating; short pulses
Electromagnetically driven shock tubes	Influence of the walls - low density	

Table V

	photons		electrons
	Glass laser 1.06 μm	Gas laser 10.6 μm	200 keV-10 MeV
Beginning of significant interaction experiments	1963	1970	1971
Pulse duration	<1 nsec	1 nsec-100 nsec	$\gtrsim$50 nsec
Maximum energy	300 J	20 J-300 J	~1 MJ
Efficiency	<1%	5-10%	50%
Maximum flux density on target	10^{17} W/cm^2	10^{12} W/cm^2	10^{14} W/cm^2 (10^6 -10^7 A/cm^2)
Maximum neutron yield	10^7 [50]	-	10^{11} [44]
Geometry	multi-beam spherical	single beam	single beam
Predictable advances (1975)	10 kJ-1 nsec	10 kJ-1 nsec	

Obviously in the case of the most powerful lasers known today, the neodymium glass laser (λ = 1.06 μm) and the CO_2 TEA gas laser (λ = 10 μm), predictable advances are compatible with the compression scheme only. Furthermore the efficiency of the glass laser is very low, so only feasibility experiments can be made with this laser. This approach to fusion could also use electron beam. Then higher current densities and much shorter pulses would be required. One of the major drawbacks might come from unwanted preheating of the core by the high energy incoming electrons.

Fusion by driven shocks could use either long wavelength high energy lasers or electron beams. However, the mechanisms, laser driven flame, ignition after local deposition of electron energy are different. But due to the high energies already achieved and to the better efficiency, conditions in target irradiated by relativistic electrons rather than by laser light might be closer to thermonuclear requirements.

The breakthrough in laser technology which would render both approaches to fusion practical would be a significant increase in energy and efficiency. Gas laser pumped by electrical discharges or electron beam with super radiant emission could be the solution. Chemical or gas dynamic lasers could also be used to drive thermonuclear flames.

Acknowledgments

The author is indebted to D. Colombant, F. Floux, D. Cognard, D. Redon, A. Saleres and G. Tonon for their contribution to the results quoted in this paper.

Thanks are also due to Professor R. A. Gross for communicating details about the schemes given in Section VI.

References

1. R. G. Sachs, Phys. Rev. 69 (1946) p. 514.

2. R. A. Gross, Rev. Mod. Phys. 37 (1965) p. 724.

3. J. D. Somon in Physics of High Energy Density, Academic (1971) p. 189.

4. J. H. Nuckolls, L. Wood, A. Thiessen, G. Zimmerman, Nature 239 (1972) p. 135.

5. J. G. Linhart, Nucl. Fus. 10 (1970) p. 24.

6. Ya. B. Zeldovich, Yu. P. Raizer, Physics of Shock Waves and High Temperature Hydrodynamic Phenomena, Academic 1 (1966) 2 (1967).

7. J. P. Somon, Nucl. Fus. 12 (1972) p. 461.

8. R. A. Gross in Physics of High Energy Density, Academic (1971) p. 245.

9. M. S. Chu, Phys. Fluids 15 (1972) p. 413.

10. J. M. Dawson, C. Oberman, Phys. Fluids 5 (1962) p. 517, A discussion of some nonlinear situations is found in R. E. Kidder in Physics of High Energy Density, Academic (1971) p. 306.

11. C. Fauquignon, F. Floux, Phys. Fluids 13 (1970) p. 386.

12. J. L. Bobin, Phys. Fluids, 14 (1971) p. 2341.

13. R. E. Marshak, Phys. Fluids 1 (1958) p. 24.

14. A. Caruso, A. de Angelis, G. Gatti, R. Gratton, S. Martelluci, Phys. Lett. 29A (1969) p. 316.

15. R. Sigel, K. Buchl, P. Mulser, S. Witkowski, Phys. Lett. 26A (1968) p. 498.

16. J. L. Bobin, F. Delobeau, G. de Giovanni, C. Fauquignon, F. Floux, Nucl. Fus. 9 (1969) p. 115.

17. C. Yamanaka, T. Yamanaka, T. Sasaki, K. Yoshida, M. Waki, H. Kang, Phys. Rev. 6A (1972) p. 2335.

18. J. W. Shearer, S. W. Mead, J. Petruzzi, F. Rainer, J. E. Swain, C. E. Violet, Phys. Rev. A6 (1972) p. 764.

19. N. G. Basov, O. N. Krokin, G. V. Sklizkov, in Laser Interaction and Related Plasma Phenomena, Plenum (1972) p. 389.

20. M. Lubin, J. Soures, E. Goldman, T. Bristow, W. Leising in Laser Interaction and Related Plasma Phenomena, Plenum (1972).

21. F. Floux, D. Cognard, A. Saleres, in Laser Interaction and Related Plasma Phenomena, Plenum (1972).

22. H. M. Thompson, R. G. Rehm, J. W. Daiber, Journ. Applied Phys. 42 (1971) p. 310.

23. S. D. Zakarov, O. N. Krokhin, P. G. Driukov, E. L. Tiurin, Quantum Electronica (in Russian) 2 (1971) p. 104.

24. F. Floux, J. F. Benard, D. Cognard, A. Saleres, in Laser Interaction and Related Plasma Phenomena, Plenum (1972) p. 409.

25. J. L. Bobin, M. Decroisette, B. Meyer, Y. Vitel, Phys. Rev. Lett. 30 (1973) p. 594.

26. A. Saleres, D. Cognard, D. Redon, F. Floux, J. L. Bobin Congrès du Centenaire S.F.P. Vittel, 28 mai-2 juin 1973. To be published in Journal de Physique.

27. J. D. Lawson, Proc. Phys. Soc. London 70B (1957) p. 6.

28. M. Born, E. Wolf, Principles of Optics, 4th ed. Pergamon (1970).

29. J. M. Dawson, A. Hertzberg, R. E. Kidder, G. C. Vlases, H. G. Ahlstrom, L. C. Steinhauer, in Plasma Physics and Controlled Nuclear Fusion Research, IAEA, Vienna (1971) p. 673.

30. A. D. Sakharov, Soviet Phys. Doklady 10 (1966) p. 1045.

31. S. B. Alikhanov, J. Scient. Instrum. Series 2 1 (1968) p. 543.

32. J. W. Shearer, Journ. Applied Phys. 40 (1969) p. 4490.

33. N. G. Kovalski, P. P. Pashinin, M. I. Pergament, private communication.

34. R. G. Fowler, W. R. Atkinson, W. D. Compton, R. J. Lee, Phys. Rev. 88 (1952) p. 137.

35. A. C. Kolb, Phys. Rev. 107 (1957) p. 345, 107 (1957) p. 1197.

36. R. M. Patrick, Phys. Fluids 2 (1959) p. 589.

37. R. A. Gross, Y. G. Chen, E. Halmoy, P. Moriette, Phys. Rev. Lett. 25 (1970) p. 575.

38. C. K. Chu, R. A. Gross, in Advances in Plasma Physics, 2 A. Simon, W. B. Thompson ed. Wiley (N.Y.) (1969).

39. R. A. Gross, Nucl. Fus. 13 (1973) p. 293.

40. An unspecified reference is made in: H. V. Babykin, Ye. K. Zavoiski, A. A. Ivanov, L. I. Rudakov, Plasma Physics and Controlled Nuclear Fusion Research, IAEA Vienna (1971) p. 635 to the initial proposal by Zavoisky.

F. Winterberg Phys. Rev. 174 (1968) p. 212.

41. D. L. Morrow, J. P. Phillips, R. M. Stringfield, W. D. Doggett, W. H. Bennett, Appl. Phys. Lett. 19 (1971) p. 441.

42. G. Yonas, K. R. Prestwick, J. W. Poukey, J. R. Freeman, Phys. Rev. Lett. 30 (1973) p. 164.

43. J. R. Kerns, C. W. Rogers, J. G. Clark, Bull. Am. Phys. Soc. 17 (1972) p. 690.

44. B. Freeman, H. Sahlin, J. Luce, O. Zucker, T. Crites, C. Nelson, Bull. Am. Phys. Soc. 17 (1972) p. 1030.

45. O. Buneman, Phys. Rev. 115 (1959) p. 503.

46. L. I. Rudakov, Soviet Phys. JETP 32 (1971) p. 1134.

47. T. E. McCann, C. W. Rogers, D. N. Payton, Bull. Am. Phys. Soc. 17, (1972) p. 690.

48. F. Winterberg, Nucl. Fus. 12 (1972) p. 353.

49. J. L. Emmett, Paper presented at the Annual Meeting of the A.P.S. New York Jan. 29 Feb. 2 (1973).

50. N. G. Basov, O. N. Krokin, G. V. Sklizkov, These proceedings.

LASER COMPRESSION AND FUSION OF PLASMA*

Sylwester Kaliski

Polish Academy of Sciences

Warsaw, Poland

ABSTRACT

In the paper a short review of works on laser concentric heating and compression of plasma, from the point of view of nuclear fusion, has been given.

Then different variants of approximate, averaged description for a D-T sphere submitted to the concentric laser compression by taking into account the recovered fusion heat, were considered. Among others the integral similarity model was discussed for the kernel of the D-T sphere (Fermi liquid) and the expanding outer layer (two-temperature plasma).

Averaged description taking into account the fronts of thermal and shock waves have also been proposed. The idea of explosive precompressing of laser compression of plasma has been proposed. It allows to reduce the critical laser pulse by a few tens per cent (2-3 kJ).

INTRODUCTION

The idea of laser plasma heating, proposed in 1964[1,2], confirmed experimentally later[3-6] and in a number of other works on recovery

*Presented at the Third Workshop on "Laser Interaction and Related Plasma Phenomena" held at Rensselaer Polytechnic Institute, Troy, New York, August 13-17, 1973.

of the fusion neutrons by laser pulse gave a stimulation to a series of theoretical and experimental works in the years 1968-1973.

In particular were investigated the problems of conductivity, absorption, propagation of thermal and shock waves in the plasma heated by laser, the fusion energy recovery being taken into account[7-14]. The review of these works can be found in Refs. 16 through 20. Parallel to this were developed in theoretical aspect, the methods of averaged description of the laser plasma heating, especially for the spherical symmetry[21-28].

Computations of critical masses necessary for energy interception of α - particles[29] and time of retaining the plasma in aggregation state showed that the laser pulse energy which is to be transmitted to the D-T ball is very high and oscillates in the range of 10^9-10^8 J. A number of works were started on additional effects, allowing to prolong the time of maintaining the plasma in the aggregation state and its preliminary compression by the implosion of heavy envelopes, driven by explosives[32-36]. These methods, in spite of the fact that they gave rise to new difficulties resulting from the radiation losses of too heavy envelopes, thermal losses in the ball due to the heating of envelopes, etc., allowed to reduce the critical value of the laser pulse energy for obtaining a positive energy recovery up to the order of 10^5 J, what essentially improved the situation in relation to the perspective of the use of lasers for realization of the thermonuclear microfusion. Parallel to all this a series of comceptions was proposed, related to magnetic plasma cumulation[34-37] with the simultaneous laser heating of the plasma. They led to a similar order of the critical value of the laser pulse as mentioned above.

In 1972 the idea of laser concentric compression of the plasma was proposed parallel in Refs. 38 through 41. The properly profiled laser pulse was used for the heating of external layer of the plasma ball. This layer, undergoing the ablation and expansion, returns its pulse, in a form of the shock wave, to the central part of D-T ball, causing its compression and heating and makes possible to obtain the positive thermonuclear fusion energy recovery. These works gave a new stimulation and brought closer the problem of thermonuclear microfusion to the existing possibilties of laser technique. With the laser pulse energy from several to some tens kJ the compression of the order up to 10^4 is obtained with the critical mass of the order 10^{-5} - 10^{-4} g as well as the positive fusion energy recovery of the order from several to 10^2 (effective energy, after reduction of the fusion energy, used for expansion, heating of the plasma, etc.). In a number of later works conducted on the averaged models for the laser concentric compression of the plasma with envelopes and without them, the fundamental microfusion characteristics were additionally confirmed[42-51] and also on the basis of other model approximations[52-54].

In the author's works[55-56] on the ground of auxiliary solutions from the gas dynamics[57-61], was proposed a method of construction of the averaged equations, in which the fronts of thermal and shock waves are taken into consideration. This theory, on the ground of considerably simpler equations (the system of ordinary nonlinear differential equations), allows, with relatively small quantitative error and with conservation of qualitative properties of the strict solutions, to analyse the different variants of the laser compression of the plasma, what presents a very precious instrument of investigation.

Regardless of enormity of experimental difficulties and lots of problems of theoretical nature[62] such as: problems of hydrodynamical stability for the spherical symmetry and shock waves, problems of parametric instability of the plasma, preheating by the fast electrons, problems of reflection and absorption of the laser radiation, question of optimalization of the process, etc., - the way of laser compression of the plasma presents today an actual, contained in technical realities perspective of realization of the thermonuclear microfusion. These methods together with the competitive and similar method of the electron-beam heating of the plasma[63-66] constitute the most actual research perspective in the range of thermonuclear microfusion. At the same time are outlined many possibilitiss of applying the combined methods: laser compression of the plasma and compression by electron-beam, analogically as, for instance, cooperation of the systems of the laser-focus type[67-69] for the laser heating of the plasma or compression with the additional placing of D-T pellet in the center of plasma focus and, for example the use of heavy multilayered envelopes[70] etc.

In the author's work[71] was proposed the method of laser compression of the plasma with preliminary cumulative precompressing by the cumulative streams initiated by explosives.

This presents a possibility to obtain the initial compression of D-T of the order 10-20 (with the use of isentropy approach of the cumulation process by means of inhomogeneous fillers) at the time order 10^{-7} sec. On such a process is placed the laser plasma compression with nanosecond laser pulse. The above mentioned method makes possible to reduce, in comparison with the purely laser compression, additionally several tens per cent the critical value of the laser pulse energy, necessary for obtaining the positive fusion energy recovery, thus reducing it to the order of several kJ. This is a further essential step forward for the realization of the thermonuclear microfusion.

In this work in the next section we present a brief characteristic of the results of works of concentric plasma compression, in the third section we present the description of the concentric laser compression of the plasma on the ground of averaged models. In the fourth section we present the idea of averaged description with consideration of the structure of fronts of thermal and shock waves, and finally in the fifth one - the idea of plasma precompression by cumulative charges before the action of laser pulse. At the end we recapitulate the conclusions.

SHORT REVIEW OF WORKS ON THE CONCENTRIC PLASMA COMPRESSION AND FUSION

In Refs. 38-41 and 62 the idea was formulated and numerical computations were performed for the concentric compression of D-T plasma. In Ref. 38 was presented the averaged description of concentric plasma compression, developed later in the works of Refs. 42-51. On the basis of integral similarity description it was possible to obtain the positive thermonuclear fusion energy recovery (> 10) with the laser pulse energy 5-8 kJ and the ball mass 6×10^{-5} g and compression of the order 10^4.

In the works of Brueckner and Nuckolls[39-41] a full two-dimensional theory of the problem was formulated and computations were performed.

Applying the programmed laser pulse the compression of the order 10^4 was obtained in the centre of the ball of initial mass 6×10^{-5} g. This compression was accomplished by the heating of external ball layer and through concentric impulse of the shock waves, generated by ablation and rejection of external layer. It was shown that with the laser pulse energy of the order 60 kJ a pure recovery, in comparison with the absorbed laser energy, amounts to the order 10-30.

In these works also were analysed the Rayleigh-Taylor instability questions, parametric instabilities, questions of electron preheating, reflections and absorption; this analysis demonstrates the fundamental possibility of realization of such high compressions. A very optimistic conclusion was attained in Ref. 52, where with the laser pulse energy 1.9 kJ the authors obtained the recovery of the order 4.7.

The thorough computation analysis was performed by the author of Ref. 53 with 30 data variants, showing a series of optimalization properties of compressive solutions. A starting point for computer calculations for D-T ball, undergoing the compression, were the state equations of Fermi gas for the electrons, (for ions the compression is directly expressed by the energy density). These equations are completed by the relations for laser radiation absorption, thermal conductivity, electron-ion relaxation, fusion energy production, radiation losses, viscosity, neutron heating. On the ground of these equations the computer codes and computation results were given in Refs. 39-41 and 62. Not going into further details of Brueckner and Nuckolls and other works we will concentrate below on the review of own works, based on the averaged description. The averaged description, though, depending on the applied approximations, it loses a series of effects of qualitative nature, generally well reflects the global features of solutions and is suitable for practical appraisals.

In the case, when on the ground of averaged description the structures of the fronts of thermal and shock waves are taken into account - also the fundamental qualitative properties of solutions are retained. The averaged description by the reason of its simplicity and physical clarity allows us to follow analytically the influence of parameters of the problem on the solution and is perfectly suitable for the optimalization analyses and modifications of the problem conditions (precompressing etc.).

THE AVERAGED DESCRIPTION OF CONCENTRIC PLASMA COMPRESSION, THE THERMONUCLEAR FUSION ENERGY BEING TAKEN INTO ACCOUNT

In Ref. 33 a system of equations and solution were given for D-T ball, enclosed with the envelope of heavier material, shot into the ball by the explosive. After preliminary plasma compression the ball was submitted to the laser heating. The state equations of the ball were assumed in the form of the ideal two-temperature gas. The solution was constructed on the basis of integral similarity model.

For the positive fusion energy recovery the critical value of absorbed laser energy of the order 7×10^4 J was obtained at $\eta_2 \simeq 2.2 (\eta_3 \simeq 0.8)$

where $$\eta_1 = \frac{E_{rec}}{E_c + E_L} ; \quad \eta_2 = \frac{E_{rec}}{5(E_c + E_L)} ; \quad \eta_3 = \frac{E_{rec}}{5E_c + 100E_L}$$

This is the most advantageous value of the critical laser pulse, which was obtained with the cumulative-laser heating of the plasma. The laser worked here as the heating system, while cumulation - as the compressive factor. In the work of Ref. 38 the averaged description was proposed for the concentric laser cumulation of D-T plasma and the appraisal of global parameters were obtained, like the radius and mass of the ball, compression, critical laser pulse. These results are approximate to that in Refs. 39-40. In Ref. 38 full averaging of parameters was employed. In the successive works[43-48] was analysed a number of further improved

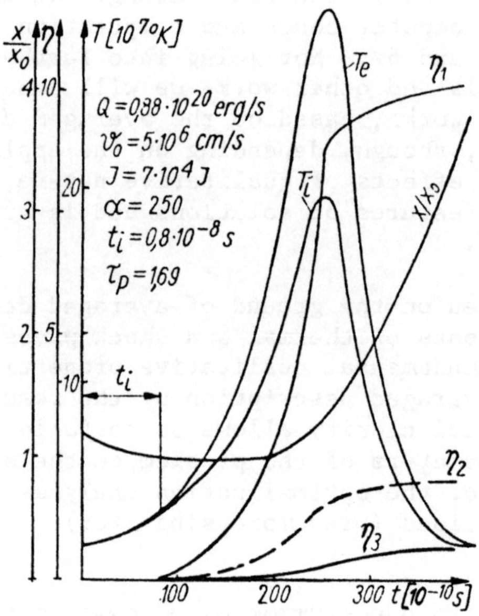

FIGURE 1.

approximations of the laser concentric compression of the plasma with the envelope and without it. In the simplest models was assumed a sudden impulse separation of the surface layer, enclosed by field temperature, (or separation of envelope), with the transmission of the compression pulse to the central part of the plasma ball. In Ref. 45 was defined the external layer, submitted to ablation on the ground of velocity of the concentric thermal wave, propagating with the variable velocity into the ball[47-48]. All the model approximations with the assumed initial mass of D-T ball of the order 6×10^{-5} g and compression $10^3 - 10^4$, led to the critical value of the laser pulse energy, necessary for the positive fusion energy recovery ($E_{rec}/E_L > 1$, where E_L - the laser pulse energy, absorbed by plasma, E_{rec} - the global thermofusion energy) of the order 5 kJ. These values in comparison with the strict ones can be slightly lowered. The 5 kJ energy was obtained after the optimalization analysis[43,50].

The most general description on the ground of the integral similarity model was presented in Ref. 45. We will present the theory of this model more thoroughly. We will namely assume that a thermal wave under the influence of the concentric laser pulse, propagates to the ball centre with the velocity, defined in an approximate way in Ref. 47. Further we will assume that the front of thermal wave divides the D-T ball into two regions: the internal ball, submitted to the compression and described by the model of Fermi liquid with assumption of integral similarity averaging and external layer, enclosed by the front of thermal wave and submitted to expansion (ablation). The external layer is treated as the two-temperature ideal gas. Both the central ball and external layer constitute, in the integral similarity description, the system of bodies with variable reduced masses, what gives a slight trouble in construction of the conservation equations, especially in conservation of energy. Let us pass to formulating the system of equations of the problem.

In Figs. 2 and 3 is presented the scheme of the system and assumed averaged distribution of velocities in the central ball and in external layer at the assumption of variable boundary dividing both regions. The velocity value of thermal wave c(t) is determined by explicit approximate formulae on the grounds of Ref. 47.

The simplified schemes of the laser radiation absorption in the external layer, discussion on the influence of bremsstrahlung radiation, linear and recombination radiation can be found in the above quoted works, therefore we omit their presentation.

Denotations assumed for the ball

$$x = \frac{\dot{r}}{\dot{R}} = \frac{\dot{r}_c}{\dot{R}_c} \tag{1}$$

and for the external layer

$$x = \frac{\dot{r}_c - \dot{R}_c}{\dot{R}_o - \dot{R}_c} = \frac{\dot{r} - R}{\dot{R}_o - \dot{R}} \tag{2}$$

where

$$\dot{R} = \dot{R}_c - c; \quad R = R_c - \int_o^t c(t_1)dt_1$$

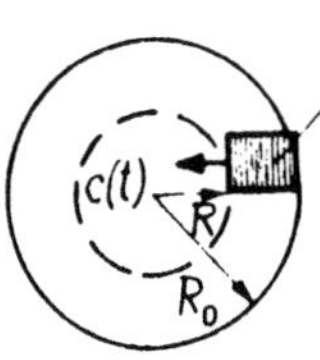

FIGURE 2.

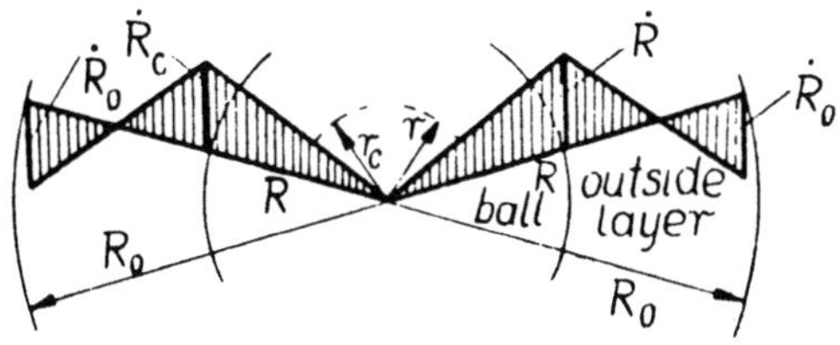

FIGURE 3.

Equations for the Central Part of D-T Ball

Assuming for the integral similarity model

$$\rho(x,t) = \rho_c(t)f(x); \quad n(x,t) = n_c(t)f(x);$$
$$p(x,t) = p_c(t)f(x); \quad T = T(t) \tag{3}$$

and the expression for energy and pressure for degenerated Fermi gas

$$E = \frac{3}{5} NE_r\left[\left(\frac{3N}{4\pi}\right)^{2/3} R^{-2} + \frac{5\pi^2}{12}\left(\frac{k}{E_r}\right)^2\left(\frac{4\pi}{3N}\right)^{2/3} R^2T^2\right]$$

$$p(x,t) = p_c(t)f(x) = \frac{2}{5} E_r n^{5/3}\left[1 + \frac{5\pi^2}{12}\left(\frac{kT}{E_r}\right)^2 n^{-4/3}\right] \tag{4}$$

where

$$E_r = \frac{\hbar^2}{2m}(3\pi^2)^{2/3}$$

we write the equations of momentum and energy conservation for a body with variable mass (the equation of energy conservation requires a modification in relation to the classical approaches):

$$\frac{2}{5} AE_r\left[C\left(\frac{N}{4\pi D}\right)^{5/3} R^{-3} + \frac{5\pi^2}{12}\left(\frac{k}{E_r}\right)^2\left(\frac{N}{4\pi D}\right)^{1/3} LRT^2\right] =$$
$$= 2B\frac{d}{dt}[M(\dot{R} + c)] - 2Bv_s\frac{dM}{dt} + \bar{A}R^2\frac{k}{V_L}(N_eT_e + N_iT_i) \tag{5}$$

$$\frac{d}{dt}\ \frac{3}{5} NE_r\left[\left(\frac{N}{4\pi D}\right)^{2/3} CR^{-2} + \frac{5\pi^2}{12}\left(\frac{k}{E_r}\right)^2\left(\frac{4\pi D}{N}\right)^{2/3} LR^2T\right] +$$
$$- A(\dot{R} + c)R^2\frac{k}{V_L}(N_eT_e + N_iT_i) + B\frac{d}{dt}[M(\dot{R} + c)^2] + \tag{6}$$
$$- 2B\left(v_s - \frac{\dot{R} + c}{2}\right)(\dot{R} + c)\frac{dM}{dt} + B_r = Q(T,R)$$

where $M = 4\pi R^3 \rho_c D$; $D = \int_0^1 f(x)x^2 dx$;

$$C = \frac{\int_0^1 f(x)^{5/3}x^2 dx}{D} \;;\quad L = \frac{1}{D}\int_0^1 f(x)^{1/3}x^2$$

$$v_s = \dot{R}_c + (\dot{R}_o - \dot{R}_c)\frac{R_s - R}{R_o - R} \;;\quad R_s = \frac{3}{4}\frac{R_o^4 - R^4}{R_o^3 - R^3} \;;\quad V_L = \frac{4\pi}{3}(R_o^3 - R^3)$$

$$\bar{A} = -4\pi \;;\quad A = -4\pi\int_0^1 \frac{df(x)}{dx} - x^3 dx \;;\quad B = \frac{1}{2D}\int_0^1 f(x)x^4 dx \tag{7}$$

we determine f(x) in accordance with the known formulae from relation of momentum conservation (it takes the exponential form). In Eqs. (5) and (6) B_r - denotes the Bremsstrahlung, Q(T,R) - characterizes the fusion energy recovery (see formulae in Refs. 25-28. For c we have the approximate expression (we write the simplified formula for q = const, where q = laser radiation intensity)

$$c = \frac{14\pi c_o q T_e^{5/2}\dot{T}_e}{(4\pi c_o T_e^{7/2} + \frac{q}{R_c})^2} \tag{8}$$

In Eqs. (5), (6) and (7) appear, besides the unknown T, R, c, the parameters of external layer T_e, T_i and R_o. We obtain them from the equations of external layer.

Equations of External Layer

For the external layer, starting from equations for two-temperature ideal gas with the distribution of velocities as in Fig. 2 and assuming f(x) = const, we obtain basing on the dynamics equations of a body with variable mass

$$4\pi(R_o^2\dot{R}_o - R^2\dot{R}_c)k(n_{oe}T_e + n_{oi}T_i) + \dot{R}_c v_s \frac{dM_r}{dt} = v_s \frac{d(M_r v_s)}{dt}$$

$$\frac{3}{2}k\frac{d}{dt}(N_e T_e) + B_L = -k\, n_{oe}T_e 4\pi(R_o^2\dot{R}_o - R^2\dot{R}_c) + \tag{9}$$

$$-\frac{3}{2}kN_e\frac{T_e - T_i}{\tau} - \eta_e\frac{dM_r}{dt}(\dot{R}_c - \frac{v_s}{2})v_s + q(t)$$

$$\frac{3}{2}\, k \frac{d}{dt}(N_i T_i) = -\, k n_{oi} T_i 4\pi (R_o^2 \dot{R} - R^2 \dot{R}_c) +$$

$$+ \frac{3}{2}\, k N_e \frac{T_e - T_i}{\tau} - \eta_i \frac{dM_r}{dt}\left(R_c - \frac{v_s}{2}\right) v_s$$

where M_r - reduced mass of the layer in relation to the point R_s; B_L - bremsstrahlung of the layer.

$$\eta_e \; ; \; \eta_i \sim \frac{T_e}{T_e + T_i} \; ; \; \frac{T_i}{T_e + T_i} \; ; \quad n = \frac{N}{V}$$

System of Eqs. (5), (6), (8) and (9) constitute the system of 6 equations of the problem in relation to the unknowns R_o, R, T, T_e, T_i, c. To this system come in addition the initial conditions, which we do not write here for the lack of place.

The non-optimalized and optimalized variants of solution of the equations system (5), (6), (8) and (9) are shown in Figs. 4 and 5.

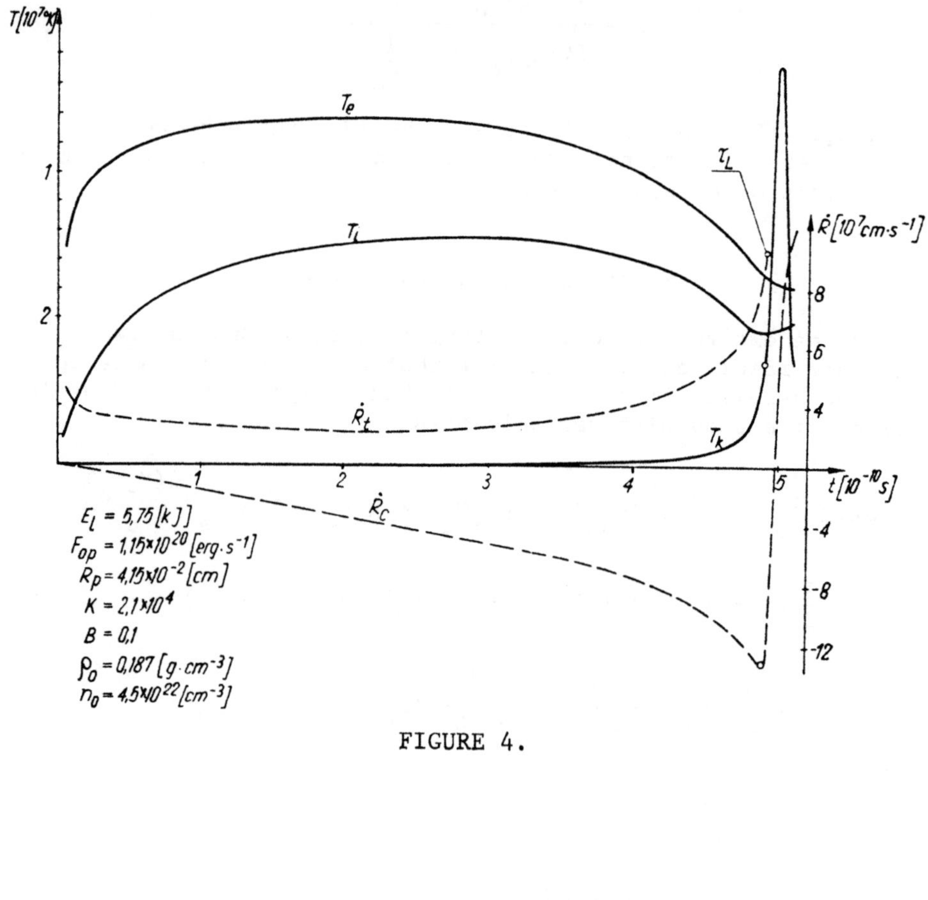

FIGURE 4.

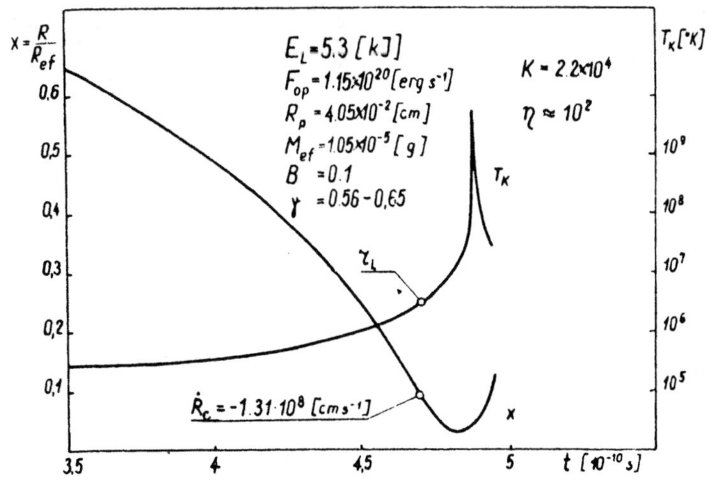

FIGURE 5.

The optimalization procedure is illustrated in Figs. 6 and 7.

We performed the optimalization procedure in Figs. 6 and 7 with some computation simplifications. As it follows from Fig. 5, with the laser pulse energy ~ 5.5 kJ we obtain the positive fusion energy recovery at the compression of the order 10^4 and initial ball mass of the order ~ 6×10^{-5} g, what is approximately in line with the results obtained hitherto in particular also by the strict methods.

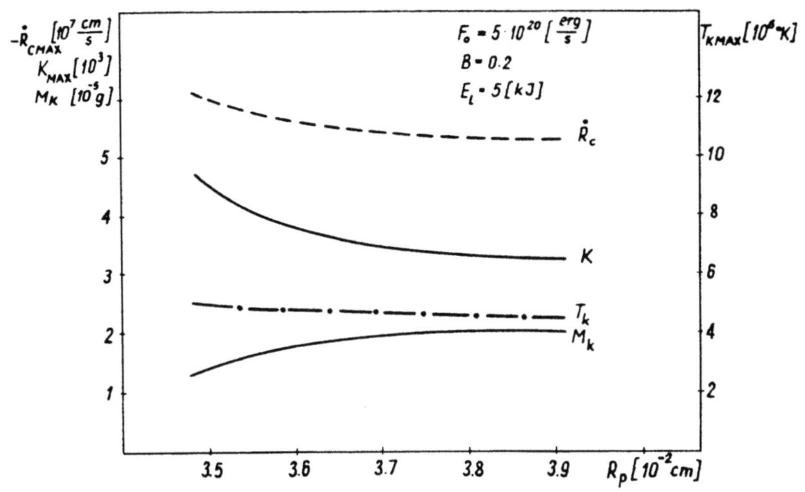

FIGURE 6.

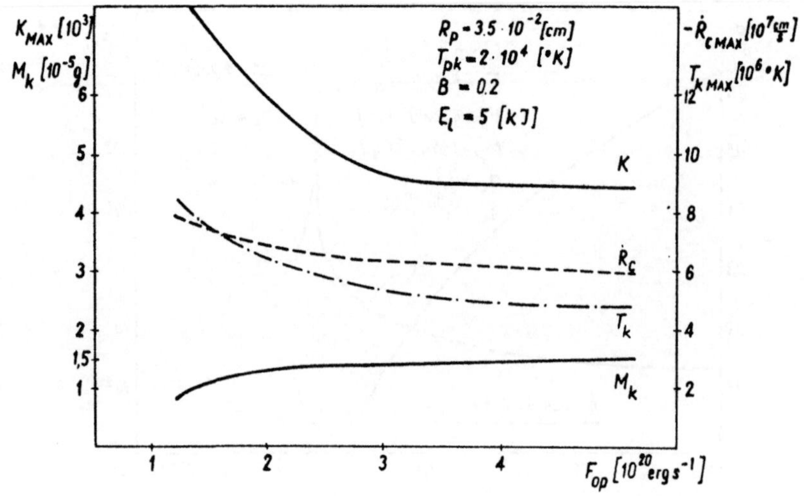

FIGURE 7.

The global recovery amounts to ~ 10^2 and the effective recovery is of the order 10. The averaged methods can lead, in appraisal of critical laser pulse energy, to a certain error in comparison with the strict solutions; it should not exceed the amount 100%. The above mentioned integral similarity model applied for the variable regions, regulated by the penetration velocity of the thermal wave into the D-T solid body, presents the most general possible description of the problem on the basis of averaging methods. A further generalization consists only in the introducing of structures of thermal and shock waves fronts into the averaged description, what will be dealt with in the next section.

AVERAGED DESCRIPTION WITH CONSIDERATION OF THERMAL AND SHOCK WAVES FRONTS

We present below the idea of constructing the averaged equations with consideration of the structure of thermal and shock waves fronts on the example of the plane wave[55] for one temperature plasma. For two-temperature plasma[56] these equations are similarly constructed, the generalization does not present any inconvenience. Moreover, we assume that the velocity of the thermal wave front c is constant. The omitting of influence of variability of c on the qualitative character of solutions is possible at the definite assumptions[61].

We will presume that the assumptions[61] are approximately satisfied. Generally they are not satisfied when c is approximately equal to $2a^*$, where $a^* = \sqrt{RT}$, i.e. in so called transshock region[59].

Then the influence of variability of c(t) is to be taken into account.

The generalization of above mentioned equations on the case of concentric spherical wave and numerical quantitative analyses are in preparation stage[73].

Let us now consider the semi-space filled by ideal gas, to the surface of which is applied the laser pulse of the constant (for simplification purposes) intensity q.

Let us assume first that $c > D$, where D - the shock wave velocity.

Then, according to Ref. 57 the hydrodynamical solution for the constant temperature field, propagating into the semi-space with the velocity $c > D$ has the form as in Fig. 8.

The course of ρ is proportional to p. For $c < D$ the solution is given in Fig. 9. The idea of constructing the averaged equations now consists in this that we assume a simplified distribution scheme of p and v as in Fig. 10 (for $c > D$). At the same time we assume that $T(t) = \text{const}$ from the surface to the wave front and depends on t only. Similarly we assume the variability of c(t), while the qualitative solution we build as for $c = \text{const}$. For the variant $c < D$ the averaged scheme is slightly more complicated[61] (see Fig. 9).

The basic system of equations for the plane wave has the form:

$$\frac{d\rho}{dt} + \rho v_{,x} = 0 \; ; \quad \frac{d(\rho v)}{dt} + p_{,x} = 0 \; ; \quad p = RT\rho = a^{*2}\rho$$
$$\frac{d}{dt}\left(\rho \frac{v^2}{2} + \rho c_v T\right) = - \frac{\partial F}{\partial x} + \Phi(T) - B \tag{10}$$

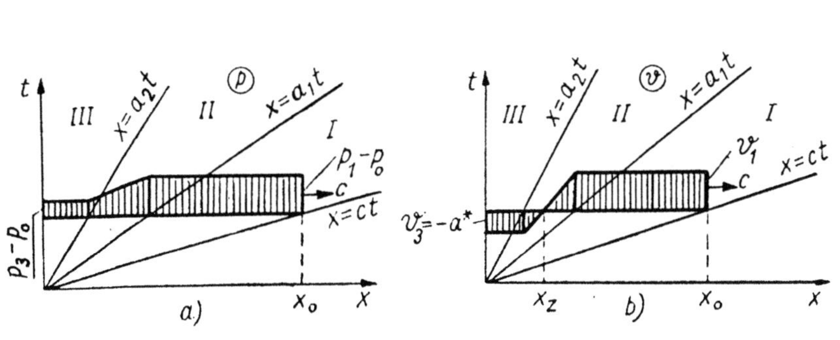

FIGURE 8.

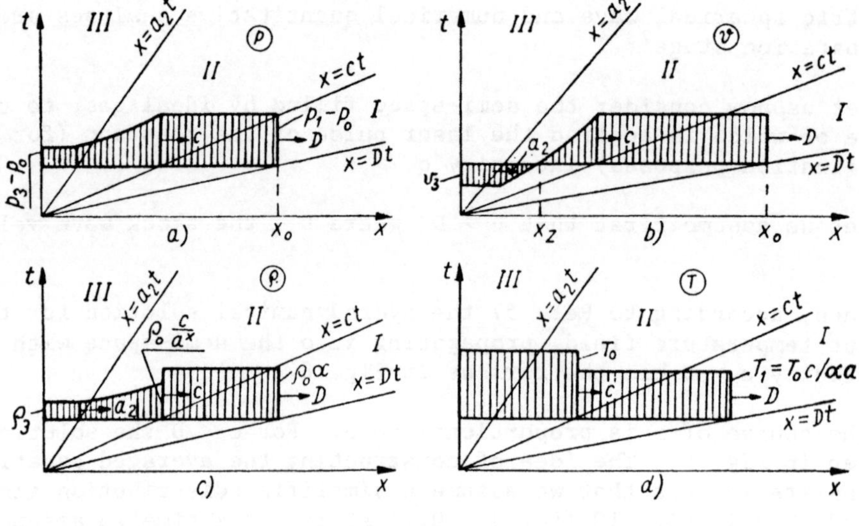

FIGURE 9.

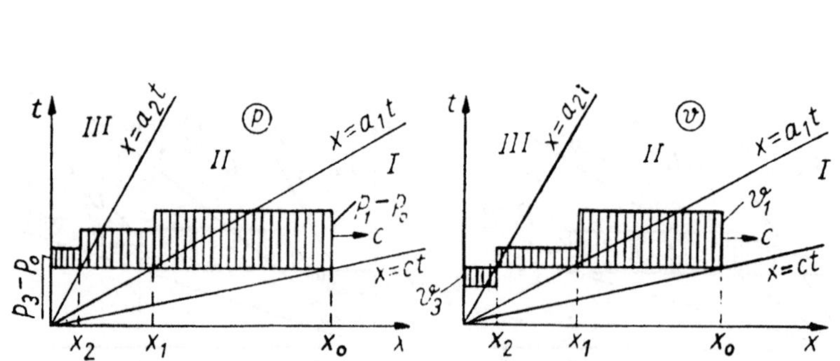

FIGURE 10.

where $\Phi(T)$ - expression defining the fusion energy output[61]; B - bremsstrahlung ~ $\rho^2 T^{1/2}$ and

$$F = -c_o \frac{\partial T^m}{\partial x} \qquad (m = \frac{7}{2}) \tag{11}$$

The equations on the thermal wave (or shock wave) front have, for the scheme as in Fig. 10, the form:

$$\begin{aligned} \rho_o \dot{x}_o &= \rho_1(\dot{x}_o - v_1) \\ \rho_o \dot{x}_o v_1 &= p_1 - p_o \end{aligned} \tag{12}$$

where

$$\dot{x}_o = c \quad (\text{or } \dot{x}_o = D \text{ for } \frac{c}{D} < 1)$$

at the same time

$$p_1 = \rho_1 RT = \rho_1 a^{*2} \quad (\text{for } c < D \quad p_1 = \rho_o c a^*) \tag{13}$$

We calculate the values of p_1 and ρ_1 from (12) and (13)

$$p_1 = \frac{1}{2}\left[p_o + \rho_o \dot{x}_o^2 - \sqrt{(p_o + \rho_o \dot{x}_o^2)^2 - 4\rho_o^2 a^{*2} \dot{x}_o^2}\right] \simeq \frac{\rho_o \dot{x}_o^2}{2}\left[1 - \sqrt{1 - \frac{4a^{*2}}{\dot{x}_o^2}}\right]$$

$$\rho_1 = \frac{p_1}{a^{*2}} \tag{14}$$

$$v_1 = \frac{p_1 - p_o}{\rho_o \dot{x}_o} \simeq \frac{p_1}{\rho \dot{x}_o}$$

On the wave $x = a_1 t$ and $x = a_2 t$ the relations occur:

$$\rho_1(a_1 - v_1) = \rho_2(a_2 - v_2) \tag{15}$$

$$\rho_2(a_1 - v_2)(v_2 - v_1) = p_2 - p_1 = \frac{p_3 - p_1}{2} = \frac{a^{*2}}{2}(\rho_3 - \rho_1);$$ (here we assumed $p_2 \simeq \frac{p_1 + p_3}{2}$)

$$\frac{\rho_3 + \rho_1}{2}(a_2 - v_2) = \rho_3(a_2 + a^*)$$
$$\rho_3(a_2 + a^*)(a^* + v_2) = \frac{a^{*2}}{2}(\rho_1 - \rho_3) \quad \text{where } v_3 = -a^* \tag{15}$$

We compute the unknowns ρ_1, ρ_3, v_1, v_2, from the system of equations (15). All the unknowns in our system of equations depend on $\dot{x}_o = c(t)$ or $D(t)$ and $T(t)$. We now use Eq. (10) employing the method of momentums. Integrating namely twice the last of equations (10) by segments in the range from zero to x_o we find:

after the first integration:

$$\begin{aligned}
&\frac{d}{dt}\left\{\rho_3 x_2 \frac{v_3^2}{2} + \rho_2(x_1 - x_2)\frac{v_2^2}{2} + \rho_1(x_o - x_1)\frac{v_1^2}{2} + \right.\\
&\qquad \left. + c_\nu T[\rho_3 x_2 + \rho_2(x_1 - x_2) + \rho_1(x_o - x_1)]\right\} = \\
&= F_o + \left[x_2\frac{\rho_3^2}{\rho_o^2} + (x_1 - x_2)\frac{\rho_2^2}{\rho_o^2} + (x_o - x_1)\frac{\rho_1^2}{\rho_o^2}\right]\Phi(T) - \\
&\qquad - [x_2 B(\rho_3) + (x_1 - x_2)B(\rho_2) + (x_o - x_1)B(\rho_1)]
\end{aligned} \tag{16}$$

after second integration:

$$\begin{aligned}
&\frac{d}{dt}\left[\rho_3 x_2^2 \frac{v_3^2}{2} + \rho_2(x_1^2 - x_2^2)\frac{v_2^2}{2} + \rho_1(x_o^2 - x_1^2)\frac{v_1^2}{2}\right] + \\
&+ c_\nu T[\rho_3 x_2^2 + \rho_2(x_1^2 - x_2^2) + \rho_1(x_o^2 - x_1^2)] = 2c_o T^m + \\
&\quad + \left[x_2^2\frac{\rho_2^2}{\rho_o^2} + (x_1^2 - x_2^2)\frac{\rho_2^2}{\rho_o^2} + (x_o^2 - x_1^2)\frac{\rho_1^2}{\rho_o^2}\right]\Phi(T) - \\
&\quad - [x_2^2 B(\rho_3) + (x_1^2 - x_2^2)B(\rho_2) + (x_o^2 - x_1^2)B(\rho_1)]
\end{aligned} \tag{17}$$

From the system of equations (16) and (17) we compute the unknowns $T(t)$ and $c(t) = \dot{x}_o(t)$ with the initial conditions:

$$x_o(0) = 0 \qquad T(0) = 0 \tag{18}$$

Having c and T we compute, on the basis of above quoted formulae, the parameters remaining. Analogically, with some additional procedure related to the discontinuity of T (Fig. 9), the case $c < D$ is solved.

A more complicated system of equations results for concentric spherical wave because of more complex purely hydrodynamical solution[60].

The method presented in this section is extraordinarily advantageous. It reduces the problem in averaged description to the system of ordinary equations, where the structures of thermal and shock waves are simultaneously taken into account. This gives a possibility of retaining the qualitative properties of the strict solutions with relatively small averaging error in quantitative aspect. It gives a very elastic and simple research instrument, allowing quick and simply to solve a series of schemes and optimalization variants.

As already mentioned, the problem of equations for concentric compression and quantitative illustrations of solutions will be given in Ref. 73.

CONCENTRIC LASER COMPRESSION OF PLASMA WITH PRECOMPRESSING BY CUMULATIVE EXPLOSIONS

At the end of this paper we will present the method of essential reduction of the critical value of laser pulse energy. This method consists in the use of D-T ball precompression by a system of cumulative streams, initiated by explosives[71]. The cumulative streams operate through inhomogeneous system of the layers to transform the precompression process in the nearly isentropic one[72] (Fig. 11).

Through the conical niches spatially dislocated, the laser beams operate, realizing the laser compression. On account of the fact that the precompression time is of the order 10^{-7} sec while laser pulse is of the nanosecond order there exists a possibility of optimal (in time) introduction of laser pulse into the D-T ball.

The precompression value oscillates in the range 10-20 (in Ref. 74 the compression of the order 21.9 was obtained nearly isentropic at 8 mbar). This allows to reduce, in essential way, the critical value of the laser pulse energy by some tens per cent. The computations were performed in Ref. 71 on the ground of averaged equations.

Thus the presented method allows to reduce the critical value of laser pulse, necessary for positive, effective energy recovery

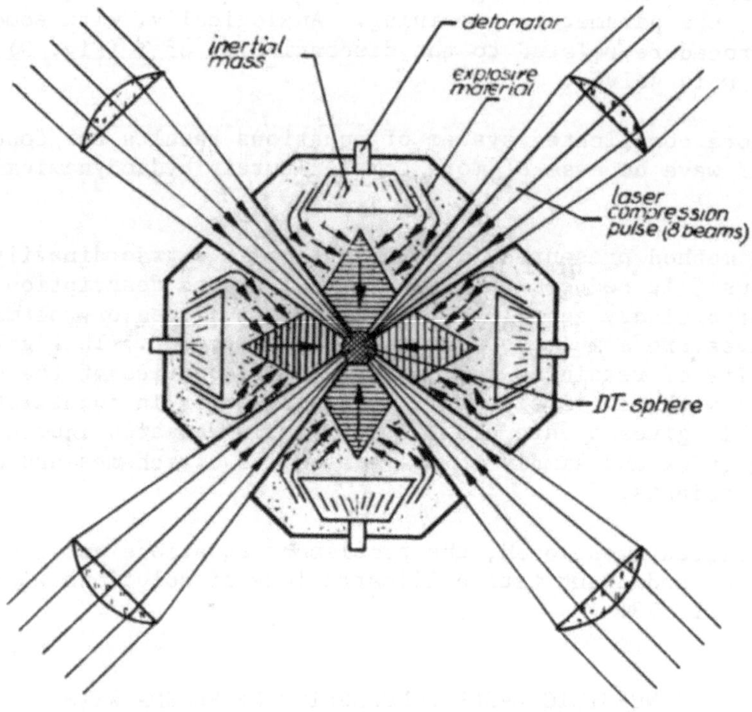

FIGURE 11a. Spherical topology of explosive precompression and laser compression.

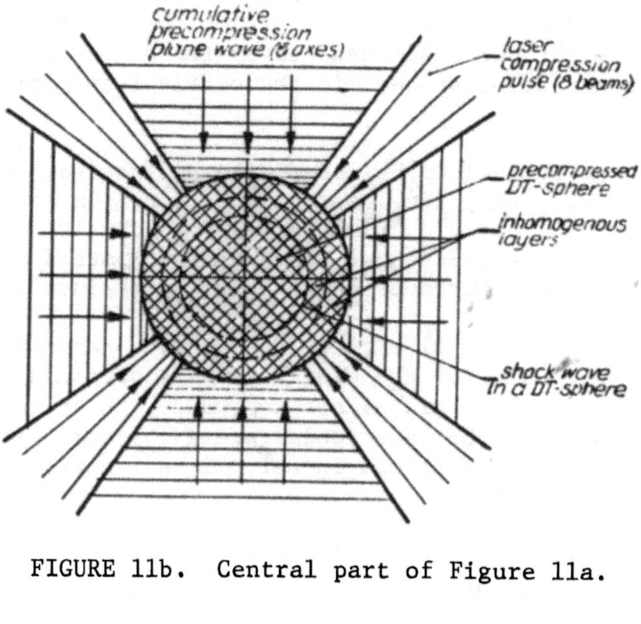

FIGURE 11b. Central part of Figure 11a.

of the order 1-10, to some kJ (2-3, assuming that without precompression $l_{cr} \simeq$ 3-10 kJ). The precompression effects can additionally be enriched by applying the compression by magnetic fields.

Analogically, it is also possible to obtain additional compression effects by employing the combined laser-electron-beam method. The explosive method in spite of definite technical difficulties, has this advantage that it is characterized by high efficiency coefficient. In the course of elaboration are also the experimental investigations.

FINAL REMARKS

Recapitulating the arguments of our paper it seems that in spite of enormity of technical difficulties, series of unexplained questions of technical nature, first of all stability type and absorption questions as well as laser radiation reflection - a perspective of realization of microfusion by the way of combined laser compression of the plasma is very promising. The critical energies of laser pulse are within the range at present technically attainable, what constitutes a fundament of all proceedings in this direction. The very problem of fusion reactors, their technical - economical aspects, will assume a real picture only when the principal question from the experimental point of view is solved namely corroboration of possibility of obtaining the positive microfusion energy recovery.

References

1. J. N. Dawson, Phys. Fluids 7, 981 (1964).

2. N. G. Basov, O. N. Krokhin, JETP 46, 1 (1964).

3. N. G. Basov, P. G. Kriukov, S. D. Zakharov, Yu. V. Senatskij and S. V. Tchekalin, IEEE. J. Quantum Electronics QE-4 (1968) p. 864.

4. F. Floux, D. Conard, J. L. Bobin, F. Delobeau, C. Fauquignon, C. R. Acad. Sciences, Paris 269, 697 (1969).

5. J. W. Shearer, W. S. Barnes, Phys. Rev. Letters 24, 92 (1970).

6. S. Kaliski et al., Bull. Acad. Polon. Sci., Ser. Sci. Techn. 21, 8 (1973).

7. N. G. Basov, S. D. Zakharov, O. N. Krokhin and others in Kvantovaja Elektronika 1, 4 (1971).

8. C. Fauquignon, F. Floux, Phys. Fluids 13, 386 (1970).

9. J. L. Bobin, Phys. Fluids 15, 3 (1971).

10. J. M. Dawson, P. Kaw, B. Green, Phys. Fluids 12, 895 (1969).

11. S. Kaliski, Proc. Vibr. Probl. 13, 1 (1972).

12. S. Kaliski et al., Bull. Acad. Polon. Sci., Ser. Sci. Techn. 20, 1 (1972).

13. S. Kaliski, Bull. Acad. Polon. Sci., Ser. Sci. Techn. 19, 6 (1971).

14. M. S. Chu, Phys. Fluids 15, 3 (1972).

15. S. Kaliski, Proc. Vibr. Probl. 14, 4 (1973).

16. F. Floux, Nuclear Fusion 11, 635 (1971).

17. P. Musler, R. Sigel and S. Witkowski, Plasma Production by Laser, Physics Reports (Section C of Physics Letters) 6 (1973) p. 187-239.

18. F. Floux, Laser Interaction and Related Plasma Phenomena, Plenum Press, New York, 447 (1971) Eds. H. Schwarz and H. Hora.

19. H. Hora, Laser + Elektro-Optik 4, 3 (1972).

20. M. J. Lubin, A. P. Fraas, Fusion by Laser, Scientific American 224, 21 (1971).

21. S. Kaliski, Proc. Vibr. Probl. 10, 4 (1969).

22. A. F. Haught, D. T. Polk, Phys. Fluids 13, 11 (1970).

23. H. Hora, Laser Interaction and Related Plasma Phenomena, Vol. 1, Plenum Press 365 (1971), Eds. H. Schwarz and H. Hora.

24. H. Hora, Laser Interaction and Related Plasma Phenomena, Vol. 1, 427 (1971), Eds. H. Schwarz and H. Hora.

25. S. Kaliski, Proc. Vibr. Probl. 12, 4 (1971).

26. S. Kaliski, Bull. Acad. Polon. Sci., Ser. Sci. Techn. 19, 5 (1971).

27. S. Kaliski, Bull. Acad. Polon. Sci., Ser. Sci. Techn. 20, 4 (1972).

28. S. Kaliski, Proc. Vibr. Probl. 12, 3 (1971).

29. S. Kaliski, S. Denus, Biul. WAT 8S, 20 (1971).

30. S. Kaliski, Proc. Vibr. Probl. 12, 2 (1971).

31. S. Kaliski, Proc. Vibr. Probl. 12, 2 (1971).

32. S. Kaliski, Proc. Vibr. Probl. 12, 3 (1971).

33. S. Kaliski, R. Swierczyński, Proc. Vibr. Probl. 13, 3 (1972).

34. S. Kaliski, Proc. Vibr. Probl. 12, 4 (1971).

35. J. P. Somon, Report L.G.I. 66, 11 (1966).

36. P. P. Pashynin, A. M. Prokhorov, JETP 60 (1971).

37. J. M. Dawson, A. Hertzberg, G. C. Vlases, H. G. Ahlstrom, L. C. Steinhauer, R. E. Kidder, W. L. Kruer, Esfahan Symposium on Fundamental and Applied Laser Physics.

38. S. Kaliski, Biul. WAT, Supplement 1972; Biul. WAT 21, 10 (1972); Proc. Vibr. Probl. 13, 4 (1972).

39. K. A. Brueckner, Laser Driven Fusion KMSF - NP 5, 1972 (paper presented at UC San Diego, May 1972).

40. J. Nuckolls, L. Wood, A. Thiessen, G. Zimmermann (Communication presented on Laser Conference in Montreal, 1972).

41. J. Nuckolls, L. Wood, A. Thiessen, G. Zimmermann, Nature 239, 139 (1972).

42. S. Kaliski, Bull. Acad. Polon. Sci., Ser. Sci. Techn. 21, 2 (1973).

43. S. Kaliski, Bull. Acad. Polon. Sci., Ser. Sci. Techn. 21, 4 (1973).

44. S. Kaliski, Bull. Acad. Polon. Sci., Ser. Sci. Techn. 21, 5 (1973).

45. S. Kaliski, Proc. Vibr. Probl. 14, 2 (1973).

46. S. Kaliski, Bull. Acad. Polon. Sci., Ser. Sci. Techn. 21, 6 (1973).

47. S. Kaliski, Bull. Acad. Polon. Sci., Ser. Sci. Techn. 21, 1 (1973).

48. S. Kaliski, Bull. Acad. Polon. Sci., Ser. Sci. Techn. 21, 1 (1973).

49. S. Kaliski, Proc. Vibr. Probl. 14, 3 (1973).

50. S. Kaliski, Proc. Vibr. Probl. 14, 4 (1973).

51. S. Kaliski, Proc. Vibr. Probl. (in print).

52. J. S. Clarke, H. N. Fisher and R. J. Mason, Phys. Rev. Letters 30, 15 (1973).

53. E. B. Goldmann, Plasma Physics 15, 289 (1973).

54. J. G. Linhard, Nuclear Fusion 13, 32 (1973).

55. S. Kaliski, Proc. Vibr. Probl. 14, 3 (1973).

56. S. Kaliski, Bull. Acad. Polon. Sci., Ser. Sci. Techn. 21, 8 (1973).

57. S. Kaliski, E. Wlodarczyk, Proc. Vibr. Probl. 14, 1 (1973).

58. S. Kaliski, Bull. Acad. Polon. Sci., Ser. Sci. Techn. 21, 8 (1973).

59. S. Kaliski, Bull. Acad. Polon. Sci., Ser. Sci. Techn. 21, 8 (1973).

60. S. Kaliski, E. Wlodarczyk, AMS (in print).

61. S. Kaliski, E. Wlodarczyk, Proc. Vibr. Probl. (in print).

62. K. A. Brueckner, S. Jorna, Laser Driven Fusion, KMS F-U97, May 10, 1973.

63. F. Winterberg, Phys. Rev. 174, 212 (1968).

64. M. W. Bobykin, E. K. Savoiskü, A. A. Ivanov, and L. I. Rudakov, Plasma Phys cs and Contr. Thermonucl. Fusion (Madison Conf) IAEA, Vien 1971, Vol. 1, p. 635.

65. L. I. Rudakov, Zh. Eksp. Theor. Fiz. 59, 2091 (1970).

66. R. V. Lovelace and R. N. Sudan, Phys. Rev. Lett. 27, 1256 (1971).

67. S. Kaliski, Bull. Acad. Polon. Sci., Ser. Sci. Techn. 20, 6 (1972).

68. S. Kaliski, Bull. Acad. Polon. Sci., Ser. Sci. Techn. 20, 6 (1972).

69. S. Kaliski, Proc. Vibr. Probl. 13, 4 (1972).

70. H. G. Ahlstrom, L. C. Steinhauer, Laser Produced Fusion with Composite Spherical Solids (submitted to Nuclear Fusion).

71. S. Kaliski, Proc. Vibr. Probl. (in print).

72. S. Kaliski, Proc. Vibr. Probl. 14, 2 (1973).

73. S. Kaliski, Proc. Vibr. Probl. (in print).

74. F. W. Grigoriew et al. JETP Letters, 16, 5, 286-290 (1972).

LASER PRODUCED PLASMAS WITH FISSIONABLE MATERIAL *

F. Winterberg

Desert Research Institute

University of Nevada System, Reno, Nevada, USA

ABSTRACT

The compression of a pellet consisting of fissionable material, such as U233, U235 or Pu239, to high densities by a laser- (or possibly relativistic electron-) beam induced implosion, will lead to very small critical masses. The critical mass can be furthermore substantially reduced by simultaneously compressing the pellet together with a neutron reflector. By this method micro-fission-explosions can be triggered which can be used for controlled power production as for a safe fast breeder reactor or for rocket propulsion. In addition, if the neutron reflector consists of thermonuclear material, such as T-D, the controlled release of thermonuclear energy becomes possible. In this case the fission chain reaction not only will assist in the ignition of the thermonuclear reaction but in turn the neutrons released in the fusion process will accelerate the fission chain reaction to a very fast pace. Both processes, the fission chain reaction and the fusion process are thus coupled in a bootstrap mode greatly increasing the energy output.

---

* Presented at the Third Workshop on "Laser Interaction and Related Plasma Phenomena" held at Rensselaer Polytechnic Institute, Troy, New York, August 13-17, 1973.

INTRODUCTION

It has been shown[1,2] that very high densities can be possibly achieved by the laser induced implosion of a spherical mass. The purpose for this proposed implosion technique was to achieve high densities in laser produced thermonuclear plasmas. In principle, the same implosion technique should be also possible with relativistic electron beams which have been previously proposed[3] as an alternative to laser beams for the controlled release of thermonuclear energy by micro-explosions. Rather detailed calculations predict for this implosion method pressures up to $\sim 10^{18}$ dyn/cm^2. If hydrogen at low temperatures with a Fermi electron distribution is subjected to this pressure its density will rise to a value which is $\sim 10^4$ times larger than its solid state density. The importance of this is that an increase in density will lead to a decrease in the required laser energy for breakeven, whereby the reduction in the required energy input is inversely proportional to the achieved density.

In this communication however, we will discuss another extremely interesting prospect which is, that by these implosion techniques the critical mass of a fissionable assembly can be reduced to such a degree that a fission micro-explosion becomes possible[4,5]. This therefore may lead to an entirely new kind of nuclear fission reactor from which the energy is extracted by a chain of small fission explosions rather than by a continuous steady state energy chain reaction. Such a reactor would be conceivable cheaper and have less serious material problems than a conventional type of fission reactor. Moreover, since the maximum fission energy released is limited by the relatively small size of the micro-fission explosion, the concept could be used for a safe fast breeder reactor. Another important application of this concept is for rocket propulsion whereby the micro-explosions would take place in the focus of a concave reflector.

Finally the possibility of such micro-fission-explosions would raise a new prospect for controlled thermonuclear energy release by surrounding the fissionable pellet with a neutron reflector of thermonuclear material to be implosion-compressed together with the pellet to high densities and subsequently ignited by the energy release of the fission chain reaction. In this case the neutrons released by the thermonuclear fusion process can greatly accelerate the pace of the chain reaction and with it the energy yield.

NEUTRON DIFFUSION EQUATION AND FISSION CHAIN REACTION

In order to simplify the problem as much as possible a mono-energetic neutron diffusion model will be assumed[6]. In this model the slowing down of the fission neutrons by scattering is neglected. This approximation puts us on the safe side since the fission cross sections tend to increase with decreasing neutron velocity. Our diffusion model is supplemented by appropriate transport corrections at the vacuum boundary of the medium in which the diffusion process takes place. This model is then equivalent to the one group model for a fast nuclear reactor.

In the diffusion approximation a scalar undirected neutron flux ϕ is defined by

$$\phi = nv_o \quad , \quad (1)$$

where n is the neutron number density and v_o the neutron velocity. For fission neutrons $v_o \simeq 10^9$ cm/sec.

In the diffusion approximation the neutron current vector $\underline{j}$ is given by

$$\underline{j} = - D \text{ grad } \phi \quad , \quad (2)$$

where $D \simeq 1/3N\sigma_s$ is the neutron diffusion coefficient, N is the atomic number density of the scattering nuclei and σ_s is the nuclear scattering cross section.

The neutron balance is governed by the equation

$$\frac{\partial n}{\partial t} + \text{div } \underline{j} = S \quad , \quad (3)$$

where S is the source or sink term for the neutrons.

In a neutron absorbing medium with a neutron absorption cross section σ_a one has $S = - N\sigma_a\phi$. With this expression for S one obtains from eqs. (1-3) the diffusion equation for the neutron flux

$$\nabla^2\phi - \kappa^2\phi = \frac{1}{Dv_o}\frac{\partial\phi}{\partial t} \quad , \quad (4)$$

with

$$\kappa^2 \equiv 3N^2\sigma_s\sigma_a \quad . \quad (5)$$

In the special case of vanishing neutron absorption and time independent neutron flux one has simply

$$\nabla^2\phi = 0 \quad . \quad (6)$$

In a fissionable medium consisting of U233, U235 or Pu239, the source term is positive. To make the treatment simple we will assume that all neutrons captured in the fissionable material will lead to a fission process. This is approximately true, especially for energetic neutrons as they are released in the fission process. The inclusion of non-fission neutron capture would leave our results essentially unchanged. If for each neutron captured by a fissionable nucleus ν fission neutrons are released, the neutron source term is given by $S = (\nu - 1)\, N\sigma_f\phi$, where σ_f is the neutron cross section for fission. With this expression for the source term one then obtains from eqs. (1-3)

$$\nabla^2\phi + B_m^2\phi = \frac{1}{Dv_o}\frac{\partial\phi}{\partial t} \quad , \tag{7}$$

with

$$B_m^2 \equiv 3\sigma_s\sigma_f\, N^2(\nu - 1) \quad .$$

B_m is called the material buckling. In case of a steady state chain reaction $(\partial/\partial t = 0)$, which is reached at the critical state one has

$$\nabla^2\phi + B_m^2\phi = 0 \quad . \tag{8}$$

For an infinitely extended medium of fissionable material $\nabla^2\phi = 0$ and one obtains from eq. (7) for the neutron flux increasing with time

$$\phi = \phi_o e^{\lambda_o t} \quad , \tag{9}$$

with

$$\lambda_o = N\sigma_f v_o(\nu - 1) \quad . \tag{10}$$

Eq. (8) for a sphere of fissionable material has the non-singular solution

$$\phi = A\,\frac{\sin B_m r}{r} \quad , \qquad A = \text{const.} \tag{11}$$

From this solution the critical equation can be derived by putting the neutron flux equal to zero at a specified distance from the surface of the fissionable sphere. Actually outside the sphere the neutron flux does not vanish and ordinary diffusion theory breaks down. From transport theory, however it is known, that diffusion theory still holds approximately by putting $\phi = 0$ at a distance d, called the extrapolation length, away from the vacuum boundary given by

$$d \simeq 0.71/N\sigma_s \quad . \tag{12}$$

Putting $\phi = 0$ at $r = R_o + d$, where R_o is the critical radius of the fissionable sphere, one then obtains R_o from eq. (11) by taking the lowest eigenvalue, hence $B_m(R_o + d) = \pi$ or

$$R_o = \frac{\pi}{B_m} - d = \frac{1}{N}\left[\frac{\pi}{(3\sigma_s\sigma_f(\nu - 1))^{\frac{1}{2}}} - \frac{0.71}{\sigma_s}\right] \quad . \quad (13)$$

In order to solve the time dependent eq. (7) we put

$$\phi = A\,\frac{\sin Br}{r}\,e^{\lambda t} \quad , \quad (14)$$

and obtain after inserting into eq. (7)

$$B^2 = \left(\frac{\pi}{R + d}\right)^2 = B_m^2 - \frac{\lambda}{Dv_o} \quad , \quad (15)$$

where R is the radius of a fissionable sphere which is assumed to be larger than the critical radius R_o. From eq. (15) we obtain, after using the expression for B_m and neglecting d,

$$\lambda = \lambda_o(1 - (R_o/R)^2) \quad . \quad (16)$$

From eq. (16) it follows that the chain reaction in a finite medium grows slower than in an infinitely extended medium ($R = \infty$) where $\lambda = \lambda_o$. For a slightly above critical mass with $R = R_o + \Delta R$, $\Delta R/R_o << 1$ one obtains for the e-fold time t_e of the chain reaction

$$t_e = \frac{1}{\lambda} \simeq \frac{1}{\lambda_o}\,\frac{R_o}{2\Delta R} \quad , \quad (17)$$

whereas for $R = \infty$, $t_e = 1/\lambda_o$. If for example $\Delta R/R_o = 0.1$, it would follow that $t_e = 5/\lambda_o$. This means that a 10% increase of the sphere above criticality would increase the e-fold time only by a factor 5 as compared to an infinite medium. Therefore, if criticality can be reached, then by making the fissionable assembly only slightly larger than critical, a fast growing chain reaction will ensue. This fact is of importance for our specific goal aimed at very small critical masses.

Smaller critical masses can be achieved by surrounding the fissionable material with a neutron reflector. We envision here the case where a fissionable sphere of radius R is surrounded by a spherical shell of a neutron reflector of thickness T. We furthermore ideally assume that this shell consists of T-D thermonuclear material. For T-D material the neutron absorption is small and eq. (6) can then be used to describe the neutron diffusion process within the reflector. We label all quantities for the fissionable core by a subscript c and for the reflector by a subscript r. In the core the neutron flux is obtained from eq. (8) with the solution

$$\phi_c = A_c \frac{\sin B_m r}{r} \quad , \quad (18)$$

and in the reflector from eq. (6) with the solution

$$\phi_r = A_r \left(\frac{1}{r} - \frac{1}{R + T} \right) \quad . \quad (19)$$

In eq. (19) the boundary condition $\phi_r = 0$ at $r = R + T$ has been already incorporated. (We neglect here the extrapolation length which for the case of a highly compressed T-D reflector turns out to be small.) At the boundary between the core and reflector, that is at $r = R$, one has to put $\phi_c = \phi_r$ and $D_c d\phi_c/dr = D_r d\phi_r/dr$, expressing the continuity of neutron flux and neutron current. By inserting eqs. (18) and (19) into these continuity conditions and after eliminating A_r and A_c one obtains the critical equation

$$\cot B_m R = \frac{1}{B_m R} \left(1 - \frac{D_r}{D_c} \right) - \frac{D_r}{D_c B_m T} \quad . \quad (20)$$

For a strongly neutron reflected fissionable mass one can expect that $R \ll R_o$, respectively that $B_m R \ll 1$, where R_o would be the critical radius without reflector. In this case then we can make use of the expansion

$$\cot B_m R \simeq \frac{1}{B_m R} - \frac{1}{3} B_m R \quad , \quad (21)$$

and after inserting eq. (21) into eq. (20) we obtain for the critical radius

$$R = \frac{3}{2} \frac{D_r}{D_c B_m^2 T} \left[1 + \left(1 + \frac{4}{3} \frac{D_c}{D_r} B_m^2 T^2 \right)^{\frac{1}{2}} \right] \quad . \quad (22)$$

For $T = \infty$ this yields

$$R_{min} = \frac{1}{B_m} \left(3 \frac{D_r}{D_c} \right)^{\frac{1}{2}} \quad . \quad (23)$$

This minimum radius is rapidly approached for a sufficiently thick reflector if one puts for example

$$\frac{4}{3} \frac{D_c}{D_r} B_m^2 T^2 \simeq 4 \quad , \quad (24)$$

which would give

$$R \gtrsim T = \frac{1}{B_m} \left(3 \frac{D_r}{D_c}\right)^{\frac{1}{2}} = [\sigma_f \sigma_{rs} N_c N_r (\nu - 1)]^{-\frac{1}{2}} \quad . \quad (25)$$

With these results we are in a position to obtain the scaling laws for the critical mass and the e-fold time of the chain reaction in dependence of a large density increase as it is expected in the implosion compression scheme.

For an unreflected spherical mass, m_o, of fissionable material the scaling law for the critical mass is obtained from eq. (13) with the result

$$m_o \propto NR_o^3 \propto \frac{1}{N^2} \quad . \quad (26)$$

Eq. (26) shows that a density increase by a factor 100 would result in a decrease of the critical mass by a factor 10^4. For an assumed critical mass of 10 kg at normal densities this would imply a reduction in the critical mass down to a value of $\sim$ 1g.

For a reflected fissionable mass m, eq. (25) gives as the scaling law

$$m \propto N_c R^3 \propto \frac{1}{(N_c N_r^3)^{\frac{1}{2}}} \quad . \quad (27)$$

If for example N_c is increased by a factor 10^2 and N_r increased by a factor 10^4, the critical mass would be decreased by a factor 10^7, that is from $\sim$ 10 kg down to $\sim 10^{-3}$ g.

The scaling law for the e-fold time of the chain reaction is obtained from eq. (16) together with the definition of λ_o by eq. (10). The result is

$$t_e \propto \frac{1}{\lambda_o} \propto \frac{1}{N} \quad . \quad (28)$$

It is instructive to compare the e-fold time with the hydrodynamic disassembly time of the exploding pellet

$$\tau \sim R/v \quad , \quad (29)$$

where v is the velocity of sound at keV temperatures. Because of $R \propto 1/N$ one obtains for the factor

$$\tau/t_e = \text{const.} \qquad \text{(independent of N)} \quad . \quad (30)$$

The significance of this factor is explained by the fact that it determines the maximum attainable neutron flux ϕ_{max}, which according to eq. (9) is given by putting $t = \tau$, hence $\phi_{max} = \phi_o \exp(\tau/t_e)$. This demonstrates that for a fission micro-explosion the same maximum neutron flux can be expected as for atomic weapon devices working at normal material densities.

CRITICALITY AND CHAIN REACTION AT HIGH DENSITIES

In the implosion scheme proposed for laser fusion[1,2] pressures up to $p = 10^{18}$ dyn/cm^2 are predicted. The pressures are generated by an ablation process whereby the high ablation product velocity creates the convergent implosion reaction pressure. Therefore, a high pressure requires a high ablation product velocity. A high ablation product velocity is more easily achieved with a low A-number material rather than with the high A-number materials of fissionable substances. But this poses no problem since one can always surround the fissionable substance to be compressed by a layer of low A-number material which ideally can also serve as a neutron reflector. Hydrogen and its isotopes seem to be especially well suited for this purpose, because of their high ablation product velocity and high neutron albedo.

The density of cold fissionable material under high pressure can be computed by a Thomas-Fermi equation of state(7). For a pressure of $p = 10^{18}$ dyn/cm^2 one finds that the densities for U233, U235 or Pu239 are increased over their solid state density values by a factor ~ 240. For hydrogen under the same pressure the density is increased by a factor $\sim 10^4$.

More precisely the atomic number density for U233, U235 and Pu239 at $p = 10^{18}$ dyn/cm^2 is $N = 1.17 \times 10^{25}$ cm^{-3} and for hydrogen $N = 5 \times 10^{26}$ cm^{-3}. The densities are given by $\rho = 4.6 \times 10^3$ gm/cm^3 for the fissionable materials and by $\rho = 7 \times 10^2$ gm/cm^3 for hydrogen. For T-D being especially considered below one would have $\rho = 1.7 \times 10^3$ gm/cm^3. The density of the compressed fissionable materials is therefore still larger than that of hydrogen even so this is not the case for the atomic number densities.

In order to estimate the critical dimensions we use the following nuclear parameters:

a) Fissionable material ($N = N_c = 1.17 \times 10^{25}$ cm^{-3})

$$\sigma_f = \sigma_s = \sigma_{cs} = 2 \times 10^{-24} \text{ cm}^2$$

$$\nu = 2.9 \quad .$$

b) Hydrogen reflector ($N_r = 5 \times 10^{26}$ cm^{-3})

$$\sigma_{rs} = 2 \times 10^{-24} \text{ cm}^2 \quad .$$

With these values one calculates with the help of eq. (13) a critical radius and mass for the unreflected fissionable pellet given by

$$R_o = 2.6 \times 10^{-2} \text{ cm} \quad ,$$

$$m = 0.34 \text{ gm} \quad .$$

For a neutron-reflected pellet the critical radius is calculated according to eq. (25) with the result

$$R = 4.7 \times 10^{-3} \text{ cm} \quad ,$$

$$m = 2.0 \times 10^{-3} \text{ gm} \quad .$$

The compression energy is given by

$$E = (2/3)\ pV \quad , \qquad (31)$$

where V is the volume of the compressed material. For the unreflected pellet $V = (4\pi/3)R_0^3$ and for the reflected pellet $V=(4\pi/3)(R+T)^3 \simeq (32\pi/3)R^3$. In the case of the unreflected pellet one obtains

$$E = 4.9 \times 10^{13} \text{ erg} = 4.9 \text{ Megajoule}$$

and for the neutron-reflected pellet

$$E = 2.3 \times 10^{12} \text{ erg} = 2.3 \times 10^5 \text{ Joule} \quad .$$

Although the energies to achieve these compressions will be actually higher since a large part of the energy goes into the ablation products, these energies are nevertheless remarkably low and within the predicted reach of future laser technology. In the case of relativistic electron beams they are within already achieved energy outputs.

Inserting the given numerical values into eq. (9) one obtains

$$\phi = \phi_0 \exp\ (4.4 \times 10^{10} t) \quad . \qquad (32)$$

From this follows as the e-fold time for an infinitely extended fissionable medium $t_e \simeq 2 \times 10^{-11}$ sec. For a finite size fissionable pellet with $\Delta R/R_0 = 0.1$ one has according to eq. (17) $t_e \simeq 10^{-10}$ sec. The hydrodynamic disassembly time τ for $R \simeq 10^{-2}$cm and $v \simeq 10^7$ cm/sec is $\tau \sim 10^{-9}$ sec, such that $\tau/t_e \sim 10$.

MICRO FISSION-FUSION CHAIN REACTIONS

If the neutron reflecting material consists of thermonuclear material, such as T-D, the heat released by the fission chain reaction can ignite a thermonuclear reaction in the T-D envelope. The T-D thermonuclear reaction produces a large number of neutrons which in turn will induce more fission reactions and thus accelerate the pace of the chain reaction. The increased rate of the chain reaction, in turn, will release even more heat, thus furthermore raising the temperature in the thermonuclear material whereby the thermonuclear reaction rate is accelerated. From this picture it is clear, that both, the fission chain reaction and the thermonuclear fusion reaction are coupled in a bootstrap mode. One can think either of a fission supported fusion reaction or a fusion supported fission reaction. As a result of this coupling a fissionable pellet surrounded by a fusionable envelope at thermonuclear temperatures may assume a smaller critical mass. Since a pellet consisting of both fissionable and fusionable material can be compressed and heated to thermonuclear temperatures by a laser- or electron beam pulse, the reduction in the critical mass as a function of the achieved temperature becomes an interesting quantity.

In order to estimate this effect we are considering a homogeneous mixture of fissionable (U233, U235, Pu239) and fusionable (T-D) material under high pressure and density, for example, under a pressure of 10^{12} atmospheres. Under a given pressure the substances would have an atomic number density of N_u for the fissionable material and N_h for the T-D thermonuclear material. We introduce the mixing parameter x, $0 < x < 1$, defined such that in the mixture of fissionable and fusionable material there are $(1 - x)N_u$ fissionable and xN_h fusionable atoms per cm^3. Under a pressure of $p = 10^{18}$ dyn/cm² the number densities would be as before $N_u = 1.17 \times 10^{25}$ cm^{-3} and $N_h = 5 \times 10^{26}$ cm^{-3}.

The neutron chain reaction for a mixture of fissionable and fusionable material of infinite extension follows the equation

$$\frac{1}{v_0}\frac{\partial\phi}{\partial t} = (\nu - 1)(1 - x)\,N_u\sigma_f\phi + S \quad , \qquad (33)$$

where S is a neutron source which in our case is provided for by the thermonuclear neutrons set free in the T-D fusion reaction. If we assume that the neutron velocity of the fission and fusion neutrons is the same, which by order of magnitude is true, the source term can be written as follows

$$S = \frac{1}{4}\,x^2N_h^2\overline{\sigma v} \quad , \qquad (34)$$

where $\overline{\sigma v}$ is the velocity averaged fusion cross section of the T-D reaction.

We are here especially interested in the temperature range from 1 keV up to 10 keV and in this range the following approximation can be used (T in keV)

$$\overline{\sigma v} \simeq 1.1 \times 10^{-20}\;T^{4.37} \quad . \qquad (35)$$

With this we have by inserting into eq. (33)

$$\frac{1}{v_0}\frac{\partial\phi}{\partial t} = (\nu - 1)(1 - x)N_u\sigma_f\phi + 2.75 \times 10^{-21}x^2N_h^2T^{4.37} \quad . \qquad (36)$$

Next we need a relation in between the temperature T and the neutron flux. This relation is obtained by taking into account the heat released by the rapidly growing nuclear reactions resulting from both the fission and the fusion processes. Part of this heat released will go into kinetic particle energy of the background plasma and part of it into black body radiation, the latter because the opacity of the plasma made up in part of fissionable material is very high. The particle kinetic energy is of the order NkT and the black body radiation energy of the order aT^4. (k, Boltzmann-constant, and a, Stefan-Boltzmann constant.) For $N \sim 10^{26}$ cm^{-3} (corresponding to the highly compressed state), and $T \sim 1$ keV, one has $NkT \sim 10^{17}$ erg/cm^3 and $aT^4 \sim 10^{14}$ erg/cm^3. At $T \sim 10$ keV both particle kinetic energy and black body radiation energy would be of the same order of magnitude. We are here interested in the temperature range below 10 keV, and therefore, can neglect the black body radiation energy as compared to the particle kinetic energy.

At an assumed atomic number density $N_h = 5 \times 10^{22}$ cm^{-3} for the T-D thermonuclear material and a neutron cross section of $\sigma_s = 2 \times 10^{-24}$ cm^2 the neutron mean free path $1/(N_h\sigma_s) = 10^{-3}$ cm is smaller than the critical radius computed above. It therefore follows that in contrast to uncompressed laser produced plasmas the released neutron kinetic energy is dissipated into the plasma rather than lost by unattenuated neutron escape. By comparison, both the charged fission- and H_e^4 fusion- products have a much smaller range and for all practical considerations completely dissipate their kinetic energy into the plasma environment.

The energy released by the fission process per cm^3 and sec is given by

$$\varepsilon_f(1 - x)N_u\sigma_f\phi \quad ,$$

where $\varepsilon_f = 180$ MeV $= 2.9 \times 10^{-4}$ erg is the fission reaction energy.

The energy released by the T-D fusion process per cm^3 and sec is given by

$$\varepsilon_\alpha S = \varepsilon_\alpha x^2 N_h^2 \overline{\sigma v}/4 = 2.75 \times 10^{-21}\varepsilon_\alpha x^2 N_h^2 T^{4.37} \quad ,$$

where $\varepsilon_\alpha = 17.2$ MeV $= 2.7 \times 10^{-5}$ erg is the fusion reaction energy.

With these heat sources the temperature increase in the fission-fusion plasma is given by

$$3k\left[g(1 - x)\ N_u + xN_h\right]\frac{\partial T}{\partial t} =$$

$$= \varepsilon_f(1 - x)\ N_u\sigma_f\phi + 2.75 \times 10^{-21}\varepsilon_\alpha x^2 N_h^2 T^{4.37} \quad , \qquad (37)$$

where $k = 1.6 \times 10^{-9}$ erg/keV is the Boltzmann constant and where g is the degree of ionization of the fissionable material such that $1 < g < 93$. $g \sim 10$ may be a reasonable estimate.

Expanding the function $f(T) = T^{4.37}$ around $T = T_o (> 1\ \mathrm{keV})$ into a Taylor series one has

$$T^{4.37} = \text{const.} + 4.37\, T_o^{3.37} T \quad . \tag{38}$$

By inserting this expansion into eqs. (36) and (37) one obtains the following set of coupled linear differential equations

$$\frac{\partial \phi}{\partial t} = \alpha_1 \phi + \beta_1 T + \gamma_1 \quad , \tag{39}$$

$$\frac{\partial T}{\partial t} = \alpha_2 \phi + \beta_2 T + \gamma_2 \quad , \tag{40}$$

where

$$\alpha_1 \equiv (\nu - 1)(1 - x)\, N_u \sigma_f v_o \quad ,$$

$$\beta_1 \equiv 1.2 \times 10^{-20} v_o x^2 N_h^2 T_o^{3.37} \quad ,$$

$$\alpha_2 \equiv \frac{\varepsilon_f (1 - x)\, N_u \sigma_f}{3k \left[g(1 - x)\, N_u + x N_h \right]} \quad ,$$

$$\beta_2 \equiv \frac{1.2 \times 10^{-20} \varepsilon_\alpha x^2 N_h^2 T_o^{3.37}}{3k \left[g(1 - x)\, N_u + x N_h \right]} \quad .$$

Furthermore γ_1, γ_2 are constants, the values of which turn out to be of no interest. The differential equations (39) and (40) can be brought into one single differential equation for either ϕ or T. The differential equation for T is given by

$$\ddot{T} - (\alpha_1 + \beta_2)\dot{T} + (\alpha_1\beta_2 - \alpha_2\beta_1)T + \alpha_1\gamma_2 - \alpha_2\gamma_1 = 0 \ . \tag{41}$$

The general solution of eq. (41) is the sum of a particular solution of the inhomogeneous equation and the general solution of the homogeneous equation. A particular integral of the inhomogeneous equation and into which the constants γ_1 and γ_2 are entering, has the form T = const. and is of no interest. The homogeneous equation

$$\ddot{T} - (\alpha_1 + \beta_2)\dot{T} + (\alpha_1\beta_2 - \alpha_2\beta_1)T = 0 \quad , \tag{42}$$

has the general solution

$$T = \text{const. } e^{\lambda t} \quad , \qquad (43)$$

with

$$\lambda = \frac{\alpha_1 + \beta_2}{2} + \left[\left(\frac{\alpha_1 + \beta_2}{2} \right)^2 + \alpha_2\beta_1 - \alpha_1\beta_2 \right]^{\frac{1}{2}} \quad , \qquad (44)$$

by taking only the physically meaningful solution growing with time.

For x = 0, corresponding to a pure fission plasma, one has

$$\lambda = \lambda_o = (\nu - 1)\ N_u\sigma_f v_o \quad , \qquad (45)$$

which is the same growth rate as it was given by eq. (10), valid for a pure fission chain reaction in a pure fission plasma. For a pure fission chain reaction in a fission-fusion plasma but where the fission chain reaction is assumed not to be coupled to the fusion neutrons one would have

$$\lambda_1 = \alpha_1 = \lambda_o(1 - x) \quad . \qquad (46)$$

In reality however, the fission chain reaction is coupled to the fusion process and the value of λ given by eq. (44) will be thus larger than the value of λ_1 given by eq. (46). For $0 < x < 1$ one properly defines a factor f

$$f = \lambda/\lambda_o \quad , \qquad (47)$$

giving the change in the rate of the chain reaction as compared to the case of a pure fission chain reaction in purely fissionable material. From this factor f an effective neutron multiplication factor ν^* to be related to f and ν can be deduced by putting

$$\nu^* - 1 = (\nu - 1)\ f \quad . \qquad (48)$$

For the following it is convenient to introduce the auxiliary function F(x) defined by

$$F(x) = \frac{x^2}{g + \frac{x}{1 - x}\frac{N_h}{N_u}} \frac{N_h^2}{N_u^2} \quad . \qquad (49)$$

With this auxiliary function we can write

$$\alpha_2 = 2.1 \times 10^8 \varepsilon_f \sigma_f (N_u/N_h)^2\ F(x)/x^2 \quad , \qquad (50)$$

$$\beta_2 = 2.5 \times 10^{-12}\ \varepsilon_\alpha T_0^{3.37} N_u F(x)/(1 - x) \quad . \tag{51}$$

One can now easily show that for $x \sim 0.5$ and for 1 keV < T < 10 keV, (and all the other numerical parameters as given before) one has

$$\alpha_1\beta_2 << \alpha_2\beta_1 \quad , \tag{52}$$

and

$$\left(\frac{\alpha_1 + \beta_2}{2}\right)^2 << \alpha_2\beta_1 \quad . \tag{53}$$

It therefore follows that approximately

$$\lambda \gtrsim (\alpha_2\beta_1)^{\frac{1}{2}} \quad . \tag{54}$$

By inspection of eqs. (36) and (37) one can see that this approximation is equivalent in neglecting in eq. (36) the fission neutron source term as being small in comparison to the fusion neutron source term, and in eq. (37) in neglecting the fusion reaction heat source term in comparison to the fission reaction heat source term.

With the definition of f (eq. (47)) one has from eq. (54)

$$f(x) \gtrsim 4.26 \left(\frac{x^2(1 - x)}{1 + 3.3x}\right)^{\frac{1}{2}} T_0^{1.68} \quad . \tag{55}$$

This function has a maximum for x = 0.57 where

$$f \gtrsim 0.94 T_0^{1.68} \quad . \tag{56}$$

Assume for example, that T_0 = 5 keV; one then has f = 14.2 and ν^* = 28.0. For T_0 = 10 keV one would even have f = 45.0 and ν^* = 86.5.

With these computed values for f one can estimate the reduction in the required critical mass due to the bootstrap coupling of the fission chain reaction with the fusion process. The critical radius of a spherical mass consisting of fissionable and fusionable material (ignoring fusion neutrons) was given by eq. (13) where B_m and d is now

$$B_m^2 = 3\sigma_s\sigma_f \left[(1 - x)N_u + xN_h \right] (1 - x)N_u(\nu - 1) \quad , \tag{57}$$

$$d = 0.71/\left[(1 - x)N_u + xN_h \right] \sigma_s \quad . \tag{58}$$

The factor $(1 - x)N_u(\nu - 1)$ in the expression for B_m^2 determines the growth rate of the fission chain reaction. In order to adjust it to the correction resulting from the fusion neutrons it has to be multiplied by the factor $\lambda/\lambda_1 = f/(1 - x)$ hence

$$B_m^2 = 3\sigma_s\sigma_f \left[(1 - x)N_u + xN_h \right] N_u(\nu - 1)f \quad . \tag{59}$$

Inserting the numbers $\sigma_f = \sigma_s = 2 \times 10^{-24}$ cm^2, $\nu = 2.9$, $N_u = 1.17 \times 10^{25}$ cm^{-3}, $N_h = 5 \times 10^{26}$ cm^{-3}, $x = 0.57$ one obtains

$$B_m = 2.8 \times 10^2 f^{\frac{1}{2}} \text{ cm}^{-1} \quad ,$$

$$d = 1.2 \times 10^{-3} \text{ cm} \quad .$$

For $\lambda = \lambda_1$, or $f = 1 - x = 0.43$, which corresponds to the case of a fission chain reaction not assisted by the neutrons of the fusion process, one obtains $B_m = 1.8 \times 10^2$ cm^{-1} and $R_o = 1.6 \times 10^{-2}$ cm. For $T_o = 5$ keV it was found that $f = 14.2$ hence there $B_m = 1.1 \times 10^3$ cm^{-1} and $R_o = 1.7 \times 10^{-3}$ cm. For $T_o = 10$ keV, $f = 45.0$, hence $B_m = 1.9 \times 10^3$ cm^{-1} and $R_o = 4.6 \times 10^{-4}$ cm. In going to $T_o = 5$ keV the critical mass is therefore reduced by a factor $\sim 9 \times 10^2$, and by going to $T_o = 10$ keV by a factor $\sim 4 \times 10^4$. These represent very substantial reductions since the input energy for compression is reduced by the same factors.

Instead of using the maximum value of the function $f(x)$ one can also ask for the maximum value of B_m^2 which would give the minimum critical radius. These two maxima will differ due to the fact that the admixture of the T-D material will not only increase the neutron production by thermonuclear processes but at the same time by its high atomic number density will reduce the neutron mean free path due to scattering. The reduction in the neutron mean free path due to scattering and which is governed by the factor $\sigma_s\left[(1-x)N_u + xN_h\right]$ will lead to a reduction in the critical mass in favor of large x values close to $x = 1$. If the material buckling for a pure fission plasma is given by $B_m^{(o)}$ we then have from eq. (59), with $N_h/N_u = 43$,

$$\left(\frac{B_m}{B_m^{(o)}} \right)^2 = (1 + 42x)f(x) \quad . \tag{60}$$

From eq. (55) it follows that the largest value of B_m^2 is given by the maximum of the function

$$y = 4.26T_o^{1.68} \left(\frac{x^2(1 - x)}{1 + 3.3x} \right)^{\frac{1}{2}} (1 + 42x) \quad . \tag{61}$$

This function has a maximum at $x \simeq 0.77$ with $y_{max} \simeq 2.8 \times 10^1 T_0^{1.68}$. The overall reduction, m/m_0, in the critical mass whereby m is the critical mass with T-D material and m_0 without it (neglecting the effect of the change in the extrapolation length) is then given by

$$\frac{m}{m_0} = \left(\frac{B_m^{(0)}}{B_m} \right)^3 = 6.7 \times 10^{-3} T_0^{-2.52} \quad . \tag{62}$$

For $T_0 = 1$ keV where $f \simeq 1$ this would already amount to a reduction in the critical mass by a factor $\sim 10^2$, which is about the same as for fissionable pellet surrounded by a neutron reflector. For T_0 = 5 keV the critical mass is reduced by a factor $\sim 9 \times 10^3$ and for T_0 = 10 keV by a factor $\sim 5 \times 10^4$.

We can now apply these calculations to estimate the effect of the fusion neutron on the heterogeneous arrangement of a fissionable core surrounded by a fusionable neutron reflector as considered above, by assuming a homogeneous mixing of both components. In this case the fissionable pellet of radius R was assumed to be surrounded by a fusionable T-D envelope of thickness $T \simeq R$. In order to extend our formulas derived for a homogeneous assembly of fissionable and fusionable material we must determine the mixing parameter x for $T \simeq R$ from the condition $x/(1 - x) = \left[(R + T)^3 - R^3 \right] / R^3 = 7$, hence $x = 7/8 \simeq 0.9$. This is close to the optimum value at $x \simeq 0.77$. From eq. (55) one obtains for $x = 0.9$ and T_0 = 5 keV, $f \simeq 9.2$ and for T_0 = 10 keV, $f \simeq 29$. The critical radius without the effect of the fusion neutrons was $R \simeq T = 4.7 \times 10^{-3}$ cm and will here be reduced by the factor $f^{\frac{1}{2}}$, that is for T_0 = 5 keV by the factor 3.0 and for T_0 = 10 keV by the factor 5.4. Accordingly, the critical masses and input energies for compression will be reduced by the factors 27 and 160.

Although the elevated temperatures promise a substantial reduction in the critical dimension of the pellet this is in part compensated by a higher compression energy resulting from the higher temperature. However, even at T_0= 10 keV the required compression energy is only $\sim$ 10 times larger to achieve the same density as for a Fermi degenerate electron distribution at $T = 0$. In contrast, the reduction in the critical mass by the fission-fusion chain reaction is governed by a much larger factor and thus the total minimum energy input can be reduced by simultaneously compressing and heating the pellet to reach temperatures where the fission-fusion chain reaction sets in. In our previous example the compression energy was 2.3×10^5 Joule. By heating the pellet to T_0 = 10 keV, 10 times more compression energy would be needed but at this temperature the critical mass would be reduced by a factor 160. It thus follows an overall reduction in the compression energy by a factor $\sim$ 10.

Finally, we would like to emphasize that the factor f also determines the change in the growth rate of the fission-fusion chain reaction as compared to the growth rate of a pure fission chain reaction. This means that the factor τ/t_e given by eq. (30), is increased by the factor f, hence

$$\tau/t_e = \text{const. } f \quad , \qquad (63)$$

where τ and t_e are defined as before, and where f can be as large as 44. Because of the exponential growth of the fission-fusion chain reaction it therefore follows that much larger energy yields can be expected than from a pure fission chain reaction. This can be most easily seen by comparing the growth rate of a pure fission chain reaction with the growth rate of a fission-fusion chain reaction. If we denote the neutron flux in the first case by ϕ and in the second case by ϕ^*, we have as the ratio for both at the time $t = \tau = R/v$, the hydrodynamic disassembly time,

$$\phi^*/\phi = e^{\lambda\tau(f - 1)} \quad . \qquad (64)$$

In eq. (64) we have inserted λ in place of λ_0 to take into account that the medium is of finite dimension. With the previously chosen numerical example $\Delta R/R = 0.1$, $t_e = 5/\lambda_0 = 1/\lambda$ we had $\lambda = 10^{10}\text{sec}^{-1}$, furthermore $\tau = 10^{-9}$ sec, and therefore have $\phi^*/\phi \simeq \exp\left[10(f-1)\right]$. Assume for example f = 10 this would then result in $\phi^*/\phi \simeq 10^{39}$. This very large ratio of course, cannot be reached due to the depletion of the nuclear fuel, but it indicates that with a fission-fusion chain reaction greatly increased energy outputs as compared to pure fission micro-explosions and probably as compared to pure fusion micro-explosions can be expected.

REFERENCES

1. J. Nuckolls, L. Wood, A. Thiessen, and G. Zimmerman, Nature 239, 139 (1972).
2. K.A. Brueckner, KMS Fusion Inc., Ann Arbor, Michigan, preprint, April, 1972.
3. F. Winterberg, Phys. Rev. 174, 212 (1968).
4. F. Winterberg, Nature 241, 449 (1973).
5. F. Winterberg, Lettere Il Nuovo Cimento 6, 407 (1973).
6. S. Glasstone and M.C. Edlund, Elements of Nuclear Reactor Theory, Van Nostrand, New York, 1955.
7. R.S. Bradley, High Pressure Physics and Chemistry, p. 253 Academic Press, New York, 1963.

SUPER-COMPRESSION OF FISSIONABLE MATERIAL BY LASERS*

M. S. Rabinovich, G. A. Askarjan, W. A. Namiot

P. N. Lebedev Physical Institute

Leninsky Prospect 53, Moscow, USSR

INTRODUCTION

Very high reactive forces may be obtained when matter is evaporated under the action of powerful light radiation or intense electron beams[1]. Recently much attention has been paid to the idea of increasing thermonuclear output by super-high compression of heavy hydrogen under the action of high intensity light irradiation causing reactive pressure[2].

In this paper we propose to use such super-compression to solve certain other problems as production of microscopic critical masses of fissionable heavy elements through more effective neutron multiplication by increasing the density of fissionable elements, production of super-strong magnetic fields and acceleration of particles.

PRODUCTION OF MICRO-CRITICAL MASSES OF FISSIONABLE MATTER

The mean-free path of neutrons before they create nuclear fission is given by:

$$\ell_f \simeq (n_i \sigma_f)^{-1} \tag{1}$$

---

*Paper scheduled for the Third Workshop on "Laser Interaction and Related Plasma Phenomena" held at Rensselaer Polytechnic Institute, Troy, New York, August 13-17, 1973, but not presented due to unforeseen circumstances preventing the authors from attending.

whereby n_i is the nuclear concentration and σ_f the effective fission cross-section. This means that increasing the density of matter may sharply increase the effectiveness of using fission neutrons resulting in a reduction of the critical dimensions

$$R_{CT} \propto \ell_f \propto n_i^{-1} \tag{2}$$

and the critical masses:

$$M_{CT} \propto n_i \ell_f^3 \propto n_i^{-2} \tag{3}$$

In the following it is assumed that the surface of a fissionable mass is irradiated from various directions by powerful radiation. In some cases it is more convenient to apply this action on the s.c. ablative layer - a layer of matter covering the fission kernel which converts the fission and leads to optical light reactive pressure. The vapor pressure[1] is given by

$$p \simeq \frac{dM}{dt} v \simeq \left[I(\lambda + 1/2v^2)^{-1}\right] v \simeq I/v \tag{4}$$

whereby λ is the sum of the heat of evaporation and ionization energy of the material, v is the velocity of the evaporating material. As an example we cite the case of 100-fold compression whereby one can obtain a pressure of $p \simeq 10^{17}$ dynes cm^{-2} with an intensity input of $I \simeq 10^{17}$ W cm^{-2} which corresponds to an energy input of $\simeq$ 100 kilojoule during one nanosecond onto an area of 10^{-3} cm^2.

Such a high pressure in the case of a degenerate electron gas corresponds to a nuclear concentration of:

$$n_i \simeq \left(p \frac{m}{\hbar^2}\right)^{3/5} Z_{eff}^{-1} \simeq 10^{25} \text{ ions/cm}^3 \tag{5}$$

Hereby Z_{eff} was entered to be approximately 10 when two upper electron shells are ionized at not very high temperatures $T_e \lesssim$ 500 eV.

At such a nuclear concentration (10^{25} cm^{-3}) the mean free path of a neutron before it splits the nucleus is

$$\ell_f \simeq (n_i \sigma_f)^{-1} \simeq 3 \times 10^{-2} \text{ cm}, \tag{6}$$

since the effective fission cross-section σ_f for fast neutrons and the type of fissionable material of interest is usually 1-2 barn ($1 - 2 \times 10^{-24}$ cm^2).

Ablative covering may be used not only for decreasing loss of fissionable matter and increasing vaporization pressure, but also for the reflection of neutrons into the material to be used for fission. The reflection coefficient α for neutrons reflected from super-compressed ablative covering (for example, from hydrogen having a sufficiently large neutron scattering cross-section 1 - 2 barn) may be close to unity: $\alpha \simeq n_i \sigma_f \ell_f \simeq 1$, where $\sigma_f \simeq 2$ barn, $n_i \simeq 10^{26}$ ions/cm^3. At the same pressure the nuclear concentration within a compressed hydrogen covering may be ten times higher than the nuclear concentration of fission material, since the compressibility of hydrogen is very high.

In order to estimate the neutron multiplication we may use a simplified equation:

$$R\frac{dn_n}{dt} \simeq R\nu n_i n_n \int [\sigma_f(v_n) - \sigma_a(v_n)] f(v_n) v_n dv_n - n_n \int v_n f(v_n) dv_n \quad (7)$$

where $\nu = \nu_o - 1$ is the coefficient of excess neutron multiplication per fission, $f(v_n)$ is the distribution function of the neutrons, σ_f the fission cross-section, σ_a the radiative capture cross-section, and R the radius of the fissionable sphere.

We can write Eq. (7) as:

$$R\frac{dn_n}{dt} \simeq R\nu n_i n_n \overline{\sigma_f v} - 3n_n v_n \quad (7a)$$

which results then in the super-critical condition for the radius of the fissionable sphere:

$$R_{CT} \geq \frac{3}{\nu n_i \sigma_f} \quad (8)$$

If one takes into account the reflection α of neutrons from the super-compressed shell the critical condition will be:

$$R_{CT} \simeq \frac{3(1-\alpha)}{\nu n_i \sigma_f} \quad (9)$$

These estimates show that one can obtain critical dimensions R_{CT} of the order of 10^{-2} cm and critical masses $M_{CT} \simeq 10^{-2}$ gr, which corresponds to an energy output of approximately 10^7 Joule assuming one percent fission efficiency. This output exceeds considerably the input energy assumed for the compression process, namely $\leq$ 100 kilojoule.

The duration τ_n for the release of energy due to the neutron avalanche is given by:

$$\tau_n \simeq K(n_i \sigma_f v_n)^{-1} \simeq 2 \times 10^{-11} K \qquad (10)$$

If, for example, the number of last neutron generations, which is the most essential for the energy output process, is K = 3, one can obtain the effective release time of neutrons and energy to be $\tau_n \simeq 6 \times 10^{-11}$ sec. The expansion time τ is expressed by:

$$\tau \gtrsim a\left(\frac{\rho}{p}\right)^{1/2} \gtrsim a(4\rho a^3 Q^{-1})^{1/2} \simeq a^{5/2}\left(\frac{4\rho}{Q}\right)^{1/2} \simeq 10^{-10} \text{ sec} \qquad (11)$$

This means that the expansion time τ exceeds the energy release time τ_n, so that no reduction in the energy release is to be expected.

Microcritical fission masses may be used for the production of super-intense neutron and neutrino beams (10^{17} neutrons within 10^{-10} sec, i.e. neutron pulses of 10^{27} neutrons per second) which should be of great interest for a number of experiments.

PRODUCTION OF SUPER-STRONG MAGNETIC FIELDS FROM SUPER-COMPRESSION OF MATTER

Super-compression may be used to compress magnetic fields, frozen in matter or in cavities of matter under ablation pressure in analogy with Zakharov's method of producing strong magnetic fields by means of a blast[3]. Super-high pressures of 10^{17} - 10^{18} dyne cm^{-2} are equivalent to magnetic fields of

$$H \simeq (8\pi p)^{1/2} \simeq 3 \times 10^9 \text{ oersted} \qquad (12)$$

However, in our case magnetic compression must be determined by taking into consideration the geometry of matter compression.

In the case of a solid body the cross-section areas S_c (compressed) and S (noncompressed) are related to the matter densities n_c (compressed) and n (uncompressed) by:

$$S_c/S \simeq (n/n_c)^{2/3} \qquad (13)$$

If the density of the material is increased by a factor of 3×10^4 and the flux kept constant, we have a field compression of

$$H_c/H \simeq S/S_c \simeq 10^3 \qquad (14)$$

which turns a field of $H \simeq 10^5 - 10^6$ oersted into a final field of

$H_c \simeq 10^8 - 10^9$ oersted. A short duration of compression ($\lesssim 10^{-10}$ sec) and high temperatures $T \sim 1 - 3$ keV will yield for the skin-time:

$$t \simeq 4\pi\sigma\, R^2/c^2 \simeq 10^{-7} \text{ sec} \tag{15}$$

which is much larger than the collapse duration of 10^{-10} sec. Hereby a conductivity $\sigma \simeq AT^{3/2} \simeq 10^{17}$ (in CGS units) and $R \simeq 10^{-2}$ cm were assumed. With this difference of skin-time and collapse time, one can certainly neglect the diffusion of the magnetic field out of the material during the duration of compression. Super-high magnetic compression may be achieved by a compression process in a hollow donut.

Strong induced electric fields E due to the fast time change of H:

$$E \simeq \frac{R}{\eta c}\frac{dH}{dt} \simeq 10^9 \text{ Volt cm}^{-1} \tag{16}$$

may result in acceleration of electrons and ions up to energies $\mathcal{E}$ in the order of 10 GeV:

$$\mathcal{E} \simeq ZeHR \simeq 3Z \text{ [GeV]} \tag{17}$$

at compressions up to a final magnetic field strength of $H \simeq 10^9$ oersted.

In conclusion the authors express their deep gratitude to academicians L. A. Artsimovich, M. A. Markov and A. M. Prokhorov for their assistance.

References

1. G. A. Askarjan, E. M. Moros, JETP 1962, Vol. 43, N6, p. 2319.

2. E. Teller Report on 7-th QE-Conference Montreal May 1972.

3. A. D. Zakharof "UFN", Progress in Physical Science, Vol. 88, 4, p. 725 (1966).
   J. P. Terlecky, JTPh 32, 387 (1957).

Note of the editors:

Recently F. Winterberg has independently worked out similar ideas as expressed in the first part of this paper. See: F. Winterberg, Lettere al Nuovo Cimento 6, 407 (1973) and in more detail in these proceedings pp. 519-535.

SUMMARY OF DISCUSSIONS AND COMMENTS

(IV. Theory of Compression and Energy Gain from Laser Plasmas)

Concerning details of his code, J. Nuckolls gave the answers to several questions. The instabilities caused by acoustic wave are not included because these are not important. The instabilities due to stimulated Compton scattering are also not important at the considered conditions. J. Nuckolls recommended unclassified near-term experiments. He said: "In that category the kind of experiments that I would like to see are some resolution of the question of what kind of electron spectra are generated. One could do that in two ways. One, where you shine the laser light on plain slabs. Two, it also needs to be done where you can shine it in an isolated pellet. I think these two ways may result in different answers. You can infer the electron spectra either by the X-rays that come out or by other experimental techniques. I would also like to see experiments where one carries out some implosions where some experimental technique of running out with a density is used as J. W. Shearer and R. Kidder have discussed. One first has to perform experiments where one gets 10-, 100- and 1000-fold compression before one can do experiments that reach 10,000-fold compression. In conclusion the discussion underlined the problem of whether the electrons are to be assumed flowing (Katz) or not (Morse).

From the discussion of K. A. Brueckner's paper it should be noted that his measured 6000 neutrons from CD_2 targets irradiated by 200 Joule - 3 nsec neodymium glass laser pulses focused down to 40 μm diameter were within his theoretically expected low values. The very strong break upward of the neutron yield will be expected only where the center ignites and the precursors of the detonation front generate self-heating. It was noted that D-D targets will not ignite and will not generate self-heating. The aim is to achieve 10^{14} neutrons from D-D with a 1 kJ laser pulse. The conversion factor for D-T from D-D without ignition is a 30 times higher yield in the number of neutrons. Brueckner explained the use of elliptical mirrors eliciting a perpendicular incidence of the laser radiation on the target following a question by J. W. Shearer on the focusing of the radiation.

R. Kidder presented his paper on the possibility of a compression up to one million times the solid state density. H. Hora asked whether this possibility agrees with the equation of state of degeneration. R. Kidder answered that there are no limitations.

A new approach in reaching the very high fusion yields is the use of the self-sustained ignition of the D-T pellet. The spherical cal ulations are new in the open literature. However, J. L. Bobin emphasized this concept of the hot spot is more than ten years old for the plane case in connection with his review on the subject.

Following F. Winterberg's theoretical result on the compression of fissionable material, G. C. Baldwin pointed out the expected very high neutron densities are suitable to invert nuclei. These high densities are selected in a way similar to those levels of the Mössbauer effect to get γ-ray lasers (grasers).

V

EXPERIMENTS FOR NUCLEAR FUSION

OBSERVATION OF THE LASER-INDUCED COMPRESSION WAVE IN A SOLID TARGET*

C. van Kessel and R. Sigel

Max-Planck-Institut für Plasmaphysik
8046 Garching, Federal Republic of Germany
EURATOM Association

ABSTRACT

The evolution of the compression wave which forms on laser impact on a plane solid target has been followed by streak- and framing photography. Using known Hugoniot-data a peak pressure of 2.5 Mbar and a density increase $\rho/\rho_o = 2.3$ has been measured in plexiglass.

As is well known laser irradiation of a solid target creates a hot and dense plasma at the surface which exerts a considerable pressure on the neighboring solid material. As a consequence the solid material is compressed and a shock wave moves into the interior of the target. The study of this phenomenon is of interest with respect to laser fusion applications and because it opens new possibilities in studying the properties of matter under extremely high pressures.

To the authors' knowledge a direct observation of laser-induced shock waves in solid targets has never been made. In an indirect manner its velocity has been determined only from the burn-through time of laser irradiated solid hydrogen foils[1] and the pressure

---

*Presented at the Third Workshop on "Laser Interaction and Related Plasma Phenomena" held at Rensselaer Polytechnic Institute, Troy, New York, August 13-17, 1973.

from a momentum balance of the expanding plasma*[2]. In the latter case the accuracy of the determination of pressure is however severely limited by the fact that the surface area where material is ablated is not accurately known. In this paper we measure the time-dependent velocity of the shock wave. If the Hugoniot-curve of the material is known this measurement then yields directly the pressure and density behind the shock front.

The experimental arrangement is shown in Fig. 1. A hot plasma is created at the concave surface of the target by the intense pulse of a neodymium glass laser. The interior of the target is imaged by a lens on the slit of a streak camera. The parallel beam of a dye laser serves as a background illumination. The shock front may then be detected either due to absorption in the compressed material behind it or by deflection of the dye laser light in the density gradient of the front.

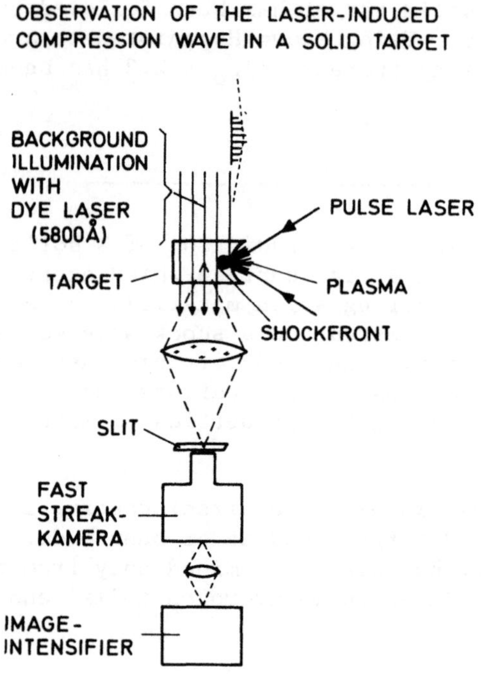

FIGURE 1. Experimental arrangement.

*The Mach-number of the shock wave produced with a relatively weak pulse laser has been determined by a similar method as in Ref. 7.

The 2 mm thick target was made of Plexiglass (solid hydrogen or other transparent materials may also be used). It has two flat and smooth side surfaces and a concave front surface. This curvature is required to allow a clear observation of the shock wave at the beginning of the irradiation where it is near the surface. Obscuring by the target edges is thus avoided. With respect to the small focal spot of the neodymium laser (~ 40 μm) this shape corresponds however essentially to a plane surface.

The dye laser (Electro-Photonics, Northern Ireland) can be operated either in the long pulse mode (~ 0.5 - 1 μs) for streak photography or in the mode-locked mode for framing photography. In the latter case the duration of a single pulse is about 5 ps[3]; the mode-locked pulse separation depends on the length of the resonator and was 4.5 ns in this experiment. The image-converter streak camera (Electro-Photonics, Northern Ireland) is equipped with an image-intensifier and is, at the fastest streak speeds (not employed here), capable of a time-resolution better than 10 ps. The neodymium laser used for plasma production was in these experiments operated typically at a pulse energy of 12 Joules with a pulse length of 5 ns. More details of the Garching neodymium laser experiment may be found elsewhere in this volume[4].

Figures 2 and 3 give experimental results. They have been obtained at a relatively low streak speed and with moderate magnification in order to give a general impression of the phenomenon. Nevertheless they demonstrate clearly that the applied experimental method is well suited for the investigation of laser induced compression waves.

Figure 2 shows a streak photograph with dye laser background illumination obtained with a plexiglass target. The vacuum-target interface remains undisturbed up to time $t = 0$ where irradiation with the neodymium laser begins. After the beginning of irradiation a dark zone extends into the interior of the material. The velocity of the boundary, which we identify as the shock front, is highest during irradiation and then decreases. Evidently the material behind the shock front is in a state where the dye laser light (5800 Å) is no longer transmitted. After about 20 ns the strength of the shock wave is already attenuated and the material behind the shock front becomes transparent again. The depth up to which the material becomes opaque (~ 300 μm) corresponds to the depth of the crater which is finally left in the target. A photograph of such a crater is shown in Ref. 4.

Figure 3 shows the same phenomenon but now in a sequence of framing photographs. These pictures have been obtained with a widely opened streak slit and with the dye laser operating in the mode-locked mode. The width of the slit corresponds to the distance

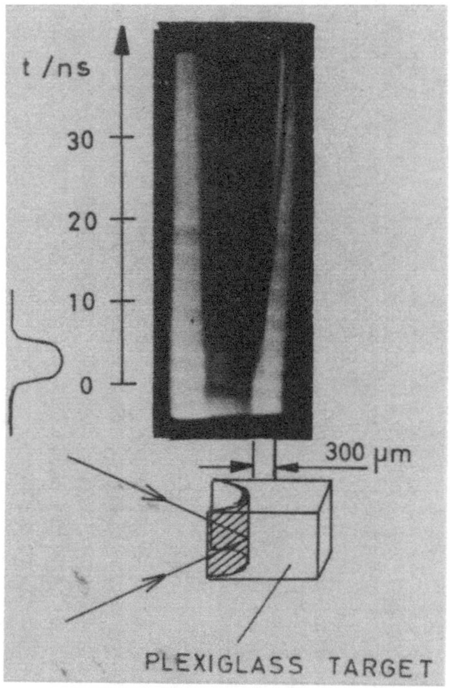

FIGURE 2. Streak photograph of the laser induced shock wave in a plexiglass target.

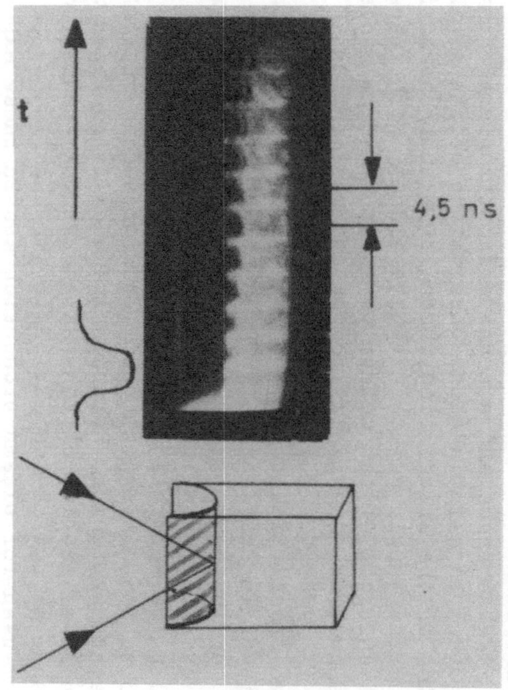

FIGURE 3. Sequence of framing photographs.

by which the electron beam of the streak tube sweeps between two subsequent pulses from the mode-locked dye laser. As the duration of such a pulse is very short (5 ps) sweeping is completely negligible during exposure and sharp pictures are obtained. From these pictures the curvature of the shock front can be determined. Its shape is roughly spherical. Again the separation of the shock front from the opaque material can be seen.

From these photographs the time-dependent velocity of the shock front has been determined. From the well known Hugoniot relations (see for example Ref. 5) and the Hugoniot curve of plexiglass[6] the pressure and density of the material behind the shock front are then readily obtained.

Figure 4 shows the pressure behind the shock front as a function of time (based on three shots). A peak pressure of 2.5 Mbar corresponding to a shock front velocity of 2×10^6 cm^{-1} is observed about 1 - 2 ns (the time resolution in this first set of experiments) after the beginning of the irradiation. The pressure drops during the irradiation and decreases rapidly after termination of the laser pulse. The peak pressure of 2.5 Megabar corresponds to a compression of ρ/ρ_o = 2.3.

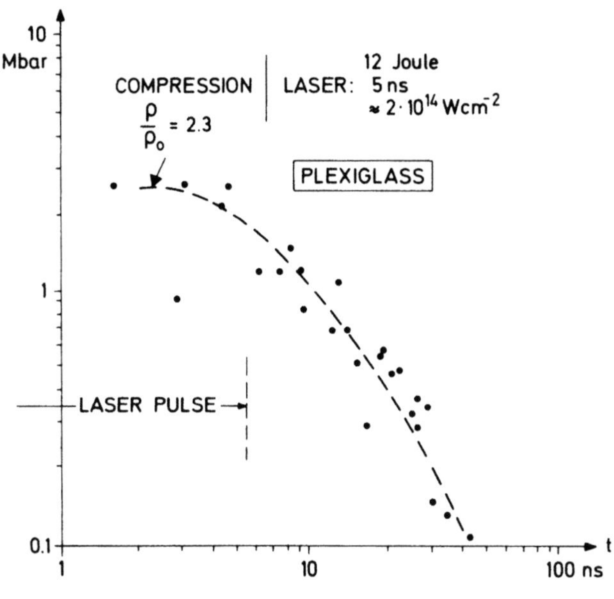

FIGURE 4. Pressure behind the shock front as a function of time.

Similar pictures with a stick of solid hydrogen as a target have recently also been obtained. In this case a maximum shock front velocity of 6×10^6 cm^{-1} corresponding to a peak pressure of 2 Mbar has been measured. A more detailed description of these experiments will be given elsewhere.

ACKNOWLEDGEMENTS

The authors would like to thank their colleagues H. Brändlein, Ch. Dorn, P. Sachsenmaier, E. Wanka and G. Wirtz for their help in conducting this experiment.

This work has been undertaken as part of the joint research program between the Max-Planck-Institut für Plasmaphysik and Euratom.

References

1. R. Sigel, Z. Naturf. 25a, 488 (1970).

2. N. G. Basov, V. A. Boiko, V. A. Gribkov, S. M. Zakharov, O. N. Krokhin and G. V. Sklizkov, Soviet Physics JETP 34, 81 (1972).

3. E. G. Arthurs, D. J. Bradley and A. G. Roddie, Appl. Phys. Lett. 19, 480 (1971).

4. See contribution of K. Eidmann and R. Sigel: "Backscatter Experiments".

5. Ya. B. Zeldovich and Yu. P. Raizer, Physics of Shock Waves and High-Temperature Hydrodynamic Phenomena, Academic Press, New York and London 1967, Vol. II, p. 705.

6. R. S. Hawke, D. E. Duerre, J. G. Huebel, H. Klapper, D. J. Steinberg and R. N. Keeler, J. Appl. Phys. 43, 2734 (1972).

7. M. H. Key, D. A. Preston and T. P. Donaldson, VI. International Quantum Electronics Conference, Kyoto, Japan, 1970.

INVESTIGATION OF PLASMA PARAMETERS AT THE SPHERICAL HEATING OF THE ISOLATED SOLID TARGET BY HIGH-POWER LASER RADIATION*

N. G. Basov, E. G. Gamaly, O. N. Krokhin,
Yu. A. Mikhailov, G. V. Sklizkov, S. I. Fedotov
P. N. Lebedev Physical Institute

Moscow, U.S.S.R.

ABSTRACT

The results of the investigation of high-temperature non-stationary plasma produced at the spherical heating of the ball-shaped $(CD_2)_n$ target by high-power irradiation of the 9-channel laser setup are reported. It had been found that the fraction of absorbed energy reached 70% of the incident radiation at flux densities ~ 10^{14} - 10^{15} W cm^{-2}. About 10-25% of the energy absorbed in the target is reemitted in the X-ray spectral range, while the fraction of energy reflected from plasma was not greater than 10%. The study of gas-dynamic parameters of the hot plasma "corona" made it possible to measure the pressure in the dense plasma core. The pressure was about 10^6 - 10^7 atm at the flux density of about 10^{15} W cm^{-2}. Experimental data on measuring density, velocity and pressure in the "corona" region made it possible to conclude about experimental realization of the compression value of the order of 30.

Neutron yield of about 10^7 neutrons per pulse was registered. Time of flight measurements of the neutron radiation indicated thermonuclear character of neutrons emitted at spherical heating of a target.

---

*Paper scheduled for the Third Workshop on "Laser Interaction and Related Plasma Phenomena" held at Rensselaer Polytechnic Institute, Troy, New York, August 13-17, 1973, but not presented due to unforeseen circumstances preventing the authors from attending.

INTRODUCTION

The up-to-date laser systems allow to obtain energy densities in the heated target up to 10^8 J cm^{-3} when the rate of energy deposition achieves 10^{17} W cm^{-3} at the nanosecond interval. The interest in lasers as a possible source of dense plasma heating for thermonuclear fusion arises from the feasibility of obtaining high energy concentrations in the matter. At dense plasma heating by high-power laser radiation one may achieve thermonuclear temperatures and produce controlled thermonuclear fusion[1]. From the estimates of Refs. 1-4 it follows that to realize a physically profitable thermonuclear reaction one needs a laser with high radiative energy and small pulse duration. In addition, efficient and symmetrical energy deposition from the laser beam into the plasma should be provided. The main mechanism regulating effective absorption of high-power laser radiation in the overdense plasma is not yet clearly understood. But from the results of this work and those of Refs. 5-6 one can consider that under some conditions a considerable fraction of laser energy is absorbed.

The construction of such lasers meets significant technical difficulties. Therefore much attention is paid to the problem of decreasing the threshold energy $E_{thresh.}$, corresponding laser driven fusion. The increase of the efficiency of energy deposition into the laser plasma due to optimization of geometry, size and contents of the target has been theoretically considered in Refs. 3, 7, and 8. Hence, it becomes however possible to reduce the value $E_{thresh.}$ to not more than an order of magnitude.

To achieve threshold of thermonuclear fusion[9] at pulsed laser heating certain conditions should be fulfilled, one of them being the Lawson criterion[10], which for a DT mixture has the following form:

$$10^{14} \leq n\tau_p \simeq n \frac{r_o}{c} \tag{1}$$

where n - density of plasma particles, τ_p - lifetime of plasma at thermonuclear temperature T ~ 10^3 eV, r_o - initial target radius at spherical irradiation, c - sound velocity in plasma. Thus, at r_o = 100 μm and c = 10^8 cm/sec the required density of the heated matter should be greater than $10^{24} cm^{-3}$. One may obtain such densities by hydrodynamic plasma cumulation[11,12] or static compression of the target matter. The experiments with laser flare collisions have shown[13-15] that the increase of substance density and plasma thermal energy occurs in the collision region. At the spherical geometry of convergent motions one may expect the density increase by 10^2-10^3 fold. Such values of matter compression have been obtained experimentally in laboratory conditions with the arrangements of plasma focus type[16,17]. Note, that when the high density targets are used in laser heating experiments one must especially investigate the efficiency of heating of the matter whose density is higher than critical.

The increase of the heated target density due to cumulative compression at spherical irradiation[18] seems to be the most promising way of achieving controlled thermonuclear fusion. Plasma density can be actually increased, in respect to the solid substance, by the factor of several orders of magnitude. In Ref. 12 matter compression could be performed at cumulation of the spherically convergent shock wave. Such a shock can be produced, for example, in the result of explosion of a heavy spherical liner[8].

Cumulation processes occurring at spherical irradiation of a homogeneous target by high-power laser radiation can be caused by a high-pressure momentum produced by the radial expansion of the ablated material - hot plasma corona. The required pulse shape of the pressure can be obtained by varying the shape of the optical pulse. One of the problems of adiabatic compression has been calculated in Refs. 19 and 21. It was indicated that at certain conditions one can achieve approximately 10^3-10^4 fold compression when the flux density at the end of the profiled pulse reached 10^{18} - 10^{19} W/cm^2.

The first experimental studies of the plasma heating in a spherical case have been realized in Refs. 18 and 21. The neutron yield of 3 x 10^6 neutr/pulse from a $(CD_2)_n$ target, 10^{-2}cm in diameter, heated by a two-nanosecond beam from the 9-channel laser, at approximately 214 J energy, was registered in Ref. 23. This value is appreciably greater than that registered at the sharply focused single-beam laser radiation[15-18].

It should be noted that gas-dynamic plasma parameters of the spherically heated isolated solid target, and some features of laser radiation absorption are not yet sufficiently studied in these experiments. As for the plasma compression at spherical irradiation the experimental data on this subject are not up to now available.

In this work we have investigated the efficiency of heating and compression of plasma at spherical irradiation of the deuterized polyethylene $(CD_2)_n$ target[28,29]. The problems of efficiency of radiation absorption in plasma at different dimensions of the target and laser pulse parameters, and the increase of target density during heating are chiefly emphasized here. In addition, the results are reported on the measurements of X-ray radiation from the plasma flare and neutron yield, which are superposed to the efficiency of energy deposition. Experimental results make it possible to determine some optimization of the conditions of plasma heating at spherical case.

I. EXPERIMENTAL SET-UP

Diagnostic Technique

Experimental set-up for high-temperature spherical plasma heating consisted of a powerful Nd-glass laser[28,29] and a vacuum chamber with diagnostic devices and registration apparatus.

Laser radiation with regulated pulse duration from 1 to 16 nsec, radiation divergence ~ 10^{-4} rad brightness ~ 5×10^{16} W/cm^2 and total energy 0.3-1.3 kJ was concentrated in the target of a conventional $(CH_2)_n$ and deuterized $(CD_2)_n$ polyethylene by means of optical system and focusing objectives. Caustic diameter in each objective did not exceed 30 μm. The focusing system permitted us to irradiate in a spherical symmetry the targets, from 5.5×10^{-3} cm to 8.5×10^{-2} cm in diameter, thus providing plasma heating at flux densities up to 3×10^{15} W/cm^2. For the targets 5.5×10^{-3} cm in diameter the rate of energy deposition of ~ 5×10^{17} W/cm^3 has been experimentally achieved. Compensation of optical paths for either light beam was realized by optical compensators with an accuracy of 1 cm, and it corresponded to retardation of pulse arrival to the target of 3×10^{-11} sec.

In this set of experiments we used a ball-shaped target. Its surface differed from the spherical one in the limits of 20%. The target was suspended in the center of a vacuum chamber with the help of a special adjustment mechanism. The target was fastened with thin $(CD_2)_n$ filament 2 mm long, and 5 μm in diameter, with a steel needle connected with a mechanism. The required matching of the laser focus with the core of the heated target was performed by shifting targets along 3 coordinates controlled by several microscopes. In the vacuum chamber various gases could be used in a wide range of pressures from 1 to 10^{-4} atm.

Vacuum chamber and some of the diagnostic devices are depicted schematically in Fig. 1. The diagnostic devices can be divided into several groups:

- apparatus for the study of energy deposition into the heated target,

- means for X-ray analysis of plasma,

- neutron detectors, including time of flight methodics,

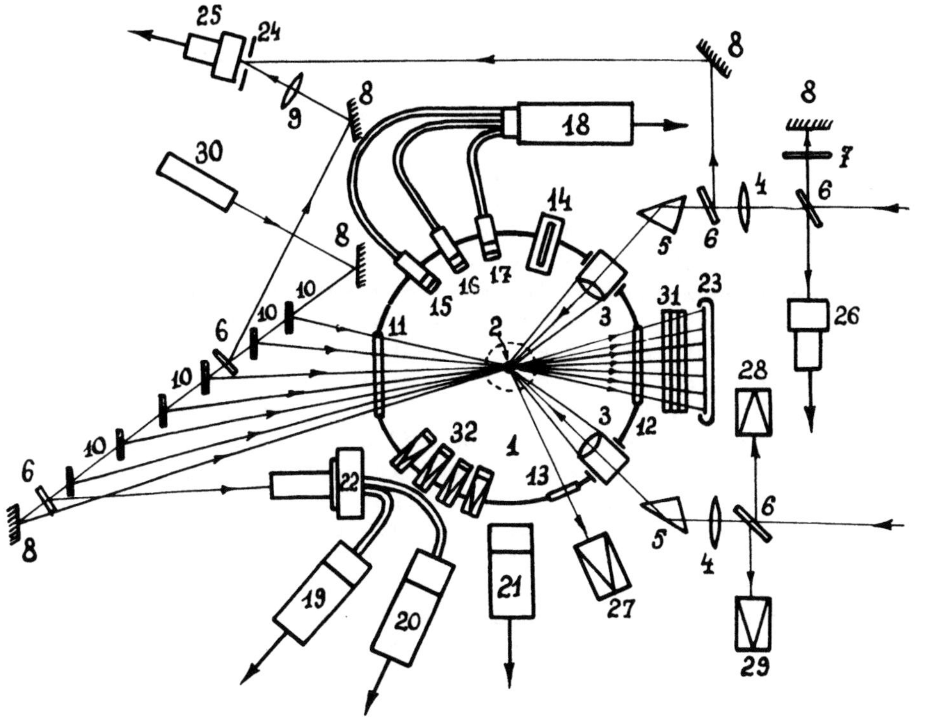

FIGURE 1. Vacuum chamber and diagnostic devices: 1 - vacuum chamber; 2 - target; 3 - short-focus component of focusing objective; 4 - long-focus lens; 5 - prism; 6 - plane-parallel plate; 7 - light filter; 8 - mirror; 9 - lens; 10 - two-component mirrors of optical delay; 11 - entrance diagnostic window of a chamber; 12 - exit window; 13 - window; 14 - cell with a nuclear emulsion; 15-17 - sensory elements of X-ray "plasma thermometer"; 18 - electron photo-multiplier; 19-21 - neutron scintillation counters; 22 - cell with filters and light pipes; 23 - photocell for shadowgraphy; 24 - diaphragm; 25-26 - coaxial photoelements; 27-29 - calorimeters; 30 - ruby laser; 31 - cell with filters; 32 - 4-channel X-ray calorimeter.

- interferometric arrangements for the study of the plasma density profile in corona.

Besides, Fig. 1 shows two of nine laser beams (elements 3,4,5).

II. SHOCK WAVE GENERATION AT SPHERICAL HEATING AND THE STUDY OF EFFICIENCY OF THE ENERGY DEPOSITION

At the spherically-symmetric heating of the ball-shaped target almost ideal conditions are realized for spherical shock wave formation in the gas medium. Therefore, wide possibilities are opened for the study of shock waves formed at the required law of energy deposition in time. In this work SWs were studied mainly to determine the fraction of energy absorbed in the laser plasma. Note, that at spherical target irradiation the only method available now for measuring energy deposition in plasma is to measure the power of SW at the later stage of gas dynamic expansion, when the most of internal energy (except for radiation losses) converts into the kinetic energy of radial motion. In spite of the SW method seems quite evident in the case of target explosion, it has some peculiarities due to laser heating, which may lead to a significant disagreement between experimental R-t curves and the theoretical calculations on instantaneous point explosion[30,31]. Thus, at the initial stage the SW expansion depends strongly on the parameters of a plasma "piston". Since energy deposition cannot be considered as point the explosion may turn out to be also not instantaneous thus leading to inhomogeneity of the target heating during radiation. In addition, in the process of SW formation the recombination may play a certain role (particularly with heavy targets) due to which the fraction of energy for ionization will be transformed into the thermal energy. The character of SW motion at the initial stage is determined by the process of energy transfer from the flare to the outer residual gas. In particular, it can cause a strong thermal wave in the residual gas. The possible mechanism of SW formation may be electron thermal conductivity. Self-similar solution of the spherically-symmetric problem on the heat propagation from the instantaneous point source has a form[31]

$$R_T \simeq 0.59E_T^{5/19}p^{-7/19}t^{2/19} \qquad (2)$$

where R_T - radius of the thermal wave front in cm, E_T - thermal energy in J, p - gas pressure in torr, t - time in nsec. For E_T = 10^2 J, p = 10 torr, t = 1 nsec we have R_T = 0.85 cm. Such a wave propagating initially with the velocity $\dot{R}_T$, which exceeds a great deal the sound speed in the heated medium, may influence sufficiently the SW motion.

Initial Stage: Measurement of the Effective Plasma Temperature at Shock Wave Formation

As it has been noted above, the heat wave moved initially in the gas surrounding the target. But its velocity sharply decreases as it expands and in 1 nsec interval SW goes out ahead of it. In the beginning of motion when the mass of the gas overlapped by SW is less than the heated mass, the SW velocity is determined by the plasma temperature at the moment of SW formation. At later stage when the heated gas mass, is rather greater than that of the target, the peculiarities of the target state are not of much importance, and the SW moves by the same law as for the point explosion. This is well confirmed by experimental R(t) diagrams. At the initial stage SW velocity dependence on the substance state behind the front for the ideal gas may be easily defined from the shock conservation laws[31]

$$\rho_o D = \rho(D - U) \tag{3}$$

$$\rho_o D^2 = p + \rho(D - U)^2 \tag{4}$$

$$\rho_o D \frac{D^2}{2} = \rho(D - U)\left[E + \frac{p}{\rho} + \frac{(D - U)^2}{2}\right] \tag{5}$$

where D, U - velocities of the shock and gas behind shock jump, ρ, ρ_o - density behind and before the jump, E, p - energy and pressure behind the jump. SW is strong, i.e. $p \gg p_o$. For the ideal gas with

$$p = A\rho T = \rho c^2, \; E = \frac{p}{\rho(\gamma - 1)} \tag{6}$$

we have

$$D = \frac{(\gamma + 1)}{[2(\gamma - 1)]^{1/2}} C = \frac{(\gamma + 1)}{[2(\gamma - 1)]^{1/2}} (AT)^{1/2} \tag{7}$$

$$\frac{\rho}{\rho_o} = \frac{\gamma + 1}{\gamma - 1} \tag{8}$$

where c - sound velocity, γ - adiabatic index, $A = 5.10^{14}$ erg/g. KeV at ρ - in [g/cm^3], T - in [keV] and p - in [dn/cm^2].

Thus, determining SW velocity from the experimental R-t diagram, we can estimate, by using Eq. 4, plasma temperature at the initial stage. Considering $(CD_2)_n$ as a fully ionized ideal gas with $\gamma = 5/3$ we have from Eq. 4

$$D = 5.17 \times 10^7 \cdot T_{keV}^{1/2} \quad [cm/s] \tag{9}$$

The shock SW velocity measured experimentally within first 10 nsec is of 5×10^7 cm/s (see the shadowgram of Fig. 2). The corresponding R-t diagram is given in Fig. 3. Comparing this value

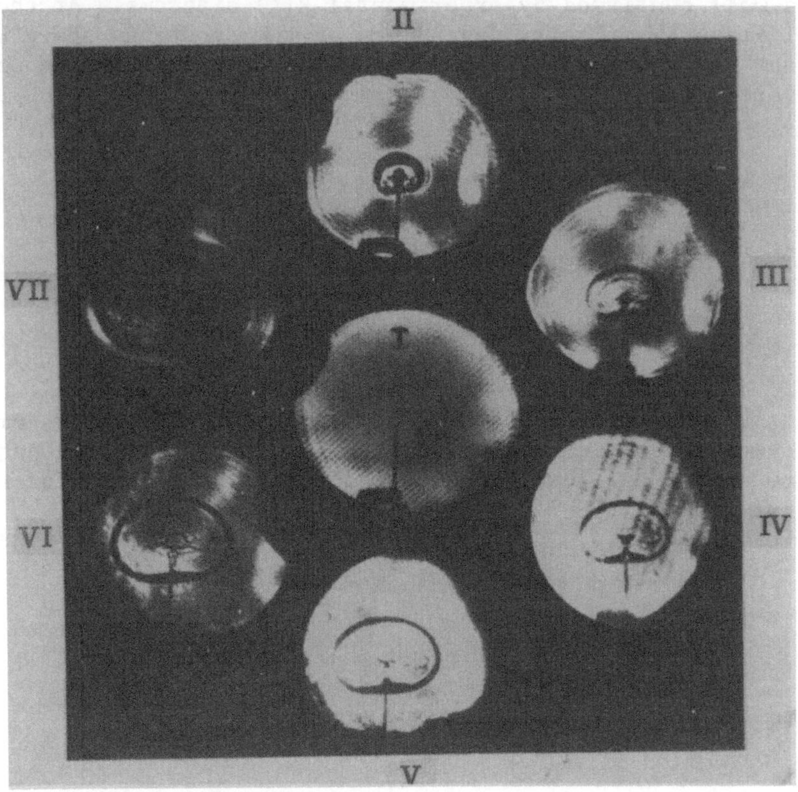

FIGURE 2. Shadowgram of a shock wave in deuterium produced by exploding $(CD_2)_n$ target at spherical irradiation from a 9-beam laser. The target is 130 μm in diameter, laser energy of 30 J at 2 nsec pulse duration. The moments of frame exposures with respect to the beginning of heating are: 24.8, 49.6, 74.4, 116, 157.7, 334.2 nsec, respectively. A frame is 57 mm in diameter. Gas pressure ~ 13 torr.

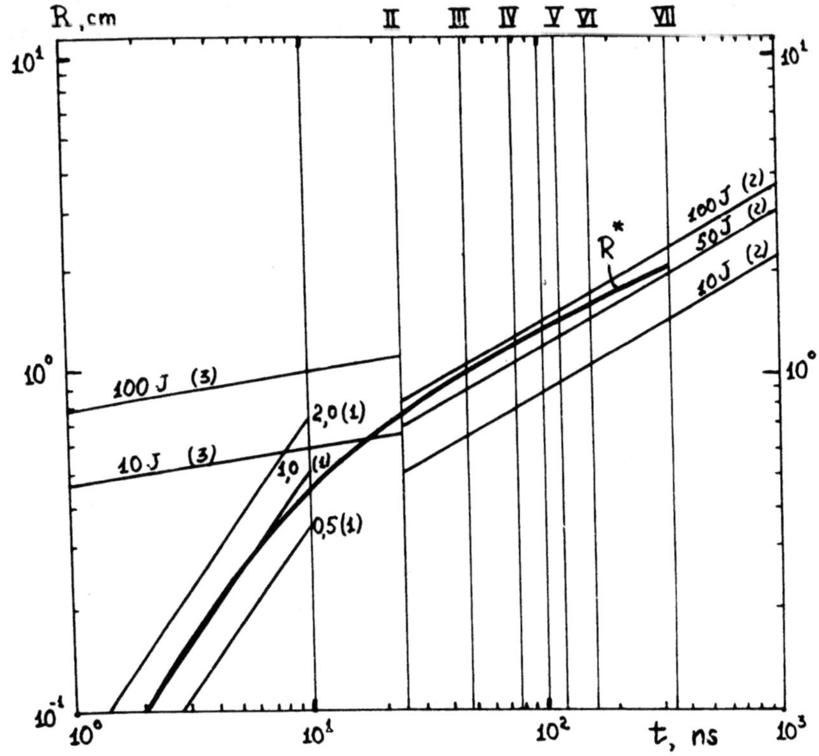

FIGURE 3. R-t - diagrams for a shock wave in the case of a spherically irradiated target by 9 beams. Numbers of frames are indicated by roman numerals (Fig. 2). Indices in brackets manifest the following processes: 1 - initial stage of SW propagation; 2 - later stage of expansion (point explosion); 3 - thermal wave.

with Eq. 9 we observe that temperature T in the target corona should be equal to about 1 keV. It is in a good agreement with other measurements and evaluations.

Self-Similar Stage of SW Motion
Measurements of Energy Deposition

If the study of SW initial stage formation permits us to determine laser plasma temperature at the end of heating, the study of SW at the later stage of expansion makes it possible to define the energy absorbed by the spherically irradiated target. Notes that due to a strong refraction it becomes actually impossible to use the methods of calorimeters and photometers for measuring absorbed energy via the reflected one. Such a method of energy deposition measurements has been proposed and developed in Ref. 21.

There has been experimentally measured an R-t diagram of the SW expansion in the rarefied gas (≃ 1 - 15 torr) by means of a laser beam high speed multi-frame photography (see Fig. 1). Then the R(t) experimental dependence at the later stage when the SW gas mass exceeds the target mass, is compared with a theoretical set of diagrams which depicts the self-similar solution for the problem of a point explosion[30]

$$R = \xi_o\left(\frac{E}{\rho_o}\right)^{1/5} t^{2/5} \tag{10}$$

where t - time from the instant of explosion of energy E, ρ_o - density of unperturbed gas, ξ_o - undimensional factor depending on the adiabatic index γ by approximation[31]

$$\xi_o(\gamma) \simeq 1.08\left[\frac{(\gamma^2 - 1)(\gamma + 1)}{(3\gamma - 1)}\right]^{1/5} \tag{11}$$

In case of a fully ionized plasma ($\gamma = 5/3$) it gives $\xi_o \simeq 1.12$.

The results of processing of SW shadowgrams in medium at 13 torr pressure are shown in Fig. 3, where the R-t diagram is for the effective SW radius

$$R^* = \left(\frac{3}{4\pi}\int_{(v)} r^2 \sin\theta \, dr d\theta d\phi\right)^{1/3} \tag{12}$$

where $\vec{r}$ - radius vector of the point in the moving gas to the given time. Here are also plotted the curves describing estimation results for the point explosion (using Eq. 10). Numerals above the curves are for the explosion energy in J. Comparing theoretical and experimental curves in the time range of 50 - 250 nsec, where the law $R \sim t^{0.4}$ (Eq. 10) is well fulfilled, we obtain the value of the absorbed energy $E \simeq 80$ J. At $t > 250$ nsec a certain slowing of the radius growth ($R \sim t^{0.2}$) is observed.

Besides Fig. 3 shows R-t curves calculated from Eq. 9, depicting SW initial stage for different values of plasma temperature (0.5, 1.0, 2.0 keV) to the moment of SW formation.

Figure 3 shows also the estimated values of the heat wave parameters, using Eq. 2, for the residual D_2 gas pressure $p \simeq 13$

torr at the instantaneous point thermal yield of 10 and 100 J. It can be easily seen the thermal wave of the second kind within the time $t_T \sim r_0/\dot{R}_T$ which is much less than 10^{-9} sec of characteristic gas dynamic time turns to the regime which is characterized with a weak time dependence.

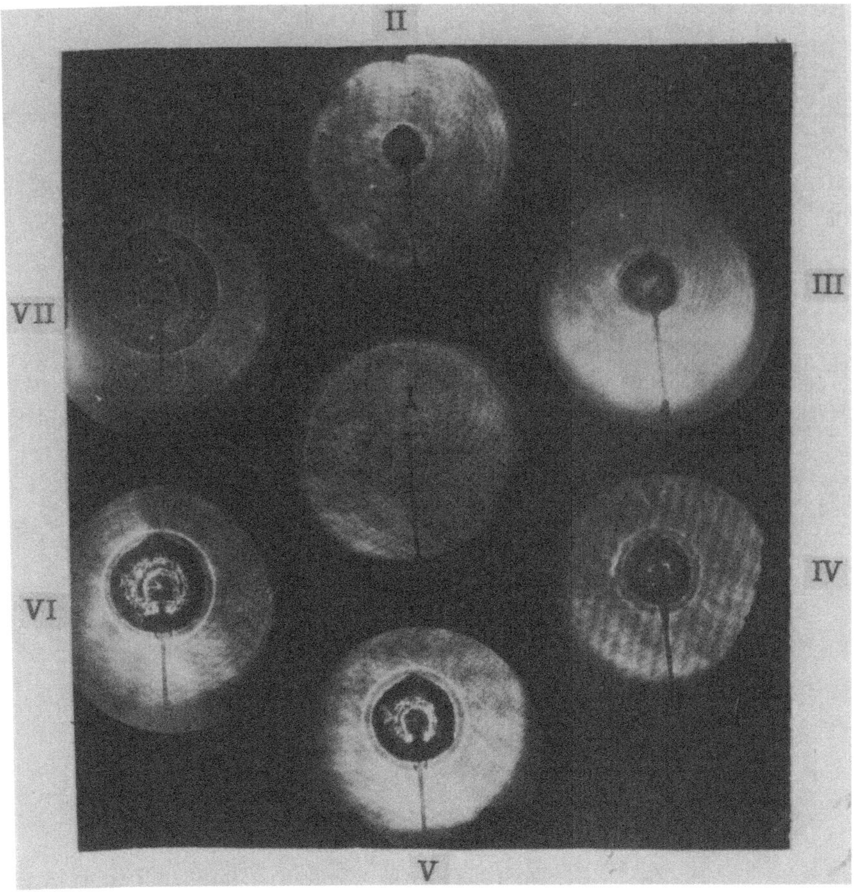

FIGURE 4. SW shadowgram. The moments of frame exposures correspond to the following delays with respect to the beginning of heating: I-2, II-20, III-45, IV-70, V-140, VI-210, VII-460 nsec. A $(CH_2)_n$ target is a ball with r_0 = 250 μm. Laser energy is ~ 560 J. Laser pulse duration is ~ 2 nsec. Pressure of the residual gas is ~ 15 torr (air). A frame is 57 mm in diameter.

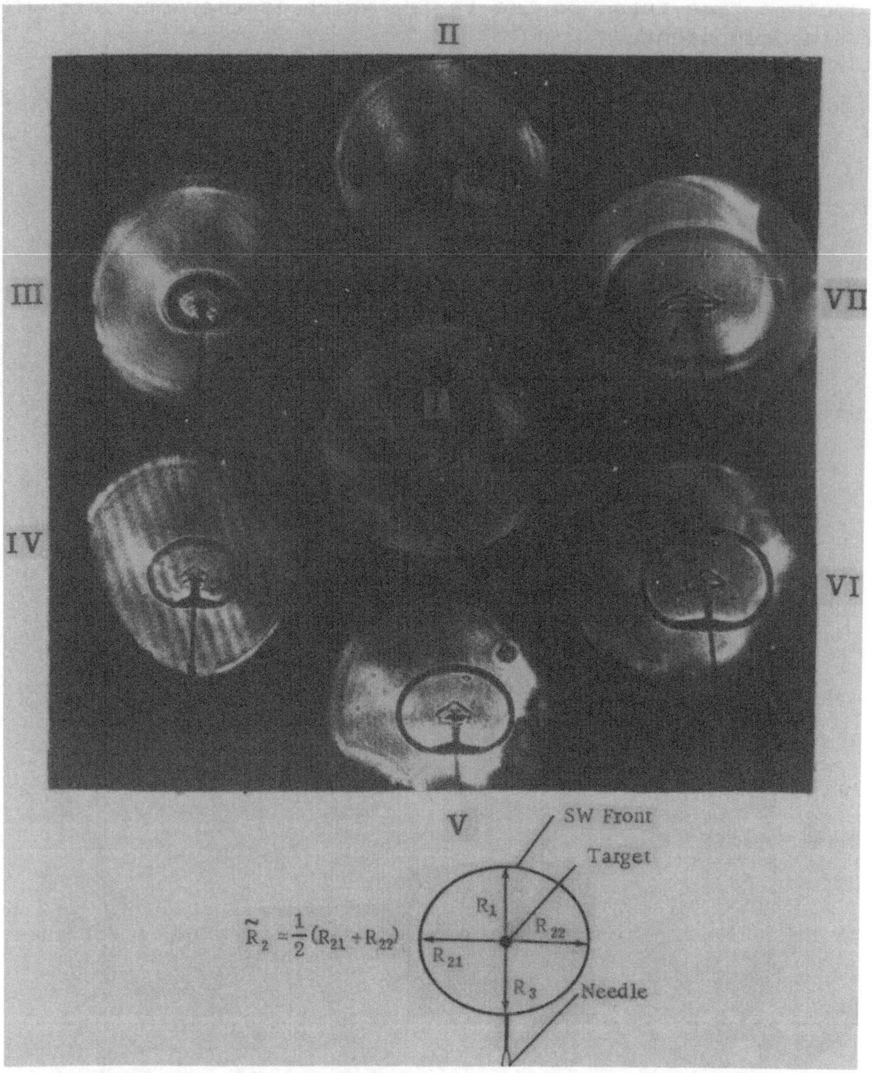

FIGURE 5. SW̃ shadowgram in deuterium. Delay of the first frame exposure with respect to the beginning of heating is 22 nsec. Time intervals between frames are the same as in Fig. 2. $(CD_2)_n$ target is a ball with r_o = 125 μm. Laser energy is ~ 310 J at pulse duration ~ 3 nsec. Pressure of the residual gas is ~ 12 torr.

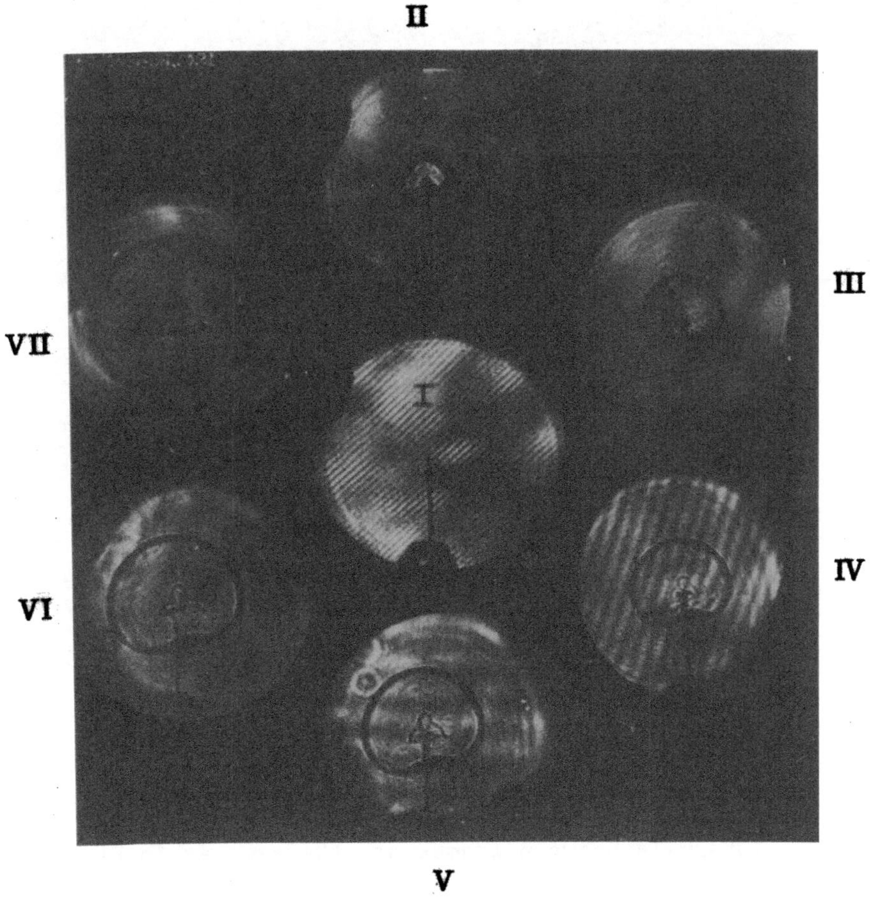

FIGURE 6. SW shadowgram in deuterium. The moment of the first frame exposure coincides with the beginning of heating with accuracy ~ 1 nsec. Time intervals between frames are 24.8, 24.8, 24.8, 41.6, 41.7, 176.5 nsec. $(CD_2)_n$ target is a ball with r_o = 250 μm. Laser energy is ~ 330 J at pulse duration 8 nsec. Pressure of the residual gas is ~ 14 torr.

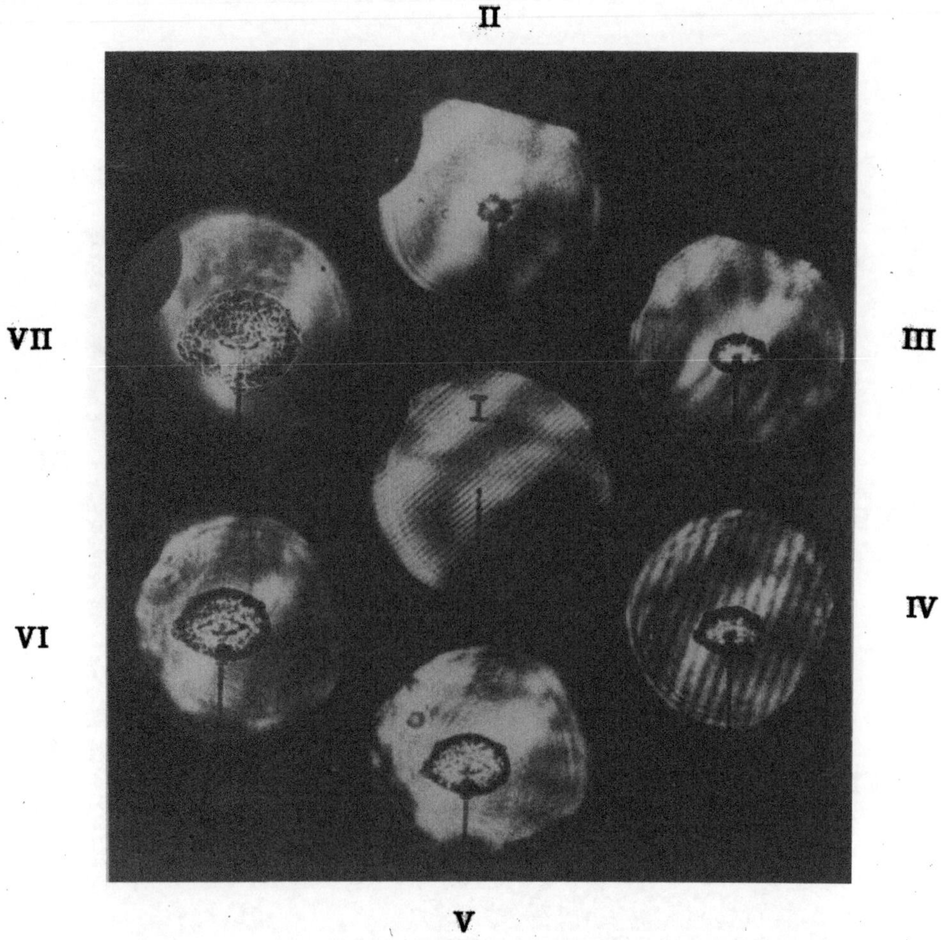

FIGURE 7. SW shadowgram in deuterium at pressure of 35 torr. Delay of the first frame exposure is 2 nsec. Time intervals between frames are the same as in Fig. 2. $(CD_2)_n$ target is a ball with r_o = 28 μm. Laser energy is ~ 360 J at pulse duration 4 nsec.

Let us discuss in detail the method of measuring absorbed energy in the target. Characteristic SW shadowgrams obtained for various experimental conditions are shown in Figs. 4-7. Figures 5, 6, 7 illustrate possible deviations of SW shape from the spherical one. Figure 8 shows R-t curves for R_1, $\tilde{R}_2$, R_3 and R* (Eq. 12) obtained by processing photographs shown in Figs. 4-7 (denotations are explained by the scheme in the same figure).

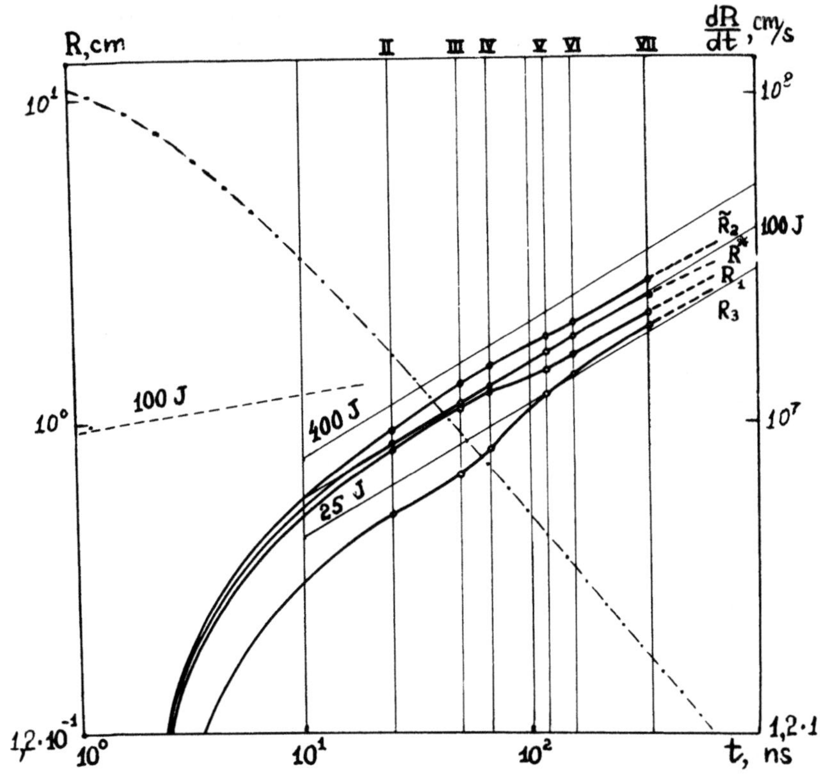

FIGURE 8. Characteristic R-t curve of SW in deuterium at laser energy of 330 J and pulse duration of 2 nsec. $(CD_2)_n$ target is a ball with r_0 = 130 μm. Deuterium pressure is 10 torr. The rate of SW front is plotted on the right of the ordinate axis. Numerals above the lines stand for the energy of implosion in J for a model of point explosion (on the right), and thermal energy for a model of momentum point thermal yield (on the left). Dashed curve illustrates the rate of SW front.

Roman numerals on the abscissa indicate the number of frame. Thin solid lines on the right correspond to the self-similar solution of the problem for point explosion of energy 25, 100 and 400 J and they determine energy absorbed by plasma. Figure 8 shows also time dependence of the effective velocity ($\dot{R}^*$) of shock.

From the noted curves it is seen that at the early stage of SW expansion the law of motion rather differs from the self-similar solution. The SW motion proceeds according to the law close to self-similar one just at the later stage beginning from the time when the SW gas mass becomes much greater than the evaporated target mass. Such SW motion is due to the fact that in case of SW formed by laser at the early stage the largest fraction of released energy converts into gas dynamic motion of the heated target. And in some time this energy is transformed to the energy of gas where SW propagates. For instance, evaluations show that by the time t_{in} when the moving gas mass is 5 times greater than the target mass, the energy contained in gas dynamic target motion is about 20%. This time is determined by experimental conditions, and when we use D_2 gas we have

$$t_{in} \simeq 1.4 \cdot 10^8 r_o^{5/2} p^{-1/3} E^{-1/2}, \text{ ns} \tag{13}$$

where r_o - initial target radius in cm, p - gas pressure in torr, E - explosion energy in J.

Another source of errors when measuring plasma energy is an intensive X-ray radiation from hot dense plasma during heating. Due to estimation and experimental results to be discussed as follows, the X-ray radiation energy may achieve ≃ 15-25% of the value of absorbed energy. This kind of radiation is slightly absorbed by the gas surrounding target, and it does not affect noticeably the process of SW formation.

Thus, the energy estimated from R-t curves gives a lower value for energy released in the target. To determine an actual value of the absorbed energy one should make corrections with regard to the mechanism of losses. In this work we measured radiation losses by means of X-ray calorimeters, and the fraction of energy in the gas dynamic flare motion was estimated from calculations. Thus, for example, the energy from R-t diagram of Fig. 8 is about 100 J, at flare motion energy of ~ 15 J, and X-ray energy 65 J. Hence, the total value for the absorbed energy is ~ 180 J. It corresponds to ~ 72% of the incident laser beam energy which is of ~ 250 J. Measuring accuracy of the absorbed energy is not worse than 15%. The rest of energy goes apparently for the reflection of laser radiation. Estimations show that coefficient of reflection in the described experiments is not greater than 10% of the incident energy.

At measuring absorbed energy by the shock wave method it is important to choose a correct time interval where one will be able to compare R-t diagrams with the law of SW motion for the point explosion in the homogeneous medium. At later stage of motion the law of SW propagation may differ from the point explosion due to the increased losses for dissociation and ionization of gas molecules, and because of radiation from the gas in SW. Sum energy losses on ionization and dissociation to the moment t are described by:

$$E_{LOS} \simeq 1.4\times10^{8}E^{3/5}p^{2/5}t^{6/5} \qquad (14)$$

at $E = 10^2$ J, p = 10 torr, t = 3×10^2 nsec, losses E_{LOS} are of ~ 25 J.

Gas radiation at the later stage may be evaluated supposing the total gas mass concentrated in the layer of $\delta R \simeq R(\gamma - 1)/3(\gamma + 1)$ thickness at the SW front. Using a quasiclassical equation for the power of bremsstrahlung and recombination radiation from the fully ionized D plasma, energy dependence on time can be expressed as:

$$E_{Rad} \simeq E^{ff}_{Rad} + E^{fb}_{Rad} \simeq \frac{3\times10^{7}}{(\gamma - 1)^{1/2}} E^{4/5}p^{6/5}t^{8/5} + \frac{1.05(\gamma + 1)^{2}\cdot10^{19}}{(\gamma - 1)^{3/2}} E^{2/5}p^{8/5}t^{14/5} \qquad (15)$$

where the first term of the equation corresponds to bremsstrahlung radiation, and the second to recombination. At the given values of E, p and $\gamma = 5/3$ for the time t = 500 nsec this equation gives $E_{Rad} = 10$ J; when ≃ 1.9 J for bremsstrahlung radiation and ≃ 8.1 J for recombination radiation.

It should be noted that Eq. 15 is valid for the totally ionized plasma. In the case of D it occurs at temperature of several eV. Plasma temperature in the radiating layer near the point of explosion is described by

$$T \simeq \frac{4.92\times10^{-7}(\gamma - 1)}{(\gamma + 1)^{2}} E^{2/5}p^{-2/5}t^{-6/5} \quad \text{eV} \qquad (16)$$

from which it follows that at time variation from 10^2 to 10^3 nsec temperature falls from 30 to 2 eV (at t = 500 nsec, T ≃ 4.5 eV). At such a temperature and pressure determined by limiting compression in the SW front the continuous radiation is of ~ 50% of the total adiative losses in lines[32]. Approximating the ratio of the linear radiation power to the sum radiative power by hyperbolical dependence for a total radiated energy we obtain at p ≤ 50 torr and T ≃ 3-30 eV:

$$E_o = E_{Lin} + E_{Rad} \simeq \frac{3\times10^7}{(\gamma - 1)^{1/2}} E^{4/5}p^{6/5}t^{8/5} \tag{17}$$

$$\frac{[(\gamma - 1)E^{2/5} + 3.5\times10^7(\gamma + 1)^2 p^{2/5}t^{6/5}]}{[(\gamma - 1)E^{2/5} - 6.1\times10^6(\gamma + 1)^2 p^{2/5}t^{6/5}]}$$

From Eq. 17 it is seen that radiation losses sharply increase with SW motion. It leads, consequently, to the variation of SW behavior and to the greater disagreement with the self-similar solution. Thus Eqs. 14 and 17 together with Eq. 13 define the time intervals, in which the motion of shock wave corresponds to the theory of point explosion.

Note, that at the fixed time of observation of SW motion the value of gas pressure is restricted. Equations 13, 14 and 17 define the lower and upper limits of pressure respectively. In our experiments at t_{max} = 334 nsec for r_o = 100 μm and $E_{abs} \simeq$ 100 J the value of pressure could vary in the limits from $p_{min} \simeq$ 5 torr to $p_{max} \simeq$ 15 torr.

Now we consider experimental results. Absorption of the optical radiation by dense plasma was investigated at laser energy of ≃ 300 J and pulse duration of ≃ 1.5 - 16 nsec. The size of deuterized polyethylene target r_o varied from 30 to 250 μm. Efficiency of energy absorption into the laser plasma versus light pulse parameters was experimentally studied. Efficiency is

$$\eta = \frac{E_{abs}}{E_{incid}} = \frac{E_{abs}}{E_L - E_{refl}} \tag{18}$$

where E_{abs} - laser beam energy absorbed by the target, E_{incid} - target incident energy, E_L - laser radiation energy, E_{refl} - energy losses in the focusing system. Figure 9 illustrates the dependence of energy deposition efficiency on the target radius. Laser radiation pulse duration τ_L is for a parameter of curves. From the given curves it is seen that for a target with $r_o \gtrsim$ 100 μm efficiency of energy absorption reduces with the increase of pulse duration. Thus, for instance, for a target with r_o = 120 μm at τ_L = 1.5 nsec efficiency of absorption is of ≃ 75%, and at the increase of duration up to 8 nsec its value decreases down to ≃ 40%. Perhaps such dependence is associated with the target bleaching, when beginning from a certain moment the optical radiation penetrates plasma without noticeable absorption. This effect can be explained by a rapid decrease of plasma electron density at gas dynamic expansion to the value less than the critical density

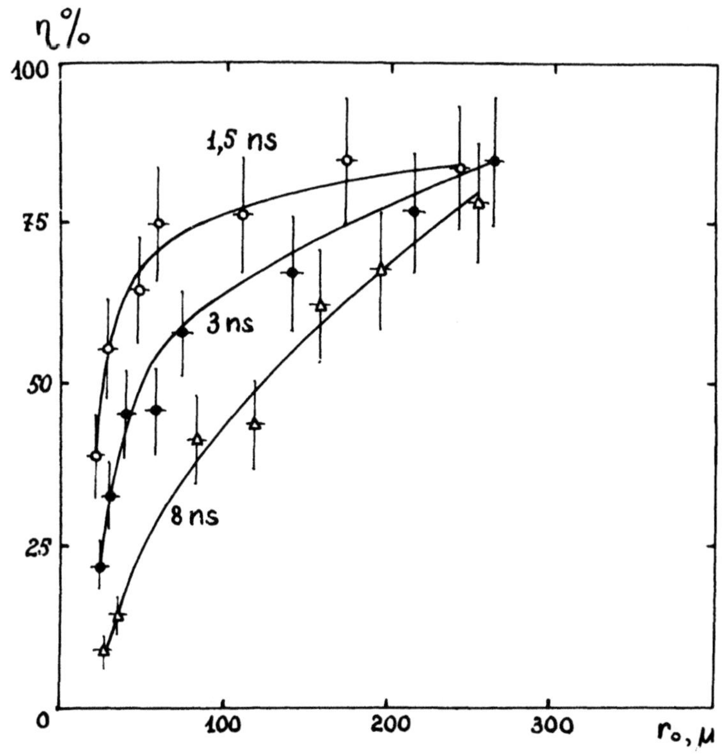

FIGURE 9. Efficiency of energy deposition into plasma versus initial radius of the target at various laser pulse durations. Pressure of deuterium is ~ 8-12 torr. Laser energy is ~ 290 - 350 J.

for the heating radiation frequency. Sharp decrease of absorbed energy for the targets with $r_0 \leq 50$ μm is apparently due to the errors of laser radiation target focusing, since the target size becomes in this case comparable with the size of a focal spot. And the probability for a fraction of radiation to propagate without absorption increases.

Figure 10 shows the dependence of radiation absorption efficiency on pulse duration.

The size of a target is for a parameter of curves. One can pay attention on a weaker decrease of absorption at great target dimensions. Such a behavior of curves is connected with the increase of characteristic gas dynamic time $t \sim r_0/c$ (c - sound velocity in plasma) when we substitute small target by large one and, consequently, with the decrease of "bleaching" influence.

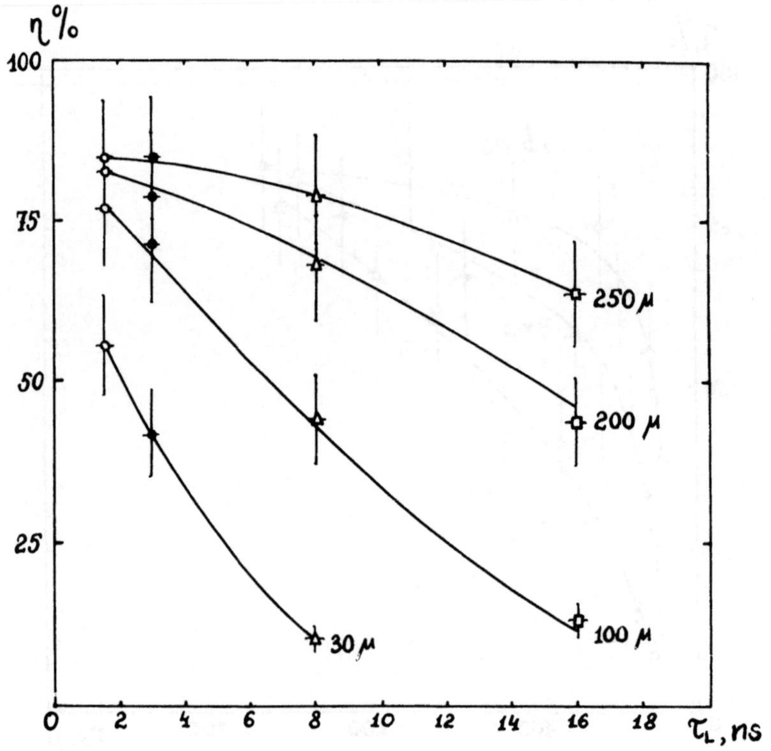

FIGURE 10. Efficiency of energy deposition versus pulse duration at various dimensions of the target. Pressure of residual gas (deuterium) is ~ 8 – 12 torr. Laser energy is ~ 320 J.

The data on the energy deposition into plasma at the known composition and size of the target allow to determine average energy per particle $\bar{\varepsilon}$. The related results are represented in Fig. 11 like $\bar{\varepsilon}$ dependence on the radiation flux density. Flux density was changed by the target size variation and the quality of beam focusing which provides homogeneous target irradiation. Pulse duration was fixed in these measurements and it was ≃ 1.5 nsec. From this curve it is seen that average energy per particle increases nearly proportionally to the flux variation $\bar{\varepsilon} \sim q$. The maximum value is obtained for $q \simeq 10^{15}$ W/cm^2 and is of about 10 keV.

Figure 12 illustrates the dependence of the efficiency on flux density at various pulse durations. The increase of flux density was realized by the decrease of target dimensions, with consequent decrease of energy deposition. From the curves it is seen that the most effective plasma heating occurs at target irradiation with the minimal pulse duration (≃ 1.5 nsec).

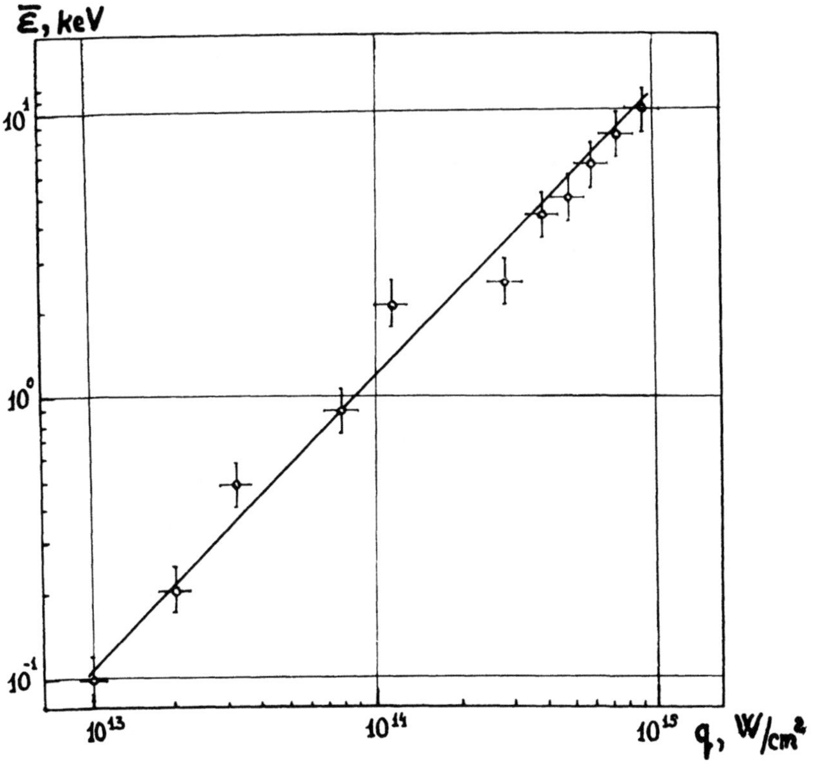

FIGURE 11. Average energy per particle in plasma versus laser flux density. Laser energy is ≃ 330 J at pulse duration of ~ 1.5 nsec. Pressure of deuterium in the chamber is ≃ 10 torr.

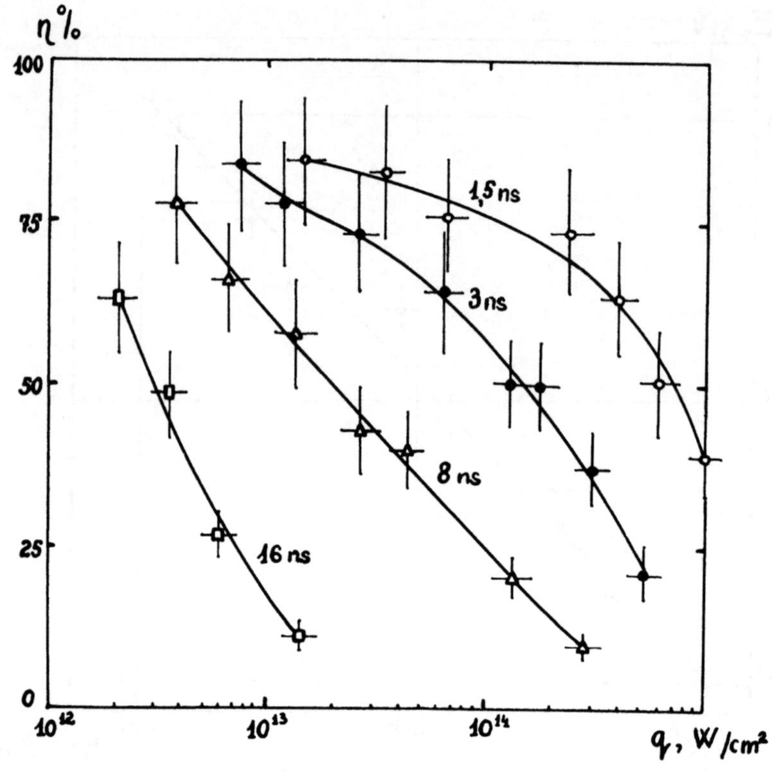

FIGURE 12. Efficiency of energy deposition versus laser flux density. Pressure of deuterium is ~ 8 - 12 torr. Laser energy is ~ 330 J. Dimensions of targets are from 30 to 250 μm.

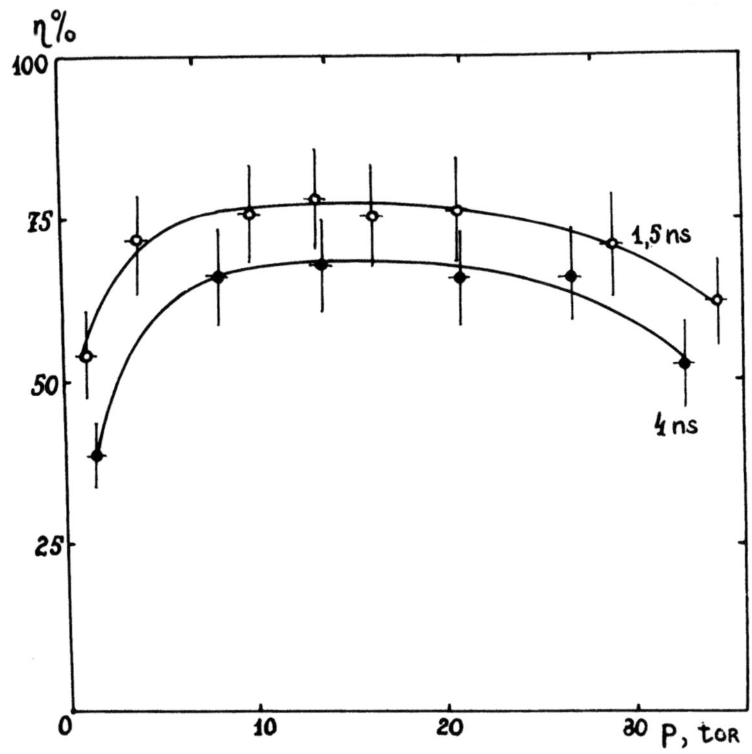

FIGURE 13. Experimental dependence of energy deposition efficiency on the pressure of residual gas at pulse duration of 1.5 nsec and 4 nsec. Laser energy is ≃ 330 J. Initial radius of the target is ≃ 120 μm.

In conclusion, consider the experimental curve of Fig. 13 illustrating the experimental dependence of absorption efficiency of laser radiation with 1.5 and 4 nsec pulse duration into the target with r_o = 120 μm on D_2 gas pressure. This curve illustrates the errors arising when determining absorbed energy by incorrect selection of the gas pressure. Note, that intolerable errors of the method of point explosion arise at small pressures ($p < 5$ torr) when due to Eq. 13 within the total time of registration the mass of moving gas has no time to exceed the target mass. In case of too large pressures ($p > 15$ torr) the figures of absorbed energy are also decreased due to the reasons noted above.

III. X-RAY PLASMA DIAGNOSTICS BY CONTINUOUS RADIATION

In our experimental conditions the temperature of plasma produced at spherical irradiation of the isolated $(CD_2)_n$ solid target with mass 10^{-6}-10^{-7} g at ~ 10^2-10^3 J energy achieves ≃ 1-10 keV. At such temperatures plasma becomes fully ionized and it radiates intensively in the X-ray spectrum. Note, that in the case of the optically thin plasma the radiation occurs from the total volume with maximum wavelength intensity $\lambda_{max} \simeq 6.2\ T^{-1}$ (T in keV). The main mechanisms of radiation are bremsstrahlung (ff) and recombination (fb) emission with electrons recombinating on the ground (n = 1) and first (n = 2) excited levels. Spectral power of this radiation is described by:

$$I_\lambda \simeq 6\times10^3 \frac{\tilde{n}_e^2(\bar{Z}^2+2)}{(\bar{Z}+2)\lambda^2 T_e^{1/2}} \exp\left(-\frac{12.4}{\lambda T_e}\right)\cdot\Big\{\bar{g}^{ff} + \tag{19}$$

$$+ \frac{3.3\times10^{-3}\bar{Z}^2}{T_e}\left[\alpha\bar{g}_1^{fb}\exp\left(\frac{1.36\times10^{-2}\bar{Z}^2}{T_e}\right) + \beta\bar{g}_2^{fb}\exp\left(\frac{3.4\times10^{-3}\bar{Z}^2}{T_e}\right)\right]\Big\}$$

$$\mathrm{GW/cm^3\,\AA}$$

where $\tilde{n}_e = 10^{-21}\ n_e$ - reduced electron density, $\bar{Z}$ - mean ion charge, λ - radiation wavelength in Å, T_e - electron temperature in keV, $\bar{g}^{ff}$ - Gaunt factor, averaged by Maxwellian distribution, $\bar{g}_n^{fb}$ - Gaunt factor averaged by the orbit quantum number for n = 1; 2; coefficients α and β have regard to deposition of recombination radiation at $\lambda \leq 912\ n^2\bar{Z}^{-2}$ Å wavelengths and correspond to:

$$\alpha = \begin{cases} 8, & 0 < \lambda \leq 9.12\bar{Z}^{-2} \\ 0, & \lambda > 9.12\bar{Z}^{-2} \end{cases}$$

$$\beta = \begin{cases} 1, & 0 < \lambda \leq 3.65\bar{Z}^{-2} \\ 0, & \lambda > 3.65\bar{Z}^{-2} \end{cases}$$

One can thus determine the electron plasma temperature by the dependence of radiation intensity on wavelength. The most widespread method of X-ray spectrum study is the method of filters[33] according to which the intensities of radiation from various absorbed filters are compared with different curves of transmittance.

But one should remember that at high flux densities $q \simeq 10^{15}$ W/cm^2 there can occur deviations of the electron velocity distribution from Maxwellian one. It may lead to significant errors in determining electron temperature T_e in such a way[34].

In our experiments we measured the ratios of intensities of X-ray radiation transmitted through Be-filters of different thickness. Light scintillation pulses were registered by a photomultiplier. The required synchronization of light pulses was provided by a delay produced on the basis of a filament optics. Characteristic pulse oscillogram is represented in Fig. 14.

The measurements of the absolute X-ray intensity in various spectral regions were simultaneously made. For this purpose a 4-channel X-ray calorimeter was used. Each channel registered radiation energy through beryllium filters of 10, 200, 400 and 800 μm thickness. It enabled us to determine the total emitted energy E_{iR} from:

$$E_{iR} = \xi_c E_c \tag{20}$$

where E_c - energy in one of the channels, ξ_c - the factor allowing for geometry and spectral content of radiation

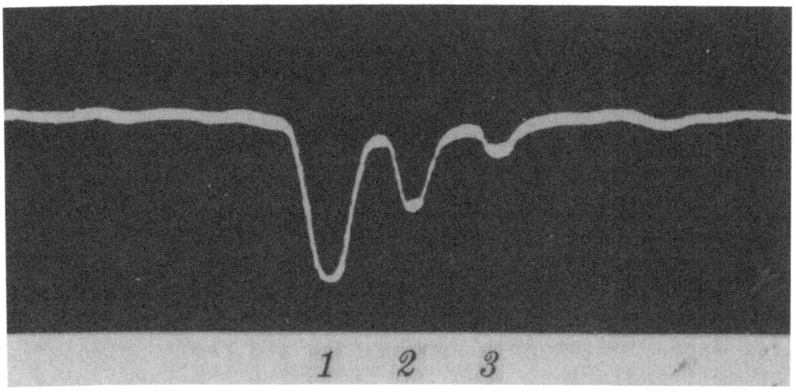

FIGURE 14. Oscillogram of X-ray pulses transmitted through Be-filters of thickness 1-100 μm, 2-200 μm and 3-500 μm. Lengths of light pipes are, respectively, of 1 m, 4 m and 7 m.

$$\xi_c = \frac{4\pi R_c^2 \int_0^\infty \int_0^t r^3(t') I_\lambda(t') d\lambda dt'}{S_c \int_0^\infty \exp\left[-\mu_c(\lambda)\delta_c\right] \int_0^t r^3(t') I_\lambda(t') d\lambda dt'} \tag{21}$$

Here R_c - distance from the target to radiation detector, S_c - square of detector μ_c and δ_c - mass coefficient of absorption and filter thickness.

Radiative power from the plasma volume unit in continuous spectrum for the spectral density (Eq. 19) may be described as:

$$P_{rad} \simeq 4.86\times10^2 \tilde{n}_e^2 \frac{\bar{Z}^2 + 2}{\bar{Z} + 2} T^{1/2}\left(1 + \frac{0.03\bar{Z}^2}{T_e}\right) \quad \text{GW/cm}^3 \tag{22}$$

Estimations of radiation from spherical plasma were done at the following assumptions: plasma density and temperature are constant all over the volume, $\rho(\vec{r})$ = const, $T(\vec{r})$ = const, and the velocity of gas dynamical expansion is linearly changed with radius

$$r\frac{dr}{dt} = v\frac{r}{r_o}$$

And the system of equations describing plasma parameters in time has a form:

$$\begin{gathered} T = \alpha_o r \frac{d^2 r}{dt^2} \\ \frac{dT}{dt} = -\frac{dr}{dt}\frac{d^2 r}{dt^2} + \beta_o(P_L - P_{Rad}) \\ T(0,r) = 0; \ \frac{dr}{dt}(0,r) = 0; \ r(0) = r_o \end{gathered} \tag{23}$$

where T - temperature in keV, P_L - laser radiation power in GW, P_{Rad} - radiative power in GW, $\alpha_o = 3.02 \times 10^{-2}$, $\beta_o = 4.2 \times 10^{12} r_o^{-3}$. At earlier expansion stage plasma was considered to be optically thick for self-radiation, when the radiative power P_{Rad} was described by

$$P_{Rad} \simeq 1.3 \times 10^9 r^2 T^4 \quad \text{GW} \tag{24}$$

For the heating radiation plasma was considered to be entirely absorbing to the moment when the laser radiation quantum run in plasma becomes comparable with characteristic plasma dimensions. At later stages in the system of equations (23) it was assumed that $P_L = 0$.

Figure 15 illustrates power and energy dependences on time obtained from the solution of a system (Eq. 23). Radiative power was assumed constant for various pulse durations which were introduced as curve parameters. Initial target radius was of 10^{-2} cm. From these curves it is seen that total energy losses due to radiation may achieve ≈ 25% of the incident energy.

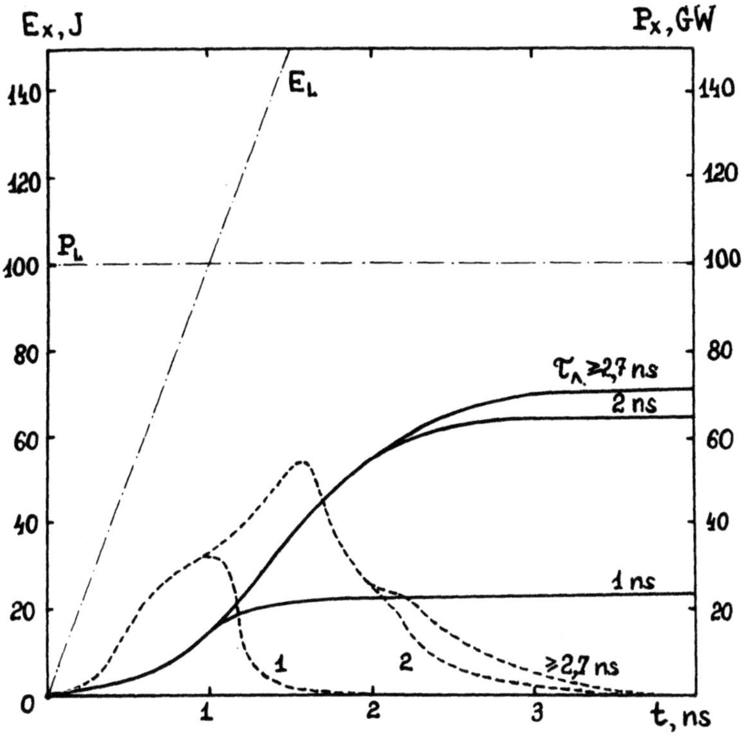

FIGURE 15. Power (dotted curves) and energy (solid curves) dependences of X-ray radiation from plasma. Estimation for P_L = 100 GW, $r_o = 10^{-2}$ cm, and τ_L = 1, 2, 2.7 nsec. Numerals above curves indicate laser pulse duration in ns.

Figure 16 shows the dependence of X-ray radiation energy from plasma on the target dimensions for the same laser pulse durations. To each duration there corresponds a certain target radius at which maximum of energy is radiated. The same figure illustrates the experimentally measured X-ray radiation from plasma, at laser pulse duration of 2 nsec and radius r_o = 130 μm. This value of 65 J exceeds calculations for the same laser parameters and target radius.

Measurements of plasma electron temperature for this case by the method of absorbing filters give the value of 1 keV, which is in an agreement with calculation results. Temperature estimations were carried out in all other experiments. They indicate that plasma electron temperature decreases rapidly at the increase of

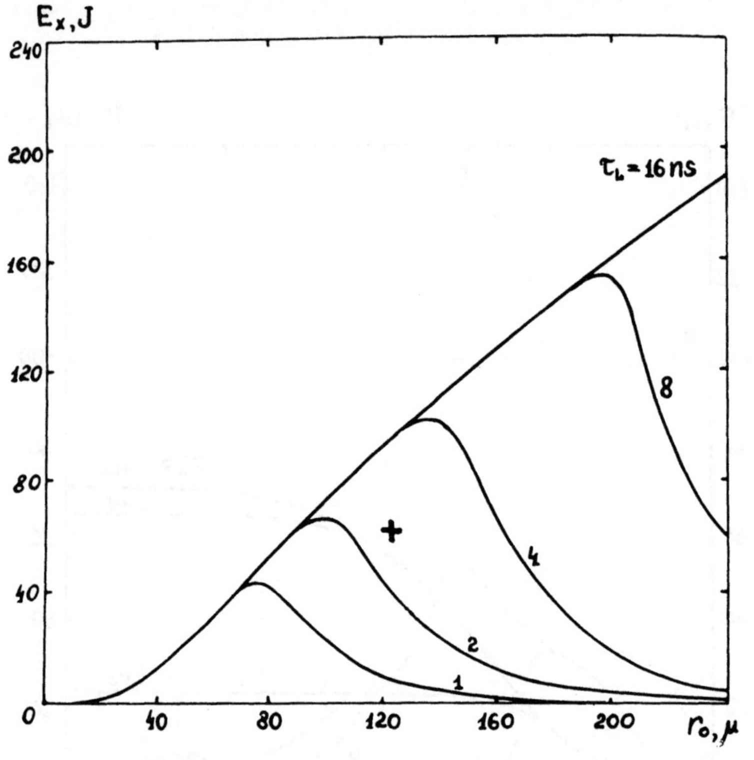

FIGURE 16. X-ray energy irradiated from plasma versus initial target radius at laser pulse duration of 1, 2, 4, 8 and 16 nsec (numerals above curves). Experimental value for energy irradiated from the target with r_o = 130 μm at laser energy of 270 J and pulse duration of 2 nsec is marked by "+".

target size and the decrease of laser power. Maximum temperature was achieved at irradiation of the target $2r_0 = 60$ μm in diameter, pulse duration $\tau_L \simeq 1.5$ nsec and radiative energy $E_L \simeq 250$ J, and it was $T_e = 6$ keV.

IV. INVESTIGATION OF NEUTRON YIELD

The neutrons from plasma were registered by three scintillation counters located at different distances from the target. Taking into account time resolution of 5 nsec it became possible to perform time of flight measurements of neutron energy. Each photomultiplier was connected with light pipe. Auxiliary laser light pulse synchronized with the beginning of heating went into light pipe. Typical oscillograms are shown in Fig. 17.

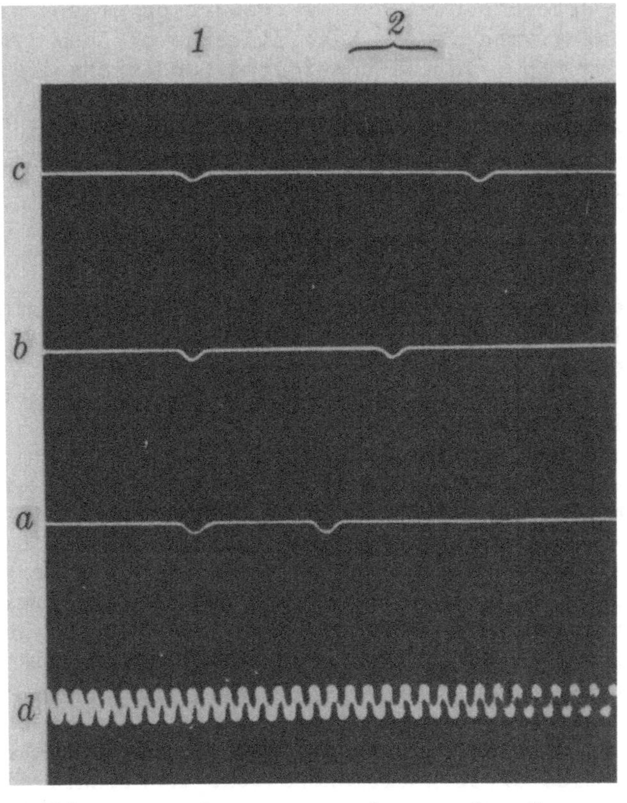

FIGURE 17. Oscillograms of neutron pulses. 1 - laser pulse, 2 - neutron pulse, a - pulse from a photomultiplier located 1.15 m from the target, b - pulse from a photomultiplier at the distance of 1.95 m, c - pulse from a photomultiplier at the distance of 3 m, d - marks of time of 10 ns.

Neutron energy determined from the time of flight was of 2.45 MeV. This value corresponds, to energy of neutrons produced by D-D reaction with temperature of several keV. It should be also noted that no anisotropy of the neutron emission was observed. These data evidence for thermal origin of neutron emission.

In this work there have been also measured the width of neutron energy spectrum by signal broadening from the neutron scintillation detector located at the distance L = 3 μm from the target. Pulse duration of the corresponding neutron signal from the detector Δt_n is connected with the neutron pulse width ΔE_n by relation

$$\Delta E_n = 2.74 \times 10^{-2} E_n^{3/2} \frac{\Delta t_n}{L} \quad \text{MeV} \qquad (25)$$

where E_n - neutron energy in MeV, Δt_n in ns. The lower limit of values Δt_n is determined with time resolution of the registration system (≈ 5 nsec), and at the base distance of 3 μm it gives an upper value for ΔE_n. In the considered conditions it is equal to ~ 170 keV. The experimentally measured width of neutron spectrum allows to evaluate the ion plasma temperature T_d, which in the case of Maxwellian ion distribution, is approximately connected in velocities with the neutron spectrum width by the ratio:

$$\Delta E_n \simeq 82.5 T_d^{1/2} \quad [\text{keV}] \qquad (26)$$

The performed estimations show that the ion temperature of plasma produced at irradiation of $(CD_2)_n$ target, $2r_o$ = μm in diameter, by a light pulse with energy ≈ 250 J and pulse duration ≈ 1.5 ns does not exceed 5 keV. This value shows a thermal origin of neutron emission in laser plasma.

Quantitative measurements of the neutron yield were performed by means of nuclear photoemulsions at observing recoil protons[23]. A type NIKFI-R emulsion 300 μm thick was placed at 6 cm from the target. After three experiments in which neutrons were recorded by the scintillation counters, irradiated and control emulsions were developed simultaneously under the same conditions. The control emulsion of the same batch was located at a larger distance. The irradiated emulsion had 87 tracks per cm^2, corresponding to the recoil protons from the neutrons of the $D(d,n)He^3$ reaction, while the control emulsion had 48 tracks. To verify that the photographic properties of both emulsions were identical, the number of stars in the same square centimeter was measured and it turned out to be 45 and 49, respectively. It follows, therefore, that if the neutrons are assumed to be isotropically emitted from plasma, the number of neutrons within three flashes is of 10^7. The minimum neutron yield is consequently equal to ≈ 3×10^6 neutr/pulse. The fact that there were registered the $\vec{R}_i$ tracks of those recoil protons which corresponded to neutrons from the laser produced plasma was especially well manifested if we summarized projections of all tracks on the direction of neutron flux $\vec{n}$ and perpendicular to its direction $\vec{n}_\perp$ in the emulsion plane. The ratio of sums of the corresponding projections which characterizes a degree of angular anisotropy is:

$$K = \frac{\sum_{i=1}^{N} (\vec{n},\vec{R}_i)}{\sum_{i=1}^{N} (\vec{n}_\perp,\vec{R}_i)} \qquad (27)$$

where N - the number of tracks per square centimeter of emulsion turned out to be equal to K_{irrad} = 1.28 for irradiated emulsion and K_{cont} = 0.98 for the control emulsion.

It should be noted that neutron emission under experimental conditions has nonregular character. Thus in some experiments, at a considerable energy deposition into the target and plasma temperature of the order of several keV the neutron yield, in the limits of experimental accuracy, has not been recorded. Such non-regular experimental results may be due to the deviations from the spherical shape of targets and a certain inhomogeneity of its irradiation.

V. REFLECTION OF LASER RADIATION

A fraction of reflected from the target radiation during target heating was measured in two channels of a multi-beam laser. In one of them there have been measured amplitude-time characteristics of the reflected signals by means of a coaxial photodiode with $\simeq$ 0.5 nsec resolution. Small portions of the incident and reflected radiation separated by divider, with time delay which provided separation of signals (see Fig. 1) which went from coaxial diode. Typical oscillogram is shown in Fig. 18. The shape of reflected signal is seen to differ from that of the incident one. Duration of its leading edge is varied most essentially. It is due to the fact that coefficient of reflection depends sharply on the gradient of electron density the value of which varies rapidly in time during plasma heating. Measurements of absolute energy of reflected radiation in another laser channel were performed by the method of calorimeters. They were chiefly to define efficiency of high-power laser radiation absorption by dense plasma. The measurements have indicated that when flux density at the target surface varies in the range of 10^{13} - 10^{15} W/cm^2 the coefficient of reflection is not greater than 5 - 10% in the solid angle.

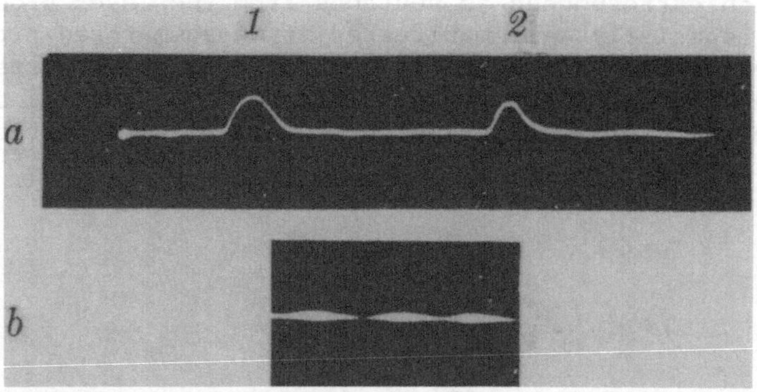

FIGURE 18. Oscillograms of laser radiation incident on the target (1,a) and reflected from it (2,a); b - marks of time of 10 ns. $(CD_2)_n$ target with r_o = 100 μm; laser energy of 330 J, at pulse duration of 4 nsec, and residual gas pressure of 6 torr (D_2).

VI. GAS DYNAMIC PRESSURE OF PLASMA "CORONA"

We used high speed interferometric methods to study spatial density distribution in the outwardly expanding ablated material and its time history. Two methods of photography were applied - both interferometric photography via electron-optical image converter[36,37] and multiframe interferometric photography.

In the first case for illumination of the interferometer there was used that fraction of laser radiation which after some amplification was transformed into the second harmonic with radiation wavelength λ = 0.53 μm.

Synchronization of interferometric system with the process under investigation was performed by laser triggered discharge gaps. The chief parameters of this system are as follows: time of scanning ≃ 30 ns, time resolution ≃ 0.1 ns spatial resolution at the object plane of 100 lin/mm, the highest value of the electron density determined by the aperture of optics was of 5×10^{19} cm^{-3}.

In our experiments we used Jamin interferometer which permitted us to make measurements in the zero order of interference. Registration of interferograms was realized when the input slit of image converter was oriented to the object in different ways. Symmetry of the plasma "corona" was controlled by comparing results obtained from interferograms for two opposite directions.

Multiframe interferometric photography presents a complete picture of plasma symmetry at the initial stage of expansion. These measurements have been also performed by Jamin interferometer in the zero order of interference. One could register seven successive frames.

Exposure time per frame was determined by light pulse duration of $\simeq$ 1.5 nsec used for plasma transmitting. Interval between frames was of $\simeq$ 3 nsec, accuracy of phase matching was not less than 10^{-10} sec.

Experimental results allowed to determine the value of time derivative of mass at gas dynamic expansion, and to evaluate reactive pressure on the target surface due to ablation. Thus, for a $(CD_2)_n$ target with $r_o = 6.5 \times 10^{-3}$ cm and $\simeq 10^{-4} - 10^{-3}$ g/cm^3 density at pulse duration of ~ 2 nsec and laser energy $\simeq$ 330 J the pressure was equal to $p \simeq 10^7$ atm. Temperature of plasma "corona" determined for this case by the method of filters and by initial stage of SW motion in deuterium was equal to about 1 keV. From these results one can deduce the value of compression in the spherically heated target.

Let us assume that laser energy is chiefly absorbed near the region of critical density ($\rho_{COR} = 3 \times 10^{-3}$ g/cm^3). From this assumption we find the maximum of electron temperature $T_{e\ max}$. Since grad T_e and grad ρ are of opposite signs the maximum of pressure is displaced relatively to $T_{e\ max}$ towards greater density, i.e. to the target center. The pressure in the "corona" is

$$P_{COR} \simeq A\rho_{COR}T_e \simeq 1.5 \times 10^6 T_e \quad [\text{atm}] \qquad (28)$$

from which at $T_e = 1$ keV we obtain value $P_{COR} \simeq 1.5 \times 10^6$ atm coinciding with the one measured experimentally. The pressure in the SW front of the target is of nearly the same value.

To evaluate pressures in the target center we shall use a self-similar solution for the convergent shockwave problem[39]. Numerical results for $\gamma = 5/3$ are represented in Ref. 40. The pressure of convergent shockwave front is[40] of

$$P \propto r^{-0.91} \qquad (29)$$

It is worth to calculate pressure at the distances from the center which are not less than characteristic dimensions of external asymmetry. In our case this size is $\simeq 0.1\ r_o$, and it is associated with accuracy of target preparation and geometry of irradiation. Pressure in the center of the target may, therefore, be greater than the pressure in the corona in

$$\frac{P_{centR}}{P} = 10^{0.91} \simeq 8 \text{ fold} \tag{30}$$

In our experiment we reached $P_{centR} \sim 10^7 - 10^8$ atm. Let us evaluate now the hydrodynamic velocity u and time of wave focusing t_{foc}. From the approximated relation

$$P \simeq \rho u^2 \tag{31}$$

it follows that $u \simeq 10^6 - 3 \times 10^6$ cm/sec and $t_{foc} \simeq \frac{r_o}{u} \simeq 2$ nsec. In this case a shockwave may be focused and then be reflected from the center and again it may return into compressed by the first wave substance. If the target substance is an ideal gas the shockwave could be considered strong and the compression behind the shock front would be limited by

$$\delta = \frac{\rho}{\rho_o} = \frac{\gamma + 1}{\gamma - 1} = 4 \quad (\gamma = \frac{5}{3})$$

But the target before the SW front is known to be a cold matter. Therefore, to determine about the power of shockwave one should compare the pressure in its front with the elastic pressure $P_{elast} \simeq \rho_o c_o^2$, where c_o - sound velocity in cold substance[31]. By interpolation of data of Ref. 31 between sound velocities of water and metals we assume for polyethylene $c_o = 3 \times 10^5$ cm/sec. Then $P_{elast} \simeq 10^5$ atm and $\rho_o >> P_{elast}$, i.e. even from this point of view the shockwave may be considered strong. Thus, from now on we may consider the target matter as an ideal gas (let $\gamma = 5/3$). To evaluate the achievable compressions we shall use numerical results for a strong self-similar spherical shockwave in gas with $\gamma = 5/3$ focused and reflected from the center. Hence it follows that after adiabatic precompression behind the shock front the substance might be compressed to $\rho = 15\rho_o$. The reflected from the center shockwave causes additional compression and $\rho_{lim} = 33\rho_o$. Since the wave is strong in our case, and a part of the target may be enveloped by the reflected wave the target compression is described by calculations of Ref. 40. It means that in a certain part of the target a density about 30 g/cm^3 was achieved. Temperature values in the compressed part of the target achieves only several eV ($T_e = T_i$). Thus, thermonuclear reactions in the dense nucleus could not occur, and the neutrons registered were produced in the corona.

It should be noted that asymmetry of target surface and irradiation may cause some decrease of density comparison value than one obtained from the spherically symmetrical calculations. If Δ_o is deviation of the target shape from the sphere the condition that perturbation will not influence compression at the absence of instability, is

$$\frac{\Delta_o}{r_o} \delta^{1/3} << 1 \qquad (32)$$

Then initial perturbation which deliberately distorts strongly the compression is

$$\left(\frac{\Delta_o}{r_o}\right)_{lim} \sim \frac{1}{\delta^{1/3}} \qquad (33)$$

Thus, at $\delta \simeq 30$ relative perturbation of the order of 0.3 should lead to sufficient distortions. It is not without exception that in our experiments the sum perturbation (of the target surface and radiation source) had such a value. This circumstance will apparently explain nonregular character of reactional pressure of the corona and neutron yield from shot to shot.

The authors thank V. M. Groznov, A. A. Erokhin, N. N. Zorev and N. V. Novikov for help with the work and V. B. Rosanov and Yu. V. Afanasyev for useful discussions.

References

1. N. G. Basov, O. N. Krokhin, JETP 46, 171 (1964).
N. G. Basov, O. N. Krokhin, Vestnik Ac. Sci. USSR, N 6, 55 (1970).

2. O. N. Krokhin, In Physics of High Energy Density, Academic Press, 1 hc. N.Y.-L., 1971, p. 278.

3. H. Hora, Laser Interaction and Related Plasma Phenomena, ed. by H. Schwarz and H. Hora, Plenum Press, 1971, p. 383.

4. P. P. Pashinin, A. M. Prokhorov, JETP 60, 1630 (1971).

5. F. Floux, J. F. Benard, D. Cognard, A. Saleres, Laser Interaction and Related Plasma Phenomena, ed. by H. Schwarz and H. Hora, Plenum Press, 1972, p. 409.
E. D. Jones, G. W. Gobeli, J. N. Olsen, Laser Interaction and Related Plasma Phenomena, ed. by H. Schwarz and H. Hora, Plenum Press, 1972, p. 469.

6. V. A. Boiko, O. N. Krokhin, G. V. Sklizkov, Preprint N 121, FIAN, Moscow, 1972.
A. A. Rupasov, G. V. Sklizkov, V. P. Tzapenko, A. S. Shikanov, JETP 65, 11, 1898 (1973).

7. Yu. V. Afanasiev, E. M. Belenov, O. N. Krokhin, I. A. Poluektov, JETP Lett. 13, 5, 257 (1971).

8. S. Kaliski, Proceeding of Vibr. Probl. 12, 3, 243 (1971).
S. Kaliski, Proceeding of Vibr. Probl. 12, 3, 231 (1971).

9. L. A. Artsimovich, Controlled Thermonuclear Reactions, Moscow, Fizmatgiz, 1963.

10. J. D. Lawson, Proceeding of the Phys. Soc., 70, Part 1, N 445B, 11 (1957).

11. N. G. Basov, O. N. Krokhin, G. V. Sklizkov, JETP Lett. 6 6, 683 (1967).

12. J. W. Daiber, A. Hertzberg and C. E. Wittliff, Phys. Fluids, 9, 3, 617 (1966).

13. N. G. Basov, D. N. Krokhin, G. V. Sklizkov, Quantum Radiophysics, Trydi FIAN 52, 171 (1970).

14. V. A. Gribkov, V. Ya. Nikulin, G. V. Sklizkov, Short Communications in Physics, FIAN, N 2, 45 (1971).

15. H. Puell, H. Opower and H. J. Neusser, Phys. Lett. 31A, 1, 4 (1970).

16. N. V. Filippov, V. I. Agafonov, I. R. Belyaeva, V. V. Vikhrev, V. A. Gribkov, L. G. Golubchikov, V. F. Dyachenko, V. S. Imshennik, V. D. Ivanov, O. N. Krokhin, M. P. Moiseeva, G. V. Sklizkov, T. I. Filippova, Plasma Physics and Controlled Nuclear Fusion Research, Vol. I, Madison, USA, 1971.

17. N. J. Peacock, M. G. Hobby and P. D. Morgan, CN-28/D-3, IAEA 4th Conf. on Plasma Physics and Controlled Nuclear Fusion Research, Madison, USA, 1971.

18. N. G. Basov, O. N. Krokhin, G. V. Sklizkov, Laser Interaction and Related Plasma Phenomena, ed. by H. Schwarz and H. Hora, Plenum Press, 1972, p. 389.

19. J. Nuckolls, L. Wood, A. Thiessen and G. Zimmerman, "Laser Compression of Matter to Super-High Densities Thermonuclear (CTR) Application". Paper presented at Conference on Quantum Electronics, Montreal, Canada, 1972.

20. K. A. Brueckner. Preprint KMS Fusion, Inc., KMSF-NRS, 1972. See also these proceedings.

21. N. G. Basov, O. N. Krokhin, G. V. Sklizkov, S. I. Fedotov, A. S. Shikanov, JETP 62, 1, 202 (1972).
N. G. Basov, O. N. Krokhin, G. V. Sklizkov, S. I. Fedotov, A. S. Shikanov. Preprint N 123, FIAN, Moscow, 1970.

22. S. W. Mead, Phys. Fluids 13, 1510 (1970).

23. N. G. Basov, Yu. S. Ivanov, O. N. Krokhin, Yu. A. Mikhailov, G. V. Sklizkov, S. I. Fedotov, JÈTP Lett. 15, 10 (1972).
N. G. Basov, V. A. Boiko, S. M. Zakharov, O. N. Krokhin, Yu. A. Mikhailov, G. V. Sklizkov, S. I. Fedotov, JETP Lett. 18, 5, 184 (1973).
V. A. Gribkov, O. N. Krokhin, G. V. Sklizkov, N. V. Filippov, T. I. Filippova, JETP Lett. 18, 9, 541 (1973).

24. N. G. Basov, V. A. Boiko, S. M. Zakharov, O. N. Krokhin, G. V. Sklizkov, JETP Lett. 13, 691 (1971).

25. S. W. Mead, R. E. Kidder, J. E. Swain, "Preliminary Measurements of X-ray and Neutron Emission from Laser-Produced Plasmas", Lawrence Radiation Laboratory (University of California), Preprint UCRL-73356, 1971.

26. C. Yamanaka, T. Yamanaka, T. Sasaki, K. Toshida, M. Waki, H. Kang, "Plasma Generation and Heating to Thermonuclear Temperature by Laser", Institute of Plasma Physics (Nagoya University). Preprint IPPJ-117, 1972. See also these proceedings.

27. F. Floux, D. Cognard, L. Denoeud, G. Prior, D. Parisot, J. Bobin, F. Delobean, C. Fauquignon, Phys. Rev. A. (General Physics) 1, 821 (1970).

28. N. G. Basov, O. N. Krokhin, A. A. Rupasov, G. V. Sklizkov, S. I. Fedotov. Preprint N 47, FIAN, Moscow, 1973.

29. N. G. Basov, O. N. Krokhin, G. V. Sklizkov, S. I. Fedotov, Preprint, FIAN, Moscow, 1974.

30. V. P. Korobeinikov, N. S. Melnikov, E. V. Ryazanov, Theory of the Point Explosion, Moscow, Fizmatgiz, 1961.

31. Ya. B. Zeldovich, Yu. P. Raizer, Physics of Shock Waves and High Temperature Gas Dynamic Phenomena, Moscow, 1966.

32. L. M. Biberman, V. S. Vorobiev, G. E. Norman, Optics and Spectroscopy, XIV, 3, 330 (1963).

33. F. G. Iahoda, E. M. Little, W. E. Quinn, G. A. Sawyen, T. F. Stratton, Phys. Rev. 119, 3, 843 (1960).

34. K. Büchl, K. Eidman, P. Mylser, H. Salzmann, R. Sigel, S. Witkowski. Paper CN28-D-II, Proc. of IAEA 4th Conf. on Plasma Physics and Controlled Nuclear Fusion Research, Wisconsin, USA, 1971.

35. N. G. Basov, V. A. Boiko, V. A. Gribkov, S. M. Zakharov, O. N. Krokhin, G. V. Sklizkov, JETP Lett. 9, 520 (1969).

36. N. G. Basov, V. A. Goiko, V. A. Gribkov, S. M. Zakharov, O. N. Krokhin, G. V. Sklizkov, JETP 61, 154 (1971).

37. L. D. Landau and E. M. Lifshitz, Theoretic Physics, Vol. V, Moscow, 1964.

38. K. P. Stanvukovich, Nonstationary Motions of a Continuous Medium, Moscow, 1971.

39. E. B. Goldman, Plasma Physics 15 4, 289 (1973).

HARMONIC GENERATION AND PARAMETRIC EXCITATION OF WAVES IN LASER CREATED PLASMAS*

J. L. Bobin

C.E.A. - Centre d'Etudes de Limeil
Boîte Postale N° 27
94190 - Velleneuve-Saint-Georges

ABSTRACT

When focusing an Nd glass laser (frequency ω_o) onto a cryogenic target (D_2 or H_2) quite intense lines with frequencies $2\omega_o$, $3\omega_o/2$, $\omega_o/2$ are found in the reflected light in both backward and 45° directions. The lines at $2\omega_o$ and $3\omega_o/2$ are broadened on the low frequency side only. The occurrence and broadening of such lines may be related to parametric excitation of waves.

I. INTRODUCTION

The absorption of large fluxes of laser light by high temperature plasmas is a crucial problem. Every application needs an efficient absorption. From the form of the usual (collisional) absorption coefficient:

$$K_\omega \propto \frac{Z^2 n^2}{\omega^2 T^{3/2}(1 - \omega_p^2/\omega^2)^{1/2}} \tag{1}$$

it is readily inferred that high temperature, low Z plasmas poorly absorb the radiation even at cut-off. Furthermore, the factor $n^2/T_e^{3/2}$ in K_ω comes from the collision frequency which involves

---

*Presented at the Third Workshop on "Laser Interaction and Related Plasma Phenomena" held at Rensselaer Polytechnic Institute, Troy, New York, August 13-17, 1973.

the thermal velocity $v_{th} \propto T^{1/2}$. Then, if the maximum oscillatory velocity:

$$v_o = \frac{eE_L}{m_e\omega} = \frac{e}{m_e\omega}\left(\frac{4\pi I}{c}\right)^{1/2} \tag{2}$$

where E_L is the electric field associated with the laser intensity I, happens to be greater than v_{th}, the coefficient becomes:

$$K_\omega \propto \frac{\omega Z^2 n^2}{I^{3/2}\ (1 - \omega_p^2/\omega^2)^{1/2}} \tag{3}$$

However, this behavior is restricted to the off-resonant case $\omega \neq \omega_p = 4ne^2/m_e$. In the resonance case, collective effects should be taken into account. They lead to an anomalous absorption of the laser light through the onset and saturation of instabilities.

Many theoretical studies have been devoted to these phenomena[1-5]. Of course, they are not restricted to the case of visible e.m. waves. Actually, some were observed using microwaves interacting with low density plasmas[6,7]. They were also suspected to play a role in the high flux interaction of laser light with matter[8].

Besides these plasma nonlinearities, optical nonlinearities such as harmonic and subharmonic generation on electron density gradients were also predicted[9,10] and observed[11]. This paper is devoted to the coupling of the two classes of nonlinearities.

II. EXPERIMENTAL

Unusual generation of harmonics was observed at Limeil in two different experiments. Both use a neodymium glass laser and a cryogenic target. The beam is tightly focused with a f/1 plane aspheric lens. But the parameters and observation conditions are quite different as shown on Table I.

Lines at $2\omega_o$ and $\omega_o/2$ were predicted by theoretical studies. On the contrary, the line at $3\omega_o/2$ and the asymmetric spectrum of lines at $2\omega_o$ and $3\omega_o/2$ which are displayed on Fig. 1 were quite unexpected.

Spectral observations were made using a Czerny Turner grating spectrometer with a mean dispersion of 8 Å/mm in the first order. For recording in the visible range a multichanneltron image amplifier fitted with an S-20 photocathode was used. Its maximum gain is 10^3. For wave lengths greater than 1 μm the spectrometer was fitted with a 50 CVP photomultiplier connected by an optical

Table I

Experiment	Target	Pulse Duration and Power Actually Impinging on Target	Direction of Observation	Observed Lines	Spectral Observation	Polarization
I[12]	H_2	10–15 nsec	45° from the beam axis	$2\omega_o$	asymmetric spectrum	none
		1–4 GW		$3\omega_o/2$	asymmetric spectrum	
				$\omega_o/2$	symmetric	
II[13]	D_2	3 nsec	back reflected light	$2\omega_o$	none	-as incident line
		15–25 GW		$3\omega_o/2$	none	

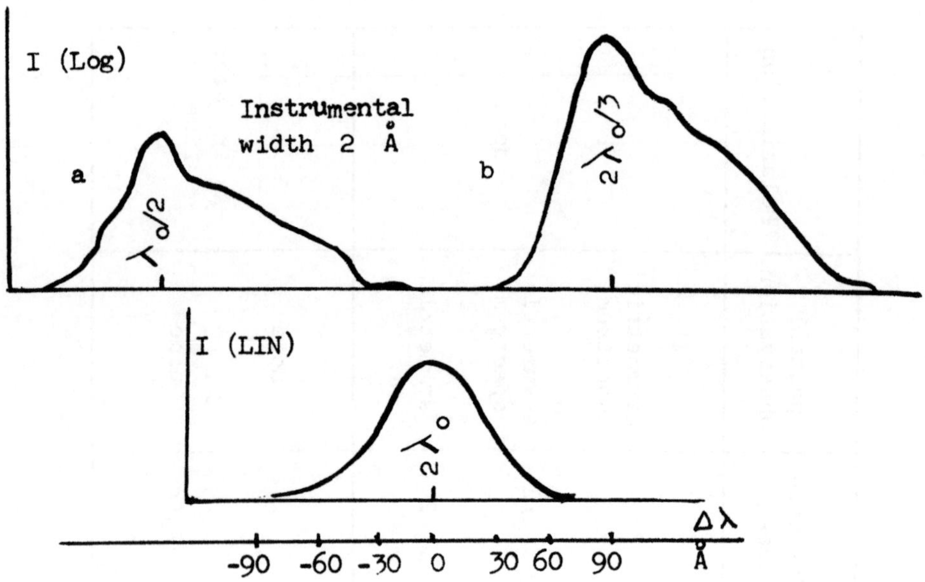

FIGURE 1. Spectral profile of the three observed harmonics.

fiber. Typical profiles are displayed on Fig. 1. The lines at $2\omega_o$ and $3\omega_o/2$ are time integrated and recorded in a single shot. On the contrary, the profile of $\omega_o/2$ is investigated on a shot to shot procedure. The incident radiation was composed of about 45 lines spaced 0.95 Å apart. The envelope is nearly symmetrical with 30 Å half width. It can clearly be seen on Fig. 1 that the broadening on the low frequency side of the lines at $2\omega_o$ and $3\omega_o/2$ is anomalously large.

The temporal behavior of the lines was also investigated. In both experiments the $3\omega_o/2$ line is contemporary to the leading edge of the incoming pulse. The pulse is shorter and looks unstable. There exists a definite threshold for the occurrence of this line. However, the threshold flux cannot be evaluated since no measurements of the focal spot in the presence of a plasma are available.

The second harmonic has a different behavior according to the direction of observation. At 45° it behaves as the $3\omega_o/2$ line. On the contrary, it is only slightly shorter than the incoming pulse in the backward reflected light. In this case, the dependence on the location of the focal spot with respect to the surface was accurately determined. The results are plotted in Fig. 2 for reflected light at incoming frequency, the number of neutrons and the two lines $2\omega_o$ and $3\omega_o/2$. There is a correlation between the number

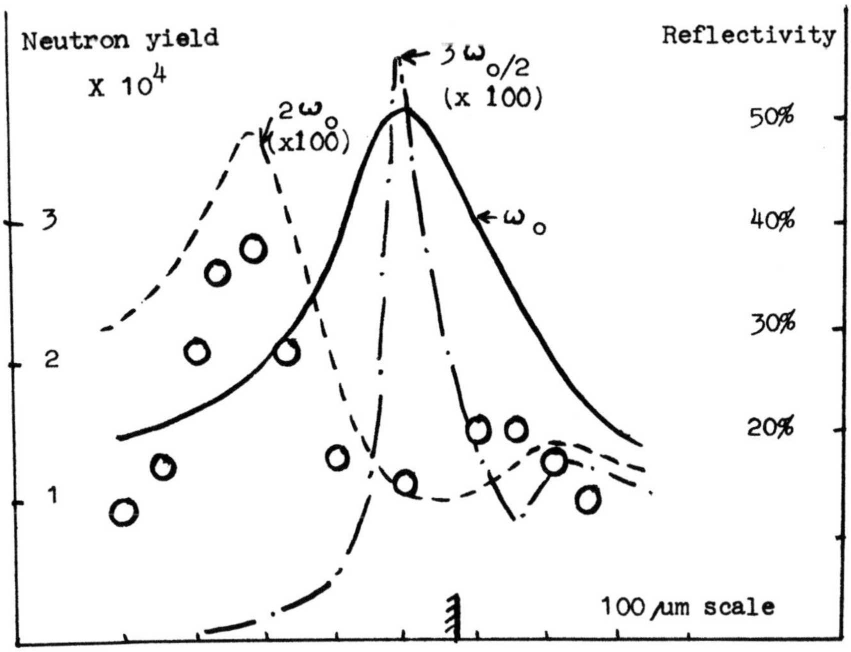

FIGURE 2. Neutron yield o and intensities of lines in the backward reflected spectrum as a function of the location of the focal zone with respect to the surface.

of neutrons and the intensity of the second harmonic whose maximum values occur for the same location of the focus, definitely different from the location which gives the peak values of the reflected light at ω_0 and of the line $3\omega_0/2$.

Furthermore, it was seen that hard X-rays are correlated with a strong reflection of light and the occurrence of the harmonic $3\omega_0/2$ [13].

III. HARMONIC AND SUBHARMONIC GENERATION

Due to the v/c term in the Lorentz force e.m. radiation at frequency ω_0 impinging on an electron gas induces polarization of higher orders. However, anisotropy might come from various origins; an axially symmetric field surrounds the focal spot in uniform plasmas[14]. In a nonuniform plasma the density gradient provides the necessary anisotropy. In the experiments described in Section II, the density gradient is weak according to numerical simulations and interferometric density measurements. The density decreases by a factor 1/e over about 100 μm, i.e., 100 wavelengths for Nd glass light. So:

$$\eta = \frac{1}{n_e} \nabla n$$

can be considered as a small parameter. Another small parameter is, of course, v/c. Expansions will be made in the following in first order in η and in higher orders in v/c.

The plasma is conveniently described by hydrodynamic equations. Since one has $k \cdot v_{th} << \omega$, one may use the cold plasma approximation. Furthermore, it is allowed to neglect $k \cdot v_o$ in the calculations, since the amplitude of the field is not high enough to make v_o greater than v_{th}.

Pushing the nonlinear terms in the right hand side, the equations read:

$$\frac{\delta n}{\delta t} + n_o \ \nabla \cdot v = -\varepsilon \ \nabla \cdot nv$$

$$\frac{\delta V}{\delta t} + \frac{e}{m} E = -\varepsilon[(v \cdot \nabla)v + \frac{e}{mc} v \times B] \tag{4}$$

in which ε is an order of magnitude label which will eventually be set equal to 1. The plasma equations should be completed by Maxwell equations:

$$\frac{\delta B}{\delta t} - c \ \nabla \times E = 0$$

$$\frac{\delta E}{\delta t} - c \ \nabla \times B - 4\pi e n_o V = \varepsilon 4\pi e n v \tag{5}$$

Now consider a propagating transverse external e.m. wave:

$$E = E_o \cos (\omega_o t - k_o \cdot r) \tag{6}$$

The solution of the momentum equation (quadratic terms in v are disregarded) is an expansion in v/c for v:

$$v_o = v_o^{(0)} + \varepsilon v_o^{(1)} + \varepsilon^2 v_o^{(2)} + \ldots$$

$$v_o = -\frac{e}{m\omega_o} E_o \sin (\omega_o t - k_o \cdot r)$$

$$- \varepsilon \frac{e^2}{4cm^2\omega_o^2} (E_o \times B_o) \cos (2\omega_o t - 2k_o \cdot r) \tag{7}$$

$$+ \varepsilon^2 \frac{e^3}{24c^2m^2\omega_o^3} (E_o \times B_o) \times B_o \sin (3\omega t - 3k_o \cdot r)$$

The particle conservation equation then yields (expansion in both η and v/c):

$$n_o = n_o^{(0)} + \varepsilon n_o^{(1)} + \varepsilon^2 n_o^{(2)} + \ldots$$

$$= N_o \{1 - \varepsilon\eta\cdot E_o \frac{e}{m\omega_o^2} \cos(\omega_o t - k_o\cdot r)$$

$$- \varepsilon\eta\cdot(E_o\times B_o) \frac{e^2}{8m^2 c\omega_o^3} \sin(2\omega_o t - 2k_o\cdot r) \tag{8}$$

$$+ \varepsilon^2\eta\cdot[r - \frac{e^3(E_o\times B_o)\times B_o}{72m^3c^2\omega_o^4} \cos(3\omega_o t - 3k_o\cdot r)] + \ldots\}$$

where N_o is the reference density at r = 0. It is furthermore assumed that the region of interest is smaller than the electron excursion in the applied field. Note also that in plasmas created by laser interaction with solid targets, the gradient is in the direction of propagation of the light, so one has:

$$\eta\cdot E_o \sim 0$$

The corresponding term is of the same order of magnitude as the first order term in v/c (2nd harmonic).

The solution of Equations 4 and 5 can be sought by the multiple time scale expansion[15-17] setting:

$$\frac{\partial}{\partial t} = \frac{\partial}{\partial t} + \varepsilon\frac{\partial}{\partial t_1} + \varepsilon^2\frac{\partial}{\partial t_2} + \ldots$$

We will restrict ourselves to the Bogolubov approximation using only the slow time:

$$t_1 = \varepsilon t$$

Fourier transforming Equations 4 and 5 with respect to both space and time, the solution is sought for $\hat{E}$. Both longitudinal and transverse fields are found. Disregarding the former, one gets in first order:

$$D(\omega_1 k_1)\hat{E}_1 = 0 \tag{9}$$

The solution is nontrivial if and only if:

$$D(\omega_1 k_1) = \omega_1^2 - \omega_p^2 - k_1^2c^2 = 0 \tag{10}$$

i.e., the transverse wave fulfills the usual plasma dispersion relation. The solution is a combination of exponentials:

$$E_1 = 2\mathrm{Re}\,\{A_1 \exp[i(\omega_1 t - k_1\cdot r)] + B_1 \exp[i(\omega_1 t + k_1\cdot r)]\}$$
$$v_1 = \frac{2ie}{m\omega_1}\,\mathrm{Im}\,\{A_1 \exp[i(\omega_1 t - k_1\cdot r)] + B_1 \exp[i(\omega_1 t - k_1\cdot r)]\} \tag{11}$$

A_1 and B_1 are functions of the slow time t_1 which are determined together with the possible frequencies ω_1 by going to the second order. The result for $\hat{E}$ is:

$$D\hat{E}_2 = S \tag{12}$$

with the source term:

$$S = \int dt\,\underline{dr}\; e^{i(\omega t - k\cdot r)}\,[-4\pi e n_o \frac{\partial v_1}{\partial t_1} - ick\times \frac{\partial B_1}{\partial t_1} - i\omega \frac{\partial E_1}{\partial t_1}$$
$$-4\pi i e\omega n_o^{(1)} v_1 + 4\pi i\omega n_o^{(2)} v_o - \frac{\omega_p^2}{i}(v_o\times B_1 + v_1\times B_o)] \tag{13}$$

Due to the form of Eq. 11 the source term leads to delta functions involving combinations of frequencies and wave numbers. Since permitted frequencies should verify Eq. 10, the solution for $\hat{E}_2$ is nondivergent if and only if the right hand side also vanishes. This implies the equality of different combinations of frequencies and the sum of coefficients should be zero. Hence, differential equations linking A_1, B_1 and their complex conjugates are found.

Such calculations predict the occurrence of the second harmonic $2\omega_o$ and of subharmonic $\omega_o/2$ [18]. There is a simple connection between the width of such parametrically excited lines and the width of the incident line. No asymmetric broadening can be found. Furthermore, no combination of frequencies yield $3\omega_o/2$.

These can be obtained only if the coupling with longitudinal waves is considered.

IV. EXCITATION OF LONGITUDINAL WAVES IN A PLASMA

An external e.m. wave with frequency ω_o impinging onto a plasma acts on a charged particle through the Lorentz force. Due to the mass difference, electrons oscillate with respect to the ions. Accordingly, any mode with frequency ω is perturbed and acquires frequency components:

$$\omega + n\omega_o$$

Note that these harmonic components which come from the Bessel expansion of the density fluctuations are different from the usual harmonic generation by free electrons which generate from the v/c expansion in the Lorentz force.

Now, we are mainly interested in resonances. Two kinds exist:

- the high frequency resonance for which

$$\omega \sim \omega_{pe} = \frac{4\pi n_e e^2}{m_e}$$

$$\omega_o < \omega_{pe}$$

- the low frequency resonance for which

$$\omega \ll \omega_{pe} \quad , \quad \omega \ll \omega_o$$

It turns out that only n = 1 is permitted in this case. Thus, one has $\omega_o \sim \omega_{pe}$ and the characteristic low frequencies are ion-sound frequencies:

$$\omega = \Omega_S \lesssim \omega_{pi} = \frac{4\pi n_i e^2}{m_i}$$

and:

$$\omega_o = \omega_{pe} + \Omega_S > \omega_{pe}$$

If no damping occurs, these regimes are unstable. The high frequency mode causes an oscillating two stream instability whereas in the low frequency case one has the parametric decay instability since an electronic Langmuir wave (near resonant) is simultaneously excited. The name comes from the way it can be diagramatically figured (Fig. 3). The e.m. wave decays into a Langmuir wave and an ion-acoustic wave. Conservation of energy and momentum implies:

$$\omega_o = \omega_L + \Omega_S$$

$$(\omega_L = \omega_{pe}) \tag{14}$$

$$k_o = k_L + k_S$$

When damping is taken into account, a threshold is found for the onset of the instability. Furthermore, it can be shown that linear

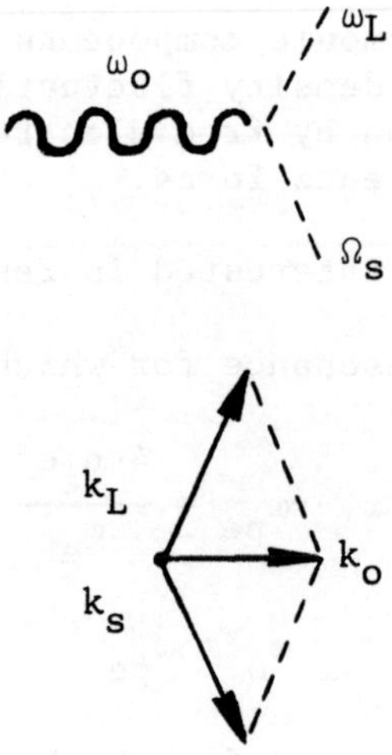

FIGURE 3. Parametric decay instability.

Landau damping cannot balance the growth rate of the instability. On the contrary, nonlinear Landau damping can. However, the amplitude of the triggered wave should be high enough. So it needs time before damping balances the growth and the instability saturates.

It is also found that the instability is easier to excite if one has:

$$T_i << T_e$$

which is the case at the beginning of the interaction process.

Other instabilities are possible in the interaction of laser beams with plasmas. Since their threshold is significantly higher, they will not be considered here.

A last point is of interest. A density gradient is expected to cause an increase of the threshold[19], except for waves propagating along a direction perpendicular to that of the gradient. Indeed, in presence of a gradient the wave vector k should be replaced by:

$$k(1 + \frac{\eta \cdot k}{k^2})$$

The results obviously do not change when $\eta \cdot k = 0$. So looking back at the Fig. 3 which shows that longitudinal waves are excited along such a direction, the thresholds to be expected are close to the one corresponding to a uniform plasma.

V. COUPLING OF PLASMA WAVES TO HARMONIC GENERATION

In the calculation Section III, no account was taken of density fluctuation due to plasma waves. These can be introduced very simply replacing n_o by:

$$N_o[1 + \eta_S(\Omega_S) + \eta_L(\omega_o - \Omega_S)]$$

where η_L and η_S are periodical perturbation of zeroth order with respect to expansion in v/c or η with frequencies Ω_S and ω_L respectively. This change only modifies the source term S_2 which keeping only those terms leading to lowest order harmonic or sub-harmonic generation now reads ($\alpha = e/m_e\omega_o$)

$$\begin{aligned} S = \int dt\underline{dr}\, e^{i(\omega t - k\cdot r)} \{ & A(\omega, \omega_1, k, k_1) + \\ & + i \frac{\alpha}{8c\omega_o^2} \omega_p^2 \eta\cdot(E_o\times B_o)(1 + \eta_S + \eta_L) \sin(2\omega_o t - 2k_o\cdot r)v_1 - \\ & -4\pi i e N_o \eta\cdot[r - \frac{\alpha^3}{72c^2\omega_o}(E_o\times B_o)\times B_o(1 + \eta_S + \eta_L)\cos(3\omega_o t - 3k_o\cdot r)v_o \\ & + \ldots]\} \end{aligned} \tag{15}$$

where A is an algebraic function. Equation 15 leads to δ-functions of combinations of frequencies different from those due to Eq. 13. One finds for instance:

On the one hand,	On the other hand,
$\delta(\omega - \omega_1)\delta(k - k_1)$	$\delta(\omega - 3\omega_o + \Omega_S + \omega_1)\delta(k - 2k_o - k_L + k_1)$
	$\delta(\omega - 2\omega_o \pm \Omega_S)\delta(k - 2k_o \pm k_S)$
	$\delta(\omega - \omega_o - \Omega_S + \omega_1)\delta(k - k_o + k_L + k_1)$

In order to get finite solutions one should have:

$$\omega - \omega_1 = \omega - 3\omega_o + \Omega_S + \omega_1$$

$$\omega_1 = \tfrac{1}{2}(3\omega_o - \Omega_S)$$
$$2k\left(\frac{3\omega_o - \Omega_S}{2}\right) = 2k_o + k_L \tag{16}$$

or by the same token

$$\omega_1 = 2\omega_o \pm \Omega_S$$
$$k(2\omega_o \pm \Omega_S) = 2k_o \pm k_S \tag{17}$$

or others (up to the 5th harmonic).

Note that due to the plasma dispersion relation (Eq. 10) one cannot simultaneously have Eq. 16 and the + sign in Eq. 17. Since the harmonic $3\omega_o/2$ has been observed, only the - sign is allowed in Eq. 17. The occurrence of the line at $3\omega_o/2$ can be physically interpreted as the parametric excitation of transverse waves due to the coupling of a Langmuir wave to the oscillation of electrons with second harmonic frequency.

Equations 16 and 17 keeping the - sign only show an asymmetric broadening of the line corresponding to the ion acoustic spectrum. The broadening of the $2\omega_o$ and $3\omega_o/2$ lines look approximately equal on the wavelength scale. Converting to the frequency scale one finds that the width at $2\omega_o$ is almost twice the width at $3\omega_o/2$ in agreement with Eqs. 16 and 17.

Finally, from Eq. 15 making coefficients vanish yields differential equations for A_1 (t_1) and B_1 (t_1). Then one finds, for instance, in the case of the line at $3\omega_o/2$ a growth rate:

$$\gamma_1\left(\frac{3\omega_o}{2}\right) = \frac{\alpha}{16c\omega_o^2}\,\omega_p^2\eta\cdot(E_o\times B_o)\varepsilon_L \tag{18}$$

where ε_L is the amplitude of the Langmuir wave. In the case of instability it has a growth rate γ_L. Thus one has

$$\gamma_1 \propto e^{\gamma_L t_1}$$

and one expects a very fast growth of the $3\omega_o/2$ line. Moreover if the Langmuir wave saturates, the harmonic still grows although at a slower rate.

Finally Eq. 18 shows that the growth is maximum when the gradient has the direction of propagation of the external wave.

VI. CORRELATION WITH HARD X-RAYS

It is well-known that the Landau damping corresponds to the trapping of electrons by the longitudinal wave[20]. Those electrons in the thermal distribution with velocities around the phase velocity

$$\frac{\omega_L}{k_L}$$

are caught up by the wave. The instability causes an amplification of the process, more and more electrons constituting a high energy tail with velocity:

$$v_{trap} \sim \frac{\omega_L}{k_L}$$

Now a minimum threshold for instability is found when one has[4]:

$$0.1 < k\lambda_D < 0.2$$

where λ_D is the Debye length:

$$\lambda_D = \frac{v_{th}}{\omega_{pe}} = \frac{v_{th}}{k_L v_{trap}}$$

Thus one finds:

$$10 > v_{trap}/v_{th} > 5$$

and for the energies:

$$100 > \varepsilon_{trap}/\varepsilon_{th} > 25$$

In the experiment of Section II the electron temperatures range between 100 eV and 1 keV. Then a high energy tail of 10 - 100 keV electrons is associated with the onset of the parametric decay instability. By bremsstrahlung such electrons generate hard X-rays which were actually recorded and are correlated with the $3\omega_o/2$ line.

VII. CONCLUSION

The occurrence of the $3\omega_o/2$ harmonic, an asymmetric broadening of lines $3\omega_o/2$ and $2\omega_o$ and the correlation with hard X-rays are in our opinion direct evidences of the parametric decay instability during the interaction of laser beams with hydrogenoid targets.

The contribution of D. Cognard, M. Decroisette, F. Floux, B. Meyer, A. Saleres and Y. Vitel to the results quoted in this paper is gratefully acknowledged.

References

1. V. P. Silin, Survey of Phenomena in Ionized Gases, IAEA, Vienna, 205 (1968).
2. R. E. Kidder, Physics of High Energy Density, Academic, 306 (1971).
3. E. A. Jackson, Phys. Rev. 15, 235 (1967).
4. K. Nishikawa, J. Phys. Soc. Japan 24, 916 (1968), 24, 1152 (1968).
5. P. K. Kaw, J. M. Dawson, Phys. Fluids 12, 2586 (1969).
6. K. Kato et al., J. Phys. Soc. Japan 20, 2097 (1965).
7. R. A. Stern, N. Tzoar, Phys. Rev. Lett. 17, 903 (1966).
8. C. Yamanaka et al., Phys. Rev. A6, 2335 (1972).
9. N. Bloembergen, Nonlinear Optics, Benjamin (1965).
10. V. L. Ginsburg, A. V. Gurevich, Soviet Phys. Uspekhi 3, 115 (1960), 3, 175 (1960).
11. M. Decroisette, B. Meyer, G. Piar, J. de Phys. 32, (1971) Colloque 5b.119.
12. J. L. Bobin, M. Decroisette, B. Meyer, Y. Vitel, Phys. Rev. Lett. 30, 594 (1973).
13. A. Saleres et al., Société Française de Physique, Vittel (1973).
14. G. Mayer, Optics Comm. (1973).
15. D. C. Montgomery, Physica (Utrecht) 31, 693 (1965).

16. E. A. Frieman, J. Math. Phys. 4, 410 (1963).

17. G. Sandri, Ann. Phys. 24, 332 (1963).

18. C. S. Chen, J. Plasma Phys. 5, 107 (1971).

19. F. W. Perkins, J. Flick, Phys. Fluids 14, 2012 (1971).

20. R. Z. Sagdeev, A. A. Galeev, Nonlinear Plasma Theory, Benjamin (1969).

SHORT PULSE LASER HEATED PLASMA EXPERIMENTS*+

J. Soures, L. M. Goldman** and M. Lubin

Laboratory for Laser Energetics
College of Engineering and Applied Science
University of Rochester
Rochester, New York 14627

ABSTRACT

The interaction between time tailored, high intensity, laser radiation and spherical light element targets has been investigated. Simultaneous measurements of the laser plasma electron temperature, charged particle distribution, absorbed energy, neutron yields and backscattered light intensity and spectra were made. For 100 psec. wide laser pulses focused to intensities ranging between $3x10^{15}$ and $2x10^{16}$ watts/cm^2 on 150-200 μm diameter LiD and $C_{36}D_{74}$ targets, plasmas with electron temperatures up to 1.2 keV and target generated yields of up to $4x10^4$ neutrons are produced. The absorption can be accounted for by inverse bremsstrahlung and no significant contributions due to parametric instabilities are observed. No more than 10^{-1} of the incident laser energy is back reflected into spectral components at ω_L, $3/2\omega_L$, and $2\omega_L$.

*Presented at the Third Workshop on "Laser Interaction and Related Plasma Phenomena" held at Rensselaer Polytechnic Institute, Troy, New York, August 13-17, 1973.

+This work was supported by the Laser Fusion Feasibility Project at the University of Rochester.

**Permanent address: General Electric Corporate Research and Development Center, Schenectady, New York.

INTRODUCTION

In this paper we report on laser produced plasmas from solid spherical targets, which produce neutrons, have a high electron temperature (≃1 kev), and produce no evidence for anomalously high electron or ion velocity distributions. The absorbed energy and the measured electron temperatures can be accounted for by assuming only collisional absorption. These results are in contrast to a number of other experiments on laser produced plasmas [1-5] in which evidence of anomalously large numbers of energetic electrons and ions are reported.

Recently a number of stimulated interaction mechanisms between the laser radiation and the dense plasma have been proposed, which, if dominant, could limit the absorption of laser light in the dense target. These include stimulated Raman, Brillouin, and Compton scattering. In addition, enhanced absorption of the laser radiation as a result of the excitation of certain parametric modes in the plasma has been predicted. [6]

The experiments reported here were conducted to investigate the dominant radiation absorption mechanisms in the range of 10^{15} to 10^{16}W/cm^2 at a wavelength of 1.06μm corresponding to intensities near or beyond the theoretical threshold for nonlinear absorption effects. This is part of a series of experiments leading up to power densities on target required for thermonuclear breakeven conditions; i.e., 10^{16} - 10^{17} W/cm^2 [7]. Emphasis has been placed on obtaining the widest variety of experimental information from the heated plasma in order that a self-consistent picture of the laser radiation plasma interaction can be formulated. Simultaneous measurements include plasma ion energy spectra, time integrated x-ray emission spectra (electron temperature), neutron flux measurements as well as incident and backscattered optical spectra.

EXPERIMENTAL CONFIGURATION

The laser system used in this work consists of a Nd:glass oscillator followed by five rod amplifiers, producing a pulse of 1.06μm radiation of 10^{-10} second duration. Energy contrast ratios (pulse to baseline energy) of better than 10^5 at the output are characteristic of this system, described in reference [8].

The temporal shape of the pulse may be varied using a system of beamsplitters and attenuators. In these experiments a prepulse containing 0.5 percent of the total energy precedes the main pulse or pulses by 6 nanoseconds. In short pulse experiments, an appropriately timed prepulse, designed to produce a sufficiently long density gradient, leads to stronger absorption of the main heating pulse [9]. Previous experiments in this laboratory have indicated that irradiation of targets without a prepulse for 10^{-10} second pulses leads to reduced absorption and increased reflected energy. In some cases, experiments have been conducted using two main heating pulses separated by 10^{-8} seconds. The beam is focused onto the target with a 20cm focal length f/2 lens. The operating characteristics of the system are shown below

LFFP LASER DRIVER

Wavelength	1.06 m
Bandwidth	1 to 4Å
Energy per Pulse	15-50 joules
Pulse Duration	10^{-10} seconds
Peak Power	$>5 \times 10^{11}$ watts
Exit Aperture	50mm
Peak Intensity (Focussed)	$>2 \times 10^{16}$ watts/cm^2
Beam Divergence	300 radians
Brightness	$>10^{17}$ watts/cm^2/ster
Effective Black Body Temperature	≃1300 eV
Focal Spot Diameter	<60 m

Spherical targets of LiH, LiD, $C_{36}D_{74}$ varying in size between 150 to 250μm diameter are placed in the focal spot of the lens.

The target chamber, shown schematically in Fig. (1), is a stainless steel sphere 72cm in diameter. All the diagnostics with the exception of the neutron detectors and the optical backscatter apparatus are mounted inside the chamber which operates at a vacuum of 3×10^{-7} torr.

DIAGNOSTICS

Six diagnostic techniques are used simultaneously on each laser shot. These are:

1. Charged particle flux measurements,
2. Time-of-flight mass spectroscopy
3. Neutron flux and time-of-flight measurements,
4. X-ray spectra measurements,
5. Optical spectroscopy of incident and backscattered radiation from the target,
6. Backscatter radiation intensity and polarization

A brief description of each technique follows:

Charged Particle Collectors. The charged particle emission from the plasma is monitored at five different directions relative to the laser beam-target axis. The collectors are basically screened Faraday cups biased to collect ions; each collector

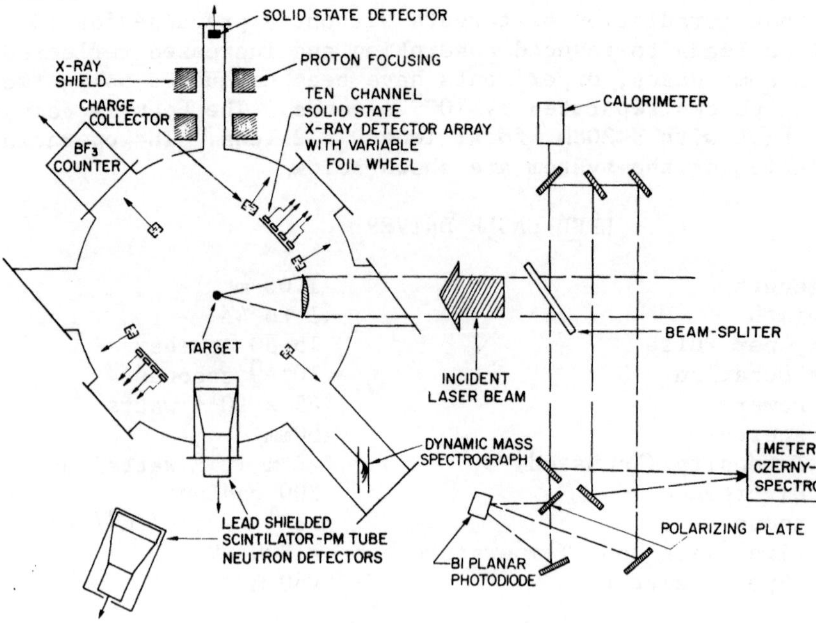

Fig. 1 Target chamber and interaction diagnostics.

subtending a solid angle of $4x10^{-4}$ sterradians. The ion collectors are located 22cm and 100cm from the target which is a sufficient distance to assume that the plasma is freely expanding so that the ions contain almost all the energy. Wide bandwidth (150MHz) current probes measure the ion current. Given the knowledge of the relative ratio of the charge particle species (i.e., D^+, Li^+, L^{++}, and L^{+++} for LiD), and their relative velocities, obtained from the mass spectrograph described below, it is possible to reconstruct the angular dependence of the expanding plasma.

A typical charged collector trace is shown in Fig. (2). Since the laser plasma is a point source emitting for a time very short compared to the collection time, the total energy incident on ion collector at a distance d from the plasma source is given by

$$E = \frac{Md^2}{2Ze} \int_0^{\infty} \frac{I(t)}{t^2} \, dt$$

where M is the ion mass, Ze is the ion charge, and I(t) is the measured collector current.

Computer processing of both oscilloscope photographs and real time digitizers yields the angular distribution of both the particle and energy flux also shown in Fig. (2). From plots such as this, the total number of particles and the total expansion energy are readily calculated.

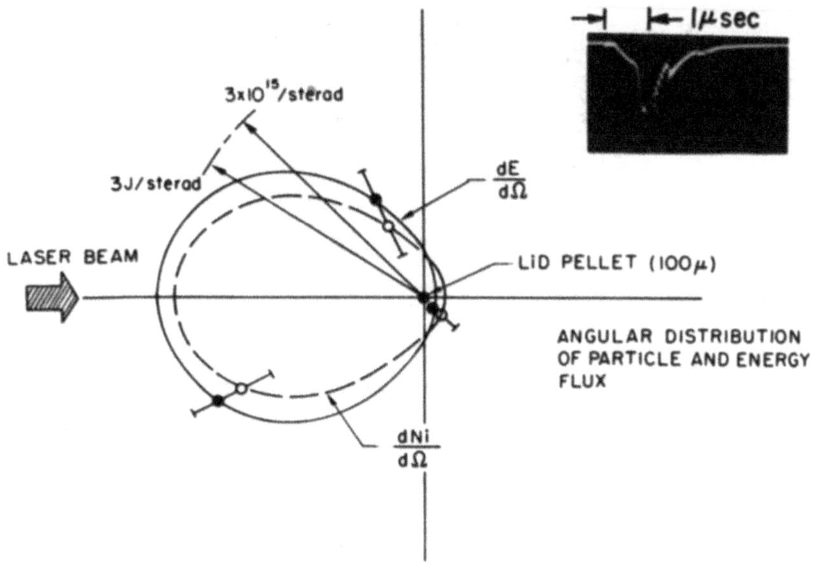

Fig. 2 a) Typical charge collector trace.
b) Angular distribution of particle and energy flux.

Time-of-Flight Mass Spectrometer. A quasi-electrostatic analyzer, whose potential varies as a function of time, is used to determine the velocity distribution and identify the principal ionized components of the expanding plasma. The ions are analyzed in a time-of-flight tube followed by the dynamic, capacitor like analyzer. This combination enables one to determine simultaneously the relative yield and energy distribution of all the ions created in one laser shot [10]. Typically in the case of oxygen doped LiD spherical targets, we examine L^+, L^{++}, L^{+++}, D^+, and O^{+7}. A sample of the data obtained is shown in Fig. (3).

Neutron Detection. Three neutron detectors are used: a 7 cm diameter, 2.5cm thick plastic scintillator, 2.5cm thick placed at 100cm and a BF_3 counter imbedded in a paraffin moderator located 40cm from the target. The pulse height spectrum for DD neutrons was measured for both plastic scintillator-photomultiplier systems and the neutron sensitivity of all three counters was determined using a Kaman Nuclear calibrated DD source. The time of arrival of the neutron signal from the plastic counters was determined relative to the laser light pulse by introducing a laser light signal on the same oscilloscope trace as the neutron signal. The time delay is determined with an estimated error of ± 4 nano-

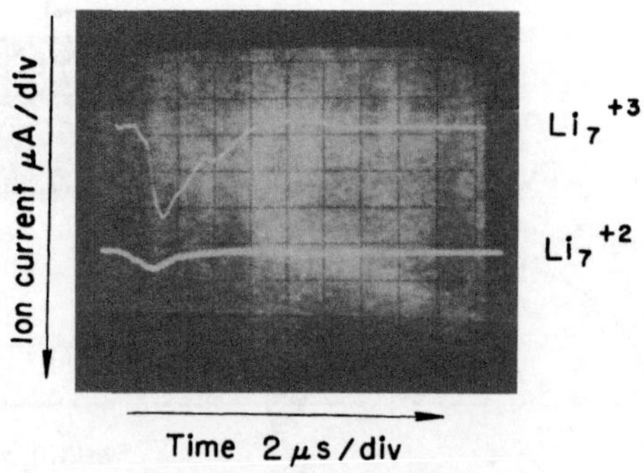

Fig. 3 Li^{+++} and L^{+++} current waveforms from dynamic mass spectrograph.

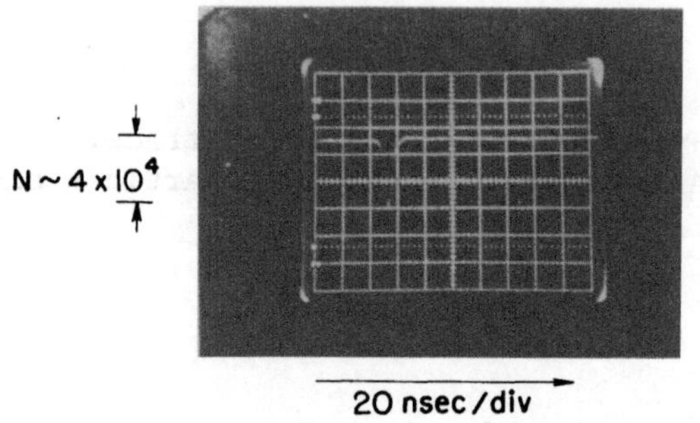

Fig. 4 Sample of neutron signal on smaller neutron detector.

seconds due to the rise time of the photomultiplier and the reading error on the oscilloscope. From these measurements, hard x-ray signals can be clearly separated from the neutron signals, since at 100cm the time-of-flight for a DD neutron is 47 nanoseconds. No hard x-ray signals were observed on the neutron detectors in any of the experiments. Both scintillators were surrounded by 6mm. of stainless steel. In addition the smaller counter was shielded by 6mm. of lead. A sample neutron signal from the 7cm detector is shown in Fig. (4). This signal represents a yield of $(4 \pm 2) \times 10^4$ neutrons.

X-Ray Temperature Measurements. Electron temperatures of the laser plasmas have been deduced from the relative transmissions of the continuum x-ray radiation through absorbing aluminum and nickel foils. Ten solid state PIN x-ray detectors with a response time of 10^{-9} seconds are arrayed at 45° and 135° to the incident beam. The detectors are arranged in two groups of four and six respectively. The aluminum foils ranged from 5 to 38 mg/cm^2. The output of each detector is integrated in a LeCroy -227 gated intergrator and then stored in the HP 2100 computer. An example of the computer output data is shown in Fig. (5a). The solid lines represent values of the relative intensity for Maxwellian distributions at several temperatures. The experimental points for some of the data do not lie along any single Maxwellian temperature curve. An example of a non-Maxwellian data is shown in Fig. (5b). The non-Maxwellian distributions have been fit by Bristow [11] assuming the superposition of two Maxwellian sources. As is seen in Fig. (6) the addition of a very small number of emitters (.1 - .2 percent) at 3 to 4 times Te is sufficient to produce a satisfactory fit to the observations.

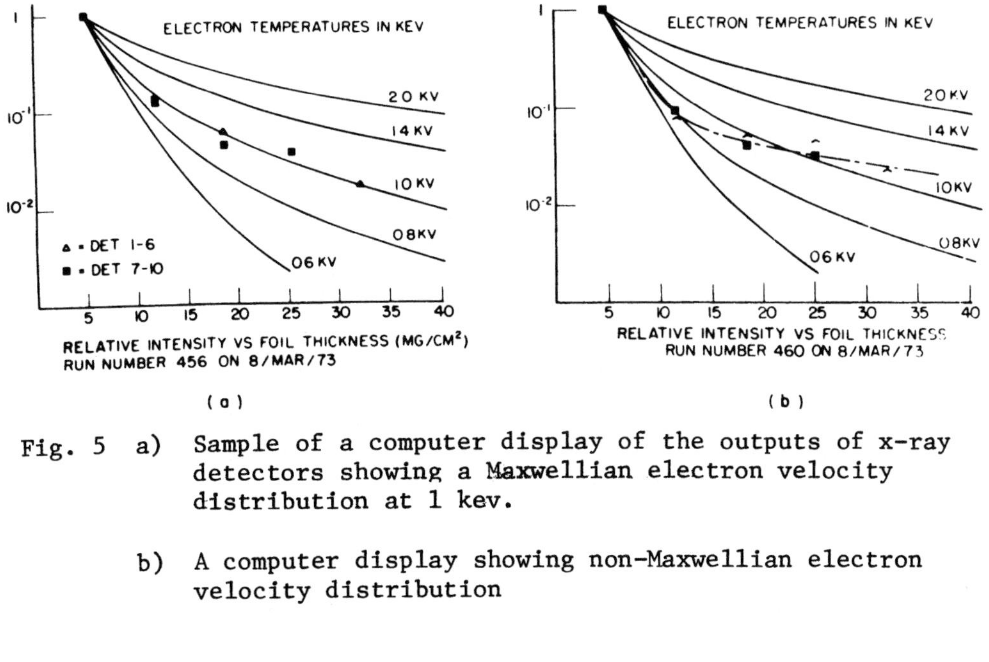

Fig. 5 a) Sample of a computer display of the outputs of x-ray detectors showing a Maxwellian electron velocity distribution at 1 kev.

b) A computer display showing non-Maxwellian electron velocity distribution

Optical Diagnostics. One of the goals of the present experiment is to utilize the backreflected light from the plasma as an indication of the possible onset of anomalous effects. The optical diagnostics shown in Fig. (1) are used for measuring the intensity, polarization, and spectrum of the backreflected light. Light reflected back along the laser optical axis is sampled by a beam-splitter and analyzed by a coated glass plate polarizer. The two component polarizations are time delayed with respect to each other and with respect to a third beam which represents a known fraction of the incident laser intensity. All three beams are directed into a single vacuum photodiode, thus facilitating the measurement of the backreflected light intensity and polarization, Fig. (1).

Part of the incident and reflected light is directed into a

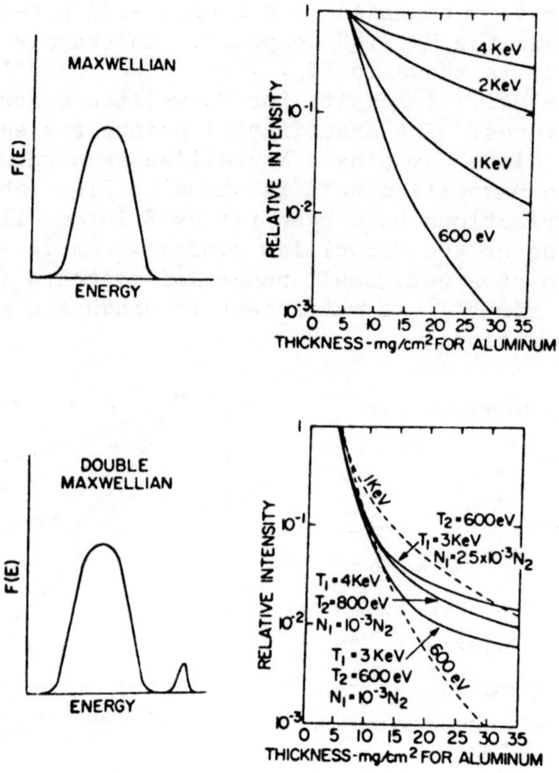

Fig. 6 A comparison of the transmission function of a double peaked Maxwellian distribution with a comparable single Maxwellian distribution function.

one meter spectrograph which simultaneously monitors both spectra. At the focal plane of the spectrograph a sensitized I-Z plate is used for monitoring the incident laser spectrum and the reflected light near the incident laser frequency,ω_L. A second plate of Tri-X Pan film is placed behind the I-Z film to photograph the reflected light spectrum in second order near the second harmonic of the incident beam. It was verified that the 1.06μm light did not expose the Tri-X Pan film at typical intensities produced at the focal plane of the spectrograph.

EXPERIMENTAL RESULTS

1. Charged Particle Data

In the free expansion phase of the laser plasma, the plasma energy is primarily in the form of directed kinetic energy of expansion of the ions. Measurements made with the dynamic mass spectrometer indicate that for LiD targets, Li^{+3} and D^+ are expanding with the same velocity. Simultaneous measurements made with the array of ion collectors show an angular distribution of both charge and energy which is strongly peaked in the forward direction. The maximum ion energies indicated by the charge collector measurements are 40 kev for Li^{+3} and about 11 kev for D^+. A typical energy spectrum deduced from the charge collector traces is shown in Fig. (7). In general, the energy spectra are peaked between 3 kev and 6 kev for Li^{+3} and .9 and 1.8 kev for D^+. Smaller secondary peaks in the energy spectra are typically seen at two to three times the main energy peak. In none of the shots have there ever appeared any of the fast ion componenets reported by other authors. [3,5] The initial small signal present in the charge collector trace, coincident with the arrival of the laser pulse on the target, was found to be due to photoelectrons and could be eliminated by the addition of a small magnetic field across the charge collectors.

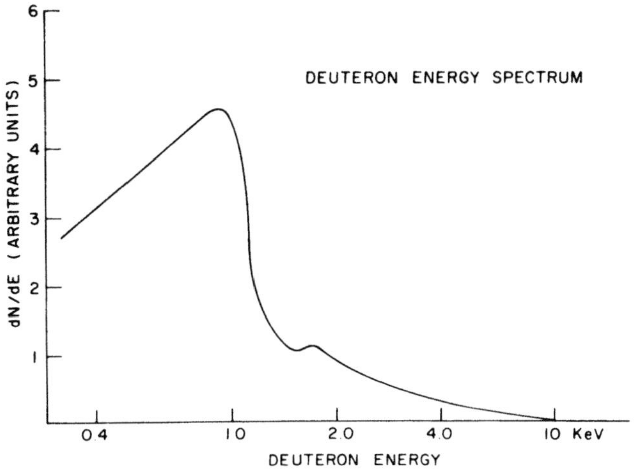

Fig. 7 A sample deuteron energy spectrum deduced from charge collector trace such as shown in Fig.

The total energy of expansion of the ions is typcially 10% of the incident laser energy in the range of incident intensities used in this set of experiments, i.e., 10^{15} to 10^{16} watt/cm^2. We refer to this as the "absorbed energy". It is a lower bound on the total energy absorbed by the target, and it does not include energies of vaporization, ionization, or reradiation. The correlation between the total number of charge particles and the "absorbed energy" is shown in Fig. (8). The linear relation between these two parameters is to be expected since the laser spot size is smaller than the target. An increase of laser energy

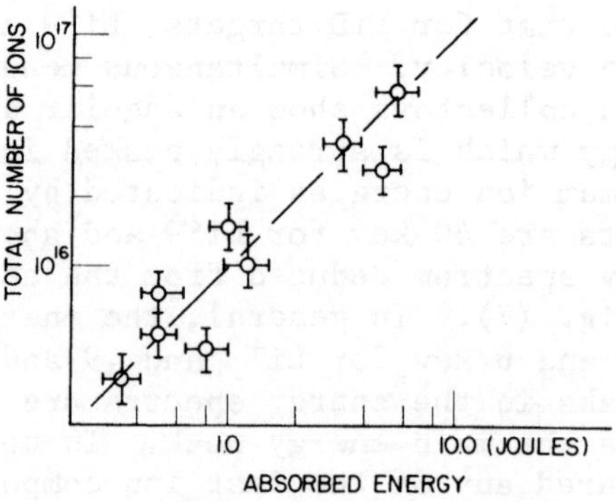

Fig. 8 Experimentally measured relationship between total number of ions created and "absorbed energy" for LiD targets.

Tends to increase the quantity of plasma but not necessarily the average particle energy.

2. Electron Temperature

The x-ray spectrum, as duduced from the foil transmission measurements, has two typical shapes. On roughly half of the shots a simple, single temperature, bremsstrahlung spectrum is obtained with temperatures ranging from 600 eV to 1.2 kev. On the remaining shots the curves could be satisfactorily fit by assuming two Maxwellian distributions with .8 kev and 2.0 kev respectively with the ratio of 2.0 kev to .8 kev radiators chosen to be 10^{-3} [11]. Fig. (5) shows examples of each type of distribution. There was no correlation between the shape of the spectra and the neutron yield or backscatter intensity. On a small number of shots, signals were recorded on the thick (11.5 mg/cm^2) Ni foil detector which might indicate components up to 2.8 kev (approximately 3.5 Te). In contrast to other measurements [3,4,5] no hard x-rays have been observed. Fig. (9) shows the measured electron temperature plotted vs. "absorbed energy" indicating a very limited dependence for "absorbed energies" above .5 joule. This is consistent with the above mentioned relation between the amount of plasma produced and the absorbed energy.

3. Neutron Yield

In the majority of the shots, in which relatively pure LiD was used as the target and the "absorbed energy" was higher than .4 joule, neutrons were produced. Targets which were contaminated with a large percentage of O_2 as measured by the mass spectrometer,

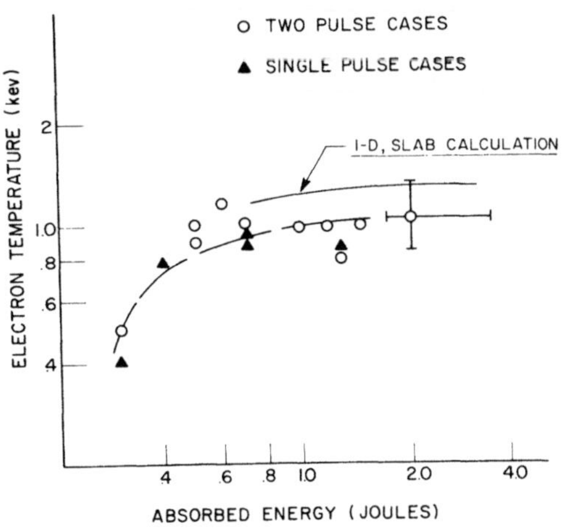

Fig. 9 Measured electron temperature versus "absorbed energy" for LiD targets.

did not produce any measurable neutron yield even though absorbed energies higher than 5 joules were recorded on some of these shots. Neutrons were observed also with $C_{36}D_{74}$ targets; no neutrons were observed with LiH or $C_{36}H_{74}$ targets. The verification that the recorded signals were indeed neutrons was primarily by their time-of-flight. All distinguishable signals on the 1 meter neutron detector had a delay of 47 x 10^{-9} seconds. On no shots was a prompt signal corresponding to an x-ray measured. Neutron production from anywhere but the target can be excluded by the time-of-flight measurements and the absence of very fast ions in the expanding plasma. The neutron yield versus "absorbed energy" is shown in Fig. (10). The data can be best fit by $N \simeq E_{ab}^{1.5}$ curve.

4. Backscattered Light Measurements:

(a) intensity. It is convenient to show separately the measured reflectivity vs. absorbed energy for the two experimental conditions reported here (i.e., one heating pulse or two heating pulses), see Fig. (11). The reflectivity of the second pulse in the two pulse cases averaged 9% while the reflectivity of the first energetic pulse averaged 3%. We note that a Lambertian reflector, reflecting 90% of the incident energy, would result in about 3.0% of the incident energy being directed back along the laser optical axis. This leaves up to 6.0% to be accounted for by other effects. The possibility of a correlation between neutron yield and reflectivity was investigated but none was found.

In addition to measuring the reflected intensity, the depolarization of the reflected beam was also monitored. It was found

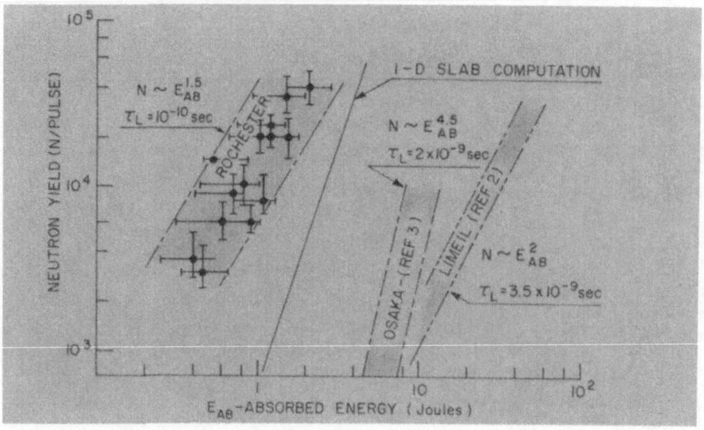

Fig. 10 The measured neutron yield versus "absorbed energy" compared to the computer calculation and to previous measurements. Note that the absolute value of absorbed energy for the various experiments is difficult to compare because of difference in measuring technique.

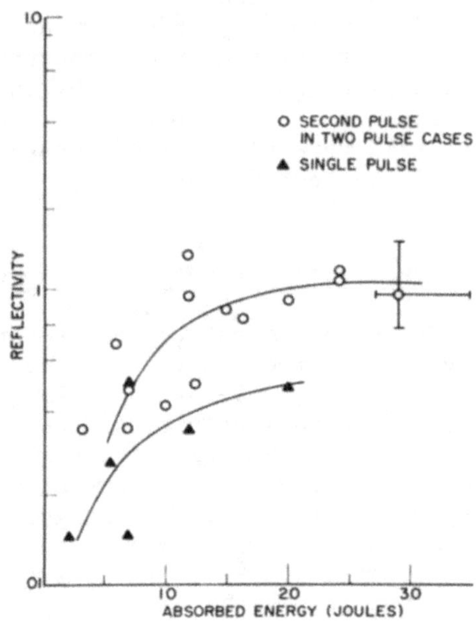

Fig. 11 Percentage of laser energy reflected back into laser optic axis, (reflectivity), versus absorbed energy, for LiD targets.

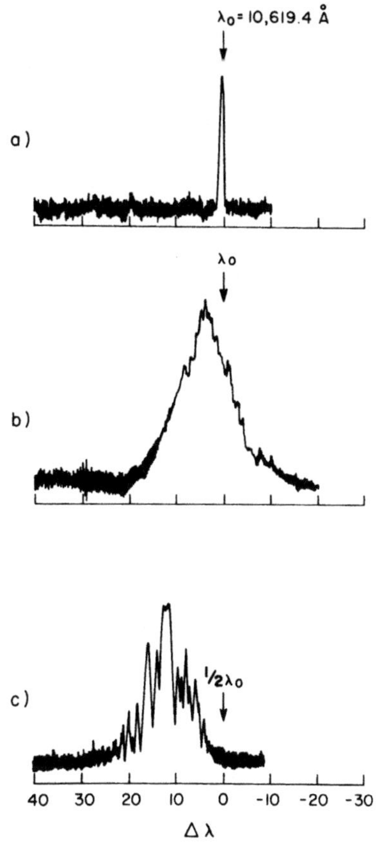

Fig. 12 a) Incident laser spectrum.
b) Spectrum of backscattered light near ω_L.
c) Spectrum of backscattered light near $2\omega_L$. All spectra were taken simultaneously under the following conditions: single 30 joule, 100 psec wide, laser pulse with prepulse incident on 150μ LiD target.

that the initially plane polarized incident beam was slightly depolarized on reflection; the degree of depolarization varied from 10% to 30%

(b) Spectral Studies. Comparisons of the spectra of the backreflected light from the targets to both the incident laser frequency band (ω_L) and its second harmonic ($2\omega_L$) indicate a consistent broadening and shift to longer wavelengths. A typical backscattered spectrum is shown in Fig. (12). Backreflected light at $3/2\omega_L$ was also monitored. The intensity was less than half of that observed at $2\omega_L$. Approximately 0.1% of the backreflected light was in the second harmonic while less than 0.05% was measured at $3/2\omega_L$.

The shift to longer wavelengths of the green light was three to four times larger than the shift of the backscattered light near 1.06μm. Typical red shifts for the 1.06μm component of the reflected light were between 2 and 6Å, while shifts as much as 15Å were recorded for the green component of the reflected light towards the long wavelength side of $2\omega_L$. The broadening ranged between 2 and 15Å for both 1.06μm and .53μm reflected light. No obvious dependence of shift was noticed on incident or absorbed energy.

DISCUSSION

The characteristic expansion time for these plasmas under the present experimental conditions is given by

$$\tau_E \simeq 2.5\text{x}10^{-13} E_{AB}^{1/2} R^{-1} \text{ (sec)}$$

where E_{AB} is the absorbed energy in joules and R the focal spot radius in centimeters [12]. For $R \simeq 3\text{x}10^{-3}$cm and $E_{AB} \simeq 1$ joule $\tau_E \simeq 8\text{x}10^{-11}$sec. As the laser pulse width is $\tau_L \simeq 10^{-10}$sec., a knowledge of the plasma dynamics during the heating phase is important in estimating the conditions under which neutron emission occurs.

A Lagrangian computer code is used to model the evolution of the laser heated plasma [13]. It treats the target as initially vaporized but cold and neutral. The effect of the prepulse in preionizing the target is calculated using the experimental ratio of prepulse to main pulse energy and the proper delay between the two pulses. The code contains the following assumptions:

1. Light absorption is due only to inverse bremsstrahlung in the weak field limit; (single photon); i.e., the absorption coefficient is that given by Dawson [14] with the appropriate correction.

2. The dynamics of the neutral and charged particles species are followed, requiring that all heavy particles have the same macroscopic velocity and temperature. The electrons are treated as having a separate temperature.

3. The electron thermal conductivity is handled in the diffusion approximation using the Spitzer expression for thermal conductivity [15]. The maximum heat transfer is flux limited using the limiter form given by Shearer [16]. Thermal conductivity of the ions and neutrals is treated in a similar manner with the state of ionization taken into account at each point of time.

4. The ionization of the heavy species can be treated either in the equilibrium approximation using the Saha equation or by integrating the full rate equations. Non-linear interactions, such as decay instability heating or stimulated Brillouin or Raman scattering, are not included in the computational work in its present form.

Since the spot size of the focused radiation is always smaller than the target diameter, a slab version of this code was used to correlate absorbed energy, neutron yield and temperature with the experimentally measured parameters. A typical set of results for a sample calculation is shown in Fig. (13).

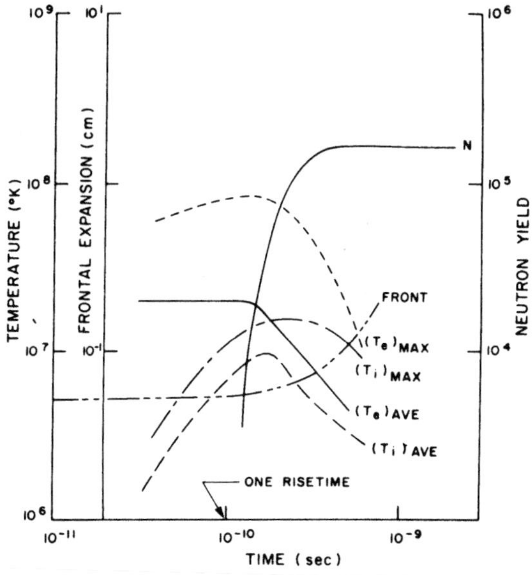

Fig. 13 Computer calculated laser produced plasma development, for the case of a 20 joule 100 psec, laser pulse incident on a square, 40 x 40 μm, of LiD, prepulse is also included.

compared the measured absorption in Fig. (14). Note that the measured fractional absorption falls rather rapidly with incident intensity while the computed vale, assuming the weak field approximation change slowly. At very high electric fields classically, one expects the absorption to be proportional to (ω_L/E_o), thus higher field strengths lead to lower oabsorption.

The parameter which determines the relative importance of weak or strong field collisional absorption (single or multiphoton absorption) of the incident radiation is the ratio of the electron oscillatory velocity to the electron thermal velocity:

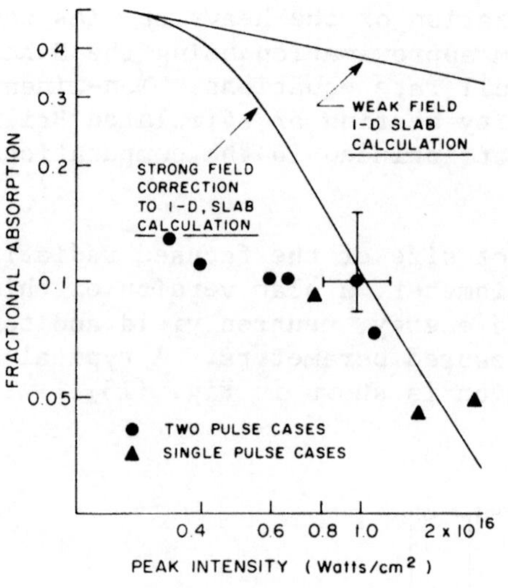

Fig. 14 Measured fractional absorption compared to the weak field, computer calculated absorption, and the strong field corrected absorption.

$$\frac{v_o}{v_T} = \frac{eE_o/m\omega}{(kT_e/m)^{1/2}}$$

$$= 6.6\text{x}10^{-7}(I/T_e)^{1/2} \text{ at } \lambda = 1.06\mu\text{m}$$

where I is the focused radiation intensity in watts/cm^2, and T_e is the electron temperature in eV.

For the conditions of these experiments ($I \leq 2\text{x}10^{16}$ w/cm^2; $T \leq 1$ kev)

$$0.1 < \frac{v_o}{v_T} \leq 2.9$$

The intensity in the plasma I
The intensity in the plasma I(Z) can be written as

$$I(Z) \quad Io \exp(-\int k d\ell)$$

where k is the absorption coefficient given by

$$\frac{\nu_{SF}}{\nu_{WF}} = \frac{3}{2} (2\pi)^{3/2} \left(\frac{v_T}{v_o}\right) \frac{1}{\ell n\Lambda} \simeq 0.1$$

This would result in approximately 91% less absorption in the plasma volume than would be predicted by weak field absorption. Thus strong field corrections are important at these field strengths. The measured fractional absorbed energy shown in Fig. (14) exhibits a fall off at intensities greater than $1\text{x}10^{-16}$watt/cm^2 on target.

Neutron Yield

A comparison of the predicted vs. observed neutron yields as a function of absorbed energy is shown in Fig. (10). We note that the predicted neutron yields are less than the measured yields by more than an order of magnitude. Part of the difference may be due to the manner in which the absorbed energy is determined. Since only the energy in the expansion is used, this number represents a lower limit on absorbed energy. Errors in the ion measurement, plus inclusion of the energy of ionization and recombination, could increase the true absorbed energy by as much as 50%. Possibly more important is the difference between experiment and theory in the scaling of neutron yield with absorbed energy. For comparison, other neutron scaling results reported in the literature are also shown in Fig. (11). Experimentally we observe $N \alpha E_{AB}^{1.5}$ whereas the predicted dependence is $N \alpha E_{AB}^{3.0}$. There is a scaling similarity between the experimental results reported here and those reported by Floux et al. [2]. A full treatment of the two-dimensional character of the expansion would, in all likelihood, reduce the discrepency between calculated and measured neutron yield scaling.

At these neutron yields it is difficult to determine whether or not these are truly thermally produced neutrons. For example, the charge collector measurements sometimes indicate a finite number ($\simeq 10^{14}$) of deuterons at approximately 15 kev. If one assumes a density profile, $n(x)=n_c(1 - \frac{x}{\ell})$, (produced for example by the prepulse) then the neutron yield resulting from $N_D \simeq 10^{14}$ deuterons traversing the underdense portion of the plasma will be:

$$N = \frac{1}{2} n_c N_D \sigma_{DD} \ell \ .$$

If $N_D \simeq 10^{14}$; $n_c = 10^{21}\text{cm}^{-3}$; $\ell = 10^{-3}\text{cm}$; and $\sigma_{DD} \simeq 10^{-28}\text{cm}^2$,

then $N \simeq 5\text{x}10^3$, which is of the same order of magnitude as the

experimentally measured neutron yields. Thus the detailed origin of the neutrons in laser produced plasmas will not be truly clarified until the plasma temperatures are high enough so that thermally produced neutrons are dominant.

There is no evidence, however, for the neutrons being produced by a beam of ions striking deuterated surfaces at a distance from the target in the manner reported by McCall et al. [5].

In the experiments reported here, v_o/v_T varies from .1 to 2.9 so that one might expect to exceed the threshold if the scale length is as long as 50μm. Even if the threshold is exceeded, it is not clear what magnitude of reflection might be expected from stimulated Brillouin scattering because of the complicated interaction between this and the other wave-wave instabilities. For example, particle in cell computer simulations indicate a reduction in backscattered intensity because of electron heating[19].

The experimental results are consistent with the assumption that part of the reflection is due to stimulated Brillouin scatterings saturating at a level of approximately ten percent. The backscattered spectra at both ω_L and $2\omega_L$ are always shifted to longer wavelengths. The magnitude of the shift varies from shot to shot and shows no correlation with the measured electron temperatures in contrast to the Brillouin shift ($\Delta\nu$) which might be expected to vary at $\sim(T_e)^{1/2}$ since

$$\frac{\Delta\nu}{\nu} = \frac{2n}{c}\left[\frac{kT_e}{m_i}\right]^{1/2}$$

Here n is the refractive index of the plasma.

It is possible to explain the red shift and broadening at on the basis of a moving reflective interface at $\omega_L=\omega_p$. The larger shifts encountered at $2\omega_L$ however cannot be accounted for in this fashion and as yet we do not have a satisfactory explanation for this behavior. Generation of frequencies at $\frac{1}{2}\omega_L$ or $\frac{3}{2}\omega_L$ could be evidence for the stimulation of parametric instabilities [20]. The low level of $\frac{3}{2}\omega_L$ and $2\omega_L$ together with the absence of energetic electrons indicate that parametric instabilities do not dominate the radiation plasma interaction.

An additional source of enhanced reflectivity could be stimulated Compton scattering. According to Ref. [21] the total reflection from a plasma layer of scale length ℓ is given by the relation:

$$R = R_o \exp \ I_o\{\beta(1-R)\}$$

where

$$\beta = \frac{8\pi^2 c r_o^2}{\omega_L^{3=}} \frac{V_o n_e}{kT_e} \left(\frac{mc^2}{2\pi kT_e}\right)^{1/2} \exp\left(\frac{mV_o}{2kT_e}\right) .$$

In fact, the time delay of the neutron signals rules out the production of neutrons from any place but the target. The two surfaces likely to produce neutrons at a distance are the focusing lens and one of the neutron counters. Both are more than 15cm. from the target and would, therefore, introduce a signal delay of at least 50×10^{-9} sec., even for a 100 kev ion.

Temperature

To compare the computed temperatures to the measured x-ray transmission curves, the computer code calculated a time and space integrated spectral distribution. Folding this predicted spectral absorption through the transmission functions of the aluminum foils leads to a predicted transmission curve for each calculated incident energy. These predicted curves are very nearly Maxwellian and we can thus assign a single temperature to each calculated case. These calculated temperatures are shown in Fig. (9) and compared with the observed values. For the aluminum foils used in the experiment only the highest temperature region of the plasma contributes to the measurement and one is therefore only measuring the temperature at the absorption region. This temperature is predicted to be relatively independent of absorbed energy which is confirmed by the measurements.

Reflection and Non-linear Processes

The intensity and the spectral distribution of the back-scattered radiation has been used as a measure of presence of stimulated Raman and stimulated Brillouin scattering. An increase in the reflectivity $R \simeq .18E_{inc}^{0.3}$ has been reported by Floux [2]. Shearer and Yamanaka also indicate a dependence of the reflectivity with incident energy [3,4].

The present set of observations show only a rough dependence of the reflectivity on the incident energy. However, if one plots the reflectivity against absorbed energy as in Fig. (11), a pattern is discernible. As discussed earlier, there is a different dependence for the single or first main pulse than for the second pulse. The reflectivity in the single pulse case increases rapidly to 5%, whereas the reflectivity for the second pulse starts at 5% and very rapidly saturates at 10%.

According to Rosenbluth [18], the threshold for stimulated Brillouin scattering, taking into account the inhomogeneous density of the plasma, occurs when $v_o^2/v_T^2 > 4/k_oL$. L is a characteristic density length in the plasma. For $L \simeq 50\lambda_o$ the threshold is given by $v_o/v_T > 0.28$.

R_o is the initial reflection coefficient in the absence of Compton effects, r_o is the classical electron radius, n_e the electron density, and V_o is the plasma velocity. At typical experimental conditions, $kT_e \simeq 1$ kev,

$v_o \simeq 2\text{x}10^7$c/s, and $n_e \simeq 10^{21}\text{cm}^{-3}$, $\beta \simeq 2\text{x}10^{14}$cm/w. For $I_o \simeq 10^{16}\text{w/cm}^2$ and $\ell \simeq 10^{-3}$cm., this predicts a 22% increase in the total reflection coefficient. Our experimental observations, however, indicate a leveling off of the backreflection at a level of about 10%. Stimulated Compton scattering has an exponential dependence on intensity and thus should be quite apparent when present. Experiments with greater incident intensities, on the order of 10^{16} to 10^{17} w/c^2 will be crucial in determing the significance of stimulated Compton scattering to the laser radiation plasma interaction problem.

SUMMARY

In these experiments using a 10^{-10} sec., time tailored laser pulse to irradiate spherical targets, extensive diagnositcs have provided us with simultaneous information on absorption, reflection, charged particle distribution, neutron yield, and electron temperature of the plasmas produced. Our conclusions can be summarized as follows:

1. Absorption of the radiation up to incident intensities of 2×10^{16} watt/cm^2 can be accounted for by classical inverse bremsstrahlung absorption, taking into account the strong field corrections.

2. No evidence has been found for anomalously high charged particle energies.

3. Backreflection saturates at 10^{-1} of the incident beam and is not greatly depolarized. This radiation has frequency components near ω_L, $\frac{3}{2}\,\omega_L$, $2\omega_L$, and the amount of reflected energy at $\frac{3}{2}\omega_L$, $2\omega_L$ is small, amounting to less than 2×10^{-4} of the incident beam.

4. Good agreement between predicted and measured temperatures (0.8 - 1.2 kev) is obtained with some distributions indicating an energetic tail at $3v_T$.

5. Neutron yields ($N \sim 4\text{x}10^4$), larger than the predicted values are observed to originate in the target.

Saturation of the reflectivity and the generally good agreement with computation gives us encouragement to continue these studies at higher energies.

ACKNOWLEDGEMENTS

We have had many helpful discussions with Drs. E. Goldman R. Kinsinger. Dr. E. Goldman has carried out the numerical work. X-ray diagnostics were conducted by T. Bristow. The time-of-flight mass spectrograph was developed and operated by M. Oron and Y. Paiss. The assistance of S. Kumpan in obtaining the optical spectra is gratefully acknowleged. This work was supported by the Laser Fusion Feasibility Project.

REFERENCES

1. Basov, N.G., Kruikov, P.G., Zakharov, S.D., Senatski, Yu. V., and Tchekalin, S.V., I.E.E.E. J. Quant. Electronics QE-4 (1968) 864.

2. Floux, F., Cognard, D., Denoeud, L.G., Piar, G., Parisot, D., Bobin, J.L., Delobeau, R., and Fauquignon, C., Phys. Rev. A 1 (1970) 821.

3. Yamanaka, C., Yamanaka, T., Sasaki, S., Yoshida, K., Waki, M., and Kang, H., Phys. Rev. A 6 (1972) 2335.

4. Shearer, J.W., Mead, S.W., Petruzzi, J., Rainer, F., Swain, J.E., and Violet, C.E., Phys. Rev. A 6 (1972) 764.

5. McCall, G.H., Yound, F., Ehler, A.W., Kephart, J.F., and Godwin, R.P., Phys. Rev. Lett. 30 (1973) 1116.

6. Kaw, P.K. and Dawson, J.M., Phys. Fluids 12 (1969) 2586.

7. Nuckolls, J., Wood, L., Thiessen, A., and Zimmerman, G., Nature 239 (1972) 139.

8. Soures, J., Kumpan, S., and Hoose, J. Report No. 20, Laboratory for Laser Energetics, University of Rochester (1973), to be published, Appl. Optics.

9. Soures, J., Bristow, T., Lubin, M., Oron, M., and Goldman, L. Bul. Amer. Phys. Soc. 17, 972 (1972) Abstract 1B8; Brauer, A., and Goldman, E.B., Bul, Amer. Phys. Soc. 17, 1000 (1972), Abstract 3C7, Jour. Appl. Physics (March 1974).

10. Oron, M. and Paiss, Y., "A Dynamic Mass Spectrometer for the Study of Laser-Produced Plasmas", Rev. Sci. Instr. 44, 9 (Sept. 1973). 1293-96.

11. Bristow, T.C., "X-ray Emission and Ionization Dynamics of Laser Produced Plasmas", Ph.D. Thesis, University of Rochester, Rochester, N.Y. (1973).

12. Gratton, R. in "Physics of High Energy Density" edited by P. Caldirola and H. Knoepfel, Academic Press, N.Y. (1971).

13. Goldman, E., Plasma Physics 15 (1973) 289; also Report No. 16, Laboratory for Laser Energetics, University of Rochester (1973).

14. Dawson, J.M., Phys. of Fluids 7 (1964) 981; also Johnston, T.W., and Dawson, J.M., Phys. of Fluids 16 (1973) 722.

15. Spitzer, L., "Physics of Fully Ionized Gases", Wiley, New York (1967).

16. Shearer, J., Lawrence Livermore Laboratory Report UCID-15745 (1970).

17. Seely, J.F. and Harris, E.G., Phys. Rev. A 7 (1973) 1064.

18. Rosenbluth, M.N., Phys. Rev. Lett. 29 (1972) 565.

19. Kruer, W.L., Eastabrook, K.G., Sinz, K.H., Lawrence Livermore Laboratory Report UCRL-74676 (1973); also Forslund, D.W., Kindel, J.M., and Lindman, E.L., Phys. Rev. Lett. 30 (1973) 739.

20. Bobin, J.F., Decroisette, M., Meyer, B., and Vitel, Y., Phys. REv. Lett. 30 (1973) 594.

21. Krasyuk, I.K., Pashinin, P.P., and Prokhorov, A.M., JETP Lett. 17 (1973) 92.

THERMONUCLEAR FUSION PLASMA HEATED BY LASERS*

Chiyoe Yamanaka and Tatsuhiko Yamanaka

Institute of Laser Engineering

Osaka University, Osaka 565 JAPAN

Hyung-Boo Kang

Institute of Plasma Physics, Nagoya University

Nagoya 464 JAPAN

ABSTRACT

The interaction between laser and plasma has been investigated using the irradiation of YAG-glass and glass-glass laser systems on the solid hydrogen and deuterium target. The broad band laser system, the latter, was much more effective for heating which might be attributed to a double resonance absorption. Induced back scattering of SHG and Brillouin was studied. Isotope effect of the ion wave was observed. In Appendix the properties of high power disc laser and the results of a computer experiment on pellet compression were presented.

*Presented at the Third Workshop on "Laser Interaction and Related Plasma Phenomena" at Rensselaer Polytechnic Institute, Troy, August 13-17, 1973.

INTRODUCTION

As laser has attained an ability to deliver a large amount of energy very rapidly to matter, one can produce a plasma of thermonuclear temperature by this means[1]. Quite recently much attention has focused on laser fusion[2] and the feasibility test has come on the time schedule. Anomalous heating physics, the development of high power laser and the high compression scheme are very important and concomitant for the laser fusion project.

The laser interaction with plasma[3] has been investigated. Parametric excitation of collective process of plasma is very effective to plasma heating. Near the cut off region, the decay instability[4] (photon-plasmon-phonon interaction) plays an essential role which has been testified not only in laser plasma but also in ionosphere and microwave simulation experiments. Various kinds of theoretical treatment[5] have been prevailed for the excitation of instabilities. But the quantitative treatment for absorption is still in a preliminary stage. In the underdense region, nonlinear optical phenomena such as SHG, Raman, Brillouin induced scatterings[6] (Photon-photon-plasmon or phonon interaction) have been investigated. These two are competitive processes in the interaction with plasma either of which earlier reaches the instability.

To clarify these situations, we have performed the laser bombardment experiment on solid hydrogen and deuterium, using a YAG oscillator-glass amplifier system and a glass oscillator-glass amplifier system. Isotope effect of backscattering has been observed. And the double resonance heating process[7] can be expected by using a broad band laser system. Experiments up to now show the source of the neutron was mainly the deuterium target. The doping of high Z materials was also tested.

In the appendix, the detail properties of laser system and the results of computer experiment on high power bombardment to a pellet are presented.

EXPERIMENTAL ARRANGEMENT

The laser system was composed of an oscillator section, an amplifier section and a booster section. The whole system installed at Osaka is shown in Fig.1. The detail data of it are given in Table 1. The oscillator section has two oscillators YAG and glass, either of which alternatively used for the change of the laser light spectral width. The TEM_{00} oscillator signal produced by the KDP Pockels cell Q switching was fed to the 2 stage preamplifier and then it was tailored by the pulse transmission mode technique. Ordinary pulse width was 2-10nsec. The signal of 2 MW level was supplied to the amplifier section which was described before[1]. The output pattern of this section is shown in Fig.2. The power level was 25 GW and the beam divergence was less than 1 m radian. The anomolous heating and induced scattering experiments were mainly performed by this power level. The booster section has one preamplifier and three stages of disc amplifier, each of which contains 4 black glass coated oval discs[1,3]. The output was about 250 joules in 2 nsec. The detail properties of the high power laser system are given in Appendix.

For the experiments on the nonlinear optical phemomena of plasma, we mainly used solid deuterium and hydrogen stick, the dimension of which was 2 mm × 2 mm × 1 cm. The laser beam was focused by an aspherical lens, f = 0.83. The focal point was checked by the 10 times magnified projection pattern of the target. The accuracy of the focal adjustment was less than 50 μm. The backscattered spectrum was measured using Czerny-Turner grating spectometer with a mean dispersion of 8 A/mm in the first order. The X-ray from plasma was detected by the 4 channel plastic scintillator. The particle measurement was performed by the time of flight method. The neutron source of the laser irradiated plasma was located by the two neutron detectors. The magnetic probe was also used to catch the laser induced current near the target.

EXPERIMENTAL RESULTS

As laser plasma exists for a very short time, the experimental measurement should be developed to get the transient informations of the process in plasma.

(1) Reflection

The backward reflection of laser light from plasma was measured. Figure 3 indicates the spectrum of inci-

Table 1. Properties of laser system

	Oscillator Section			Amplifier Section					Booster Section	
	TEM_{00} Osc	Preamp I	Preamp II	I	II	III	IV	V	Preamp	Disc Laser
Glass	LCG11	LCG11	LCG11	LCG11	LCG11	LCG11	LCG11	LCG11	LSG91	LSG91 H
Nd^{3+} (wt%)	3.5	3.5	3.5	3.5	3.5	3.5	3.5	3.5	1.5	3
Glass size	$10^{\phi}\times150^{\ell}$	$10^{\phi}\times150^{\ell}$	$15^{\phi}\times330^{\ell}$	$20^{\phi}\times330^{\ell}$	$20^{\phi}\times330^{\ell}$	$30^{\phi}\times330^{\ell}$	$30^{\phi}\times330^{\ell}$	$40^{\phi}\times600^{\ell}$	$60^{\phi}\times400^{\ell}$	$80^{\phi eff}\times25^{t}\times12$
Flash Lamp	1	1	1	4	4	6	6	10	10	30
Pump. Energy (kJ)	0.8	1	4.3	11	11	17	17	72	40	140
(Max. Pump. E.)	(2)	(2)	(6)	(20)	(20)	(30)	(30)	(200)	(100)	(300)
Gain		3	6	16	14	10	6	5	2.4	2
Power (MW) Inp.		0.3	0.9	2	20	180	1.5×10^{3}	5×10^{3}	2.5×10^{4}	6×10^{4}
Power (MW) Outp.	0.3	-0.9	5.4	32	280	1.8×10^{3}	6.7×10^{3}	2.5×10^{4}	6×10^{4}	1.2×10^{4}

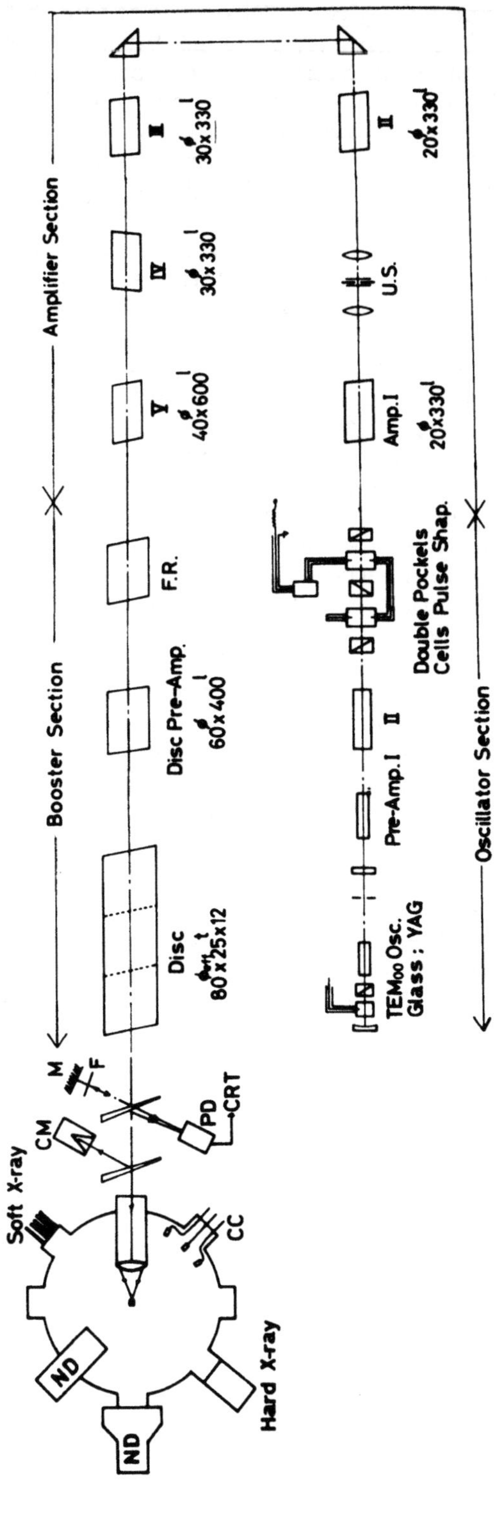

Fig.1. Schematic diagram of whole laser system.

dent and reflected light. The spectrum of the reflected light near the incident spectrum became broad about 10 A for YAG and 100 A for glass laser and the peak showed red shift about 5 A. The total reflectivity versus glass laser intensity is shown in Fig.4. The reflectivity due to glass laser was at most 18%-20% and it due to YAG laser was 20%-25%. The focal point on the target is very critical for the reflectivity with the distance from the surface. The back scattered light appeared in the various spectrum range. The second harmonics in back scattered light was mainly measured. Figure 5 indicates the SHG characteristics which accorded with the square law of incident laser power. The energy of the second harmonic light has linear dependence on that of the reflected fundamental light as shown in Fig.6. The second harmonic scattering showed very interesting multistructures of spectrum especially clear using YAG laser. Figure 7 indicates these behaviors in hydrogen and deuterium. There were four scattering peaks. The nearest side peak to the second harmonic light seems to be due to Brillouin scattering. The others were attributed to the over tones. The difference shift between H and D was due to the isotope effect. In the case of glass laser, the spectrum was a comvolution of the incident broad spectral light and the density profile of plasma, so the detail structure was hardly recognized except 2 peaks. But the growth rate was much larger than the YAG.

(2) Electron Energy

The X-ray measurements give us the information of electrons in plasma. We used a 4 channel soft X-ray detector to clarify the difference of heating process between glass and YAG lasers. Figure 8 showes the typical responce of the X-ray detector and estimations of electron tempereture. They show the deviation from the Maxwellian distribution of electron. Using the two combination of the absorption foils, the results represent the two groups of temperatures. As shown in Fig.9 and 10, the glass laser was much more powerful for heating of electrons than the YAG laser. As the origin of X-ray has been studied in Garching[8], the isotropic emission of X-ray means the heating of plasma by laser. The anisotropic expansion of plasma detected[9] was caused by the hydrodymamic effect appeared in a certain distance of the flight. The hard X-ray above 42 KeV was also observed. Intensity of hard X-ray was shown in Fig.11.

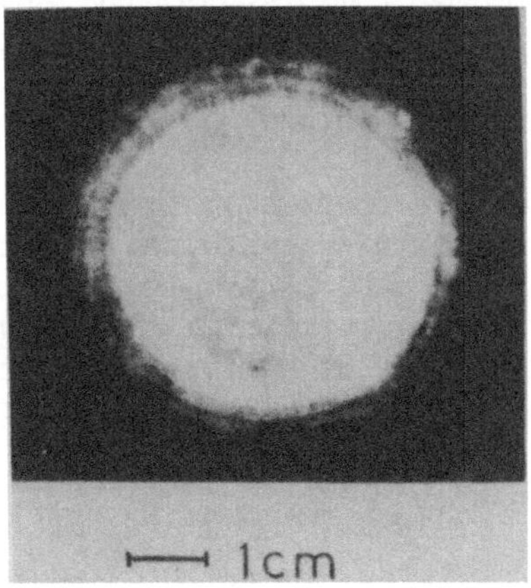

Fig.2. Output pattern of amplifier section.

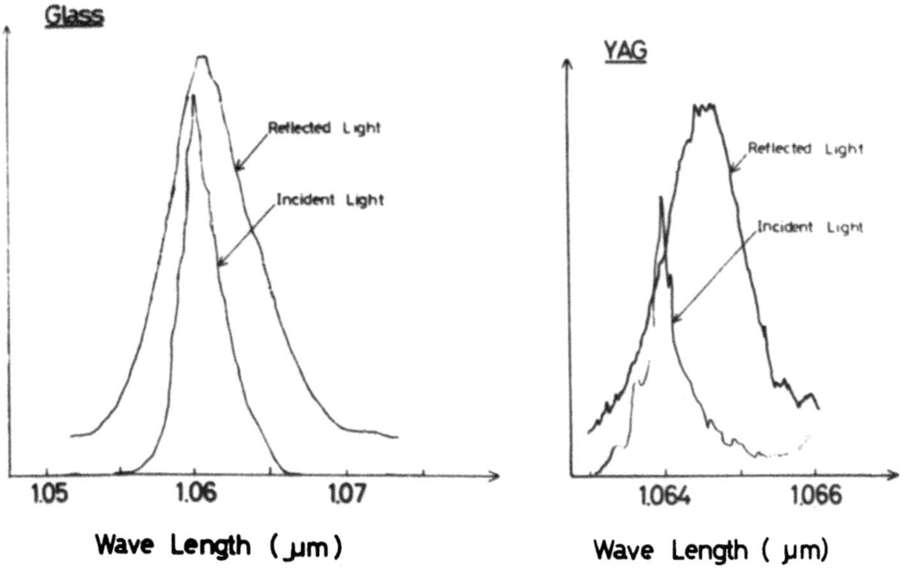

Fig.3. Spectrum of incident and reflected laser light.

(3) Ion Energy

The ion energy was measured by the time of flight method. The ion has two groups of velocity[3] above the critical laser intensity. The kinetic energy dependence on the laser power was shown in Fig.12.

The threshold laser intensity for glass and YAG was measured. The former was one order smaller than the latter. The magnetic flux was measured by a samll pick up one turn coil at the point 7 mm from the target forward about 45 degree to the incident direction. The intensity was about 30 gauss. The responce of probe is shown in Fig.13.

(4) Neutrons

The neutron measurement has been performed by glass and YAG lasers. Figure 14 shows the neutron yield versus laser energy relations. The neutron production was assumed to be isotropic for the whole solid angle. To get the same amount of the yield, the YAG laser needed twice energy of the glass. The dependence of the neutron yield to laser energy accorded with 5 powers of the energy in the low energy region and as pointed by the Limeil group[10] tended to saturate in the high energy region.

The origin of the neutron source was tried to locate by the time of flight with the two detectors as shown in Fig.15. The main source of the neutron was the target itself within our experimental error.

DISCUSSION

Laser plasma has two interesting features, one of which is the individual motion of collision free constituent caused by the thermally dissociated origin and another is the collective wave motion due to the long range force of interaction. The coupling between laser light and plasma is one of the most interesting topics of plasma physics. As well known, classical theory predicts such a long absorption distance to the laser plasma that anomalous process has been attracted the attention. At the under dense region, stimulated Raman and Brillouin scattering can be induced. According to Manley-Rowe relation, the most part of laser energy goes to the backscattered light and not to the plasma waves. While at the cut-off region, we can expect to

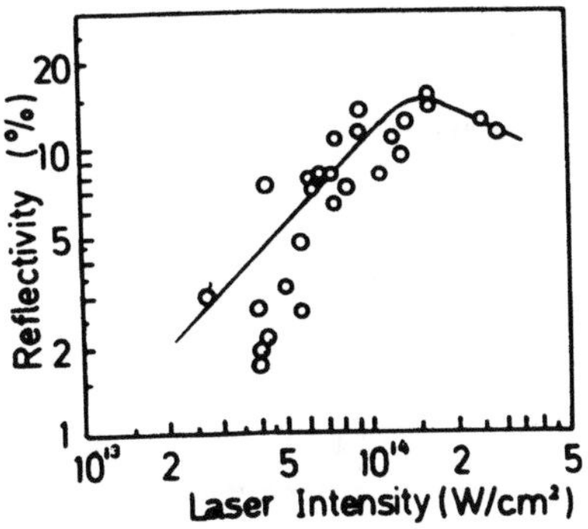

Fig.4. Reflectivity of plasma by glass laser.

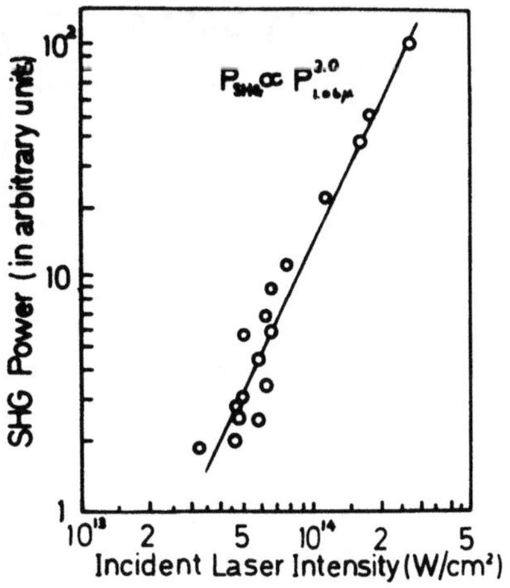

Fig.5. Power dependence of second harmonic generation.

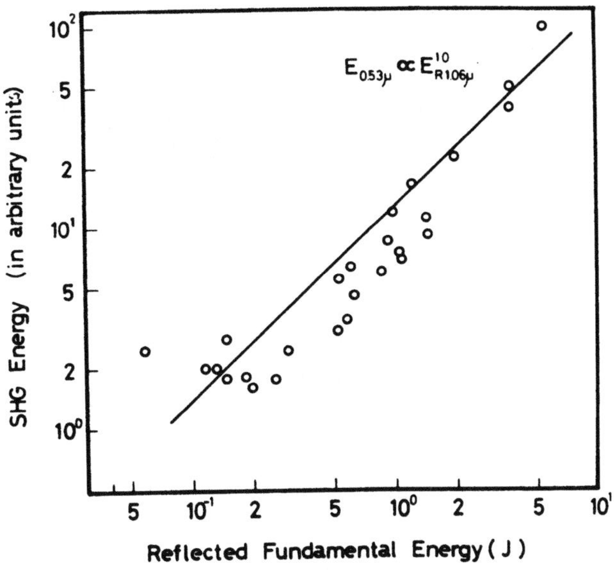

Fig.6. Relation between reflected fundamental light and second harmonics.

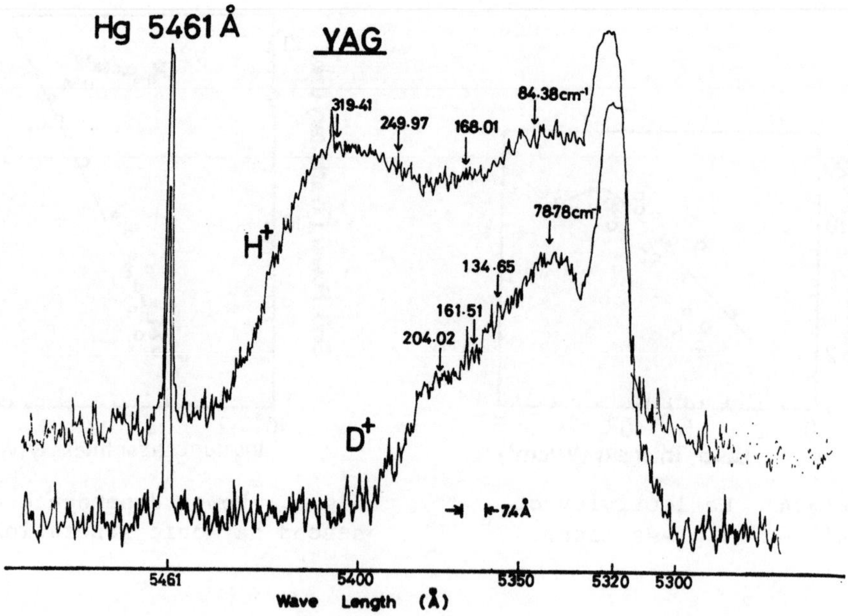

Fig.7. Spectrum of second harmonics of hydrogen and deuterium plasma with Brillouin shifts.

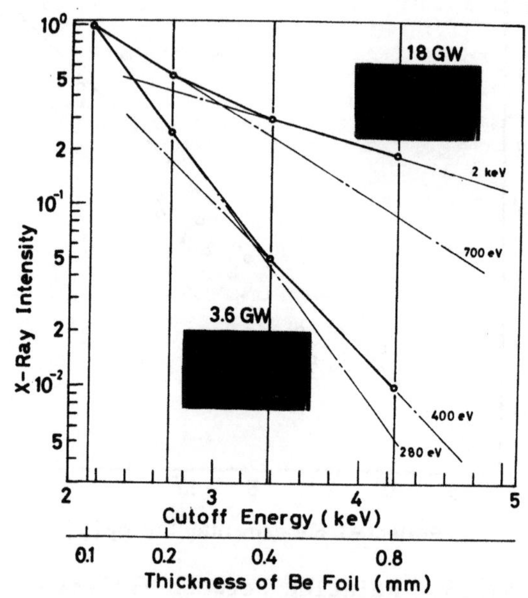

Fig.8. X-ray intensity measured by 4 channel X-ray detector.

rise the electron and ion waves. In a homogeneous plasma, the threshold of these instability is determined by the damping of the excited waves. The threshold energy[11] density $E_0^2/4\pi$ normalized by the electron thermal energy NT_e is given by Λ^2

$$\Lambda^2 = \frac{E_0^2}{4\pi NT_e} = \alpha\frac{\Gamma_1\Gamma_2}{\omega_1\omega_2}$$

where Γ_1 and Γ_2 are the damping rates of the excited waves of frequency ω_1 and ω_2 respectively and α is the couping factor of the laser light and the excited waves. Table 2 gives some typical values of α which is inversely proportional of coupling strength.

Table 2. Coupling factor α of parametric excitation

plasma		Excited process		α
under dense		Stimulated Raman Scatter		
		Stimulated Brillouin Scatter		$(c/2\,V_e)^2$
Near cut-off	under dence	Parametric decay instability	$\omega_1, \omega_2 >> \Gamma_1, \Gamma_2$	4
			$\omega_2 >> \Gamma_2 >> \omega_1 >> \Gamma_1$	$\frac{16}{9}\sqrt{3}\,\frac{\Gamma_2}{\omega_1}$
	over dence	Oscillating two stream instability	$\omega_1 = 0$ growing mode	ω_1/Γ_1 (Virtual)

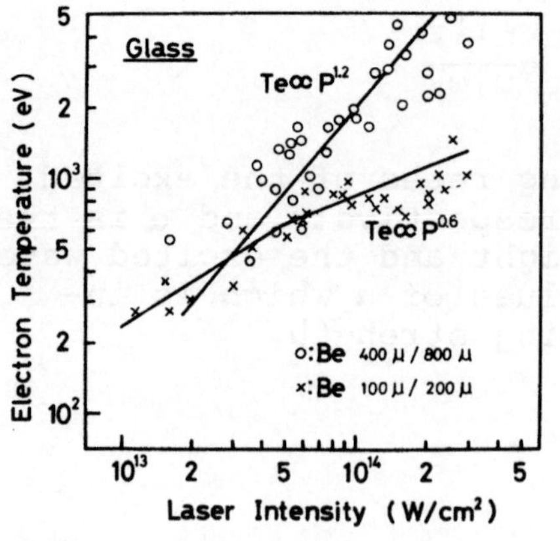

Fig.9. Power dependence of electron temperature by glass laser.

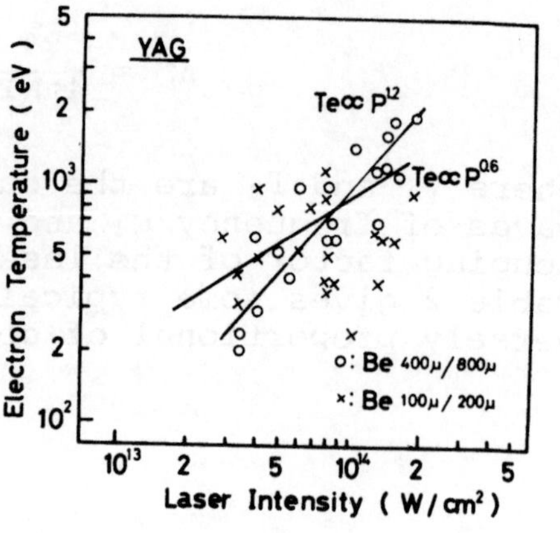

Fig.10. Power dependence of electron temperature by YAG laser.

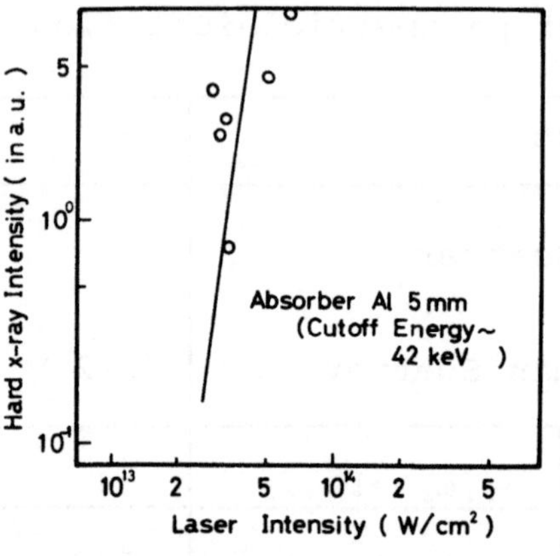

Fig.11. Power dependence of hard X-ray.

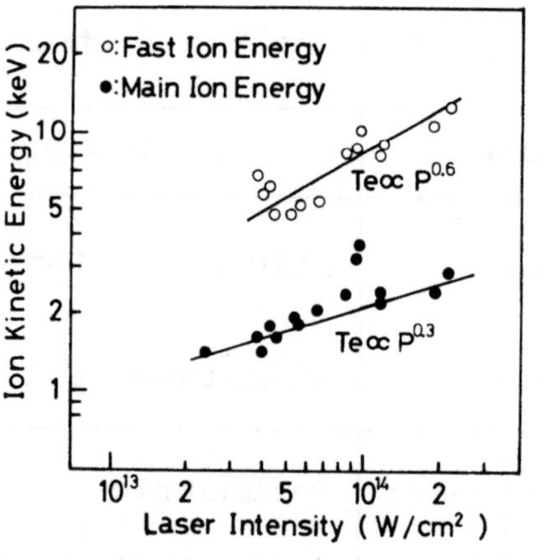

Fig.12. Ion kinetic energy versus laser power; glass laser.

The threshold of light scattering by glass laser for 1 KeV spacialy homogenious plasma is 10^{11} W/cm^2, that of the decay instability is 10^{12} W/cm^2 and that of the oscillating two stream instability is 10^{13} W/cm^2 respectively.

Our experimental results indicated that the laser beam could reach the cut-off density region without heavily stimulated scattering in the underdense region. As the laser plasma is inhomogenious in space, the excited waves propagate away from the interaction region before the waves grow to a substantial level and decay in outside. The convective loss for excitation should be considered. In the stimulated scattering process, the scattered electromagnetic wave has the light speed, while in the longitudinal waves the group velocity is small. The convection loss is not so important for the latter comparing the former. If the convection loss is dominant, the threshold condition is roughly given by $\gamma_0 > v_g/L$ where γ_0 is the growth rate in homogenious plasma, v_g is the group velocity of excited wave and L is the length of the interaction region. For stimulated Raman scattering $\gamma_0^2 = e^2E_0^2(\omega_0\omega_{pe})/(2m\omega_0c)^2$, so the threshold power becomes $10^{13}/L$ W/cm^2 for Nd glass laser and for decay instability case

$$\gamma_0^2 = \frac{e^2}{4mM}\frac{(kE_0)^2}{\omega_{pe}\Omega_k}$$

so the threshold is about $10^9/L$ W/cm^2. Comparing two these process, the interaction region length $L = (\frac{1}{n}\frac{dn}{dx})^{-1}$ is different each other and also in the latter case swelling effect of the electric field near the cut off becomes dominant. So in our plasma the threshold of decay instability becomes several orders smaller than that for the stimulated scattering.

At first we discuss the parametric instability in the cut off region. The detail features of this anomalous effect had been published before[3]. Now we present the difference of heating effect between the YAG system and the glass system as shown in section 3 which deliver the same amount of energy in TEM_{00} mode but the different spectrum width. Nishikawa[7] et al have presented theoretical treatment of the double resonance of parametric excitation using two pumps, both being in resonance with the electron plasma oscillation and at the

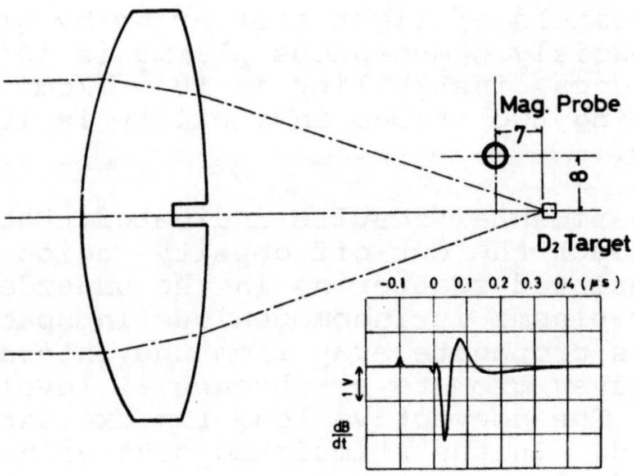

Fig.13. Response of magnetic probe.

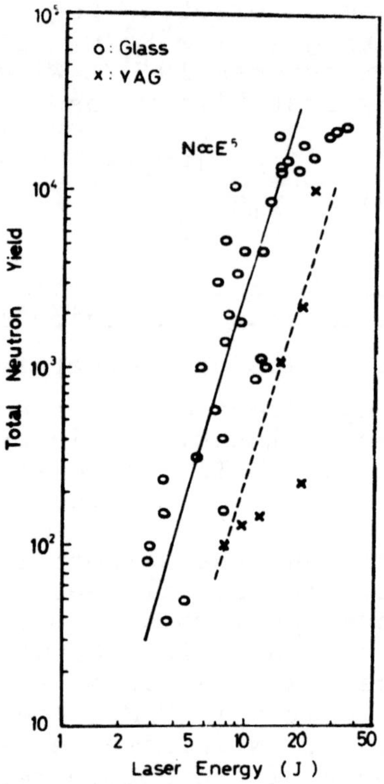

Fig.14. Neutron yield versus laser energy.

same time their beat frequency being resonance with the ion acoustic frequency. According to it, (1) When the beat frequency Δ is tuned to the twice of the ion acoustic frequency 2Ω, the total threshold power for excitation of the ion-wave is

$$\Lambda_1^2\Lambda_2^2 = \left(4\frac{\Gamma_i\gamma_e}{\omega_{pe}\Omega}\right)^2 ,$$

where Γ_i and γ_e are the damping rate of ion wave and electron wave respectively and $\gamma_e > \Omega$. A single pump threshold is

$\frac{16\sqrt{3}}{9}\frac{\gamma_e}{\Omega}\frac{\gamma_e\Gamma_i}{\omega_{pe}\Omega}$ for $\gamma_e >> \Omega$ is much higher than the double pumps. And also the growth rate

$-\Gamma_i + \frac{\Lambda_1^2}{4}\frac{\omega_{pe}\Omega}{\gamma_e}$ for double pumps is larger than

$-\Gamma_i + \frac{9\Lambda_1^2}{16\sqrt{3}}\frac{\omega_{pe}\Omega^2}{\gamma_e^2}$ for a single pump. If $\gamma_e < \Omega$, a single pump and double pumps give the same threshold. (2) When the first pump intensity is marginal for the decay instability, the very small second pump red shifted by Ω can produce the oscillating two stream instability.

$\Lambda_1^2\Lambda_2^2 = 8\frac{\gamma_e^2\Gamma_i}{\omega_{pe}{}^2\Omega_i}$, $\Omega << \gamma_e$. If the marginal pumping is Λ_c for Λ_1,

$$\mathrm{Min}\Lambda_2^2 = \frac{\Omega_i}{2\Gamma}(\Lambda_c^2 - \Lambda_1^2).$$

As for the experiment, the glass laser has the half intensity width of spectrum 60 A, 1.6×10^{12} HZ and the peak at 1.06 μ, while the YAG laser had 6 A, 1.2×10^{11} HZ and 1.064 μ respectively. The plasma density range 1.13×10^{19} cm^{-3} around 1.0043×10^{21} cm^{-3} can be resonanced with the glass laser and the range 2.44×10^{18} cm^{-3} around 0.9989×10^{21} cm^{-3} with the YAG laser. The case of $\Omega_i < \gamma_e$, according to reference (3)-Fig.13, Ω_i is estimated as 10^{12} HZ. The beating condition for the glass laser is hold. The laser plasma is inhomogenious in space, so the clear double resonance cannot be expected. But the experimetnal data on electron

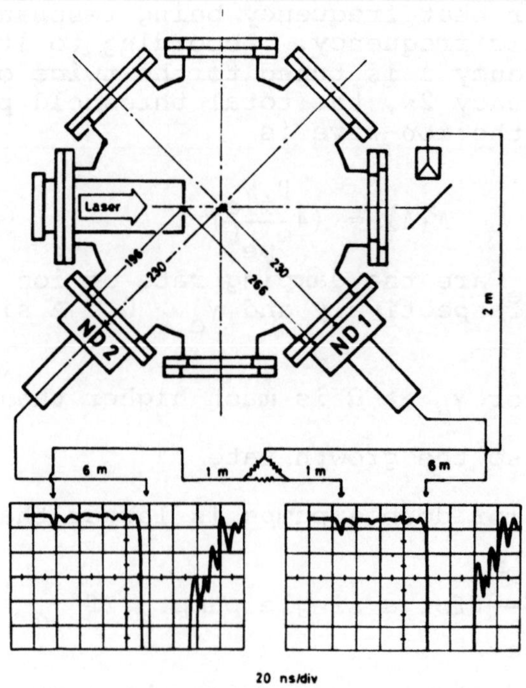

Fig.15. Location of neutron source.

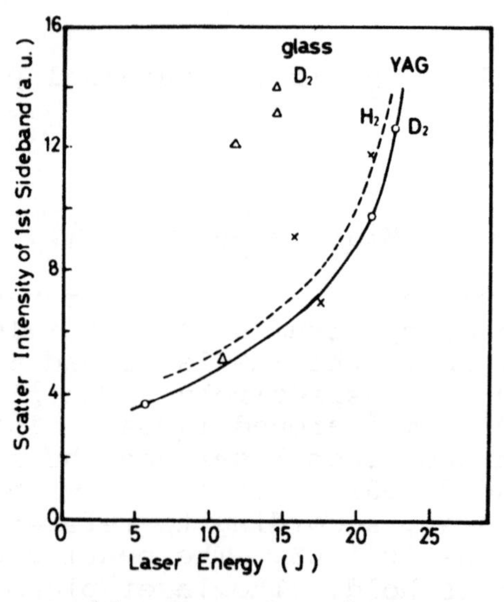

Fig.16. Power dependence of side band spectrum of second harmonics.

temperature, ion energy and also the yield of neutron indicate the effectiveness of heating by broad spectrum of the glass laser.

We tried to get several insights on these situation by scattering from a plasma. We used mainly the induced SHG from the plasma. In glass case, multi-coupling of laser and plasma, scattering spectrum[12] has a broad band which can only devide two peaks: SHG and red sifted side band. Comparing to YAG, it showed the appearance of the side band at the half intensity of the laser power as shown in Fig.16. This is also interpreted as the easiness of heating by glass laser. Using YAG laser, the induced scattering has four peaks on the red side of SHG. For deuterium plasma from the center of the second harmonics, the first peak 78.78 cm^{-1}, the second peak 134.65 cm^{-1}, the third peak 161.51 cm^{-1} and the fourth peak 204.02 cm^{-1} shifts were observed. For hydrogen plasma their shift were 84.38 cm^{-1}, 168.01 cm^{-1}, 249.97 cm^{-1} and 319.41 cm^{-1} respectively. The shift due to deuterium and hydrogen ion-acoustic waves are estimated 66.66 cm^{-1} and 93.99 cm^{-1} respectively. They are almost accorded with the first peak shift. The series of peak shift is the over-tone of the ion waves which indicates the cascade scattering process of the SHG light by the ion wave. As shown in Fig.16, the scattered intensity was increased beyond the square of the energy of laser light. The glass laser also shows the stronger scattering than the YAG laser.

The mechanism of neutron emission by laser has been discussed[13]. For the picosecond laser pulse, a requisite condition for neutron emission was said the existence of a precurser bombardment which suggested the collision of pre-exist deuterium plasma with the accelerated ions. We tried to locate the position of the neutron source using two neutron detectors by nanosecond pulse laser. As shown in Fig.15, the sequential annalysis indicated the origin of neutron was the target itself within an error of 2 ns. The energy of neutron was 2.45 Mev. This means that the main emission of neutron comes from the heated plasma stick, though it does not deny the small yield of neutron from the vessel wall.

The magnetic field was not so large as expected before[14]. The coupling between electron and ion was too strong to cause the appreciable charge separation.

In conclusion, parametric decay process for absorption and induced scattering have an important role for the laser fusion scheme. The double resonance absorption was qualitatively testified. The SHG scattering from plasma had the spectral structure to show the cascade process of ion wave generation. The main source of neutron was the deuterium target itself. The magnetic field induced in the plasma was small, which means the charge separation was small.

As shown in Appendix, a new high power system of disc lasers have been installed.

The computer experiment of a pellet compression scheme indicated that a very high laser intensity caused the fast propagation of heat wave into the pellet to prevent the shock compression. There was an optimum laser intensity for the effective neutron yield.

ACKNOWLEDGMENTS

The authors wish to thank Dr. K. Nishikawa for stimulating discussion and Dr. H. Takuma of JEOL, Dr. T. Izumidani of Hoya Glass Works and Mr. G. Ushio of Ushio Electrics for the cooperation in constructing the high power glass lasers. The authors also thank to the Ministry of Education and the Science and Technology Agency, Japan for scientific research funds.

REFERENCES

1. C. Yamanaka: Laser Interaction with Matter, Japan Society for the Promotion of Science, 1973
2. Laser Fusion Program, Semianual Report, Lawrence Livermore Laboratory, UCRL-50021-72-2, 1973
   J. Nuckolls et al: Nature 239, 139 (1972)
   K. Boyer: Progress Report Laser Program at LASL, LA-5251-PR, 1973
3. C. Yamanaka et al: Phys. Rev. A 6, 2335-2342 (1972)
4. K. Nishikawa: J. Phys. Soc. Japan 24, 916 (1968), 24, 1152 (1968)
5. M. N. Rosenbluth: Phys. Rev. Letters 29, 565 (1972)
   A. N. Kaufman et al: Phys. Rev. Letters 30, 1306 (1973)
6. D. W. Forslund et al: Phys. Rev. Letters 30, 739 (1973)
   K. Eidman et al: IPP IV/46 Garching (1972)

7. K. Nishikawa: Private communication
8. K. Eidman et al: European Conf. Control Fusion Plasma Phys. Moscow (1973)
9. C. Yamanaka: Progress Reports of Plasma Electronics (1968)
10. F. Floux et al: Laser Interaction and related Plasma Phenomena vol.2 page 426, Plenum Press, 1972
11. K. Nishikawa: Reference-1, page 272
12. J. L. Bobin: Phys. Rev. Letters 30, 594 (1973)
13. G. H. McCall et al: Phys. Rev. Letters 30, 1116 (1973)
14. J. A. Stamper: NRL Report 7411, (1972)

Appendix A

High Power Glass Laser System

C. Yamanaka, T. Yamanaka, T. Sasaki, K. Yoshida

M. Hongyo and K. Suzuki

Osaka University

The high power glass laser is composed by three sections: the oscillator section, the main amplifier section and the booster section. It has been specified to be a four beam system with a total output of 1 kJ in a 100 picosecond pulse. One system has been installed in Osaka and is being adjusted. The properties of this system are presented.

(a) Laser Glass

We used HOYA LSG-91H glass for the new laser system instead of LCG-11. It has a large stimulative emission cross section 2.5×10^{-20} cm^2 and low loss coefficient 0.001 cm^{-1} with the antisolarization improvement which is suitable for the picosecond operation. The properties of this glass is presented in Table 1A. In order to avoid parasitic oscillation, the disc glass was coated on a edge surface with a black solder glass. The bonding property has been improved using the compression stress in the coated layer. The transmittance of the black layer was shown in Fig. 1A. The damage strength of LSG-91H is almost same as LCG-11. The solarization of LCG-11 and LSG-91 glass is effectively protected by using filter glass L-39.

(b) Nanosecond Oscillator and Two Spectrum Beam Generation

Fig. 2A shows the detail arrangement of nanosecond oscillator which was reported before[1,2]. The master oscillator is either a glass laser 150 mm in length, 10 mm in diameter, spectral range 50 Å, or YAG laser 50 mm in length, 5 mm in diameter, spectral range 2 Å. The cavity is 140 cm in length with the end mirror a concave, reflectivity 98 %, a radius of curverture 8 m, KDP Pockels cell Q-switching, a mode selection of afocal system 1 m and 70 cm, a pinhole 2 mm in diameter and an output mirror for glass reflectivity 40 % and a sapphire etalon 3 mm of separation for YAG reflectivity 30 %. The TEM_{00} output of these oscillator was 10 mJ in 30 nsec. These pulses were fed to the two stage rear amplifiers: glass laser 150 mm in length 10 mm in diameter and 300 mm in length 15 mm in diameter. The output was 100~300 mJ. Then the pulse transmission mode method of double Pockels cells 14mm × 2 and the laser triggered spark gap 0.1 mm with 20 kV and 12 atmospheric pressure of nitrogen

Table 1A

Properties of Laser glass LSG-91H

Glass type			LSG-91H
Neodium doping			3.0 wt.
Fluorescent life time			(300 μsec.)
Fluorescent wave length			(1.062 μ)
Fluorescent line width			(260 Å)
Cross section for stimulated emission			(2.5×10^{-20} cm^2)
Specific gain			(0.133/cm/joule)
Loss coefficient (at 1.060 μ)			(0.001 cm^{-1})
Lasing threshold (10 x 160 mm Rod, normal osc. 60 % output mirror)			(110 J)
Slope efficiency			(2.0)
Refractive index	365 mμ	(n_i)	1.58552
	435.8	(n_g)	1.57346
	486.1	(n_F)	1.56804
	587.6	(n_d)	1.56115
	656.3	(n_c)	1.55812
	768	(n_A)	1.55471
	1014	(nt)	1.55024
	1060	($n_{1.06}$)	1.54980
Abbe's number	ν_d		56.56
Density			2.81 (g/cm^3)
Young's Modulus			8,890 (kg/mm^2)
Shear Modulus			3,590 (kg/mm^2)
Poisson's Ratio			0.237
Knoop Hardness (100 g)			590
Brewster's Angle			57°10'
Linear Coefficient of thermal expansion (25 - 300 °C)			105×10^{-7} ($°C^{-1}$)
Softening temperature (point) Sp			505 °C
Transformation temperature (Tg)			465 °C
Thermal conductivity	(25 °C)		0.89 kcal/m.h.°C
Specific Heat	(50 °C)		0.15 cal/g°C
	(122 °C)		0.17 cal/g°C
	(246 °C)		0.19 cal/g°C
Chemical durability (weight loss)			
H_2O (100 °C 1hr.)			0.036 %
HNO_3·Ph2.2 (100 °C 1hr.)			0.039 %

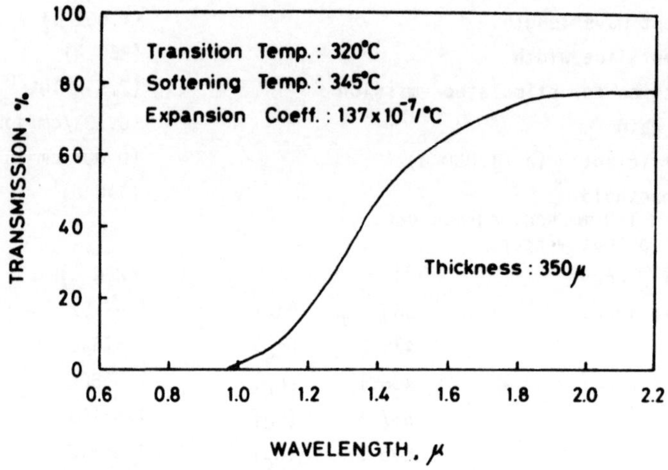

Fig. 1A Transmission curve of Black Solder Glass

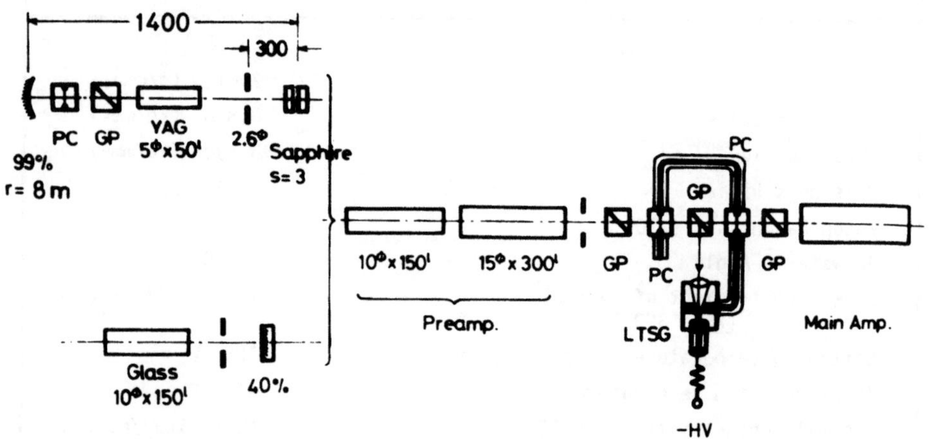

Fig. 2A Schematic diagram of the oscillator and pulse transmission mode method.

was used. This extinction ratio was 1000. The output was about 2 MW. The pulse duration was variable from 2 to 10 ns.

Two spectral beams for the doule resonance heating have been demanded. The simple way was to use an etalon in the cavity which restricted the oscillation spectrum by the separation of a free spectral range. The experimental results are shown in Fig. 3A.

(c) Picosecond Pulse Formation

The ultra-short pulse from self-mode locked laser by a saturable absorber is very important for investigation of light interaction with matters. When a condition $\Delta\nu < 1/\tau$ is fullfiled, a stable mode locking can be expected[1A], where $\Delta\nu$ is the spectrum width of the laser light and τ is a life time of the upper level of a bleachable absorber. Other important point in order to get stable mode locked pulses is to prevent nonlinear effects in a cavity such as self trapping and self phase modulation. The smaller value of nonlinear refractive index of active medium is desirable. The YAG or $YAlO_3$ laser fullfils these condition and about 100 psec can be obtained stably. In order to get narrower pulse, for example 10 psec pulse, the combination of a saturable absorber and an amplifier gives a stable method[1A,2A].

We have developed stable mode-locked oscillator system for plasma generation.

The picosecond pulse former consists of three parts, the oscillator, the single pulse switchout and the amplifier. Fig. 4A shows the schematic diagram of the system.

The oscillator was Nd:$YAlO_3$ rod which was brewster cut, 5 mm in diameter and 75 mm long. This was cooled by water. Q-switching and mode locking was performed by a saturable absorber, Kodak 9860 or NDL-101 (Japan Research Institute for Photosensitizing Dyes Co. LTD). The deterioration characteristic due to ultraviolet light (500 W, Xe arc lamp) is shown in Fig.5A. The better performance of NDL-101 was obtained. The flowing dye cell was attached to the total reflective mirror. The effective thickness of the dye was 500 μm.

The pulse width of mode locked laser was controlled by the thickness of a sapphire etalon which was inserted in the cavity. The slit whose diameter was 2 mm was inserted in the cavity to get a smooth spatial distribution of laser light. Fig.6A shows the relation between the pulse width obtained and the thickness of etalons. The measurement was performed by TPF (Two Photon Fluorescence)

1A) P.G.Kryukov and V.S.Letokov, IEEE J. Quant. Elect. QE-8 (1972) 766.

2A) A.Penzkofer et al., Appl. Phys. Lett. 20 (1972) 351

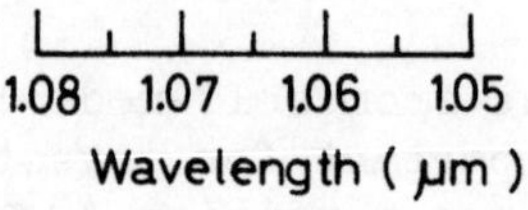

Fig. 3A Oscillation spectra in Q-swithched mode from glass oscillator. Thickness of the etalon in the cavity is 25 μm.

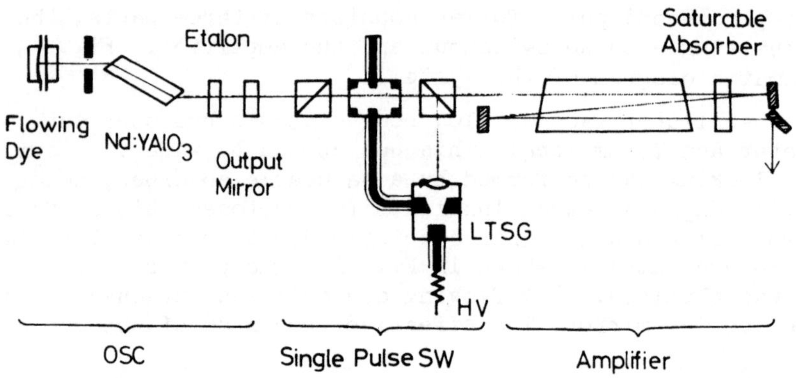

Fig. 4A Schematic diagram of the picosecond pulse former.

method and the high speed streak camera whose resolution time was about 10 psec.

The switchout method of a single pulse from mode locked pulse train was an usual combination of Pockels cell and laser triggered spark gap. But the whole system was modified into a very compact one.

The single pulse was sent into the combination system of an amplifier and a saturable dye cell. When 100 psec pulse is required the dye cell is removed. The rod dimension of the amplifier is 15 $mm^{\phi} \times 300\ mm^{l}$ and three path system is adopted. The amplifier characteristics is now under investigation.

(d) Main Amplifier and Isolation

The five stage main amplifier was reported before[3)]. The output energy was 50 J and gain 10^5. The isolation between stages was mainly kept by the aperture limitters provided at every stage, the enough distance of the amplifiers and an uniguide slit which is shown in Fig. 7A. The uniguide slit was composed of afocal system and a pinhole covered with a thin plastic film situated at the f. focal point. The action of the uniguide slit was freely to pass the outgoing beam of diffraction limitted and to stop the backward going reflected beam by producing plasma on a plastic covered slit. The uniguide was set between the first and second stage which can effectively protect the system of pulse tailoring. The Faraday rotator switch using HOYA FR-4 was checked but not necessary for this scale of lasers.

(e) Booster Laser

The booster section has been newly developed to enforce the laser power from the main amplifier 50 J in 2 ns up to the energy 250 J. The computer code which was compiled in ILE Osaka University was used. The pulse shape was a gaussian; rise time 1.5 ns and half intensity width 2 ns. The amplification factor for the rod type preamplifier was 2.4, the input 50 J to the output 120 J and that for the disc amplifier was 2, the input 120 J to the output 240J. The input energy density for the preamplifier was 1.77 J/cm^2, the diameter of which was 6 cm. For the disc amplifier it was 1.69 J/cm^2, the effective diameter 7 cm (actual diameter 8cm). Pumping efficiency was assumed from the previous experiments 6.9×10^{-3} for the preamplifier and 2.8×10^{-3} for the disc amplifier. The life of flash lamp was estimated by the formula of Emmett

$$N = \left(\frac{E_{exp.cavity}}{E/N_{lamp}} \right)^{8.8}$$

where N is the shot number, the explosion energy is $E_{exp.cavity} = E_{exp.air}/1.5$, $E_{exp.air} = 2 \times 10^4\ dl \cdot \sqrt{T}$ (J), d is the inner diameter of flash lamp, l is the lamp length, T is the lamp period $\sqrt{LC}$. We used d = 1.8 cm, T = 300 μs, $E_{exp.air} = 0.624 \cdot l$ (kJ). The design

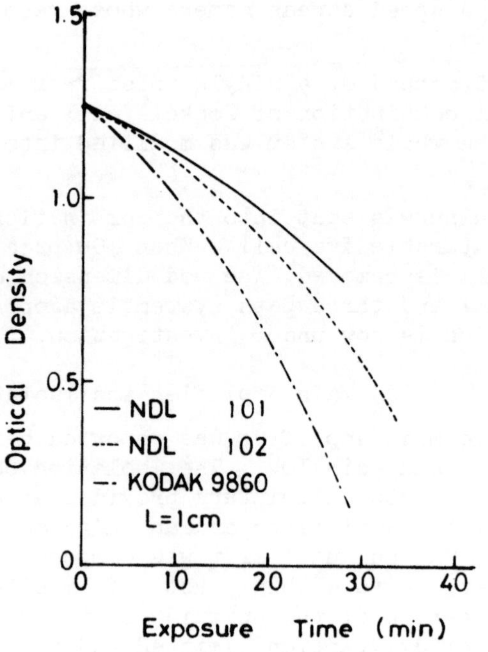

Fig. 5A Deterioration characteristics of various saturable absorbers due to ultraviolet light. The light source is 500W arc lamp.

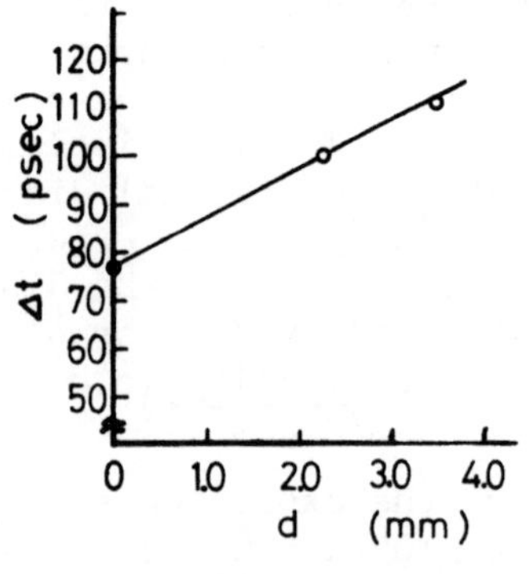

Fig. 6A Relation between the pulse width obtained and the thickness of etalons.

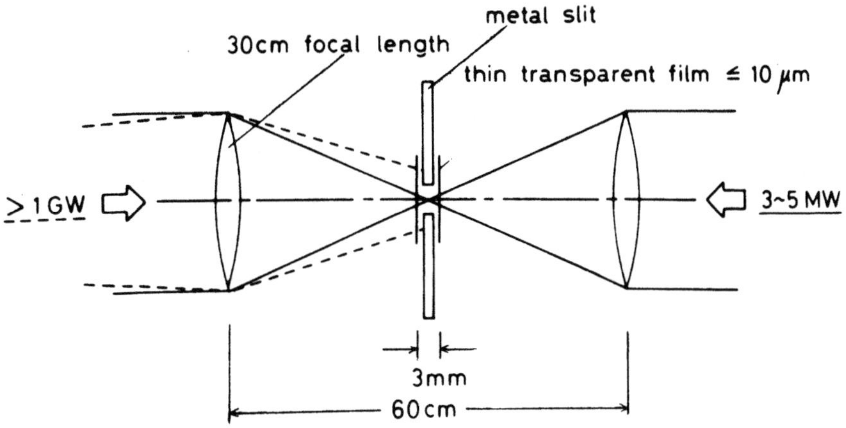

Fig. 7A Schematic diagram of the uniguid slit.

Table 2A Design data of the preamplifier. Input energy density is 1.77 J/cm^2 in 2 ns. The diameter of the rod is 6 cm.

Active Medium Length 36 cm

Net Gain	Stored Energy Density	Pumping Energy	Shot Number till Explosion	Output energy Flux
G_0	n_0 (J/cm^3)	E (kJ)	N	(J/cm^2)
2.4	0.326	48.1	2.2×10^4	4.2
3.7	0.377	55.6	6.1×10^3	4.8
3.0	0.416	61.4	2.6×10^3	5.3

Table 3A Design data of the disc amplifier. Total length of active medium is 36 cm.

Disc Diameter $\Phi = 80$ mm

Net gain G_0	Stored Energy Density n_0 (J/cm^3)	Pumping Energy E (kJ)	Shot Number till Explosion N	Output Energy Flux (J/cm^2)
2.0	0.308	305	1.6×10^4	2.6
2.5	0.394	390	1.8×10^3	3.3
3.0	0.475	471	3.4×10^2	3.9

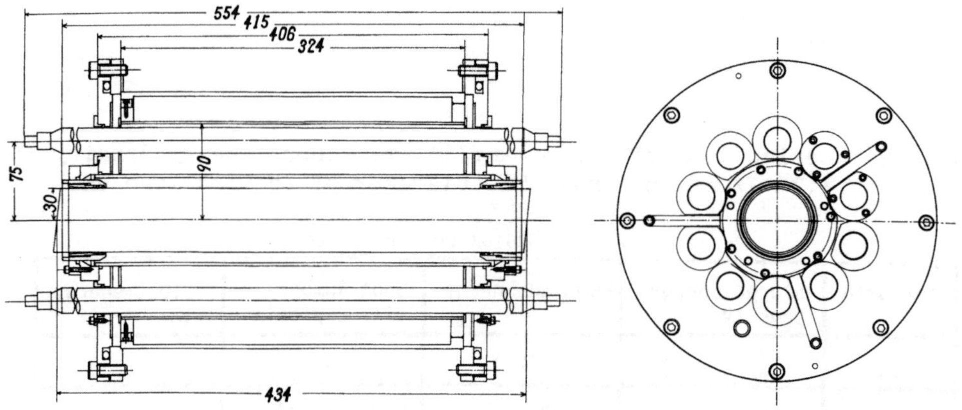

Fig. 8A Dimension of the preamplifier. The glass rod is 6 cm in diameter and 40 cm in length. The concentration of Nd_2O_3 is 1.5 wt %. Ten flash tubes are set aroud the rod.

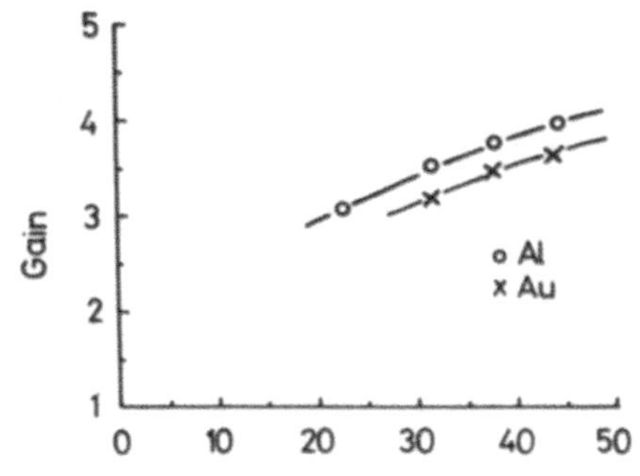

Fig. 9A Small signal gain of the preamplifier. 20 ns pulse was used. The alminum reflector showed better result than gold plated reflector.

Fig. 10A Photograph of the preamplifier.

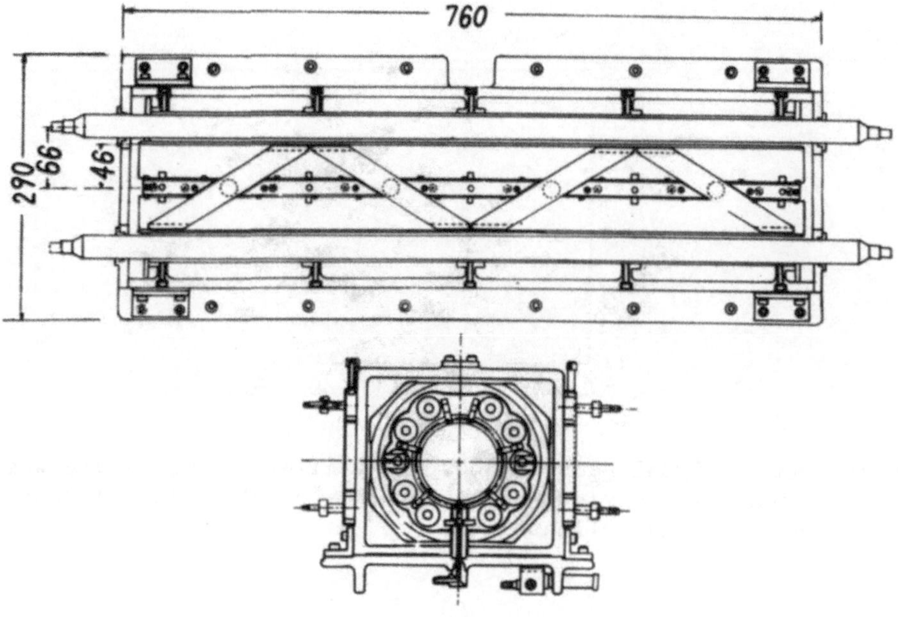

Fig. 11A Dimension of the disc laser. Disc laser system consists of three laser heads.

Fig. 12A Photograph of disc amplifier

data is given in Table 2A and Table 3A. The temperature rise in discs under the pumping of 400 kJ was estimated less than 3 °C. According to the calculation on the lamps and discs, the air cooling system was adopted. The disc amplifier was devided into three sections composed of four oval discs each in the brewster arrangement. The illumination calculation concluded to use ten flash lamps 90 cm in length arranged on circuler in the distance of 6.6 cm from the center axis for each section.

(I) Preamplifier

The dimension of the preamplifier is shown in Fig. 8A. The experimental properties are shown in Fig. 9A. The small signal gain was in good accordance with the calculated values. Fig. 10A is the view of the preamplifier.

(II) Disc Amplifier

Fig. 11A shows the dimension of the disc laser. Fig. 12A shows the construction of disc amplifier.

1. Black coated disc

The black coated glass was checked out. It prevented the parsitic oscillation in the disc. The dimension and the view of the disc are shown in Fig. 13A. Fig. 14A shows the characteristics of the discs with different edge treatments.

2. Flash lamp and reflector

Flash lamp in the main amplifier was immersed in the cooling water. This condition was very effective for the protection of explosion. But in the disc amplifier, the lamp was in the nitrogen atmosphere. The several explosion tests had been performed. The flash lamp of the electrode distance 67 cm, the whole length 90 cm, inner diameter 20 mm, tube thickness 3 mm, Xenon pressure 600 torr was used and the maximum input energy 20 kJ, pulse duration 1 msec. The reflector was investigated in the form and also the surface quality, gold, silver and alminum. The individual eliptic mirror for the lamp gold plated was adopted.

3. Disc laser

One stage of disc lasers was fed 100 kJ, capacity 3130 μF, series inductance 150 μH, applied voltage 8 kV. The amplification properties was measured in detail. The data are given in Fig.15A. The pumping efficiency was two times better than the designed value. The expected properties have been attained.

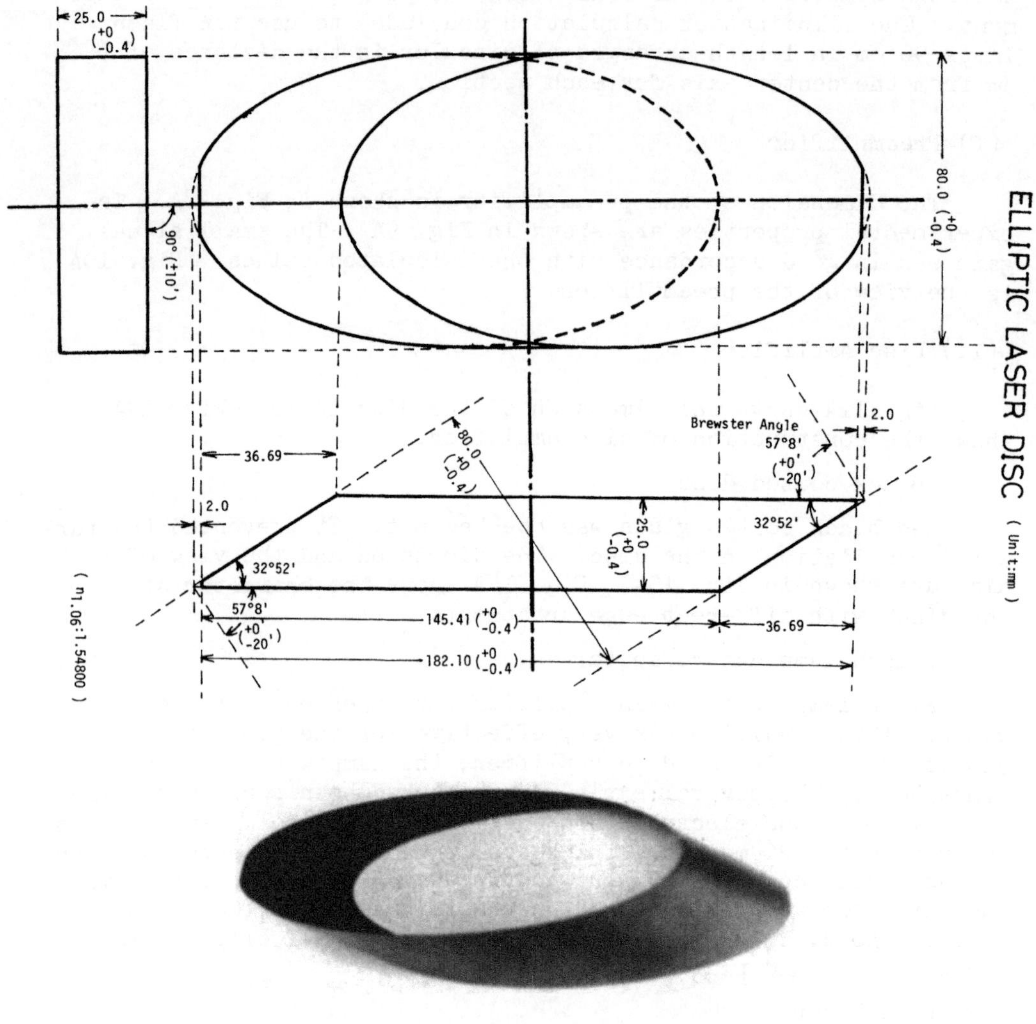

Fig. 13A Dimension and photograph of a glass disc. The Concentration of the Nd_2O_3 is 3 wt%.

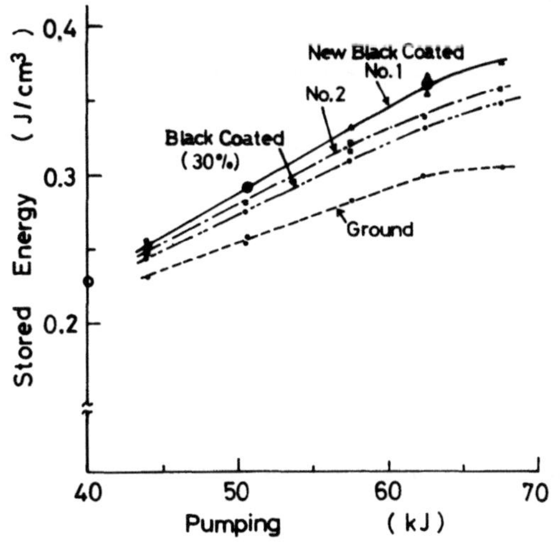

Fig. 14A Fluorescent intensity from glass discs whose edges have different treatment. Black coated (30%) shows that absorption on the black coated layer is 30%. New black layer are durable against pumping energy 72 kJ per one disc laser head. Stored energy was estimated from small signal gain (Large circles).

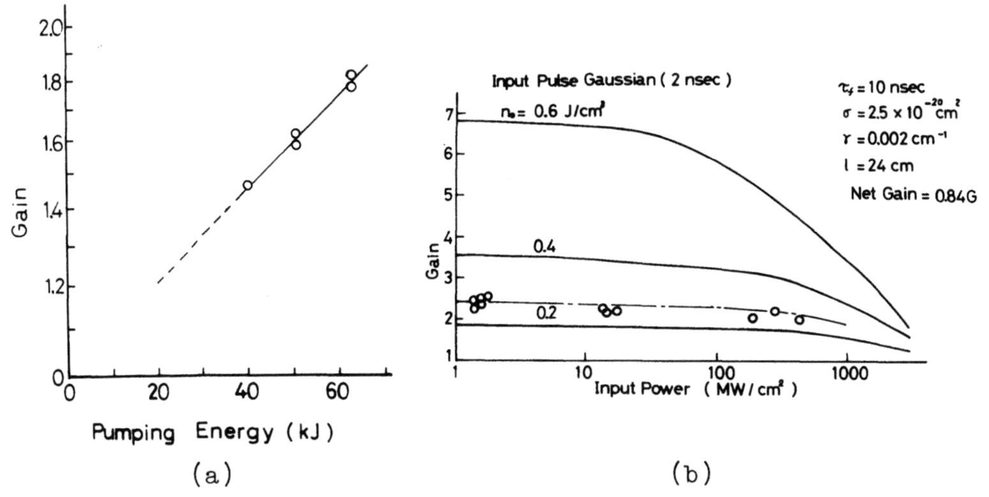

Fig. 15A Amplification properties of the disc laser. 2 ns pulse was used. (a) Small signal gain of one disc laser head. The length of active medium is 12 cm. (b) Large signal gain of the disc laser heads. Solid lines show the results of calculation. The loss due to reflection and absorption of two disc heads (12 discs) is 16 %. **Pumping is 100 kJ for the experiment.**

Appendix B

Computer Experiments "CLAB-H" on a Pellet Compression

C. Yamanaka, J. Mizui and T. Yamanaka

Osaka University

A computer experiment "CLAB-H" on a pellet compression has been performed assuming the classical absorption and hydrodynamic behaviors. The plasma was assumed as fully ionized, quasi-neutral and two temperature components[1B]). The hydrodynamic equations included classical absorption, electron heat conduction, ion viscosity, relaxation between ion and electron, bremsstrahlung and thermonuclear energy generation terms. Initial solid D-T cores of radius, r_c = 50 μm (0.048 μg) and 100 μm (0.384 μg) with low density tail, which was proportional to r^{-3} beyond r_c, were irradiated by laser pulse whose shape was a step function of time and wavelength 1.06 μm. Our results have been checked by the energy balance equation with accuracy better than 0.4 %.

Fig. 1B shows the power dependence of the maximum temperature and the absorption efficiency of the laser energy. Maximum electron temperatures at the low density tail Teo were proportional to $P^{0.37}$ and the efficiency $\eta \propto P^{-0.52}$, where P was the laser power. The electron absorbed the laser energy mainly near and under cut off region and $\eta \propto KP/P = K \propto T_{eo}^{-3/2}$, where K was the classical absorption coefficient. (The reflected light also contributed to the absorption.) From the Fig. 1B Teo $\propto P^{0.37}$ and therefore $\eta \propto P^{-3/2 \times 0.37} = P^{-0.55}$. This agree with the numerical result. The maximum of the ion temperature at the center of the core showed a maximum value at a certain input power; P = 2.5 × 10^{16} W/cm^2 at r_c = 50 μm ($T_{i\ max}$ = 2.1 keV) and P = 1.8 x 10^{17} W/cm^2 at r_c = 100 μm ($T_{i\ max}$ = 4.8 keV). This optimum condition was realized when the imploding shock wave was reached to the center of core before the heat conduction front arrived. If the excess laser power was applied the fast heat front propagated to the center of the core to decrease the compression.

In the case of the excess laser power, at the initial period, 0 ~ 10 psec, (Mach number M < 1) there was no dynamic motion and the most absorbed energy transformed to the electron thermal energy (3 keV) with the good absorption efficiency η = 9.1 % at M = 1. In the next phase, the shock was appeared. But the electron heat front broke the shock formation at 50 psec. In this case the plasma could not be highly compressed and the ion of core was adiabatically heated. Total absorption efficiency was η = 3.2 % when the ion temperature

1B) M. S. Chu, Phys. Fluids. 15 (1973) 413.

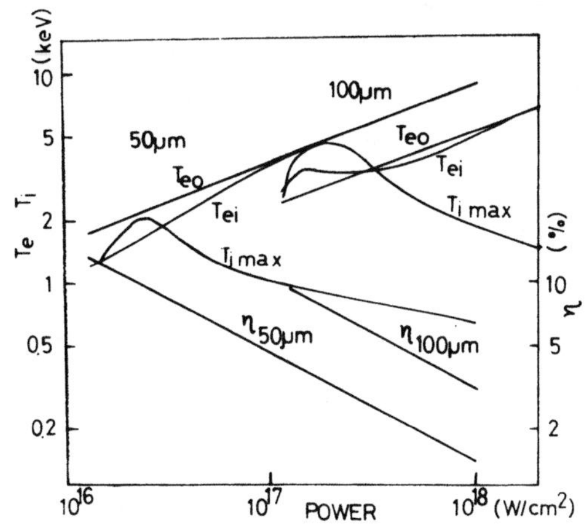

Fig.1B Power dependence of the maximum temperature and the total efficiency for 50μm and 10μm D-T pellets. T_{imax} and T_{ei} are the maximum temperature of ion and electron respectively at the dense core. T_{eo} is the maximum electron temperature at the low density tail.

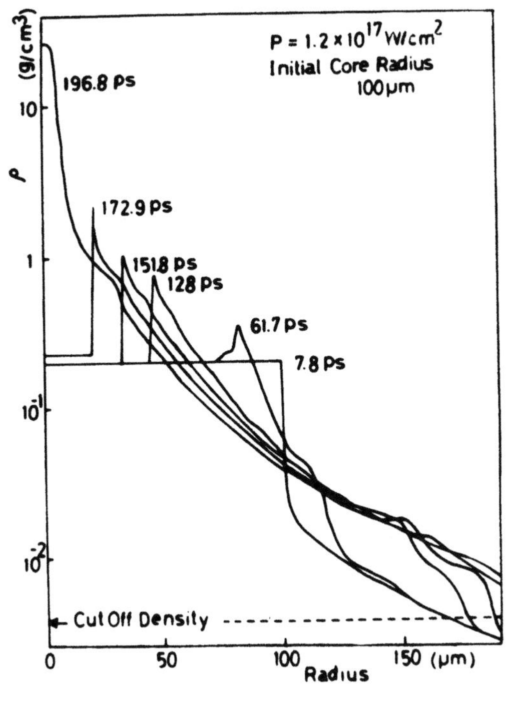

Fig.2B Density distribution as a function of radius at various time for $P = 1.2 \times 10^{17}$ W/cm^2.

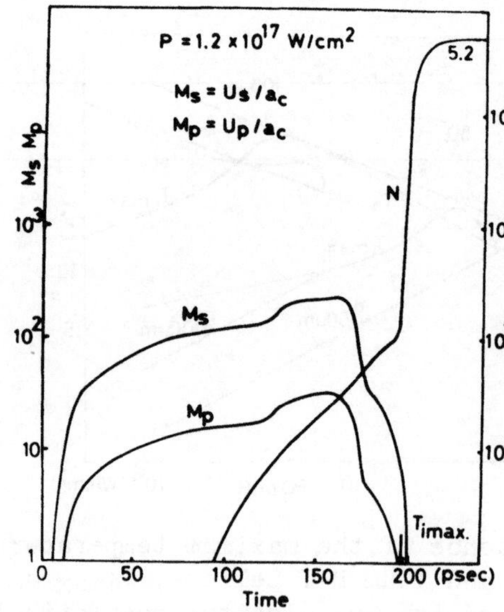

Fig. 3B Neutron production and Mach number as a function of time for the case of $P = 1.2 \times 10^{17}$ W/cm^2 (100 μm pellet). U_s, U_p and a_c are the shock speed, the particle convergence velocity and the sound velocity at the center of core respectively.

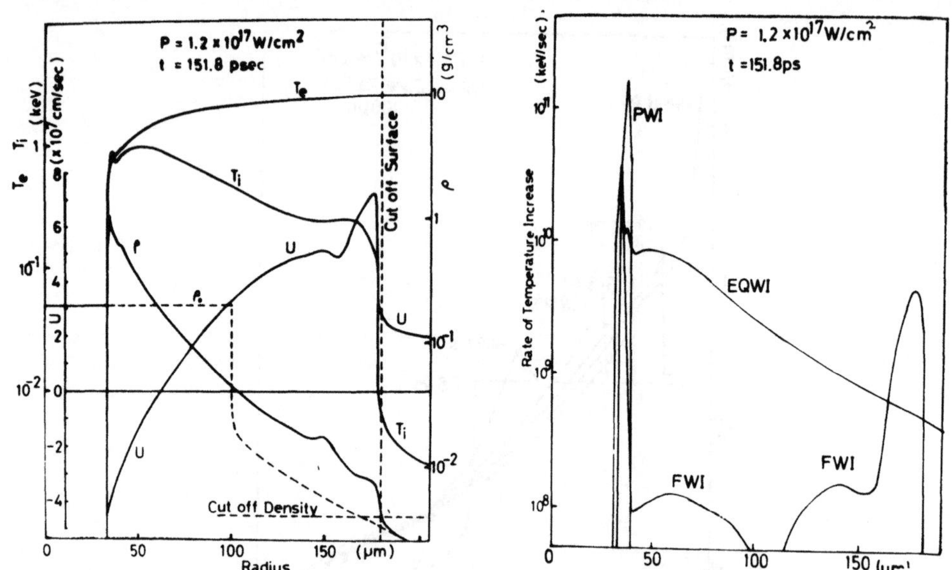

(a) (b)

Fig. 4B (a) Density, ion temperature and electron temperature distribution at 151.8 psec. (b) Different mechanisms in ion heating for $P = 1.2 \times 10^{17}$ W/cm^2. PWI, FWI and EQWI are respectively the work due to pressure, the ion viscosity and the electron-ion relaxation.

became the maximum and total neutron production was $N = 3.1 \times 10^{10}$.

In the case of optimum heating condition was shown in Fig. 2B, 3B. The shock strength and steepness of the front increased at 120 psec when the exploding wave reached to the cut off surface of the laser light. This case is shown in Fig. 4B(a),(b). The width of imploding shock front is 5.0 ~ 8.0 μm where the ion was heated by the work due to pressure, the electron-ion relaxation and the ion viscosity. The ion temperature exceeded the electron temperature. The Mach number was M = 220 and the density jump agreeded with $\rho_1/\rho_0 = \gamma+1/\gamma-1 = 6$ where $\gamma = 5/3$ is the specific rate. The spherical convergence effect appeared later. In this front the mean free path is 0.35 μm and so the shock wave was collision dominant. But the width of the exploding front is 20 μm which was comparable or smaller than the mean free path. Considering the problem of compression and explosion, we must use the equations of the electron and ion motion respectively estimating the charge separation effect which causes the acceleration. The energy dispatch between the acceleration and the thermalization will be investigated. This scheme is now going on concerning with the McCall's experiment[13]. At 196 psec the shock collapsed at the center and the density became 26 g/cm^3, the ion temperature 2.98 keV and neutron production increased remarkably. This final process was very different from that of the excess laser power. Total neutron production is 5.2×10^{12}, irradiated laser energy 3.0×10^4 J, absorbed energy 2.8×10^3 J and total efficiency $\eta = 9.2$ %.

BACKSCATTER EXPERIMENTS+)

K. EIDMANN and R. SIGEL

Max-Planck-Institut für Plasmaphysik

8046 Garching, FRG

ABSTRACT

On irradiation of a solid target with a Neodymium laser it has been found that part of the incident radiation is backscattered with a frequency near the laser frequency, and to a smaller fraction with twice the laser frequency (second harmonic). Extensive experiments with various target materials give strong evidence for stimulated Brillouin backscattering of laser radiation and frequency conversion into the second harmonic due to the presence of a density gradient in the plasma.

+) Presented at the Third Workshop on "Laser Interaction and Related Plasma Phenomena" at Rensselaer Polytechnic Institute, Troy, New York, August 13 - 17, 1973

I. INTRODUCTION

Irradiation of a solid target by an intense laser beam with a wavelength in the visible or infrared creates a hot and dense plasma at the surface of the target. The density in this plasma may exceed the so-called critical density where the plasma frequency equals the laser frequency. In the most simple (linear) treatment, the absorption of the laser radiation is assumed to be determined by electron-ion collisions (inverse bremsstrahlung) in which case the absorption coefficient is independent of the intensity of the incident radiation. An analysis of the reflection problem under these circumstances shows that with increasing intensity of the laser radiation and hence increasing temperature and decreasing collision rate more and more of the incoming laser radiation should be reflected from the plasma, like from a highly conducting metal, and therefore be lost for the heating of the plasma (see for example /1/).

During the last few years it has been shown theoretically by many authors that at the high laser intensities in laser-fusion experiments nonlinear phenomena will become important and finally will determine absorption and reflection from a laser produced plasma. Stimulated backscatter of laser radiation by instabilities of the Brillouin- and Raman-type /2/ can lead to enhanced reflection of laser radiation. Through further nonlinearities of the laser plasma interaction, harmonics of the incoming radiation may be created which then leave the plasma /3/. On the other hand parametric instabilities of the absorptive type /4/ can cause enhanced absorption of a sizable fraction of the laser radiation even in a collisionless plasma /5/. Depending on the conditions, these effects may become practically important for flux densities of the order $\varnothing \gtrsim 10^{11}$ Wcm^{-2}. For $\varnothing > 10^{15}$ Wcm^{-2} (depending on wavelength) relativistic effects must also be considered /6/. The present situation is however characterized by the fact that the different modes of interaction have mostly been treated separately and under idealized conditions. It is therefore still an open question how the various processes affect the interaction of the laser light with the plasma in an actual experiment.

Experimental observations of absorption and reflection in the case of laser irradiated solid targets have been reported in the literature several times /7/. Unfortunately the experimental conditions in present experiments are far away from the idealized conditions of theoretical analysis. The two-dimensional situation in the experiments, due to the focussing of the beam, the hydrodynamic motion of the plasma during irradiation, and the intrinsic dependence of the plasma properties on the interaction mechanism lead to a highly complex situation in actual experiments. In fact an analysis of the above mentioned experiments shows that in none of them the interpretation of the results is unequivocal and free from arbitrariness.

This paper summarizes and extends the backscatter investigations which have been performed in the Garching Neodymium laser experiment /8/. Essentially new aspects are a systematic investigation of backscatter from targets with low and high Z-number and a photographic study of the angular distribution of the backscattered laser light and its second harmonic. In particular the latter shows strikingly a strong collimation of backscattered laser light and a systematic dependence of the intensity and polarization characteristics of the second harmonic on the direction of the polarization vector of the incoming laser light. The measurements presented below are based on several thousand laser shots which have been performed under fairly constant and well defined conditions. All the results have been found to be entirely reproducible. Nevertheless - as will become clear from the discussion given below, even with the information available now, the interpretation of the experiments still presents problems in some respects. Therefore we present the measurements here mainly as experimental facts, with a final and more detailed interpretation to be published elsewhere.

Other new aspects of this experiment, namely an investigation of the isotropy of X-ray emission /9/ and the observation of the laser induced compression wave in solid hydrogen and plexiglass targets /lo/ will be published elsewhere.

II. EXPERIMENTAL ARRANGEMENT

Most decisive for the properties of the plasma produced with a high power laser is the intensity distribution of the laser radiation near the focal point of the focusing lens. This has been measured by a photographic technique and is represented in Fig.1. Within the resolution of the film (lo μ) the intensity in the central maximum is about $4 \cdot 10^{14}$ Wcm^{-2} (within an

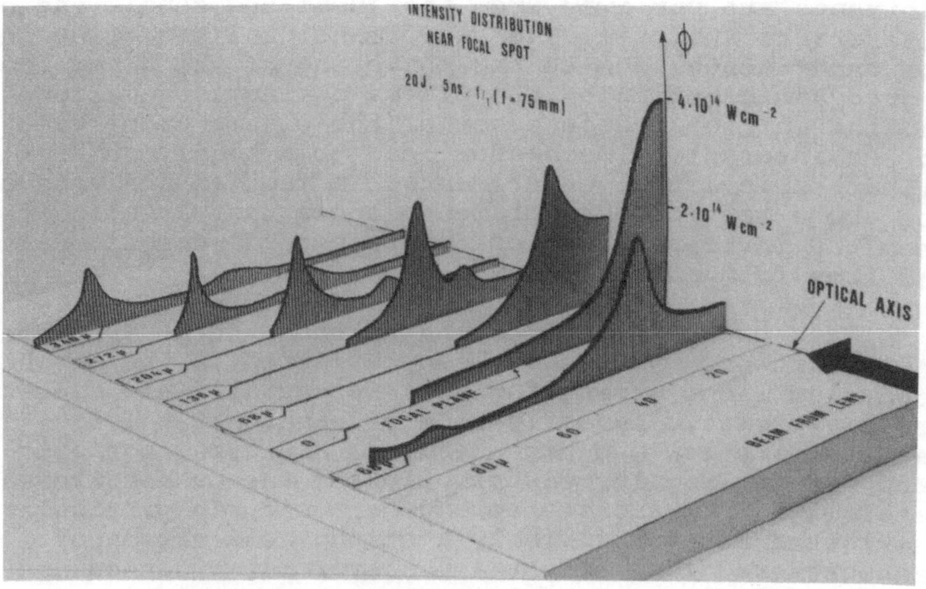

Fig.1 Measured intensity distribution near the focus of the laser beam

accuracy of 2). In front of and behind the focal plane a fairly complicated, hollow structure of the beam is resolved. This resembles the intensity distribution at the periphery of the last amplifier (64 mm Ø) of the laser and shows higher intensities in the periphery than on the axis. The standard operating conditions of the Neodymium laser (λ = 1,o6 μ)/11/ are given in Fig.1.

Fig.1 represents a time-average of the laser intensity. Recently it has been detected with a fast streak camera (Electro-Photonics LTD) that the laser pulse (which looks smooth on a nanosecond time scale, see Fig.3) consists of irregular bursts of light typically separated in time by about 2o ps (see Fig.2). It should be noted that according to the bandwidth of the laser ($\Delta\lambda \sim$ 3o Å) fluctuations with a duration of 1 ps could be present in the beam which would not be resolved in Fig.2.

As targets either sticks of solid deuterium with a square 2 x 2 mm cross section /12/ or thick rods (6 mm Ø, hence essentially plane targets in our case) of carbon, plexiglass, glass, copper and tungsten have been used. Their position with respect to the focal plane (Fig.1) could be controlled with an accuracy of about ± 1o μ.

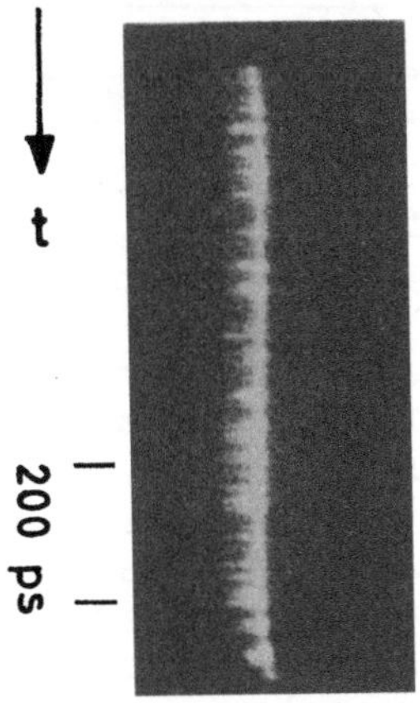

Fig.2 Fast streak photograph, showing fluctuations in the laser beam

III. INTENSITY AND SPECTRAL DISTRIBUTION OF BACKSCATTERED RADIATION

The time behaviour of the incident laser radiation (a in Fig.3), the laser radiation (c) and the second harmonic (b) scattered back through the focusing lens, and of the laser radiation scattered in backward direction through a window in the experimental chamber at an

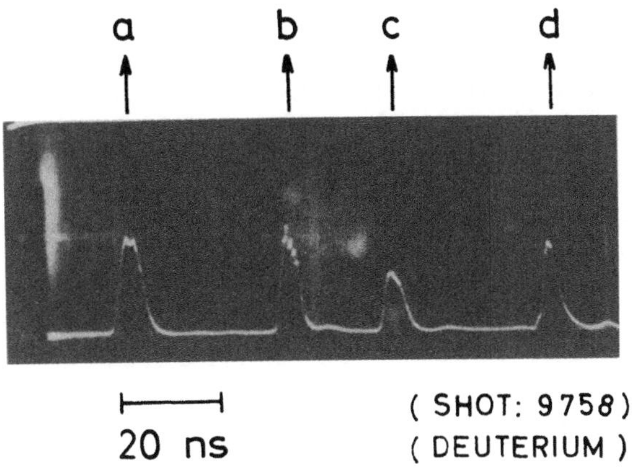

Fig.3 Oscilloscope trace, showing signals of 4 photodiodes. a) laser pulse, b) second harmonic, c) backscattered laser light (observed through the focusing lens), d) backscattered laser light (observed under 45° to the optical axis)

angle of 45° to the optical axis (d), were registered simultaneously on a fast oscilloscope (Tektronix 519). The photodiodes (a) and (c) were calibrated by calorimeters, whereas the sensitivity of the photodiode (b, second harmonic) was calculated from the spectral sensitivity of its photocathode as given by the manufacturer. The signal of photodiode (d) is later in Fig.4 given in arbitrary units. Parameters of experimental variation were the position of the target and the laser energy. For the latter case the pumping of the laser was always held constant and the laser energy was attenuated by inserting filters into the laser beam after the last amplifier. Hence we can expect that we changed only the absolute intensity, but not the intensity pattern near the focal plane by this procedure. During the course of these experiments - several months - some minor changes in the behaviour of the plasma could be observed. For example the absolute value of backscattered energy varied in different series of experiments by about 3o %. Also, for example, the right minimum in the reflection versus target position curve (Fig.4a) was not present in all of the series. The width of the reflection curve, however, was essentially constant. Both effects may be attributed to new alignments of the laser which slightly changed the intensity pattern near the focal spot. Those changes were felt however to be under control and do not affect the results presented below.

The signals of the three photodiodes in Fig.3 observing the scattered radiation vary in shape slightly, but nevertheless distinctly, when the target is moved in subsequent shots through the focal plane. We do not, however, consider this effect here and turn immediately to Fig.4 a - c where the maximum signal amplitude has been plotted versus the target position.

Fig.4a summarizes the behaviour of the various target materials with respect to the reflection of laser light. The halfwidth of the reflection curves decreases with increasing Z-number of the target material:

deuterium	45o μ	
carbon	25o μ	(35o μ with second max.)
plexiglass	26o μ	
copper	19o μ	
tungsten	17o μ	

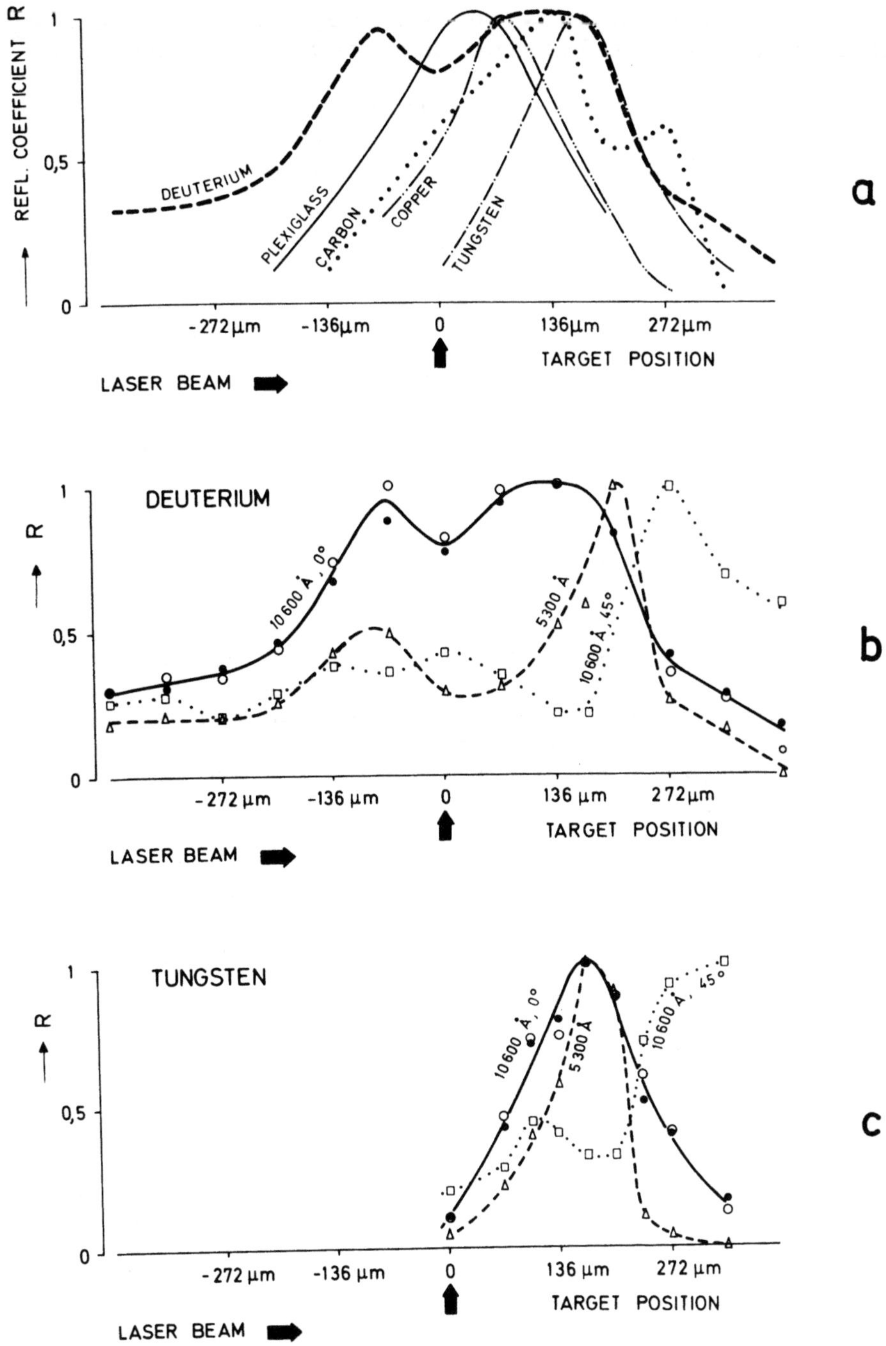

Fig.4a-c Reflection versus target position. Reflection normalized for all curves.

For a given target material the width of the reflection curves depends on the incident laser energy. For tungsten the following data have been measured:

Attenuation	Incident Laser Energy (Joule)	Halfwidth	Target Position for maximum reflection (see Fig.1)
0 db	13	170 μ	+ 170 μ
3 db	8,8	147 μ	+ 140 μ
7 db	3,0	120 μ	+ 153 μ
10 db	1,9	128 μ	+ 136 μ
17 db	0,44	75 μ	+ 136 μ

With decreasing laser energy the halfwidth of the reflection curves decreases and, in the case of tungsten, becomes as small as 75 μ. Also the target has, for maximum reflection, to be positioned nearer to the focal plane by about 30 μ. This effect is, however, not very pronounced and is near the limit of detectability (step width of target position 34 μ).

It appears also from Fig.4 that in order to observe maximum reflection we have to shift the focus to slightly in front of the target surface. The optimum shift increases with heavier target material. We note, however, that any unwanted motion in the projection system used for target positioning after the measurement of Fig.1 had been performed would be misleading in this context and hence the exact position of the focal plane should be reconfirmed before making use of this information. Nevertheless we have no reason to doubt the relative position of the maxima of the different target materials.

Fig.4b correlates the scattered laser intensity in backward direction, at 45° and at the frequency of the second harmonic for the case of a deuterium target. Whereas the backward reflection curve and the second harmonic curve have corresponding maxima, the laser radiation scattered at 45° has always a minimum when the backscattered radiation has a maximum. This indicates that under the condition of maximum reflection in the backward direction the reflected laser light must be well collimated: a fact which will become quite obvious from the experiments presented in the next section. According to Fig.4c a tungsten target behaves basically the same way as one of deuterium, except that all the curves have a narrower halfwidth. In both cases the second harmonic curve has its maximum with the target somewhat further away from the focal plane than for maximum backscatter, and then drops very steeply.

Fig.5 shows absolute values of the backscattered laser energy (integrated output of photodiode c) versus the incident energy (integrated photodiode a) for deuterium, carbon and tungsten. The incident energy is given rather than the intensity, since the former is the more accurately known experimental value. (The energy scale could be replaced by an intensity scale with the help of Fig.1.) In the case of solid deuterium

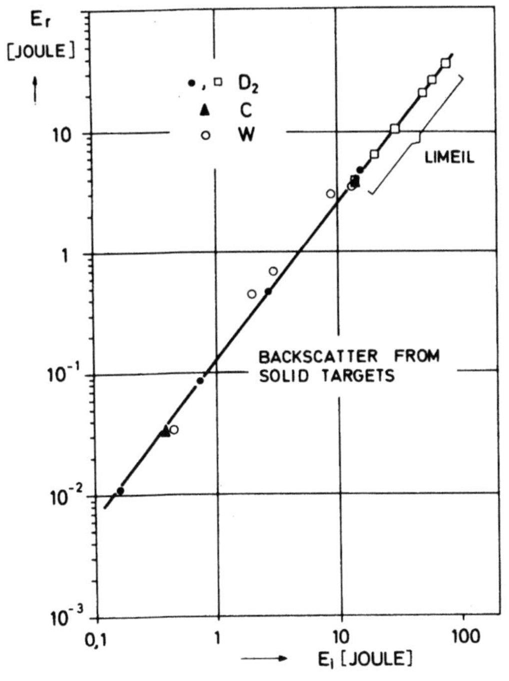

Fig.5 Incident versus reflected laser energy

the incident laser energy has been varied from $2 \cdot 10^{-1}$ - 85 Joule (i.e. over nearly three orders of magnitude (including the Limeil measurements /7a/)). In this range the reflected versus incident laser energy varies according to $E_r \sim E_i^{1,28}$. The reflection coefficient $R = E_r / E_i$ varies accordingly as $R \sim E^{0,28}$ i.e. in that energy range the reflection coefficient increases with incident laser energy.

Since the collision frequency increases with increasing ion charge Z one would expect enhanced absorption and hence less reflection from the high Z materials. It is surprising therefore that the reflection measured from carbon is the same as for deuterium (plexiglass, glass and copper give the same result). In the case of tungsten reflection, in the range 1 - 1o Joule, is even somewhat higher than with the other materials. Around 1o Joule the reflection curve of tungsten bends and its slope drops below one thus indicating that the reflection coefficient begins to decrease with increasing laser energy. Since each point in Fig.5 is based on a full reflection versus position curve of the type shown in Fig.4 (comprising 2o - 3o shots) this fact must be considered as real. Tungsten is the only material for which, with the limited intensities available to us, we observed such a behaviour. Quite recently several authors /13/ have reported similar behaviour also for light elements. It may be interesting to note that the reflection measurements performed at Limeil /7a/ fit ours surprisingly well. This is true also for the reflection versus position curves of /7a/ as well as for the minimum observed in scattering at 45° (/7j/, compare with Fig.4). This shows that, once reproducible conditions have been achieved, similar experiments can indeed give similar results - a fact which still seems to be a curiousity in the field of laser produced plasmas.

Fig.6 shows the dependence of the radiation energy scattered back through the focusing lens at the frequency of the second harmonic as a function of the incident laser energy. For all materials the reflected energy varies roughly as $E_r(\text{harmonic}) \sim E_i^{\,2}$. Since the fraction of laser light converted into the second harmonic is not more than $1o^{-5}$ the loss of energy via the second harmonic is of no importance in this experiment.

A comparison of the spectrum of the incident and reflected laser light has been published earlier /14/. With an improved laser (shorter puls, better focusing) a slight red shift of the backscattered radiation is still observed (Fig.7). The spectrum of the second harmonic (Fig.8) shows a wing towards the red. This has been observed also by other authors /15/ and is interpreted as an indication for the presence of ion acoustic waves in the plasma.

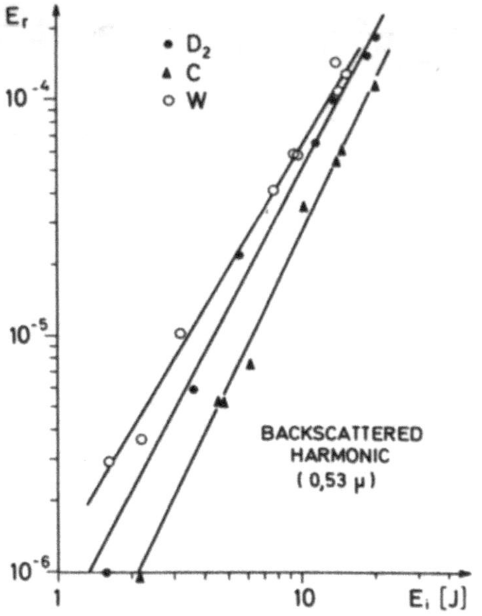

Fig.6 Incident versus reflected (second harmonic) energy

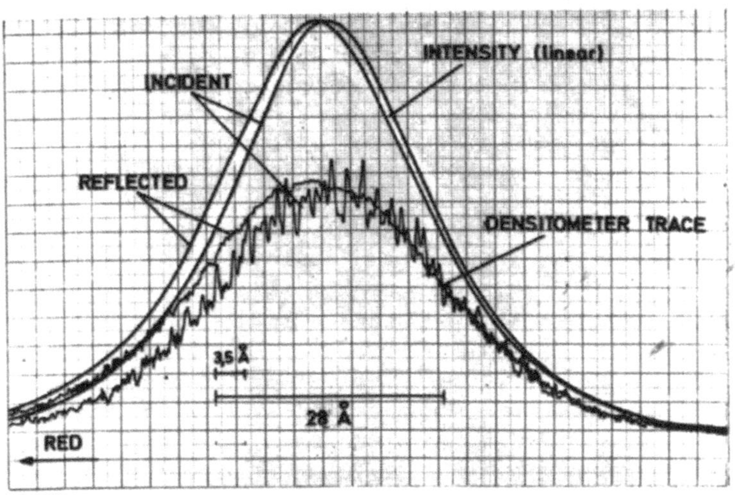

Fig.7 Spectrum of incident and reflected laser light

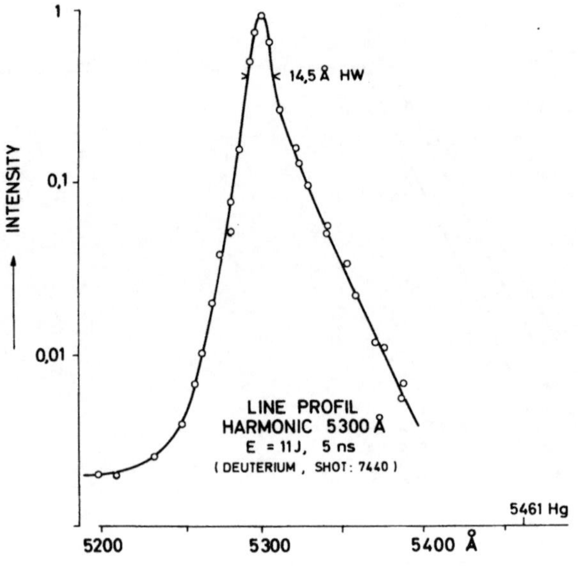

Fig.8 Spectrum of second harmonic

IV. ANGULAR DISTRIBUTION OF BACKSCATTERED LASER LIGHT AND SECOND HARMONIC

There can be no doubt that measurements with fast photodiodes as presented in Section III give quantitative and hence most valuable information on the energy losses due to backscatter from laser-produced plasmas. Their drawback lies perhaps in the fact that spatial information, i.e. on the radiation characteristic of the plasma is difficult to obtain. They are ideally completed by the photographic technique described here /16/. It immediately shows the radiation pattern of the plasma and gives a fairly direct insight into the physical phenomena occurring in the plasma.

The arrangement used for these experiments is shown in Fig.9. The intention of this experiment is to photograph the angular distribution of backscattered light by imaging a plane some distance away from the target on a photographic plate. Actually this plane is the focusing lens. For the backscattered light the focusing lens acts like a condenser and projects the light towards the imaging lens. Unfortunately the solid angle which can be observed this way is restricted to $\Omega = 1$ around the optical axis corresponding to the aperture of the focusing lens. As turns out, however, such an angle is already high enough to obtain useful infor-

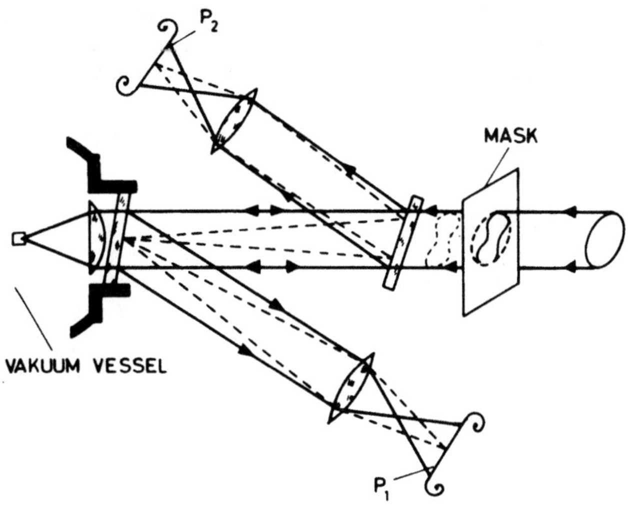

Fig.9 Experimental arrangement for photographing the angular distribution of incident and reflected laser light

mation on the radiation characteristic of the plasma. Actually two cameras P_1 and P_2 have been used in this experiment. A close inspection of the arrangement of beam splitters and lenses in Fig.9 shows that P_2 monitors the intensity distribution of backscattered light, whereas P_1 is used for a control of the intensity distribution of the incoming laser light. Insertion of an interference filter in front of P_2 allows one to photograph alternatively the reflected laser radiation or its second harmonic.

Under normal experimental conditions the incoming laser beam fills the whole cross section of the focusing lens. Any radiation scattered into larger angles than that covered by the lens would then be lost for observation. Therefore various masks have been used for the incoming laser beam; thus the beam fills only part of the lens and some aerea on the lens is left for detection of light scattered at angles larger than that of the incoming beam. Various masks have been used (holes, triangles), but for the experiments to be described in the following usually half of the beam has been masked by the straight edge of a sheet of black paper. The basic feature of backscatter observed in this experiment appears, however, most impressively in the case where a mask in the form of a question mark has been used (Fig. lo, the question mark has been cut out of a sheet of black paper). The upper half of Fig.lo shows a photograph of the distribution of the incoming beam, taken for comparison with camera P_1. Within the contours of

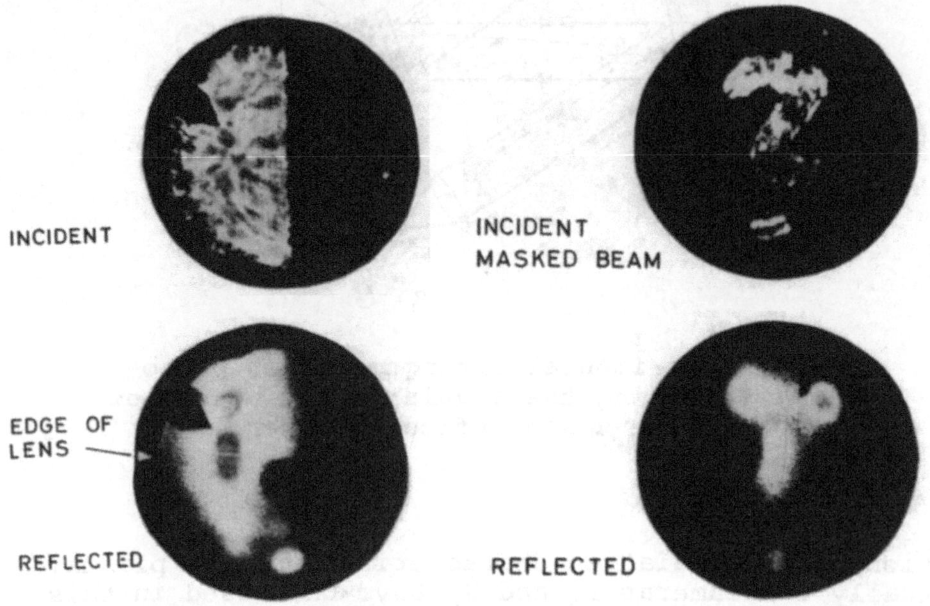

Fig.10 Photographs of incident and reflected laser light distribution on the focusing lens. Left: Right half of the lens masked. Right: Mask in the form of a question mark

the distribution hot spots are visible which - though emphasized very strongly due to the steep response characteristic of the simple burn pattern technique applied in this case - are clearly present in the incoming beam. Though such an intensity distribution is already fairly detailed it is quite surprisingly reproduced in the backscattered laser light (Fig.lo, lower half). Even the point of the question mark is clearly resolved. Also is shown the corresponding result for the case where half of the beam has been masked. Hot spots in the incoming beam are again resolved. It appears clearly from these photographs that in our experiment conditions can be achieved where to a high degree of perfection each light ray is scattered back in itself.

The results of Fig.lo were achieved with a solid deuterium target, the target being in the position of maximum reflection. Subsequently photographs were taken with heavier targets like carbon, copper and tungsten. Fig.11 shows photographs made with copper as a target material. Basically the same effect is observed: Most of the scattered laser light comes back in itself. Again

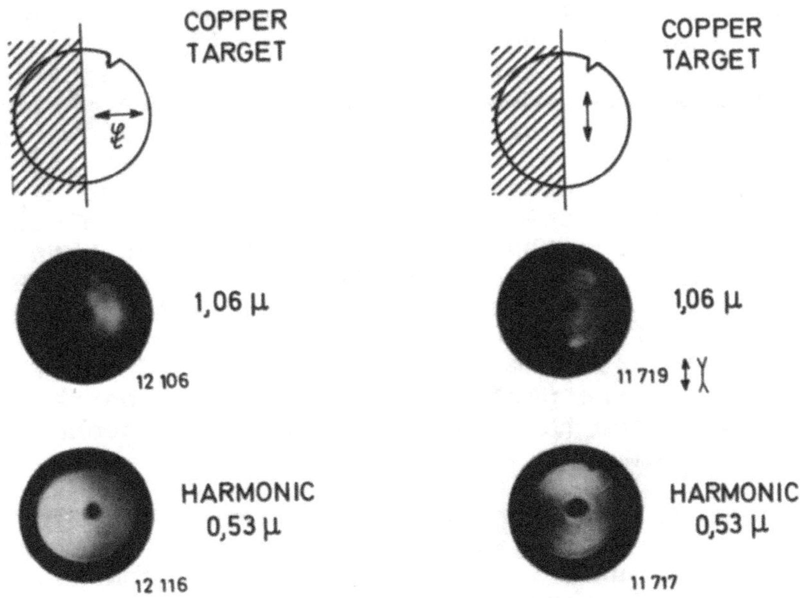

Fig.11 Patterns of reflected laser light and second harmonic. The orientation of the polarization vector of the incident laser light relative to the edge of the mask is different for the two cases

the distribution of backscattered laser light is brightest in that half of the lens which is illuminated by the incoming beam. However, with heavier targets some fraction of the laser light is also scattered into the other half of the lens which remains completely dark in the case of deuterium. Quantitative measurements on Kodak IZ plates yield the following picture of backscattered intensities in the two halfs of the lens (illuminated part/masked part).

solid deuterium	$\gtrsim$ 5o : 1
carbon	2o : 1
copper	2,5 : 1
tungsten	~ 5 : 1

There are characteristic differences between the directed and undirected parts of the backscattered light: The directed part always shows a structure corresponding to the distribution of hot spots in the incoming beam whereas the undirected part has a smooth distribution. If a polarizer is brought in front of P_2 and rotated in such a way that radiation polarized parallel to the laser beam is stopped that half of the lens which is illuminated by the incoming laser beam and which shows directed backscatter becomes completely dark. This means that the directed part of the backscattered light is linearly polarized. The degree of polarization is greater than 2o : 1 i.e. within the accuracy of the measurements no depolarization of the linearly polarized laser beam could be detected in directed backscatter. On the other hand the intensity coming back through the masked part of the lens was found to be insensitive to a rotation of the polarizer. This means that this part of scattered is not linearly polarized. Whether the directed and the more diffuse part of the pattern are emitted at the same time is not known since the applied photographic technique integrates in time.

Whether all the scattered light is directed back in itself (deuterium) or part is scattered in a geometrical manner (copper, tungsten) is not a direct consequence of the different charge of the ions in the plasma. It rather seems to depend on how the hydrodynamic flow of the target material under laser radiation develops. This results from an experiment illustrated by Fig.12. In this case the distribution of backscattered light was compared for two subsequent shots on a copper target which was not moved between shots. The first shot creates a crater in the target with a depth of several hundred microns. Such a crater+) is shown in Fig.12 where a

+)The formation of such a crater under laser impact on a solid body is of course in itself an interesting phenomenon. The instantaneous deposition of laser energy on the target surface sets up an intense shock wave which subsequently penetrates into the interior of the target. Up to a certain depth the shock wave is strong enough to vaporize the target material. This material than expands into the vacuum and a crater is left on the target surface. The phenomenon is similar to a micrometeorite impact on a solid body /17/.

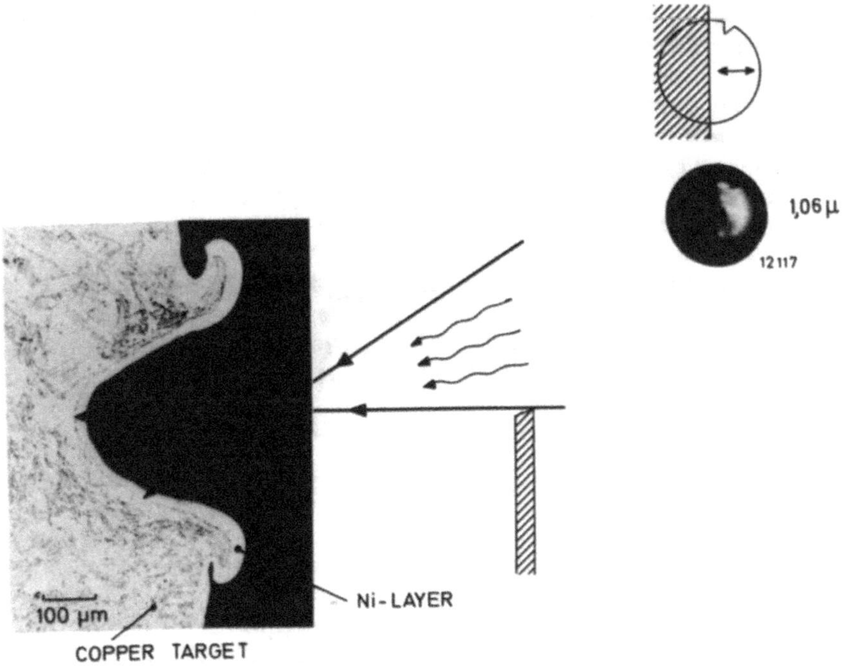

Fig.12 Laser produced crater in a copper target and reflection pattern obtained with this kind of target. (The nickel layer has been deposited onto the crater only for the purpose of polishing).

sample irradiated once has been cut through and polished. For the second shot the incoming laser beam now finds a curved rather than a plane target surface. Whereas during the first shot part of the light is scattered in a geometrical manner (Fig.11), with the crater present all of the reflected laser light is now scattered back in itself as in the case of solid deuterium (with an intensity ratio illuminated/masked greater than 5o : 1). Again the hot spots in the beam are reproduced clearly.

The behaviour of the radiation scattered back with the frequency of the SECOND HARMONIC is quite different from that observed for the backscattered laser light. This is illustrated by Fig.11. In contrast to the latter case where most of the scattered laser light is scattered back in itself the harmonic pattern in Fig.11 (left side) has maximum intensity in that half of the lens which was masked during illumination. Whereas the patterns of the backscattered laser light are apparently insensitive to the orientation of the polarization vector of the incoming beam relative to the edge of the mask, the harmonic patterns are not. If the polarization vector is rotated with respect to the edge of the mask, the harmonic pattern rotates also and takes on a figure-eight shape. This observation corresponds to the case of copper where part of the laser light is reflected from the target in a geometrical manner. Apart from some minor diffuse background the 2nd harmonic patterns have always a minimum on the optical axis and a maximum in the plane containing the polarization vector and the optical axis, located somewhere near the edge of the lens. As will be discussed below these observations are most useful in understanding the mechanism of harmonic production.

V. DISCUSSION

The authors are well aware that their present understanding of the observed phenomena is still far away from a coherent and quantitative description of the physical mechanisms leading to backscatter in such experiments. Nevertheless reproducible and relatively well controlled conditions have been established in this experiment - a situation which is not obvious in many experiments and by itself already indicates some progress. Also it is felt that some of the new aspects of backscatter reported here are elementary as well as conspicuous enough to stimulate work on a unified picture of backscatter under real experimental conditions. This may justify publication of the experimental facts in their present status.

Those of the experimental results like the time behaviour of the scattered light intensity, the width of the reflection curves (Fig.4) and also the fact that reflection increases in a quite pronounced manner with intensity (Fig.5), which are believed to be intimately connected with the two-dimensional problem of crater formation in the target under relatively long (nanosecond) irradiation are not considered in the following at all. However, some qualitative arguments concerning the angular distribution of scattered radiation and illustrating various possible interpretations are presented.

Perhaps the most striking feature of these experiments is the directed behaviour of backscatter as seen in Fig.10. For a classical interpretation i.e. where the light is reflected from the layer with $\omega_L = \omega_P$ (ω_L =

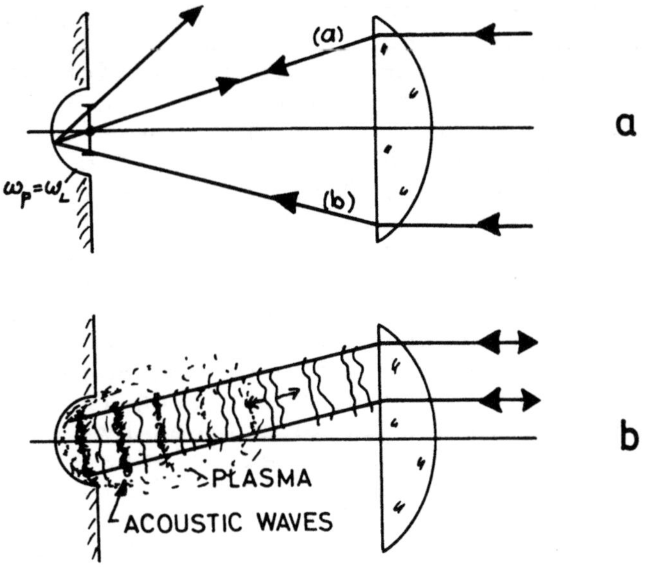

Fig.13 Two models for backscatter:
a) classical reflection from a curved mirror
b) stimulated Brillouin-backscatter

frequency of laser radiation, ω_p = plasma frequency) this layer has to be a curved mirror as shown in Fig.13 (a mirror curved outward is also possible). The experiment indeed shows directed backscatter for the case of copper when a mirror like target (crater, Fig.12) is used. In the case of deuterium which also shows directed backscatter as well it is known /lo/ that the shock wave penetrates several hundred microns into the target during irradiation and hence a similar geometry may be present. The difficulties connected with such an interpretation appear, however, immediately. Firstly the center of curvature of the mirror must coincide exactly during the whole pulse length (since the photographs show time-integrated distributions) with the focal spot of the beam. Only then a light ray (a in Fig.13) is reflected back in itself. In view of the precision required for such a mirror in order to give the high degree of directivity observed such an explanation looks quite artifical. Secondly a more close inspection of this hypothesis shows that it is valid only in the approximation where the focal spot of the laser is assumed to be a point. In reality, as is shown by Fig.1, the focal spot has a finite extent. This means there that a light ray (b) which does not pass through the (idealized) focal spot is reflected under some angle rather than back in itself. A quantitative consideration taking into account an realistic radius of the mirror and the focal spot shows that reflection should occur into angles larger than consistent with the photographed distributions. Thirdly it is quite surprising in a classical picture where absorption is determined by collisions that the measured reflection coefficients are independent of the ion charge in the plasma (which increases the collision frequency). Though the decrease of the reflection coefficient with increasing Z of the plasma seemingly has been overestimated in the literature a distinct decrease of reflection is to be expected for the heavy materials /18/.

With respect to these arguments we favour another explanation which is different in its physical nature. We attribute the observed directed backscatter to stimulated Brillouin backscatter. In this case the reflected laser radiation is scattered from ion acoustic waves. Since these waves in the case of stimulated scattering are excited by the beating of the two electromagnetic waves (incident, reflected) directivity is an intrinsic property of the scattering process itself. Though oblique backscatter is possible in principle, the most extended length of the plasma along the optical axis, as well as the fact that the incoming beam is coherent over a much smaller solid angle than that of the lens, favour amplification of the scattered light in the backward direction. Thus the first two difficulties mentioned above do not exist for this interpretation. Since backscatter in this case can occur at densities lower than the critical density, the third objection (high reflectivity from high Z materials) is , at least in principle, not severe for stimulated Brillouin scattering. Unfortunately the spectral investigation of incident and backscattered laser light (Fig.7) does not allow so far a discrimination between the two possibilities. The observed redshift is consistent with reflection from a moving critical layer /14/ as well as with stimulated Brillouin scattering. In the latter case the Doppler shift from the ion acoustic waves $\Delta\lambda/\lambda = 2n(C_s - V)/c$ depends on the refractive index n as well as on the streaming velocity V of the plasma in the interaction region (C_s = sound velocity, c = light velocity). If, for example, the plasma counterstreams to the ion acoustic wave with a velocity close to the velocity of sound the wavelength shift would indeed be close to zero as observed in the experiment.

Emission of the second harmonic from a laser produced plasma has been reported by several authors /19/. The nature of the underlying production mechanisms seems however, not to be clarified. The measurements presented here give strong evidence that the harmonic is produced during oblique reflection of the laser light from the inhomogeneous plasma. The current density along the density gradient is given by:

$$\dot{j}_z = e\left(n_e + \frac{\partial n_e}{\partial z}\Delta\right)\dot{\Delta} = e n_e \dot{\Delta} + e\frac{\partial n_e}{\partial z}\Delta\dot{\Delta} = \dot{j}_\omega + \dot{j}_{2\omega}$$

where Δ is the excursion length of the electron oscillating under the influence of the laser radiation with the frequency ω . The second term (which disappears in a homogeneous plasma) varies with a frequency 2ω and gives rise to second harmonic emission. A theoretical treatment of this process has been given for a metal surface in /2o/ and for an inhomogeneous plasma in /21/. Since our results have been obtained only recently, a quantitative discussion will be given elsewhere. However, though a plane wave approximation could be expected to be a poor approximation of the experimental conditions, qualitatively the main features of harmonic emission in this case are clearly observed in the experiment. These are:

(1) The harmonic is emitted in a direction parallel to the reflected electromagnetic wave.
(2) Strong harmonic emission occurs only if the electric field vector of the incident electromagnetic wave is contained in the plane of incidence. This is due to a singularity which then occurs in the critical layer ($\omega_L = \omega_p$).
(3) Emission should occur only for oblique incidence with some optimum angle determined by the density gradient of the plasma.
(4) Harmonic emission should increase with the square of incident power.

According to this model of harmonic production we interpret(in the case of copper) that part of the backscattered light (about 1/3 of the total energy scattered back through the focusing lens) which is diffusely scattered back through the masked part of the lens (see Fig.11), as being due to reflection from an essentially plane target. This then explains readily according to (1) why the harmonic is seen also in the left (masked) half of the lens in Fig.11. The shape of the harmonic pattern is expected from (2): The light ray A (see Fig. 14) is focused by the lens, reaches the focal spot located on the optical axis c, is reflected and appears at B. According to (1) and since in this case the electric field vector is contained in the plane of incidence harmonic light should appear at B. This is in fact observed. In contrast ray A appears at B , in this case the electric field vector is perpendicular to the plane of incidence and no harmonic appears at B . Also harmonic emission is low along the optical axis according to (3) with the strongest emission being observed near the edge of the lens. If the same arguments are applied to the case where the electric field vector is parallel to the edge of the mas (right half of Fig.11) rotation and the change of the pattern into a figure-eight pattern are found to be consistent with the above interpretation. Finally as expected according to (4) the harmonic emission is found to increase with the square of the incident power.

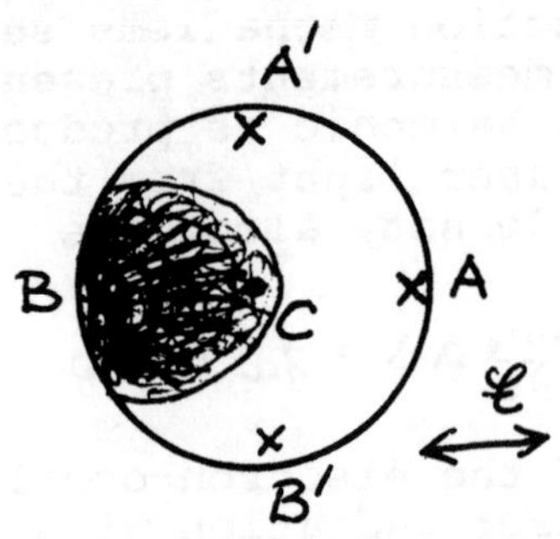

Fig.14 Schematic sketch of harmonic pattern shown in Fig.11 (lower left side)

ACKNOWLEDGMENT

The authors thank Dr. P. Mulser and Dr. H.M. Key for discussions concerning the material presented here. In particular they acknowledge the excellent work of their team: H. Brändlein (photographs and spectra), P. Sachsenmaier (laser and coordination) and E. Wanka (cyrogenics and targets).

This work was performed under the terms of the agreement on association between Max-Planck-Institut für Plasmaphysik and Euratom.

REFERENCES

1 P. Mulser, R. Sigel and S. Witkowski, Physics Rep. (Section C of Phys. Lett.) 6, No. 3 (1973), 187, North-Holland Publishing Company

2 D.W. Forslund, J.M. Kindel and E.L. Lindman, Phys. Rev. Lett. 3o, 739 (1973) and many references contained therein to earlier work on this subject

3 J.L. Bobin, M. Decroisette, B. Meyer and C. Vitel, Phys. Rev. Lett. 3o (1973) 594; K. Boyer, Progress Report, Laser Program at LASL, LA-5251-PR (1973), see also ref. /2o - 21/

4 V.P. Silin, Sov. Phys. JETP 21 (1965) 1127; K. Nishikawa, J. Phys. Soc. Japan 24 (1968) 1152; W.L. Kruer and J.M. Dawson, Phys. Fluids 15 (1972) 446 and other references contained therein

5 N.G. Denisov, Sov. Phys. JETP 4 (1957) 544; J.P. Freidberg, R.W. Mitchell, R.L. Morse and L.I. Rudsinski, Phys. Rev. Lett. 28 (1972) 795

6 P. Kaw and J. Dawson, Phys. Fluids 13 (197o) 472

7 F. Floux, J.F. Benard, D. Cognard and A. Saleres, in Laser Interaction and Related Plasma Phenomena (Acad. Press, New York 1972) p. 4o9; C. Yamanaka, ibid. p. 481; K. Büchl, K. Eidmann, P. Mulser, H. Salzmann and R. Sigel, ibid. p. 5o3; E.D. Jones, G.W. Gobeli and J.N. Olsen, ibid. p. 469; N.G. Basov, V.A. Boiko, S.D. Zakharov, O.N. Krokhin and G.V. Sklizkov, JETP Lett. 13 (1971) 691; J.W. Shearer, S.W. Mead, J. Petruzzi, F. Rainer, J.E. Swain and C.E. Violet, Phys. Rev. A6 (1972) 764; N.G. Basov, S.D. Zakharov, O.N. Krokhin, P.G. Kryokov, Yu.W. Senatskii, E.L. Tyurin, A.I. Fedosimov, S.W. Chekalin and M.Ya. Shchelev, Kvantovaya Elektronika 1 (1971) 4 (in Russian); H. Salzmann, J. Appl. Phys. 44 (1973) 113; see also ref./3b/; A. Saleres, D. Cognard and F. Floux, Proceedings V. European Conf. on Contr. Fusion and Plasma Physics, Commissariat à l'Energie Atomique, Grenoble 1972, 59; see also paper by Stamper et al. in this volume

8 K. Büchl, K. Eidmann, P. Mulser, H. Salzmann, R. Sigel and S. Witkowski, Plasma Physics and Contr. Nuclear Fusion Research, Vol. I, 645 (Int. Atomic Energy Agency, Vienna, 1971)

9 K. Eidmann and R. Sigel in Proceedings of 6th Europ. Conference on Controlled Fusion and Plasma Physics, Moscow 1973, p. 435

1o C.v.Kessel and R. Sigel, to be published (see Abstracts of the Annual Plasma Physics Meeting (Philadelphia) in Bull. Amer. Phys. Soc. (1973))

11 H. Baumhacker, P. Sachsenmaier, Laser und angewandte Strahlentechnik 3/1971; H. Salzmann, Lab. Report IPP IV/21 (1971); see also ref. /7c/ and /8/

12 H. Krause, Lab. Report IPP IV/43, to be published in Journal of Physics E: Scientific Instruments 1973 (Volume 6)

13 See papers of J.L. Bobin, K. Brueckner and C. Yamanaka in this volume

14 K. Büchl, K. Eidmann, H. Salzmann and R. Sigel, Appl. Phys. Lett. 2o (1972) 3

15 See papers of J.L. Bobin and C. Yamanaka in this volume

16 Results were first presented on the Conference Laser Interaction with Matter at Marly-le-Roi in October 1972 (K. Eidmann and R. Sigel, Lab. Report IPP IV/46 (1972))

17 Ya. B. Zeldovich and Yu. P. Raizer, Physics of Shock Waves and High-Temperature Hydrodynamic Phenomena (Acad. Press, New York 1966) p. 844

18 E. Cojocaru and P. Mulser, to be published

19 A. Caruso, A. De Angelis, G. Gatti, R. Gratton and S. Martelluci, Phys. Lett. 33A (197o) 29; see also ref. /3a,b/ and /13/

2o S.S. Jha, Phys. Rev. 14o (1965), A 2o2o; N. Bloembergen and Y.R. Shen, Phys. Rev. 141 (1966) 299

21 N.S. Erokhin, V.E. Zakharov and S.S. Moiseev, Sov. Phys. JETP 29 (1969) 1o1; Y.C. Lee et al., to be published

EXPERIMENTS WITH LASER-PRODUCED PLASMAS: ELECTRONS, IONS, AND NEUTRONS + *

R. P. Godwin

Los Alamos Scientific Laboratory

University of California, Los Alamos, New Mexico 87544

ABSTRACT

Experimental results, particularly x-ray bremsstrahlung and ion time-of-flight measurements, obtained from plasmas generated with a 10 J, 25 psec Nd:YAG-Nd:glass (1.06 μm) laser delivering focal spot intensities of greater than 10^{16} W/cm^2 are reviewed. Preliminary experiments with a 10 J, 1 nsec CO_2 (10.6 μm) laser delivering $\sim 10^{14}$ W/cm^2 to a target are briefly discussed. Both sets of experiments yield high-energy (tens of keV) electrons and ions. Experimental evidence indicates that neutron emission can be explained by electron heating and a consequent ion acceleration. Therefore, neutron production in laser-produced plasmas can be a delusive diagnostic.

+ A report of work performed under the auspices of the U. S. Atomic Energy Commission by G. H. McCall, A. W. Ehler, D. Giovanielli, R. P. Godwin, J. F. Kephart, K. B. Mitchell, T. F. Stratton, and F. Young.

* Presented at the Third Workshop on "Laser Interaction and Related Plasma Phenomena" held at Rensselaer Polytechnic Institute, Troy, New York, August 13-17, 1973.

INTRODUCTION

Very high target compressions (densities greater than one thousand times normal density) appear necessary if we are to obtain significant thermonuclear yields from laser-irradiated targets with lasers built using technology now available or expected soon.[1,2,3] While the hydrodynamics calculations of yield from laser-driven target implosions seems to be well understood, certain features of the problem are not. Major problem areas which require experimental and theoretical clarification include:

1. Laser absorption in plasmas

   Which instabilities dominate the interaction of intense laser pulses with short lengths of plasma having steep density gradients? Do these instabilities lead to enhanced absorption or reflection?

2. Hot electrons

   What is the number and energy distribution of nonthermal electrons created in laser-heated plasmas? How do these electrons influence target implosions?

3. Wavelength dependence

   How are the absorption processes and the production of "hot" electrons influenced by laser wavelength? Must we use ultraviolet lasers to achieve laser-initiated fusion?

4. Compression

   When significant target compression is achieved laser fusion will have been shown to be a viable concept.

You may notice that neutron generation has not been listed above. We do not consider neutron diagnostics a primary experimental diagnostic (at least until larger yields than have thus far been reported are achieved).

Numerous laser-plasma experiments have been performed at Los Alamos in the last year and a half. Most of our experiments have been attempts to understand the coupling of energy into a laser plasma. We have done, for example, experiments on backscattered and specularly scattered 1.06 μm radiation and upon the creation of doubled 1.06 μm, i.e., 0.53 μm light.[4,5] In concise summary, we have found in the interaction work that the light is absorbed in an extremely thin layer and that the plasma density profile is probably crucial in determining the interaction characteristics.

We shall concentrate in this discussion on electron-energy spectra and fast-ion emission. After relating these features to one another through an energy flux-limit condition, we will use them as the basis for a discussion of nonthermal neutron production in laser-produced plasmas.

A Nd:glass laser with a YAG oscillator has been used in most of our experiments. The pulse length is 25 psec as measured by two-photon fluorescence. We have delivered more than 15 J to targets. We normally operate at less than about 12 J because the lack of isolation allows backscattered light to damage the laser. The focal diameter is less than 50 μm giving us a power at focus of greater than 10^{16} W/cm^2. While the 10 to 15 J laser energy is not large, we have laser powers comparable with any being used because of the short pulse length. Figure 1 is a schematic of the small laser which is like the preliminary stages of the larger glass system (> 1 kJ) being constructed at Los Alamos. The mode-locked dye oscillator is followed by a Pockel's cell switch to select a single pulse from the Q switched train of pulses and then by two 10 mm YAG amplifiers followed by a 20 mm and finally a 51 mm glass amplifier. Note the divergence of the beam through the glass amplifiers. This divergence appears to alleviate self-focusing problems in the 51 mm rod.

An aspheric f/3.5 focusing lens with a 70 mm focal length delivers our beam to targets. Most of our experiments have been performed with thin films of CH_2 as targets; although we have also used deuterated polyethylene as well as various metal films, small carbon spheres, and metal wires.

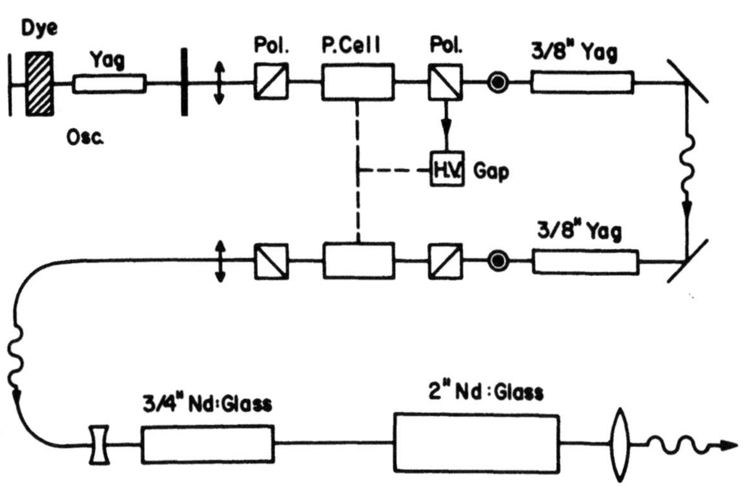

Fig. 1. Schematic diagram of the Los Alamos 10 J, 25 psec Nd:YAG--Nd:glass laser system.

ELECTRONS

At the time of the last RPI Workshop it was becoming evident that high-energy electrons were created in laser-produced plasmas.[6] The only information available was that the electrons had a very nonthermal character and that high-energy electrons were present. We believe that a small number of electrons, perhaps 10^{10}, may escape the plasma completely until an electrostatic barrier is set up to prevent further electron escape. While we have attempted to directly measure escaping electrons, we have not succeeded in doing so. Perhaps the most direct evidence of escaping electrons is the x-ray measurements performed at Garching in which x rays have been shown to be created at the walls of the target chamber.[7] Figure 2 shows what may be indirect evidence of escaping electrons. Magnetic fields[8] were being measured with a small loop probe near the laser plasma. In the figure one sees the signature of a magnetic field as the characteristic negative-positive pulse. This signal, as expected, inverts if one turns the probe 180°. The other signal early in time does not invert. We believe this early signal may be due to electrons escaping from the plasma leaving it positively charged.

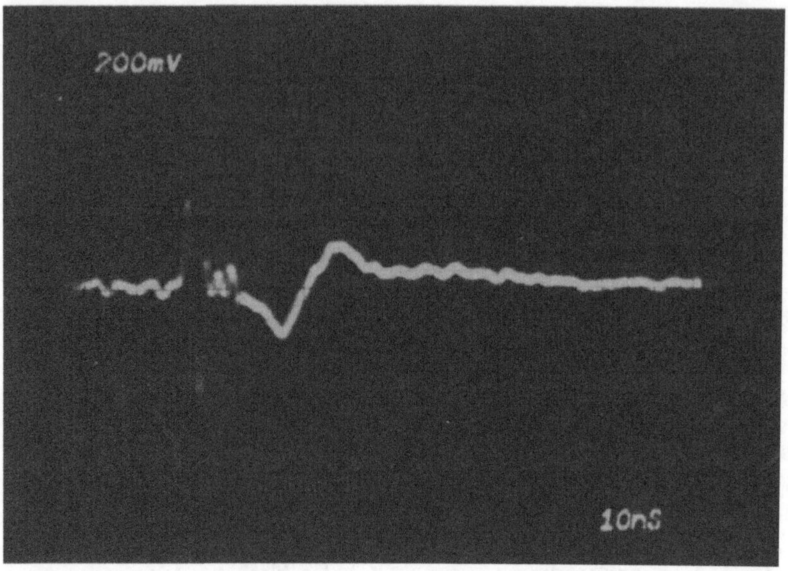

Fig. 2. Raw data obtained with a magnetic probe 1.5 cm from a laser-produced plasma in high vacuum. The negative-positive pulse represents a magnetic field at the probe of a few gauss. The early fast signal may be due to electrons escaping from the plasma.

In effect, the signal is that of an instantaneously charged capacitor. If one knew the geometry of the plasma one would be able to determine from the signal the number of electrons leaving the plasma. While we are not able to do that accurately, the signal is consistent with the interpretation above.

We turn briefly to an experiment which has been discussed quite a lot in the past year and perhaps overemphasized. We have constructed a classical x-ray polarimeter.[9,10] The polarimeter consists of a thin plastic scatterer with four detectors as nearly identical as possible located on the symmetrical spokes of a detector assembly. The detectors are NaI(Tl) scintillators coupled to conventional photomultiplier tubes. Since Compton scattering is polarization dependent, one can measure the polarization by determining the asymmetry in signals of the various detectors. Conceptually, this is a very simple experiment; practically it is quite difficult. One would like to learn from such a measurement the electron velocity distribution in the plasma as a function of angle. The interpretation of polarization measurements depends on whether the electrons are creating bremsstrahlung in what is essentially a thick or thin target, i.e., whether or not the electrons make multiple scatterings in their transit of the radiating volume. The photon-energy acceptance in the detectors relative to the maximum electron energies radiating the x-ray spectrum is particularly important. Backgrounds, experimental asymmetries, solid-angle depolarization, and multiple scattering in the scatterer must all be considered. The influence of background radiation is difficult to eliminate. Since the scatterer has an efficiency $\sim 10^{-4}$, a small number of hard-to-shield 100 keV or higher energy x rays can falsify measurements. Since we have not been able to sort all of these factors to our satisfaction, we have not formally published the results of this experiment. We do, however, believe that we have measured a polarization of the order of 15% in the laser-plasma x rays. The polarization appears to be such that electrons are moving preferentially along the k vector of the incident laser beam, i.e., not parallel to the electric vector of the incident radiation. The Garching group has attempted to find an asymmetry in x-ray emission from laser plasmas. Within the errors of their experiment, they found no asymmetry.[7] Whether or not there can be a marked asymmetry of electron excitation in laser plasmas is, in our opinion, still an open question.

Most experiments measuring x-ray spectra in laser plasmas have used filter techniques for channel energy selection. This is an easy method to implement, but is a difficult one to interpret. We have constructed a multiple-channel x-ray spectrometer[11] which is more difficult to operate than filtered detectors, but for which data are essentially trivial to interpret. The spectrometer consists of a number of flat Bragg crystals with detectors located at the proper angles to accept energy diffracted from the various crystals. The detectors are NaI(Tl) scintillator-photomultiplier combinations. While the spectrometer has the capability of eight detector channels, it is at present operating with six channels at energies of 4.3, 6.9, 11.2, 18.5, 30.3, and 50.4 keV. The resolution $\Delta E/E$ is typically $\sim$ 20%. All the channels, except that at the highest energy, have pyrolytic graphite diffraction crystals. The highest energy channel uses a 200 plane in lithium fluoride. In Fig. 3 are exhibited four x-ray spectra randomly chosen from a large number of measurements. No two measurements are identical, although all have certain similarities. The channel at 50 keV has the smallest signals and is also near the direct beam of x-rays through the spectrometer so that statistical fluctuations and background sig-

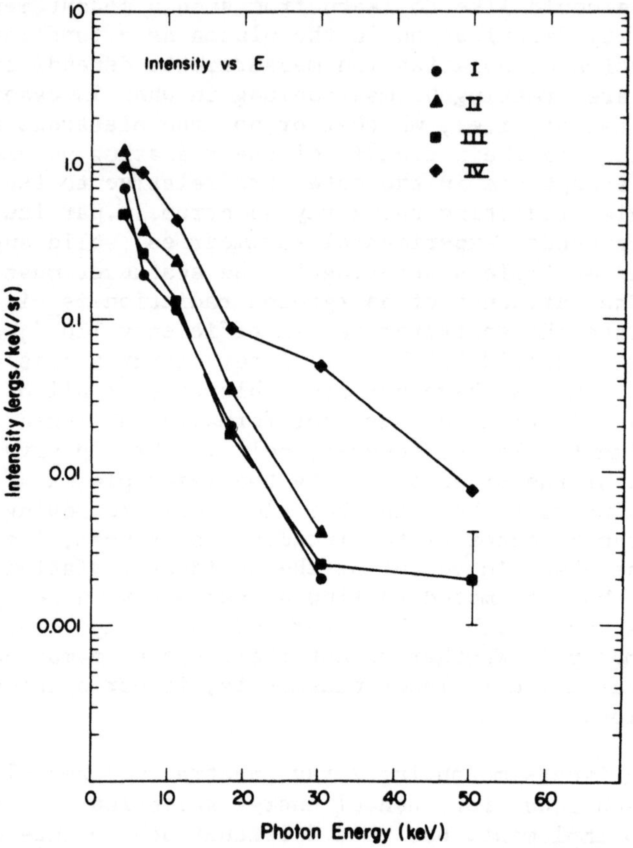

Fig. 3. X-ray spectra obtained with a multiple crystal spectrometer viewing a CH_2 target irradiated by nominally 10 J, 25 psec 1.06 μm laser pulses. The absolute intensity assumes isotropic radiation.

nals are most likely to influence it. The 4.3 and 6.9 keV channels are less accurately calibrated than are the higher energy channels. The most dependable channels are those in the 10 to 30 keV region. Note that in a single instrument measurements are obtained over four decades in intensity. The spectrometer channels are calibrated (with a conventional x-ray tube) relative to one another quite well and absolutely to about a factor of two. Although nonthermal electrons exist in these experiments, a semi-log plot of x-ray intensity vs photon energy is fairly linear. This implies that we can define a "temperature" k_bT. The slope of the curves yields an effective hot-electron temperature of about 10 keV. We should point out that the spectra also fit reasonably well a dependence $\sim \exp(v/v_o)$, where v is the electron velocity. The tail of the electron distribution suggested by Kruer would give such a spectrum.[12]

IONS

Ion emission from our laser-produced plasmas has been examined with several techniques.[13] Ion collectors constructed using commercial BNC feedthroughs with grids provided for applying various voltages to check spurious effects have been the primary diagnostic. Figure 4 shows a spectrum obtained with such ion detectors viewing a laser irradiated CH_2 foil target. It is a direct tracing of an oscilloscope record. Initially, one sees a large spike which is a photoelectron signal due to the x-rays and extreme ultraviolet radiation emitted by the plasma. Then we find a string of jagged peaks. We identify the first of these as protons and those following as

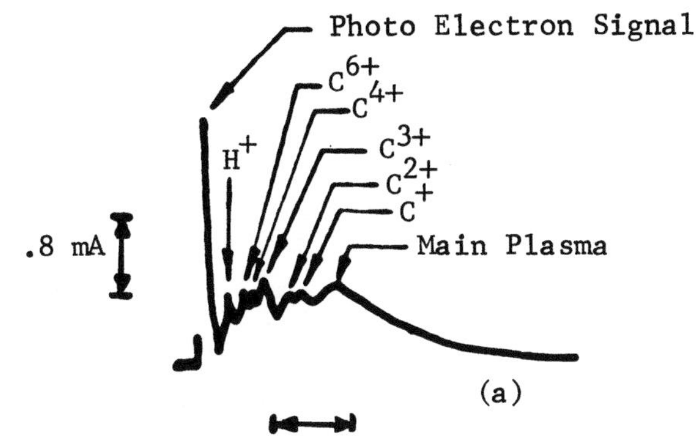

Fig. 4. Ion collector current as a function of time for laser-plasma experiments with a CH_2 target.

carbon in various ionization states. The final broad peak is due to a cool neutral plasma. On the side opposite a thin (25 μm) target from the laser, one sees a cold plasma, but none of the ion spikes. We interpret the ion-current spikes as being due to essentially electrostatically-accelerated ions associated with hot electron production. A consistent fit is obtained if one identifies the spikes with the energy being Z (the ionic charge state) times a constant voltage. The ion-collector information alone may not be convincing; but we have another detector, the Thomson parabola, which allows the unambiguous determination of the charge-to-mass ratio of plasma ions. We use a design developed by Kuswa at Sandia Laboratories. Figure 5 is a photographic record obtained with the device. The bright point in the lower-right-hand corner of the photo is a pinhole image of the laser-produced plasma. The parabolas (from left to right) are protons and three charge states in carbon. (In raw data one can often see more than three charge states.) The vertical displacement from the pinhole picture of the plasma to the start of the parabolas gives the maximum energy of ions present. For instance, the proton trace of Fig. 5 contains energies up to 50 keV. Thomson parabola records are consistent with the interpretation of the ion collector data discussed above.

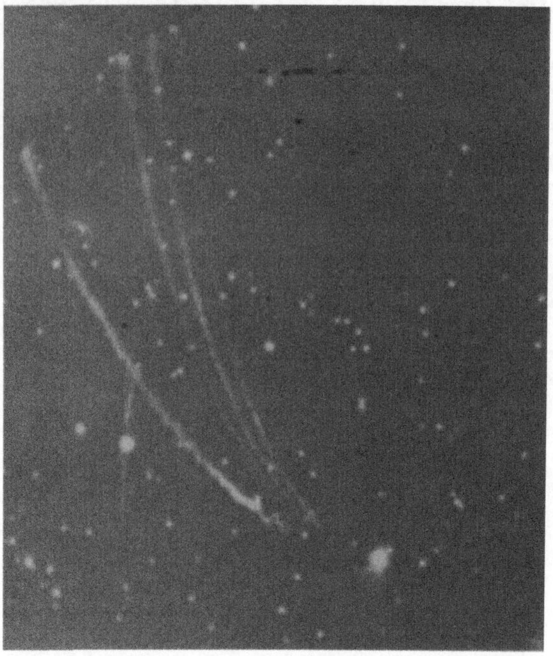

Fig. 5. Thomson parabola record obtained with laser-produced plasma from a CH_2 target. The bright spot in the lower-right-hand corner is an image of the plasma. The parabolas (from left to right) are protons and various charge states of carbon.

CO_2 LASER EXPERIMENTS

We have recently begun experiments with the short pulse CO_2 laser being developed at Los Alamos by the group of Dr. Fenstermacher. The laser has been discussed at this Workshop.[14] The laser wavelength is 10.6 μm. The pulse length is about 1.5 nsec FWHM. It has a very fast rise time and a much slower decay, as one would expect from saturated amplifiers. As yet we are not sure how much energy is in the wings of the pulse. We have delivered up to about 15 J to a target. We believe the focal diameter is about 80 μm, giving us a power upon targets $\sim 10^{14}$ W/cm^2--the highest yet achieved in 10.6 μm laser-produced plasmas. The laser-plasma diagnostics we have thus far performed are indicated schematically in Fig. 6. We are using the first three amplifier stages of the laser. The beam is deflected 90° by a turning mirror. A hole in that mirror allows a photon-drag detector to view a portion of the beam to study its time dependence. The beam is brought to focus on targets which, to this time, have been CH_2 films by an f/3.5 silver coated off-axis parabolic aluminum mirror with a 10 cm focal length.

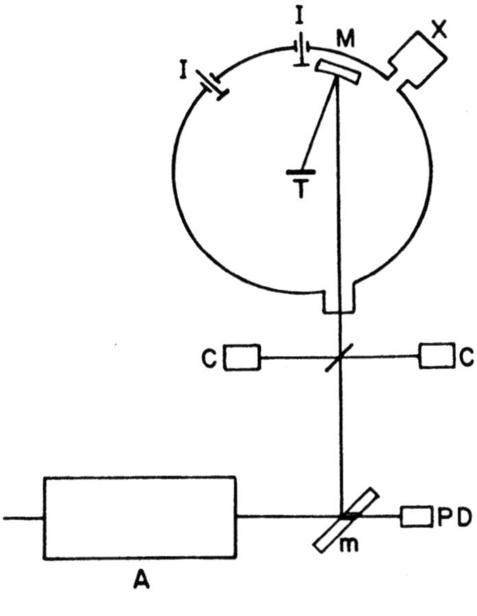

Fig. 6. Schematic diagram of the initial diagnostics of target interaction experiments with the Los Alamos 1.5 nsec CO_2 laser.

We have a NaCl pellicle for reflection of a known fraction of the beam into calorimeters for measuring the incident and backscattered radiation intensities. A number of filter type x-ray detectors and ion current collectors view the target. In these preliminary experiments we have measured backscattered fractions of 2 to 6% with a lot of scatter, perhaps due to focusing difficulties. Table I is a summary of three shots chosen from our preliminary CO_2 measurements.

TABLE I

Energy Inc. (J)	Backscatter (%)	X-Ray "Temp." (keV)	X-Ray Energy $\gtrsim$ 8 keV (erg/sr)
9.8	3.6	$\sim$ 10	45
5.4	2.8	$\sim$ 15	10
7.5	2.1	$\sim$ 10	12

The x-ray "temperature" was defined by using filtered detectors with three channels in the region of about 8 to 40 keV. The temperatures from different detector combinations are not identical. In Fig. 7 we have exhibited ion current as a function of energy per

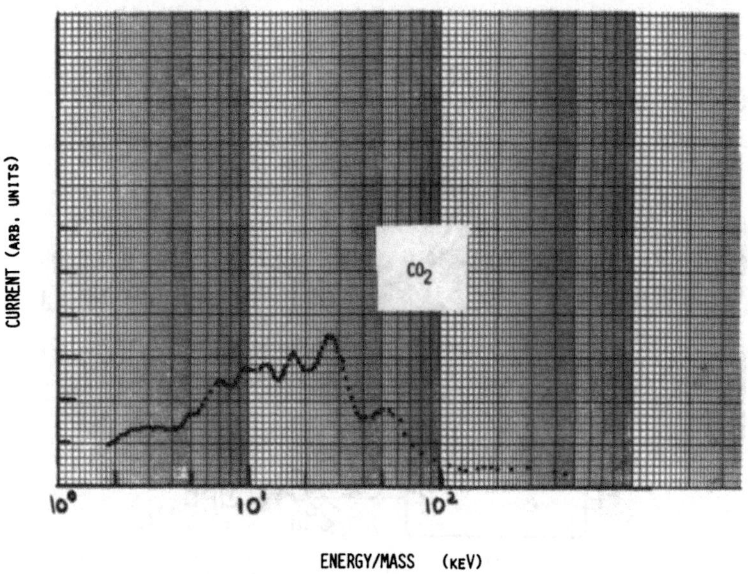

Fig. 7. Ion collector current as a function of energy per unit atomic mass for a CH_2 plasma created with a $\sim$ 10 J, 1.5 nsec CO_2 laser pulse. Note the 50 keV proton peak.

unit atomic mass for a laser plasma created by irradiating CH_2. We identify the peak at 50 keV as protons and the other peaks as various ionization states in carbon.

In summary, plasmas produced by 10 J pulses from both a 1 nsec CO_2 laser and a 25 psec Nd:glass laser contain high-energy electrons with an effective average energy of about 10 keV as determined by x-ray measurements and ions with energies of 50 keV or greater.

FLUX LIMIT FOR ELECTRON HEATING

Consider the implications for x-ray and ion measurements of the flux-limit condition of Morse and Nielson which was briefly discussed at this workshop by Morse.[15] First let us review some x-ray physics. The integrated bremsstrahlung radiation emitted by an electron in a thick target x-ray tube with an anode of atomic number Z is well approximated by[16,17]

$$I_x = kZ\,E^2, \tag{1}$$

where $k \simeq 0.7 \times 10^{-6}$ when the intensity I_x and the electron energy E are both in keV. We wish to estimate the x-ray emission from our laser plasmas. We assume that a small region at the surface of the target (at or near the critical density where the electron plasma frequency equals the laser frequency) is heated and that electrons leave that region and emit x-rays in a denser cold region behind the laser-electron interaction region--an emission model similar to conventional thick target x-ray emission. The bremsstrahlung emission can be estimated as

$$I_x = N_h\,k\,Z\,\langle E_x^2\rangle, \tag{2}$$

where N_h is the number of hot electrons and $\langle E_x^2\rangle$ is an appropriate electron energy squared averaged over the electron distribution function present in the plasma. Detailed calculations support the validity of this simple picture.[18]

The flux-limit condition of Morse and Nielson requires that the absorbed laser power is equal to the energy flux carried away from the interaction region by hot electrons, i.e.,

$$P_L = \tfrac{1}{2}\int mv^2\;v\;f(v)\;dv; \tag{3}$$

where P_L is the laser power, m and v are respectively the electron mass and velocity, and f(v) is the electron distribution function. We represent the integral by

$$P_L = \tfrac{1}{2}\,m\,n_h\,\langle v^3\rangle, \tag{4}$$

where n_h is the hot electron density and $\langle v^3 \rangle$ is an average over the distribution function. This representation of the flux limit can be used to estimate N_h and $\langle E_x^2 \rangle$; and thus the x-ray emission. We define an average energy $\langle E \rangle$ or "temperature" k_bT as

$$\langle E \rangle \equiv \tfrac{1}{2} m \langle v^2 \rangle \simeq \tfrac{1}{2} m \langle v^3 \rangle^{2/3} \quad . \tag{5}$$

Now let us assume that the number density of hot electrons n_h is equal to the critical density n_c. Then using the flux limit condition we find that the effective electron energy is

$$\langle E \rangle \simeq \left(\tfrac{1}{2} m\right)^{1/3} \left(P_L/n_c\right)^{2/3} . \tag{6}$$

Notice that the average energy contains the ratio of the laser power to the critical density. This ratio can be considered a "reduced intensity". In our 1.06 μm experiments we have a critical density of 10^{21} electrons/cm^3 with a laser power of about 10^{16} W/cm^2, while in our 10.6 μm experiments we have n_c equal 10^{19} and P_L equal 10^{14}. Both of our experiments have approximately the same "reduced intensity" and according to the flux limit, both should give effective electron energies or temperatures of ∿ 10 keV. This prediction is in qualitative agreement with the x-ray spectral measurements discussed earlier. The $\langle E \rangle$ given by Eq. (6) is not the proper energy for calculation of x-ray intensity according to Eq. (2). We require the average square of the electron energy. Since x-ray spectral measurements do not uniquely determine electron spectra, we cannot directly measure the electron distribution function with x-ray spectroscopy. However, since the x-ray measurements performed with the Bragg spectrometer are consistent with a Maxwellian energy distribution we will assume such a distribution is present. One thus finds

$$\langle E_x^2 \rangle = \int E^2 f(E)\, dE \Big/ \int f(E)\, dE = 3.75 \langle E \rangle^2 . \tag{7}$$

The simplest estimate of N_h, the number of hot electrons, is obtained by assuming all the laser energy E_L is initially deposited in electrons at the average energy $\langle E \rangle$. In that case

$$N_h \simeq E_L / \langle E \rangle ; \tag{8}$$

with the laser energy 10 J and the electron temperature 10 keV we find $N_h \sim 6 \times 10^{15}$ electrons. An alternate estimate of N_h can be obtained by assuming N_h is equal to the critical density times the volume of the interaction region multiplied by a "turnover factor." The turnover factor is necessary since, during the laser pulse, electrons are heated and leave the interaction volume to be replaced by cold electrons which are in turn heated and leave the volume. The turnover factor can be approximated by the laser pulse length divided by the time required for an electron to leave the interaction volume (roughly the interaction region thickness divided by an average electron velocity which can be estimated using the flux-limit condition). Thus

$$N_h \simeq n_c \; A \; t \; \tau_L \div (t \; / \; \langle v \rangle) = n_c \; A \; \tau_L \langle v \rangle , \tag{9}$$

where A is the focal area, t the interaction region thickness, τ_L the laser pulse length, and $\langle v \rangle$ the average hot-electron velocity. In this picture N_h is independent of the interaction thickness. We have, using the flux-limit condition,

$$N_h \simeq (n_c \; A \; \tau_L)^{2/3} \; (2 \; E_L \; / \; m)^{1/3} . \tag{10}$$

The quantity $n_c\tau_L$ is roughly the same in both our 1 and 10 μm experiments. For our Nd laser experiments N_h estimated by Eq. (10) is $\sim 4 \times 10^{15}$ electrons.

Upon inserting estimates $N_h \sim 5 \times 10^{15}$ and $\langle E_x^2 \rangle \sim 375$ into Eq. (2), we find that with 10 J incident on a CH_2 ($Z \simeq 6$) target we expect $\sim 10^{13}$ keV or $\sim 10^4$ ergs to be emitted as bremsstrahlung x rays. This is an x-ray production efficiency, $e_x \equiv I_x \; / \; E_L$, of $\sim 10^{-4}$, in qualitative agreement with our measurements of the x-ray emission intensity and with those of other laboratories.[19] It appears that, qualitatively at least, the flux-limit condition with essentially all the laser energy assumed to initially reside in hot electrons is consistent with both the effective "temperature" and absolute bremsstrahlung intensity obtained from our x-ray measurements performed with plasmas created by single short laser pulses.

The hot electrons implied by the flux limit condition lead to the creation of hot ions with energies of a few times the hot electron energy. The number of hot ions expected is a significant fraction of the hot-electron number. We measure $\sim 10^{15}$ ions with energies > 50 keV in both our Nd:glass and CO_2 experiments in rough agreement with the implications of the flux limit. We hope that with improvements in the theory of hot-ion production from electrons and in ion measurements that more useful comparisons can be made.

NEUTRONS

Let us briefly review the evidence for neutron production from laser-produced plasmas that was available several years ago.[20,21,22,23] With 1.06 μm Nd:glass laser pulses of a few nsec duration and a few joules or more of energy an enhanced reflectance and production of x rays and ions seemed correlated with the emission of small numbers ($\gtrsim 10^3$) neutrons from laser-produced plasmas. While there had been a small number of experiments performed with 10 psec pulses, no neutrons had been observed with the very short laser pulses, although apparently the x-rays, etc. associated with neutrons from longer pulses were observed. Hydrodynamics calculations have failed to give a convincing overall agreement with experimental neutron yields.

We performed some short-pulse experiments looking for neutrons at Los Alamos.[24] The experimental configuration is shown schematically in Fig. 8. We focused about 15 J onto a deuterated polyethylene film. We observed fast ions and x-ray production, but no neutrons. (Incidentally, we shielded our neutron detector with 3 cm of lead to remove hard x-ray signals from the neutron detector, while using a CH_2 target, before inserting CD_2.) After a few shots we began to measure a small number of neutrons. We carefully made a time-of-flight measurement by allowing a portion of the x-ray pulse to enter the neutron detector and then measuring the neutron arrival time relative to the x-ray signal. The time-of-flight measurements indicated that the neutrons were coming from the wall of the chamber--presumably from CD_2 deposited there. We then inserted CD_2 foils in front of and behind the CD_2 laser target. We found in

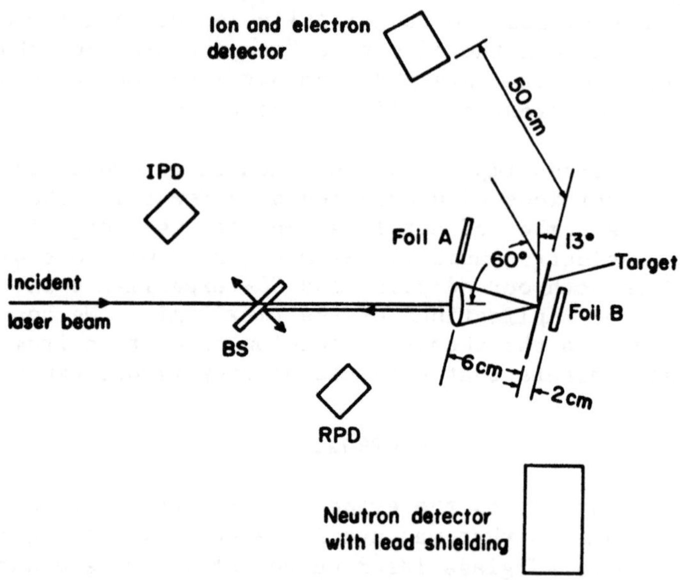

Fig. 8. Experimental layout for investigation of neutron generation with our 10 J, 25 psec Nd:glass laser. With this setup neutrons are generated at the secondary CD_2 target on the laser side of the primary CD_2 target.

this case an enhanced neutron emission as one would expect if deuterium ions formed at the CD_2 laser target interacted with deuterons in the foil in front of (i.e., on the laser side of) the laser target to produce neutrons. The presence of fast ions of an appropriate energy to explain the neutron time-of-flight measurements was confirmed by coincident measurements with an ion-current detector. Very few neutrons were emitted from the laser target itself or from the foil which was behind the target. The main features of this experiment have been reproduced by the Naval Research Laboratory.[25] We are convinced that our neutrons were not thermonuclear in origin, but were associated with fast-ion production.

The short-pulse experiments discussed above are not convincing evidence that the neutrons measured by other experimenters with longer pulse (few nsec) lasers were emitted by the same or a similar mechanism, but we became suspicious--especially since hydrodynamic code calculations of neutron emission are in poor or at least strained agreement with experiment. McCall pursued the idea of hot ions being associated with neutron emission and found a qualitatively, satisfactory explanation of several neutron production experiments for which he has been able to obtain sufficient information for comparison with his theory.[26] Several other theories of nonthermal neutron production have been proposed.[27]

We digress to remind you of the Gamow tunneling factor for D-D reactions. The D-D cross section is[28]

$$\sigma_{DD} = \frac{288}{E_i} \exp(-45.8/E_i^{1/2}), \tag{11}$$

where σ_{DD} is in barns and the deuteron energy E_i is in keV. The ion energy in the exponential is by far the dominant factor in determining the cross section. For example, using Eq. (11), the ratio of cross sections for 30 keV deuterons to that for 3 keV deuterons is greater than 10^6. It is obvious that a small number of high-energy deuterons can make a very significant contribution to neutron production.

Consider the production of neutrons by plasmas produced with nsec pulse length lasers keeping in mind the possible importance of high-energy ions. If a laser pulse is incident on a cold plasma with a steep density profile, a plasma is created of blown off material in front of the surface. Instabilities may then create hot electrons and consequently hot ions near the critical density. These hot ions stream out through the cold blowoff material. With this situation we have in effect a cold secondary ion-target material, which in the case of our shorter pulse experiment was replaced by a foil of CD_2 in front of the laser target. That the neutrons from a nsec pulse laser experiment do indeed come from the target region, as this model suggests, has been verified by Yamanaka.[29]

We expect a neutron generation rate

$$R_n = n_i v_i N_b \sigma(E_i), \tag{12}$$

where $n_i v_i$ is the hot-ion flux and N_b is the number of blowoff ions that the hot ions stream through. The neutron generation rate is independent of the distribution of plasma in front of the critical density. It depends only on the integral N_b of that distribution, which can be estimated from scaling laws. The argument of the exponential in σ_{DD} dominates Eq. (12), so we will discuss only that term here (although in the more complete formulation of these ideas[26] all the factors are considered to give not only the energy dependence of neutron production but also an absolute numerical prediction).

We return to the flux-limit condition[15] and let

$$P_L = \alpha \tfrac{1}{2} m n_h v_h^3; \tag{13}$$

a slightly different formulation of the flux limit than that we used earlier. The α has been introduced as a phenomenological parameter and v_h is an effective hot-electron velocity. McCall has introduced another factor β, where $n_h = \beta n_c$; β is essentially the fraction of hot electrons effective in producing hot ions. In this formulation

$$v_h = (2 P_L / m n_c \alpha\beta)^{1/3}. \tag{14}$$

One further finds that the energy of the hot ions produced by the electrostatic sheath due to hot electrons is proportional to v_h^2. This relation between hot electrons and hot ions gives us the dominant factor in the energy dependence of the neutron production through the exponential of the Gamow tunneling factor. The important parameter $\alpha\beta$ can be determined experimentally from the ion energy E_i measured in an experiment for calculation of neutron production in that experiment. McCall finds the neutron production/ cm^2 is given by

$$N = A\, \alpha^{1/3} \beta^{7/3} \lambda^{-14/3} P_L^{-1/3} \exp\left\{ -B(\alpha\beta)^{1/3} \lambda^{-2/3} P_L^{-1/3} \right\}, \tag{15}$$

where λ is the laser wavelength and A and B are constants.

We summarize McCall's numerical results with graphs. In Fig. 9 we have plotted measured ion energy in keV vs incident laser intensity in W/cm^2 for Nd:glass lasers. The points are ion-energy measurements which were presented by Yamanaka at the 1971 RPI Workshop.[22] The $\alpha\beta$ determined phenomenologically in this way is

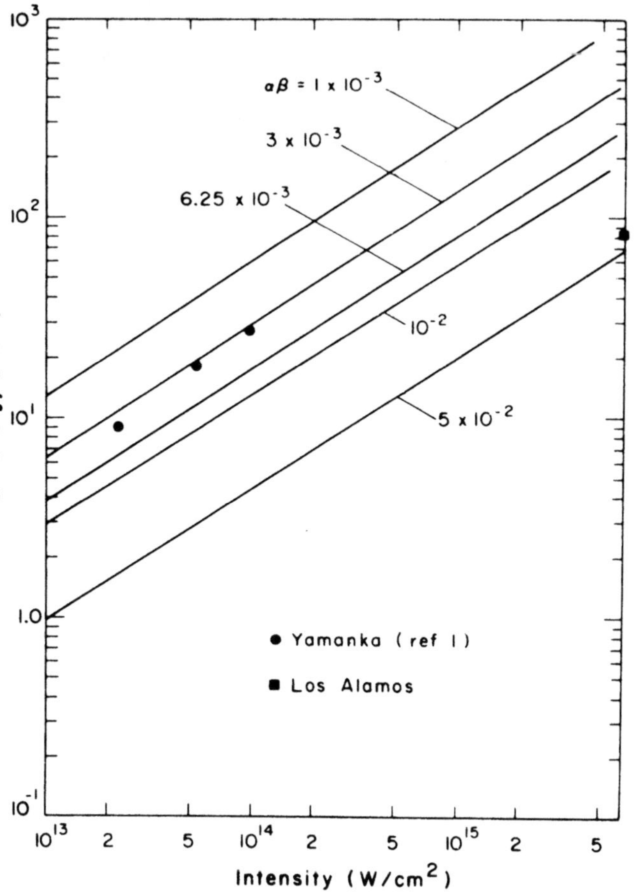

Fig. 9. Fast-ion energy as a function of laser energy for various values of the parameter $\alpha\beta$ which relates fast ions to hot electrons. Data from the work of Yamanaka and McCall are indicated.

about 3×10^{-3} in those experiments. For our 25 psec experiments we find the larger $\alpha\beta \simeq 5 \times 10^{-2}$. (We note parenthetically that these $\alpha\beta$ values appear reasonable on theoretical grounds, but we are using experiments to determine $\alpha\beta$.) Figure 10 is measured neutron number vs laser energy for the experiments of Yamanaka (triangles) and Floux (circles) discussed at the last Workshop.[21,22]

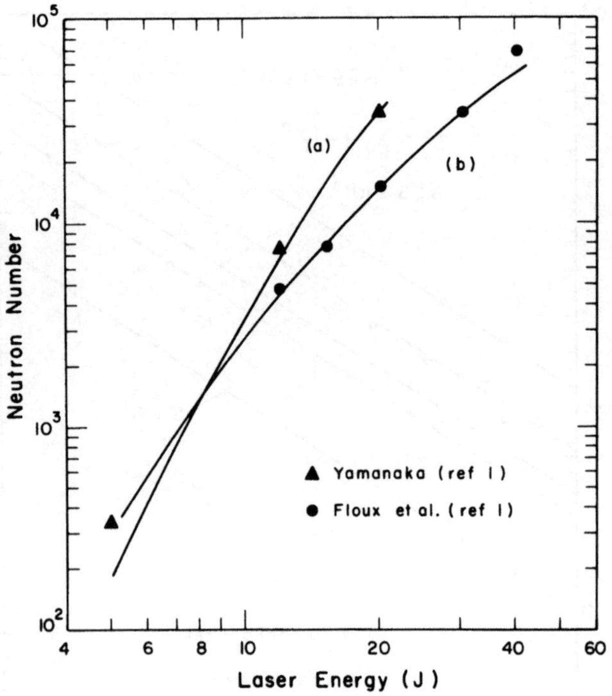

Fig. 10. Neutron yield as a function of laser energy for experiments of Yamanaka and Floux and a fit to these data using McCall's fast-ion neutron generation mechanism.

The solid lines are predictions of the model under discussion. For Yamanaka's experiment a slight fitting, by varying individually α and β, has been used to fit the absolute neutron number; while the scaling with energy is dependent only upon the $\alpha\beta$ fixed by Yamanaka's ion measurements. Since no ion measurements were reported by Floux, McCall has assumed an $\alpha\beta$ to obtain the displayed fit to Floux's experiment. This model in which hot electrons lead to accelerated ions and thus neutrons may be meaningful in the interpretation of many low-energy laser-neutron experiments.

An interesting feature of this neutron generation model is that the generation rate is given explicitly as a function of laser wavelength. Figure 11 is predicted neutron generation as a function of wavelength for various laser intensities. For Nd:glass and CO_2 irradiation at 10^{14} W/cm^2 (roughly 10 J in 1 nsec) one expects about an order of magnitude fewer neutrons from CO_2 than glass. We believe that the hydrodynamics codes, which have been fitted to 1.06 μm laser-produced neutron experiments predict much lower neutron yields than this for 10.6 μm CO_2 laser experiments. We hope soon to make measurements at Los Alamos which will check this aspect of the theory.

Even if laser-produced neutrons reported to date have not been created in the manner described above, the model points out that neutrons can be a delusive diagnostic in laser-plasma experiments aimed at fusion. We believe that experiments which give conclusive evidence of the attainment of high compressions in laser targets are particularly important.

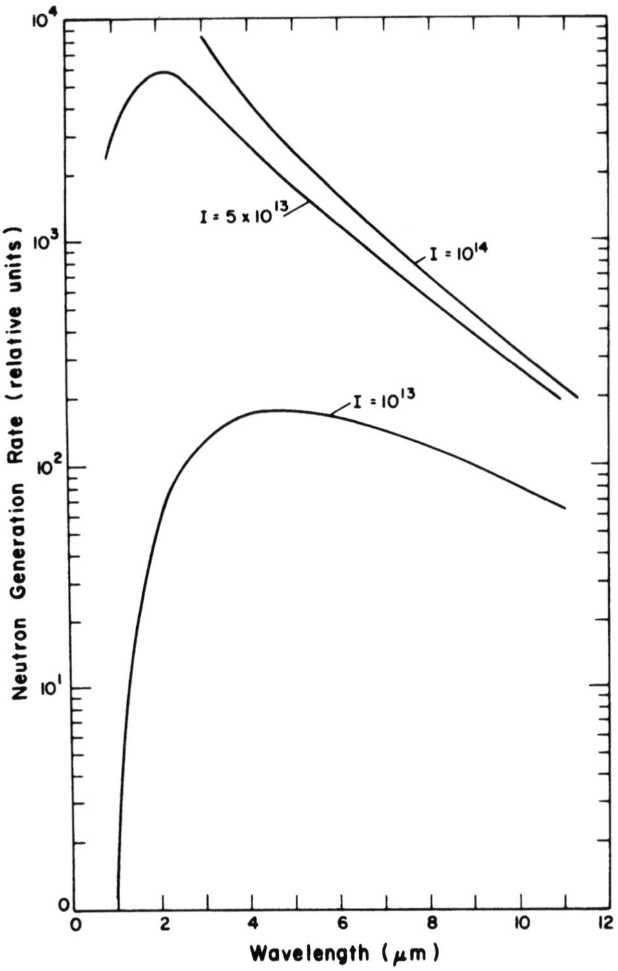

Fig. 11. Predicted neutron generation rate as a function of laser wavelength for $\alpha\beta = 3 \times 10^{-3}$.

REFERENCES

1. This Workshop published as: Laser Interaction and Related Plasma Phenomena Vol. III (H. Schwarz and H. Hora, eds.) Plenum Press, New York. See contributions by K. A. Brueckner, R. E. Kidder, R. L. Morse, and J. Nuckolls.

2. K. Boyer, Astronaut. Aeronaut. 11, 28 (1973).

3. J. Nuckolls, J. Emmett, and L. Wood, Phys. Today (August 1973) p. 46.

4. G. H. McCall, R. P. Godwin, and J. F. Kephart, Bull. Am. Phys. Soc. 17, 1044 (1972) and unpublished work.

5. Our work agrees in most aspects with that of other researchers. See, for example, Laser Interaction and Related Plasma Phenomena Vol. III, contributions by J. L. Bobin and R. Sigel.

6. See, for example: K. Büchl, et al., Laser Interaction and Related Plasma Phenomena Vol. II, p. 503 and J. W. Shearer et al., Phys. Rev. A6, 764 (1972).

7. K. Eidmann and R. Sigel, VI European Conf. on Controlled Fusion and Plasma Physics, Moscow, July 30-August 3, 1973, and R. Sigel, this Workshop.

8. D. Giovanielli, Los Alamos Scientific Laboratory (Internal report). For discussions of magnetic fields associated with laser plasmas see Laser Interaction and Related Plasma Phenomena Vol. III contributions by F. Schwirtzke and J. A. Stamper.

9. R. P. Godwin, J. F. Kephart, and G. H. McCall, Bull. Am. Phys. Soc. 17, 971 (1972).

10. A. H. Compton and S. K. Allison, X Rays in Theory and Experiment, 2nd ed. (Van Nostrand, New York, 1935), p. 93.

11. J. F. Kephart, R. P. Godwin, and G. H. McCall, Bull. Am. Phys. Soc. 17, 971 (1972).

12. W. Kruer in Laser Interaction and Related Plasma Phenomena Vol. III.

13. A. W. Ehler, D. Giovanielli, and F. Young, Anomalous Absorption Conf., Los Alamos, March 1-2, 1973.

14. R. S. Cooper in Laser Interaction and Related Plasma Phenomena Vol. III.

15. R. L. Morse and C. W. Nielson, Phys. Fluids, 16, 909 (1973) and R. L. Morse, this Workshop.

16. A. H. Compton and S. K. Allison, X Rays in Theory and Experiment, 2nd ed. (Van Nostrand, New York, 1935) p. 104.

17. R. D. Evans, The Atomic Nucleus (McGraw-Hill, New York, 1955) p. 614.

18. W. P. Gula, Los Alamos Scientific Laboratory (private communication).

19. See, for example: C. E. Violet, et al., IEEE/OSA Conf. on Laser Engineering and Applications, Washington, D. C., May 30-June 1, 1973.

20. N. G. Basov, et al., Laser Interaction and Related Plasma Phenomena Vol. II, p. 389.

21. F. Floux, et al., Laser Interaction and Related Plasma Phenomena Vol. II, p. 409.

22. C. Yamanaka, Laser Interaction and Related Plasma Phenomena Vol. II, p. 481.

23. K. Büchl, et al., Laser Interaction and Related Plasma Phenomena Vol. II, p. 503.

24. G. H. McCall, et al., Phys. Rev. Letters 30, 1116 (1973).

25. J. A. Stamper in Laser Interaction and Related Plasma Phenomena Vol. III.

26. G. H. McCall, Los Alamos Scientific Laboratory, Report LA-UR-73-791 (to be published).

27. A. Caruso, Frascati (preprint) and J. Katz, Lawrence Livermore Laboratory (preprint).

28. S. Glasstone and R. H. Lovberg, Controlled Thermonuclear Reactions, (Van Nostrand, Princeton, 1960), p. 20.

29. C. Yamanaka in Laser Interaction and Related Plasma Phenomena Vol. III.

LASER-MATTER INTERACTION STUDIES AT NRL*

J.A. Stamper, O.C. Barr, J. Davis, G.A. Doschek,
C.M. Dozier, U. Feldman, B.M. Klein, W.M. Manheimer,
E.A. McLean, J.M. McMahon, D.J. Nagel, D.A. Tidman,**
R.R. Whitlock, K. Whitney,** N.K. Winsor, and F.C. Young

Naval Research Laboratory, Washington, D.C.

ABSTRACT

Experimental and theoretical studies at NRL are directed toward understanding the interaction of a focused, high-brightness laser pulse with a solid target. The physics of laser-fusion and efficient x-ray generation are of particular interest. A Nd-doped, YAG-glass laser system is used which has rod and disc amplifiers and protection against back-reflected pulses. Experimental studies are discussed of x-ray emission, neutron production, magnetic field generation and back-reflected laser light. Theoretical studies include both analytic and numerical treatments of radiation emission and transport, magnetic field dependent plasma transport phenomena and the role of magnetic fields in plasma collective effects.

---

* Presented at the Third Workshop on "Laser Interactions and Related Plasma Phenomena", at Rensselaer Polytechnic Institute, Troy, N.Y. August 13 - August 17, 1973.

** Science Applications, Inc., Arlington, Virginia

EXPERIMENTAL FACILITY

Laser System

The design of a neodymium laser system for irradiation of targets with 10^{11} - 10^{12} watt pulses in a high brightness beam requires adequate solution of several difficult technical problems. These are primarily related to self focusing and target and inter-stage isolation in the system.

Self focusing is inextricably related to lasers of this type since the desired output levels for the laser fusion problem are 6 - 8 orders of magnitude higher than the critical power for self focusing (~ 1 MW). Since the mechanisms causing self-focusing for subnanosecond pulses appear to have a picosecond response time[1] any temporal structure on the pulse will lead to self focusing at abnormally low average intensity levels[2].

Self focusing of the entire beam can generally be avoided by either expanding the diameter at higher intensities or diverging the beam[3] but self focusing due to spatial intensity fluctuations and radiance degradation due to wavefront distortion can still occur.

Fresnel diffraction from apertures in the system has been shown to lead to downstream self focusing damage[4]. A TEM_{oo} mode oscillator does not solve this problem since the wings of the Gaussian will interact with apertures and ultimately lead to damage. It is therefore necessary to modify the TEM_{oo} oscillator mode to suppress Fresnel diffraction.

We have used a two step truncation operation in which the oscillator pulse is apertured at the e^{-4} intensity point and then apertured again in the far field at the first dark ring of the "Airey" disc. This gives a central lobe with a Gaussian like character in the center but essentially zero intensity beyond the aperture. In principle the near field amplitude ripple is $\simeq$ 1% due to this procedure; in practice ~ 3% was more generally attained.

An isolation system is also required to prevent catastrophic damage due to amplified backscatter from the target. Based on experiments at NRL and elsewhere there appear to be two components to the backscatter from targets; reflection from the critical density region in the plasma at $I < 10^{14}$ w/cm^2 and stimulated scattering processes at $I > 10^{14}$ w/cm^2. Total reflections of 10% - 40% have been reported. These lead to a requirement that the net isolation be on the order of the product of the small signal gain times the area ratio of output to input beam size times a safety factor since the back reflection may be presumed to have a poorer beam quality than the outgoing beam. Secondly, the isolation must be staged so that at all points in the amplifier train the back reflection is suppressed sufficiently.

At small apertures (25 mm or less) Pockel's cells with either dielectric or prism polarizers were used. For larger aperture sizes 45° Faraday rotators were constructed to use Owens-Illinois EY-1 terbium glass. The rotator aperture was 8 cm and the glass thickness was 2 cm to minimize requirements on magnetic field uniformity and also to suppress self focusing. Both rotators had extinction ratios greater than 30 db. They were driven by a three section PFN which produced a flat current pulse ~ 100 μsec wide to allow fail-safe circuitry to check their performance before an optical pulse was generated.

Figure 1 shows a schematic of the laser system. The Nd:YAℓG oscillator is followed by a Pockel's cell gate and two Nd:YAℓG preamps. Three CGE glass rod amplifiers are used to increase the energy to ≃ 20 J and then the disc amplifier boosts this to 60 to 125 J, depending on pump energy. With *f*/12 focusing optics energies of 50 - 100 J have been delivered on targets routinely. With *f*/2 optics energies of up to 50 J have been put on CD_2 targets.

System reliability has been reasonable. The 45mm rod amplifier accumulates self focusing and surface damage slowly and needs to be replaced or refinished after 500-600 shots. The discs in the disc amplifier slowly accumulate surface damage due to the flashlamps vaporizing dirt or dust on the surfaces but also have a relatively long lifetime before they need to be refinished. No self focusing damage has been noted in the disc amplifier. The radiance is $\sim 10^{18}$ w/cm^2-ster and is limited mainly by geometrical aberrations in the disc amplifier and stacked plate polarizer due to cumulative polishing errors.

In summary we have found it possible to design a glass system such that high radiance pulses can be routinely produced by improving beam uniformity to the point where self-focusing does not limit operation. It has also been shown that large aperture, high extinction ratio, low loss isolation systems can be designed and used to isolate backscatter from targets.

Target Facility

The target facility is located in a separate building from the laser. This separation (~ 100 feet) serves to minimize unwanted interactions between the laser and target areas while not appreciably

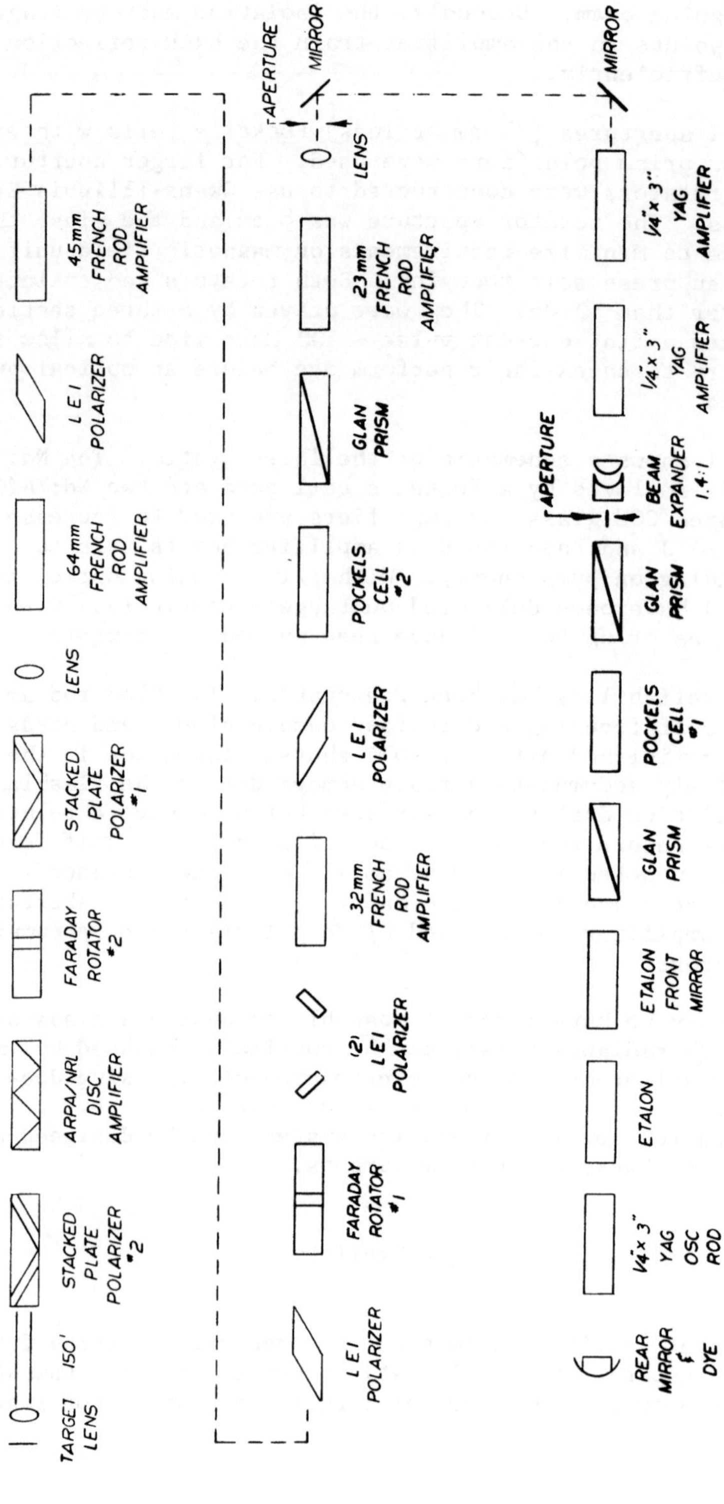

Figure 1. NRL Laser System Block Diagram

increasing the diameter of the beam. The beam propagates between the buildings in an evacuable tube.

The target area is pictured schematically in Fig. 2. Part of the beam is split-off and used to monitor the energy in the incident pulse. Temporal information is monitored by a fast photodiode which looks at light reflected from the lens and lens tube. Another monitor, which is a useful check of focusing, is a hard (> 20 keV) x-ray detector. Focusing must be correct for a beam having the same collimation as that in the experiment. Thus, when focusing is done with pulses of little amplification, care must be taken to correct for any dynamic focusing of the fully amplified pulse.

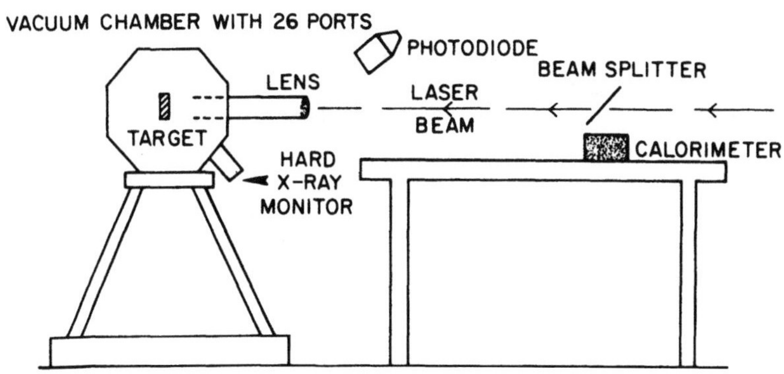

Figure 2. Schematic of NRL Target Facility

The target vacuum chamber is made of stainless steel and is basically a rhombicuboctahedron-having 18 square and 8 triangular faces[5]. All of the 26 ports are usable. An outstanding feature of the chamber is the high degree of accessibility for diagnostics. There are many pairs of mutually perpendicular ports. The chamber can be evacuated to a base pressure of a few tenths of a microtorr with a turbo-molecular pump. The target is positioned, in three dimensions, to a precision of 50 microinches by three motor-driven translators.

EXPERIMENTAL STUDIES

Neutron Studies

Part of the laser-matter interaction program at NRL includes studies of the potential for thermonuclear fusion through laser produced plasmas. Measurements of neutron production from deuterated targets have been initiated as a part of this program. Diagnostics have been developed to determine the magnitude of neutron production even at very low yields. Neutrons may be generated from laser produced plasmas by thermonuclear reactions within the plasma core, or by the acceleration of deuterons in the plasma expansion. Preliminary measurements with slab CD_2 targets and 1-nsec, 1.06-μ laser pulses have been devoted to attempting to identify the source of neutron production.

Two sensitive detection systems have been used to measure neutron yields: An array of BF_3 counters and a ^{6}Li-glass scintillator. Properties of these detectors are given in table 1. Neutron moderation is used to increase the sensitivity and to delay the interaction in the BF_3 counter or ^{6}Li-glass scintillator. The moderator is considered a part of the detector in the quoted sensitive areas and source-to-detector distances. Both detectors have been calibrated in place on the target chamber with an Am(Li) neutron source of known strength. Also measurements with a pulsed neutron source provided by an electron linac indicate that neutrons are detected by both counters for up to 300 μsec after exposure to a burst of neutrons. For this time interval there is essentially no background in either detector of neutron-like pulse height signals. Laser induced neutrons have been detected with both systems.

Measurements of laser produced neutrons have been made with both short (10.7 cm) and long (74 cm) focal length lenses. For laser pulses focused with a f/1.9 lens, neutrons have been detected with as little as 18 Joules on target. An oscilloscope display of the responses of the ^{6}Li-glass scintillator and the array of BF_3 counters to laser-produced neutrons is shown in figure 3. Neutron interactions in the detectors are delayed and dispersed in time by moderation so that the detector response to individual neutrons can be identified and counted. Also, in this way neutron interactions are separated from prompt x-rays. The initial pulse on each trace in figure 3 corresponds to x-rays. Three neutron events are evident from the ^{6}Li-glass detector (upper trace) while one event is observed from the BF_3 counter (lower trace). These events correspond to a total yield of about 400 neutrons based on an isotropic angular distribution.

Table 1

Type of Detector	Sensitive Area	Source to Detector Distance	Sensitivity (neutrons into 4π/count)
Array of BF_3 Counters	9 in. x 14 in.	14.5 in.	400
^6Li Glass Scintillator	7 in. dia.	2.0 in.	70

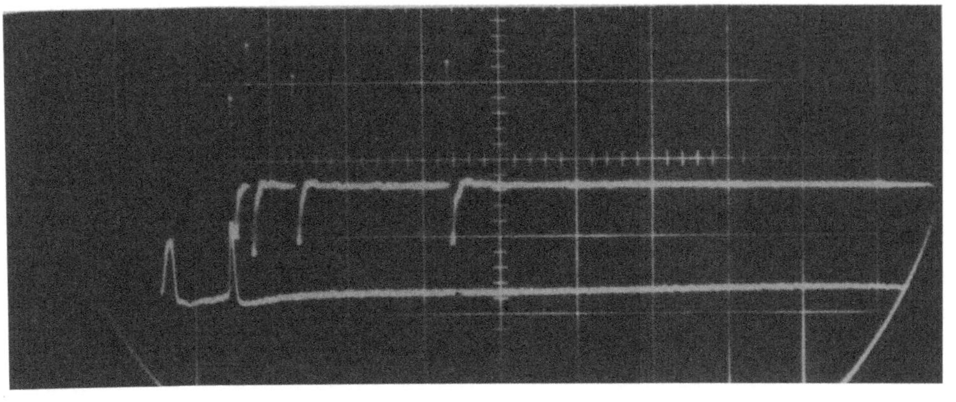

Figure 3. Oscilloscope Traces of Neutron Events

No neutrons were observed from measurements with laser pulses focused by an f/14 spherical lens onto CD_2 targets for incident energies of up to 70 Joules. Neutrons were observed under these conditions by adding a secondary CD_2 target as shown in figure 4. It is concluded that neutrons result from the acceleration of energetic deuterons from the plasma onto the secondary target. For 50 to 70 Joules on target, pulses corresponding to a total yield of about 500 neutrons were observed. This result assumes that neutrons are emitted isotropically from the secondary CD_2 target.

We have demonstrated that neutrons can be produced by accelerated ions for irradiations with a long focal length lens. The source of neutrons observed with a short focal length lens is unresolved at this time, but also may be due to an ion acceleration process. Our results are consistent with McCall's conclusion[6] "that no neutrons of thermonuclear origin are necessary to explain the results which have been obtained thus far."

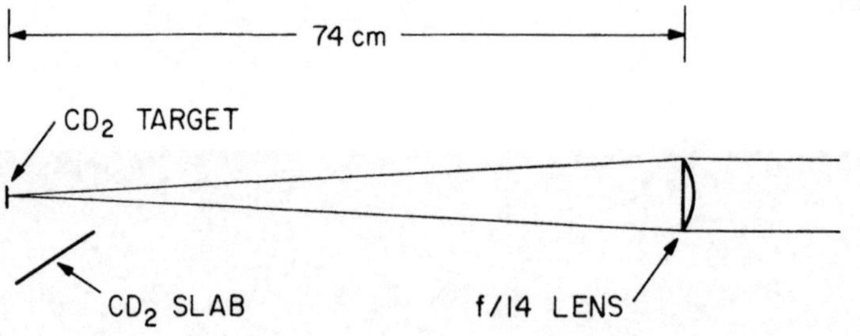

Figure 4. Experiment Showing Neutrons Due to Accelerated Ions

X-Ray and UV Measurements of Laser-Produced Plasmas

Measurement of photons emitted by plasmas is one of the fundamental diagnostic techniques for plasmas of any temperature[7]. The plasmas produced by interaction of a high-energy laser with a solid target are exceptionally hot, dense, and small. Their high temperatures (1 - 20 million $^{\circ}$K or 0.1 - 2 keV) mean that the emitted radiation largely falls in the x-ray and XUV regions of the spectrum. The dense and small characteristics of laser-produced plasmas make them a unique spectroscopic source which is relatively convenient to use.

The goal of the photon measurements is twofold: (a) to provide diagnostic data from which plasma conditions can be inferred and which can be used to test the numerical simulation work discussed below, and (b) to determine empirically the overall characteristics of the laser-plasma x-ray source as a function of laser pulse, focussing and target parameters. Laser pulse parameters include the wavelength; the pulse length, shape and integral; and prepulse characteristics (length, shape, integral and separation from main pulse). Focussing depends on lens characteristics and lens-target placement. The target parameters include composition, geometry and orientation relative to the laser beam. We would like to be able to determine an experimental intensity distribution I $(E, \bar{x}, \bar{r}, t)$ as a function of photon energy E, position within the plasma $\bar{x}$, the emission direction $\bar{r}$, and time t for various pulse, focussing and target conditions. But this cannot be obtained and we must content ourselves with certain integrals of this intensity function over its several variables:

A. Source Size. Here a pinhole photograph of the plasma, taken in some direction $\bar{r}$, records all photon energies (above the cutoff E_c of the window covering the film) which are emitted along the direction parallel to $\bar{r}$ during the entire plasma lifetime. The source size as shown by pinhole photographs has a major influence on how the other x-ray measurements are made. When it is small, high-resolution spectra can be recorded with slitless spectrographs.

B. Spectra. Crystal or grating spectrographs are used with film recording to obtain time-integrated spectra for a particular emission direction $\bar{r}$. By use of a slit perpendicular to the dispersion plane, spatial information in a direction perpendicular to $\bar{r}$ can be obtained along with the spectrum. Details of x-ray spectra such as line intensity ratios are especially useful for determination of temperature, density and equilibrium conditions of the emitting plasma.

C. Spatial Distribution. Calorimeter data and the measured spatial distribution are used to compute the total x-ray emission above E_c by integrating over the available range of $\bar{r}$. Efficiencies for the conversion of laser light into useful x-rays can then be calculated by dividing the incident laser pulse energy into the empirical total x-ray intensity.

D. Emission Time. An ultra-fast, total-absorption detector behind a light-tight window in some direction $\bar{r}$ would give the time history of emission above E_c. Available detectors only approximate this performance for laser pulses $\leqslant$ 1 nsec. A streak camera in conjunction with an x-ray monochromator might yield the temporal emission profile for a particular x-ray line. Temporal information provides a test of computed emission rates.

The following discussion is organized according to the various measurables just enumerated. Emphasis in this section will be laid on experimental results and what they mean rather than on apparatus. A less extensive review of this work has already been given[8].

A. Source Size. Figure 5 shows x-ray pinhole pictures taken at right angles to the laser beam and parallel to the target surface which was also normal to the incident beam[9]. The x-ray source obtained with the target out of focus reflects hot spots introduced into the laser beam by propogation through about 150 feet of air. An emission region of less than 100 μm in size is obtained with x-ray energies above about 0.8 keV for the best focussing conditions.

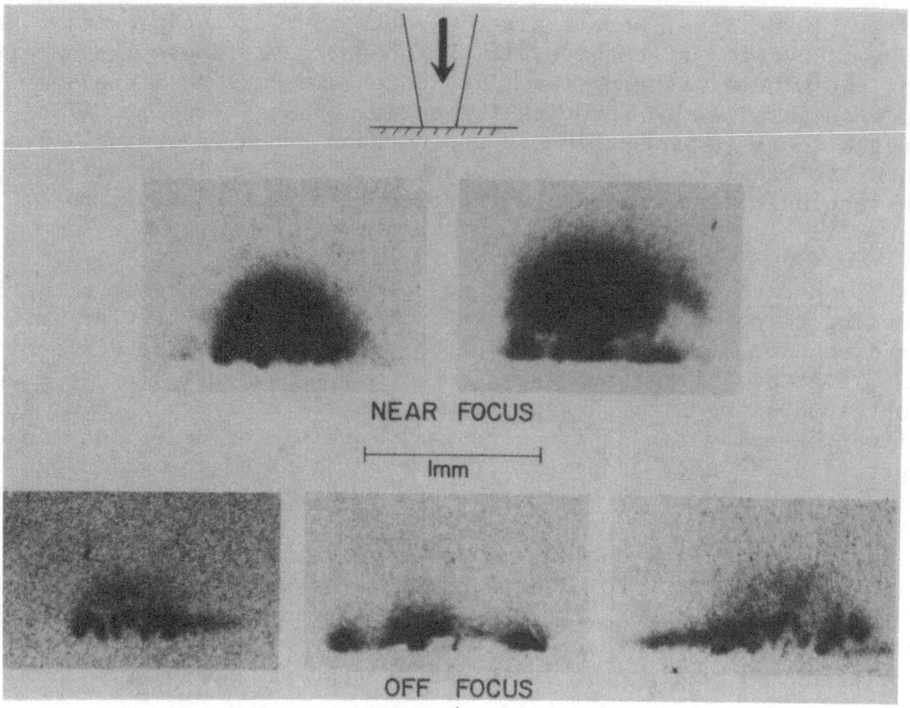

Figure 5. X-Ray Pinhole Photographs Taken on Kodax No Screen Film Behind 15 μm Be Windows.

B. Spectra. X-ray and UV spectra have been obtained over a wide range of laser and target variables. The initial work was done with 1.06 μm gaussian laser pulses of 0.9 ns duration containing a few joules. X-ray spectra in the energy range 0.8-5 keV were measured with simple flat-crystal spectrographs for many elements across the periodic table using slab targets. Less extensive x-ray spectroscopy has been done with frequency-doubled 5320 Å pulses, with pulses of shorter duration (0.25 ns), and with pulses of higher energy (up to 100 J). Also, compound (or alloy) targets have been used, and some of the shots have incorporated a 2 ns prepulse 7% as big as the main pulse. XUV spectra have been obtained in the 9-360 Å wavelength range using a grazing-incidence grating spectrograph[10] for several pulse conditions and many target materials.

Figure 6 shows spectra of highly ionized Aℓ, Zn, and Gd, in the range 1-2.5 keV with a few joules incident on the pure metal targets[11]. The Aℓ spectrum originates from lithuimlike, heliumlike, and hydrogenlike ions, i.e., ions which have lost all but 3, 2, or 1 electrons. That is, the ions are stripped to the K shell and have charges of 10-12. The Zn spectrum indicates stripping down to and into the L shell, i.e., n = 2 level, with ionization stages of 20 and greater. The Gd spectrum similarly shows ionization to and into the M shell, i.e., 36 or more electrons absent[12].

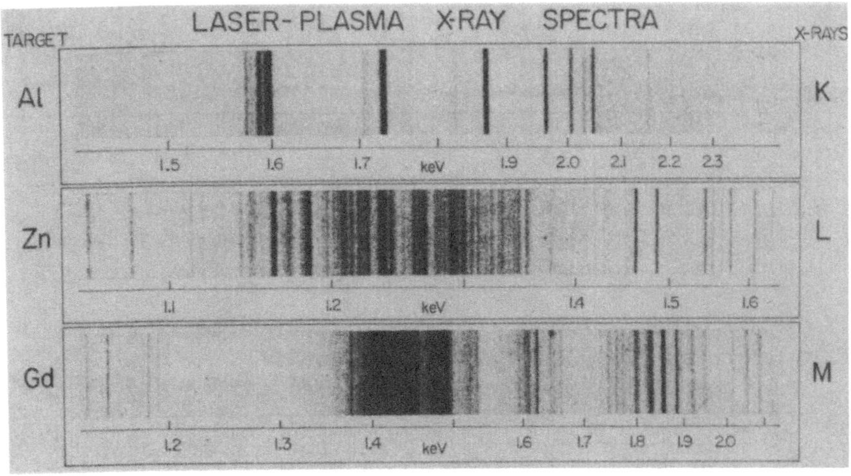

Figure 6. Aℓ k, Zn L and Gd M Spectra Produced by 1.06 μm, Few-Joule Pulses.

Similar spectra of other elements have also been measured, with most emphasis being placed on spectra of hydrogenlike and and heliumlike ions, since they are relatively easier to interpret. Although lithiumlike and heliumlike spectra up to S (Z = 16) were obtained in a single shot without use of a prepulse, as shown in figure 7[13], the addition of a prepulse was found to signficantly increase the plasma temperature. With the 7%-2 ns prepulse, similar spectra of Ti (Z = 22) were excited with sufficient intensity for measurement in three shots. Addition of the prepulse was more effective in producing higher plasma temperatures than an increase in either the total energy in a single pulse, or the energy density on target.

Spectra of 1, 2 and 3 electron aluminum have been obtained as a function of pulse conditions and target composition. Plasma temperatures were computed from observed line intensity ratios assuming coronal equilibrium[14]. The results available to date are sketchy, but they indicate the following[15]:

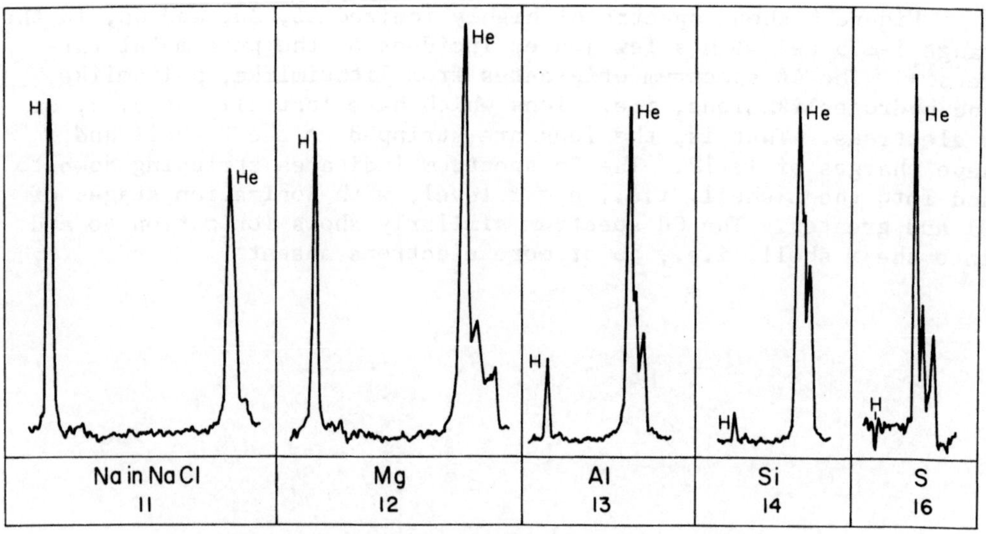

Figure 7. Densitometer traces of 2p → 1s transitions in H-like and He-like ion species produced by 1.06 μm, 2J, 0.9 ns pulses. Vertical scales are arbitrary.

a. 5320 Å light produces somewhat lower plasma temperatures than 1.06 μm light for 0.9 ns, few-joule pulses.

b. At 1.06 μm, 0.25 ns pulses yield significantly higher temperatures than 0.9 ns pulses of a few joules.

c. Temperatures range from 0.6 keV for 2 J incident to about 1 keV for 60 J incident at 1.06 μ and 0.9 ns.

d. Similar pulses on Aℓ and a 50-50 AℓNi alloy yield similar temperatures.

These results do not depend strongly on the assumption of coronal (ionization) equilibrium, which is only approximated for laser pulses ≃ 1 ns.

Spectra have been taken in the XUV range with the grating instrument for atomic numbers ranging from lithium through uranium and many laser pulse conditions. Figure 8 shows transitions in lithiumlike, heliumlike, and hydrogenlike fluorine[16]. Data such as these are rich in information about plasma conditions: (a) the intensity ratio for 2p → 1s lines in 1 and 2-electron fluorine indicates a temperature of about 4 million °K assuming coronal equilibrium, (b) line-merging near the series limits at principal quantum numbers of 9 and 10 for 1 and 2-electron species, respectively, implies an electron density near 5×10^{19} cm^{-3} (i.e., Stark broadening in the quasi-static limit determines the line width), and (c) the relative intensities of lines of lithiumlike fluorine indicate that the plasma was ionizing at the time of emission, but that departure from coronal equilibrium was not drastic.

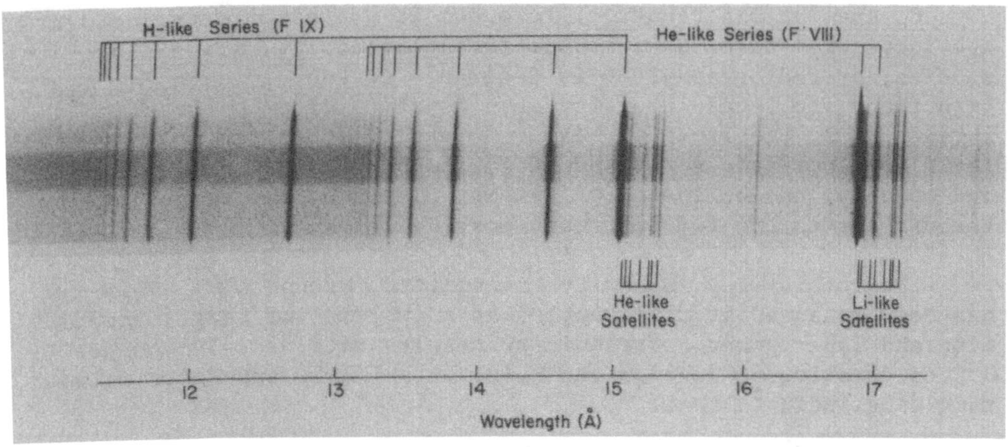

Figure 8. Fluorine Spectra from Laser-Produced Plasmas. (Entire spectrum - 4 shots; central band - 2 shots)

The laser plasma is an attractive spectroscopic source in the x-ray and XUV ranges because its emission arises primarily from a few high ionization stages which can be varied easily by changing pulse or focussing conditions. Figure 9 shows Fe spectra from vacuum spark[17] and laser-produced plasmas[18] around 100A. The spark, which is a useful source of very hot plasmas, in addition to producing lines of highly ionized stages of iron, also yields a thicket of lines from relatively low ionization stages in this region. The laser-plasma contains mainly lines from the highest ionization stages. These can be varied easily by altering the target position relative to focus of the lens. This combination of characteristics (clean and controllable spectra) has lead to identifications of lines of high ionization stages which previously were difficult to classify due to the large number of nearby lines produced by lower ionization ions.[19]

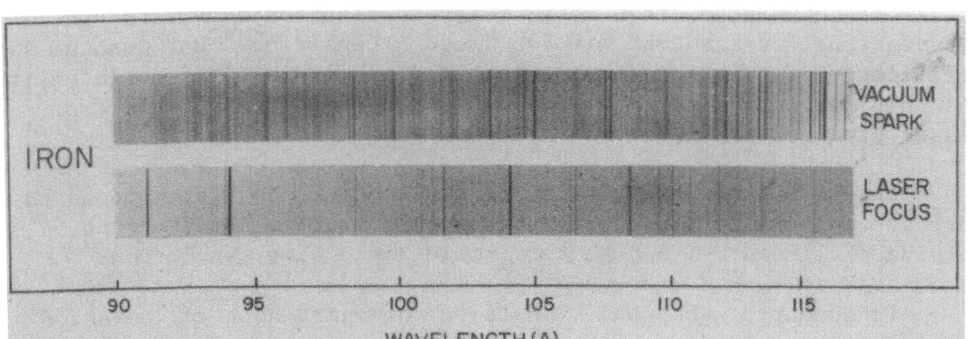

Figure 9. Iron Spectra Taken with Laser and Vacuum Spark Excitations.

C. Angular Distribution. In order to use plasmas produced at the focus of a laser as a source for spectroscopy, materials studies, or radiography, it is desirable to know the angular distribution, i.e., the intensity as a function of emission direction $\bar{r}$ relative to the incident beam. Several factors can conspire to yield non-isotropic x-ray and UV emission. Plasma non-uniformity and opacity, and eclipsing of the hot plasma by the target at low takeoff angles are two important possibilities.

The total x-ray intensity transmitted through a 25 μm Be foil has been measured at fixed angles as a function of target composition and laser power. Preliminary results obtained with pulses of 0.9 ns duration at 1.06 μm and using both film[20] and calorimetric[21] recording indicate that:

a. At 45° to the incident beam and a distance of 30 cm from the target, the emitted x-ray line energy (ergs/cm^2) varies from 7 for Mg to 0.2 for S.[15]
b. At 25° to the laser beam, the line and continuum emission (ergs/cm^2) at 30 cm from the Aℓ target varied from about 18 at 2 J to 3700 at 60 J.[13]

D. Emission Time. The temporal distribution of x-ray emission is useful to compare with the predictions of MHD and rate codes, and it limits applications of the laser-plasma source to flash radiography. Silicon x-ray detectors have rise times of about 2 ns, which are useful for some measurements of laser-produced plasmas, but not for time resolution of sub-nanosecond pulses. It was found, however, that lithium tantalate pyroelectric detectors normally used for infrared measurements[22] give 0.9 ns rise time and 1.6 ns FWHM, approximating the 0.9 ns FWHM laser pulse. Measurement with such a detector places a limit on the x-ray emission time of about 1 ns FWHM in this case.

It would be desirable to have time resolution of < 100 psec for the x-ray pulse. This could be obtained in principle by use of an x-ray monochromator, thin metal x-ray-to-electron converter and a conventional streak tube with an image intensifier. But quantum efficiency calculations[23] show that the available electron intensity is only marginal with x-ray intensities presently available from laser produced plasmas.

Lasers with 18 nsec[24] and 40 nsec[25] pulses have been found to produce measurable x-rays from Mg and Al targets. In these cases, ordinary silicon p-i-n detectors yield ample time resolution.

In summary, x-ray and UV data on the size, spectra, emission distributions and time histories of laser-produced plasmas is proving valuable in two ways: (a) as a plasma diagnostic tool and

(b) in determining the characteristics of this source for use in atomic spectroscopy, materials studies or radiography.

Magnetic Field Studies

Magnetic fields generated by laser-produced plasmas have been observed at several laboratories[26,27,28,29]. A discussion of the fields is given in a previous RPI lecture[30] and is expanded on in the theoretical sections of this lecture.

The studies reported so far have been based on measurements by magnetic probes. These measurements can give useful information about the fields as close as 1 mm to the target surface. Figure 10 is a $\dot{B}$ or unintegrated signal as recorded by a Tektronix 454 oscilloscope. A 2J, 250 psec laser pulse was focused at normal incidence onto a slab of polythylene. The probe was glass encapsulated and had an overall diameter just under 1 mm. The tip of the probe was located a little over 1 mm from the focus. The sweep speed on the oscilloscope was 20 nsec/cm and the gain, with attenuation was 50 V/cm.

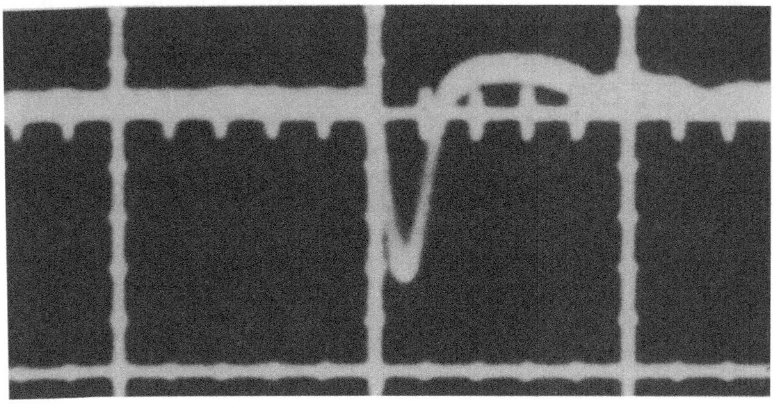

Figure 10. Oscilloscope Trace of Magnetic Probe Signal

Some useful information is obtained. The risetime is about 2 nsec and may be scope limited. Typical magnetic pulses measured a centimeter from the focus have a width around 100 nsec. The short pulse width of the magnetic fields near the focus is in agreement with the theoretical idea that the fields are generated primarily at the focus of the laser beam in the presence of laser radiation.

Using the known probe calibration, yields a value of one kilogauss for the maximum field. Several theoretical estimates, on this case, show that a much larger field is expected, probably as high as 100 kgauss. The experimental result, however, is understandable as the following discussion shows.

Calculations of the interactions of a typical magnetic probe structure with a hot, expanding plasma containing a short magnetic pulse show that reliable measurements aren't possible closer than about 3 mm from the target. The magnetic field does not have time to diffuse from the hot plasma into the probe structure during the time it takes for the magnetized plasma to flow past the probe. Thus the magnetic fields indicated by probe measurements very close to the laser focus are smaller than the actual magnetic fields in the plasma. There are also other problems which must be considered such as probe damage due to the plasma and probe disturbance of the plasma and magnetic field flow patterns.

For laser-fusion and x-ray generation, one is interested in the large magnetic fields in or very near the target. Reliable probe data can be obtained at distances greater than a few millimeters from the target and one might think that the fields near the target could be obtained by extrapolation. However, extrapolation is not likely to give useful information for short pulse high-power lasers. In general, the evolution of the magnetic pulse in space and time is very complicated. Irreversible processes (magnetic field diffusion) and incompletely understood processes such as pulse compression and coupling between the laser-plasma and ambient plasma are involved. In fact, there is experimental evidence[31] that small magnetic fields can be re-generated as the laser-plasma interacts with the ambient, well away from the focus. Thus, in general, it is not possible to infer what the large fields in the focal region are by extrapolating from the weak fields at larger distances.

In retrospect, the inference of large magnetic fields[32] by extrapolating the original data was rather risky. However, there were some features of those measurements[26] which reduced the risk. The measurements were made at a distance (~ 1 cm) which was large enough for reliable probe performance and yet small enough that both field decay and any re-generation[31] could be ignored. Laser energy (30 nsec pulse) was still incident on the target when the magnetic pulse arrived at the probe.

A reliable measurement of the magnetic fields in or near the focus should be possible by utilizing either emitted or probing electromagnetic radiation. Such measurements have not been reported to date[33]. However, some possible techniques are discussed here. Zeeman splitting or broadening of line radiation in the x-ray and VUV is one possibility. However, the strong spatial variations and competing broadening mechanisms make the interpretation very difficult. Plasma collective effects provide another possibility. The radiation emitted at the second harmonic (5320 Å) of the incident radiation would show a splitting at sufficiently strong magnetic fields.

We will try studies depending on Faraday rotation of a probing beam. A diagram of the arrangement for these studies is shown in figure 11. The main laser beam (λ = 1.06 μ) is focussed by means of lens L, onto the target. Part of the main laser beam is split off and frequency doubled to 5320 Å (green). This linearly polarized green beam is transmitted about 20 nsec ahead of the main beam into the experimental room. The green beam is delayed about 20 nsec and directed parallel to the target surface by mirrors M_1 and M_2. Positioning and localization of the probing beam is achieved by the long focal length lens L_2. The probing beam is passed to one side of the main beam axis in order to sample the azimuthal magnetic field preferentially in one direction.

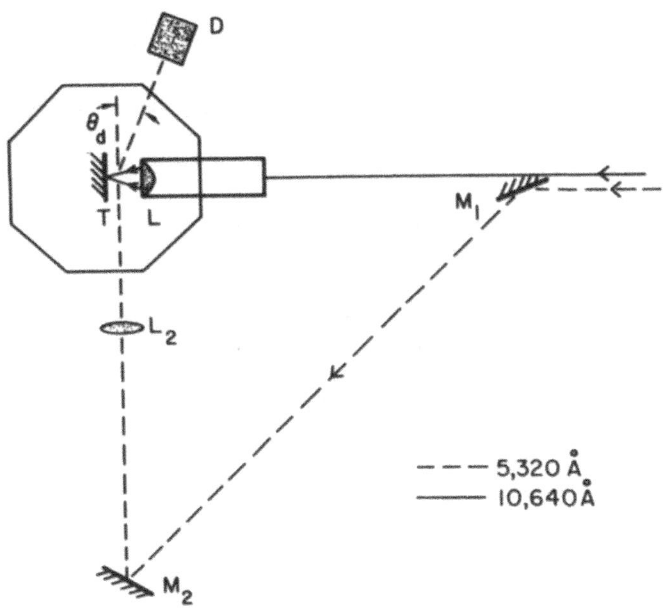

Figure 11. Arrangement for Faraday Rotation Experiment

The detector D allows a measurement of the Faraday rotation angle. This angle depends on the line integral of density times magnetic field. Thus, an accurate unfolding of the magnetic field requires a knowledge of density variations. However, the deflection angle (Θ_d) in the diagram) is itself a measure of this density variation. By measuring the rotation angle and deflection angle for several positions of the probing beam it should be possible to obtain reliable, but averaged, magnetic field information without the necessity of a detailed unfolding.

Laser Backscatter

Studies of the reflected light from laser-irradiated targets are important from two standpoints. First, a large backscatter can influence the efficiency of coupling to the target, and secondly, backscatter can have a drastic effect on the laser when the backscatter is amplified. For these reasons it is necessary to investigate the properties of the backscattered radiation and the plasma produced to determine how much of the laser radiation is coupled to the target and how much is reflected. If the reflected light indicates instability mechanism in the plasma, how does this affect the plasma temperature and density? Calculations of Forslund, Kindel and Lindman[34] and Rosenbluth and Sagdeev[35] indicate the possible presence of processes in which the electromagnetic laser wave couples to various plasma wave phenomena and possibly lead to coherent backscatter at laser intensities > 10^{14} watts/cm^2.

Experiments done at NRL[36] to date address the following properties of the backscattered laser light: polarization, spatial properties, energy dependence and spectral and angular dependence.

Polarization. The experimental arrangement to measure the polarization of the backscattered laser beam utilized two detectors, one of which detected either plane of polarization and the other detecting only the same polarization as the outgoing laser light. Alternately, an auxiliary polarizer was placed downstream between the laser and the target and the signal ratios of the detectors were compared. It was found that there was no difference between these signals, implying that the reflected light was polarized the same as the incident laser light to an accuracy of ~ 10%. The results were true out to the highest energies reported here.

Spatial Properties. To investigate the "intensity pattern" of the reflecting surface, half the laser aperture was blocked with a blackened metal plate. If the reflection was from a plane, reflecting surface, there would be image inversion and hence zero back reflection. However, if the reflection was from an appropriate spherical reflecting surface, the rays would retrace themselves giving light in one-half the aperture. The experiment showed that the backscattered light was consistent with reflection from a spherical surface. (K. Eidmann and R. Sigel,[37] at Garching, Germany have reported similar results.)

Energy Dependence. Figure 12 shows the magnitude of the back-reflected energy into the laser as a function of the energy onto both CD_2 and Cu slab targets. A short focal length lens (*f*/1.9) was used to focus the laser beam onto the target. The curvature in the semi-log plot indicates the possibility that two processes are taking place simultaneously, and further investigation showed that specular reflection, i.e., reflection from the critical density region was the dominate mechanism at lower energies and contributed 4% of the back-reflected signal. When this 4% was substracted from the values in Fig. 12 then the reflected energy values fell along a straight line on a semi-log plot with incident laser energy. This, of course, results in an exponential dependence of the reflected energy on the incident laser energy for the *f*/1.9 lens case.

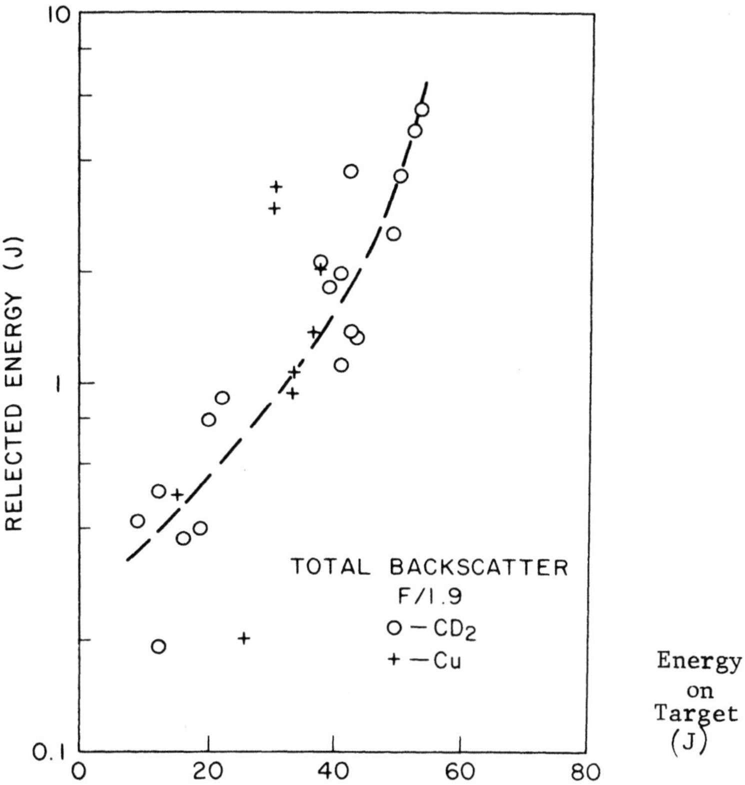

Figure 12. Back-Reflected Energy with *f*/1.9 Lens

When using a $f/14$ optical system (74-cm focal length lens), the back-reflected energy from both CD_2 and Al targets did not continue to rise with energy on target, but saturated. Such a saturation of back-reflected energy could be of key importance in laser-fusion applications and must be studied more closely. It is not known, at present, why there is such a difference between the experiments with long and short focal length lenses.

Spectral and Angular Dependence. Spectra of the backscattered laser light were obtained from $(CH_2)_n$ and various high-Z targets. The target was either oriented perpendicular to the laser beam (0 degree angle of incidence) or else rotated to a different angle of incidence. The target was in a vacuum ranging from .1-50 m Torr. For these data runs, the laser energy deposited on the target was in the range of 1-10 J and the pulse duration was 250 psec giving an intensity on the target of about 10^{15} watt/cm^2.

The light reflected from the target is intercepted by a pellicle placed near the laser and a portion of the light is focused with a cylindrical lens onto the slits of a 2-m Ebert Spectrograph, where time-integrated spectra are taken using hypersensitized, Kodak I-Z plates. There is sufficient light so that a spectrum could be taken with a single exposure.

Photographic spectra were taken with different laser pulse energies and a variation of the target angle of incidence, which proved to be two significant variables. The incident laser beam has a half-width of about 1 Å. These spectra show a strong line at the wavelength of the incident laser beam and approximately the same width, but also additional discrete satellite lines are clustered out to 20 Å on both sides of the laser line at 10,640.7 Å. Densitometer scans of 3 shots at normal incidence, but at different energies are shown in Fig. 13. Reference lines are seen on the right side of the scans. Here one can see the increasing reflected laser line intensity with increasing pulse energy. However, one can see an even greater increase of the satellite line intensities with increased pulse energy indicating a non-linear growth of these lines.

Spectra were also taken at approximately the same laser energy but at different angles of incidence. It was observed that as the angles of incidence increased from 0 degrees, the satellites shifted from the low to the high wavelengths. The explanation of this effect is unknown at this time.

Photographic spectra were taken at 5300 Å, the second harmonic of 1.06 μm, and no lines were seen. This would indicate that stimulated Raman scattering was not present at the intensities we report here.

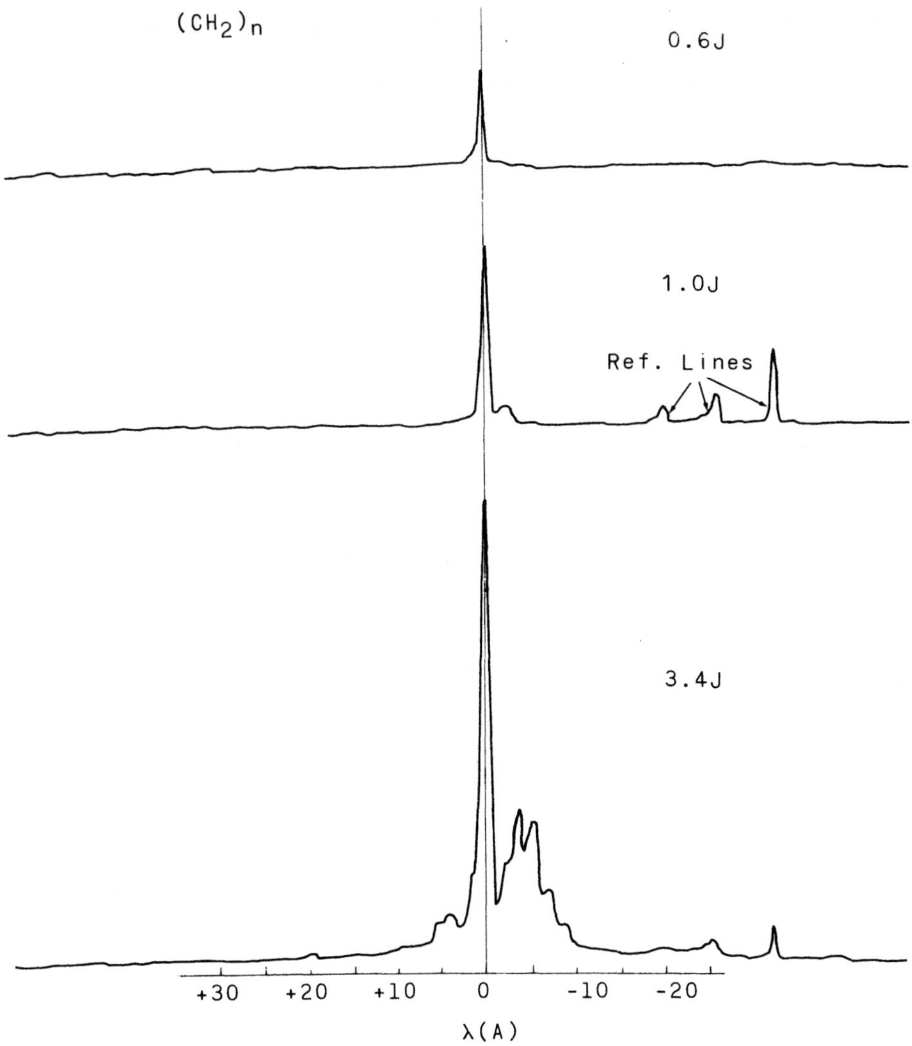

Figure 13. Back-Reflected Spectra at Different Energies

In summary back reflected laser light has all the properties expected of a stimulated process, i.e., the proper polarization, the right spatial symmetry, an apparent threshold at $\simeq 10^{14}$ w/cm^2, and exponential growth above that point. The remaining question is whether the process can go to completion and reflect all or most of the energy or whether it turns off because "phase matching" cannot be maintained.

The difference is experimentally resolvable either by:

1. going to much higher intensities and measuring the backscattered energy and

2. time-resolving the backscatter to determine whether the duration of emission is very short (Kindel) or comparable to the laser pulse duration.

THEORETICAL STUDIES

The theory of interacting plasma and laser radiation is complicated on several accounts: the occurrence of non-linear plasma and optical effects, the large spatial and temporal variations, the generation of large magnetic fields and the interaction of radiation with atomic systems. However, quantitative results can be obtained with the aid of fast computers. A discussion is given first of laser-induced magnetic fields and certain collective plasma phenomena depending on these fields. The numerical simulations are then discussed: a rate equation treatment of plasma radiation and a 2-D fluid treatment, including magnetic and radiative effects.

Laser-Induced Magnetic Fields

It is now generally recognized that large magnetic fields are generated as a result of the interaction of a powerful laser pulse with a solid target. The original explanation of the fields assumed that the magnetic fields originated from the thermal energy of the plasma and were due to solenoidal electric fields associated with electron pressure and temperature gradients[26]. Another possible source of magnetic field energy is the flow energy of the plasma. This effect, called dynamo action, has been assumed to be small but should not be ignored when complicated plasma motions develop. It has been stated that thermal sources for magnetic fields would vanish under adiabatic conditions. However, Widner and Wright[38] have pointed out that the adiabatic condition ($d(Tn^{1-\gamma})/dt = 0$ where $d/dt = \partial/\partial t + v \cdot \nabla$) does not imply, in a time-dependent problem, the vanishing ($\nabla T \parallel \nabla n$) of the thermal source term. This is due to the mixing of space and time variations in non-steady flow.

Computer calculations[39,40,41] have shown that these fields are in the megagauss range under conditions of interest to laser-fusion. Such large fields affect the physics of laser-produced plasmas in several ways: (1) They produce a hotter plasma in the central region since thermal conduction across the field is reduced; (2) Magnetic field dependent instabilities (e.g. whistler type) may occur and thus affect the absorption and reflection of the laser radiation; (3) Superthermal particle trajectories are altered by the magnetic field and the generation of these superthermals by stochastic acceleration processes is also affected[42]. These magnetic effects are of interest to laser-fusion since they can affect the heating and compression of a pellet[43].

Magnetic Field Generation by Radiation Pressure

The calculations, discussed earlier, which show that large magnetic fields are generated, were based on thermal sources[44], i.e. the energy for the fields came from the thermal energy of the plasma. However, for high laser intensity ($\gtrsim 10^{14}$ W/cm^2) there is also a direct conversion (i.e. not involving plasma thermal energy) between the laser energy and the magnetic field energy. This field generation is due to a solenoidal electric field associated with radiation pressure gradients[45,46,47]. Physically, the electric field is associated with radiative forces resulting from interactions (absorption, refraction and reflection) of the laser beam with the plasma[45].

A discussion is given in the appendix of radiative forces acting on the electron fluid. An equation of motion (Eq. A7) is derived which includes radiation forces. Solving this equation for the electric field, taking the curl, and combining the result with Maxwell's equations gives an equation governing the time-averaged or observable magnetic field,

$$\frac{\partial \underline{B}}{\partial t} = \nabla \times (\underline{V} \times \underline{B}) - \frac{c^2}{4\pi} \nabla \times \left[\underline{\underline{\rho}} \cdot (\underline{\nabla} \times \underline{B}) \right] + \underline{S} \tag{1}$$

where

$$\underline{S} = - c\nabla \times \left[\underline{\underline{\alpha}} \cdot \nabla T - \frac{1}{ne} \nabla \cdot (\underline{\underline{P}}^e + \underline{\underline{P}}^r) \right] \tag{2}$$

is the source term. The electron inertia (nmdV/dt) and field momentum ($\partial g/\partial t$) terms are ignored. Thermal sources, associated with thermoelectric power $\underline{\underline{\alpha}}$ and the scalar part of electron pressure $\underline{\underline{P}}^e$, are not considered here. The radiative source term

$$\underline{S}^r = \frac{c}{e} \nabla \times \left(\frac{1}{n} \nabla \cdot \underline{\underline{P}}^r \right) , \tag{3}$$

depends on radiation pressure $\underline{\underline{P}}^r$ which is evaluated in the Appendix (Eq. A9).

One can obtain an insight into the nature and magnitude of radiative sources by considering a special case where thermal sources would vanish. The thermal sources for a scalar electron pressure $P^e = nkT$ vanish when temperature and density vary in the same direction. However, the radiative source due to directed radiation exists even when all quantities vary in the same direction. This is due to the inherent tensor character (anisotropy) of radiation pressure associated with a directed flow of radiation.

For a plane wave traveling in the +Z direction, the time-averaged radiation pressure (Eq. A9) in Cartesian coordinates is $P^r_{xx} = P^r_{yy} = P_\perp = I\ (1 - \epsilon)/2\bar{n}c$ and $P^r_{zz} = P_\parallel = I\ (1 + \epsilon)/2\bar{n}c$ with the other components vanishing. The dielectric constant is denoted by ϵ and the index of refraction $(\sqrt{\epsilon})$ is denoted by $\bar{n}$. The anisotropy $(P_\perp \neq P_\parallel)$ is a result of the transverse nature of light and would exist for any directed flow of radiation (e.g. converging). The azimuthal symmetry $(P^r_{xx} = P^r_{yy})$ is a result of averaging over many laser periods so that polarization effects are averaged out. The main effect of $\underline{\underline{P}}^r$ on field generation can be seen by considering the special case where all quantities vary only in a spherically radial direction. Transforming $\underline{\underline{P}}^r$ and $\underline{S}^r$ into spherical coordinates shows that $\underline{S}^r$ is in an azimuthal direction about the laser beam and has a magnitude

$$S^r = \frac{c \sin 2\Theta}{er} \frac{\partial}{\partial r} \left[\frac{1}{n} (P_\perp - P_\parallel) + \frac{r}{2n} \frac{\partial}{\partial r} (P_\perp - P_\parallel) \right] . \qquad (4)$$

Note that the source term is due to the anisotropy in $\underline{\underline{P}}^r$ and is a function of $P_\perp - P_\parallel = -\bar{n}I/c$.

Using Eq. (4), it is possible to estimate the growth of magnetic fields associated with radiation pressure. A typical scale length Δr for spatial variations is 10^{-3} cm and a typical radius r of the focal region is 10^{-2} cm. The radiative source function may thus be approximated as $S^r \sim I/2ne(\Delta r)^2$, assuming $\epsilon \cong \bar{n} \cong 1$. One finds, for $I = 10^{15}$ W/cm^2 and $n = 10^{21}$ cm^{-3} that $S^r \sim 10$ MG/nsec. The fields will grow until the laser pulse is over or until they are convected, with the expanding plasma, out of the absorption region. For an expansion velocity V of 10^7 cm/sec, the growth time $\Delta r/V$ is .1 nsec. This indicates that radiation pressure can generate magnetic fields in the megagauss range.

The importance of radiative sources for magnetic fields is due both to the large fields produced in existing experiments and to the implications for future experiments. Radiative sources don't have the same limitations as thermal sources. It is quite possible, when lasers of very large energy (> 1 kJ) and short pulse width (< 10^{-10} sec) are developed, that a large fraction of the laser energy can be converted directly into magnetic field energy. This means that enormous magnetic fields can be generated and that field generation will be a significant part of the absorption process.

For linearly polarized laser radiation, the radiation pressure can produce an anisotropy in the electron pressure. This anisotropy is a preferred heating along the laser $\underline{E}$-field and can be viewed as resulting from a balance between the effects of laser driven electron oscillations along $\underline{E}$ and the collisional and collective distribution of the energy into other directions. Numerical simulations at Princeton[48] have shown such an effect. A simple argument[49] indicates that for a collisional plasma, the anisotropy $(T_{\parallel}/T_{\perp}) - 1$ may be of the order of the ratio of radiation pressure to electron pressure. The radiation induced anisotropy in the electron pressure gives a solenoidal character to the associated electric field and thus leads to magnetic field generation. It is somewhat different from the other mechanisms in that the resulting magnetic field has a component along the laser beam.

Magnetic Field Effects on Plasma Collective Phenomena

We now discuss the effects of the magnetic field on the various collective phenomena, specifically parametric instabilities, in laser produced plasmas. Most of what is discussed here is taken from a paper by one of us and Edward Ott[50]. In Ref. (50) a general dispersion relation was obtained for parametric instabilities which couple the high frequency laser light to low frequency electrostatic or electromagnetic perturbations. It is

$$1 = \frac{\underline{k}}{\omega} \cdot \left[\underline{\underline{\sigma}}_e \cdot \left\{ \frac{\underline{\underline{Q}}}{D} \cdot \frac{\omega^2}{c^2}\, i\underline{k} \right\} \right] \times$$

$$\frac{4\pi e^2 |E(\underline{\varkappa}, \Omega)|^2}{m^2 \Omega^2 c^2} \left\{ \frac{t^+}{D^t_+} + \frac{\ell^+}{D^\ell_+} + \frac{t^-}{D^t_-} + \frac{\ell^-}{D^\ell_-} \right\} \Big] \qquad (5)$$

assuming the ions to be unmagnetized. Above, $(\Omega,\underline{\varkappa})$ are the frequency and wave number of the laser light, $(\omega,\underline{k})$ that of the low frequency perturbation, $|t^{\pm} = (\underline{k} \pm \underline{\varkappa}) \times \underline{E}(\varkappa,\Omega)|^2/|\underline{k} \pm \underline{\varkappa}|^2|E(\underline{\varkappa},\Omega)|^2$, $\ell^{\pm} = |(\underline{k} \pm \underline{\varkappa})\cdot\underline{E}(\underline{\varkappa},\Omega)|^2/|(\underline{k} \pm \underline{\varkappa})|^2|E(\underline{\varkappa},\Omega)|^2$,

$$D_t^{\pm} = \frac{(\Omega \pm \omega)^2}{c^2} - |k \pm \varkappa|^2 - \frac{\omega_p^2}{c^2}, \quad D_{\ell_e}^{\pm} = \frac{(\Omega \pm \omega)^2}{c^2} - \omega_p^2/c^2,$$

$\underline{\underline{\sigma}}_e\,(k,\omega)$ is the electron conductivity tensor at $(\underline{k},\omega)$, $\underline{\underline{D}}$ is the dispersion tensor $\underline{\underline{D}} = -\,\underline{k} \times (\underline{k} \times - \frac{\omega^2}{c^2}\,\underline{\underline{I}} - \frac{4\pi i\omega}{c^2}\,\underline{\underline{\sigma}}_e - \frac{4\pi i\omega}{c^2}\,\sigma_i\,\underline{\underline{I}}$ where the ion conductivity is assumed to be a scaler since $\omega \gg \omega_{c_i}$, and $\underline{\underline{Q}}$ is the cofactor tensor of the transpose of $\underline{\underline{D}}$, and D is the determinant of $\underline{\underline{D}}$. Equation (5) contains all of the standard parametric instabilities with or without magnetic field, i.e., Raman and Brilloun scattering, the decay instability and the oscillating two stream instability. We now concentrate on effects of the spontaneous magnetic field.

The first effect which we will discuss is backscatter and absorption by whistler processes. For densities below the critical density the laser light can decay into a backscattered photon and a whistler as long as the propagation direction is neither parallel nor perpendicular to the incident field. Near the critical density the laser light may be absorbed by decaying into a whistler and an electron plasma oscillation. In either case, one may show from Eq. (5), after some manipulation that the growth rate is given roughly by

$$\gamma_w \sim (1/4)\,\sqrt{\Omega\omega_{ce}}\left(\frac{eE}{m\Omega c}\right). \tag{6}$$

One way to assess the importance of whistler processes is to compare the growth rate with that for Brilloun backscatter or the ordinary decay instability, which is

$$\gamma_B \sim (1/4)\,\sqrt{\omega_{pi}\Omega}\left(\frac{eE}{m\Omega V_e}\right) \tag{7}$$

where V_e is the electron thermal velocity. If the magnetic field is sufficiently large $\left(\omega_{ce} > \omega_{pi}\left(\frac{V_e}{c}\right)^2\right)$ that $\gamma_w \gtrsim \gamma_B$, then whistler processes will be important. Even if these processes are less important, whistler backscatter may be useful as a field diagnostic.

We have just seen that the magnetic field can introduce new modes (i.e. the whistler) which can be driven unstable by the laser light. The magnetic field can also modify the properties of existing modes and thereby affect parametric instabilities. In particular, it may lower the threshold for the standard parametric instability.

In a plasma with equal electron and ion temperature, the threshold for the decay instability is rather high because the ion acoustic wave phase velocity is roughly equal to the ion thermal velocity, so ion Landau damping is very strong. However, in a magnetized plasma, for propagation perpendicular to the magnetic field for $k \leqslant \omega_{pe}/c$ the phase speed is not the acoustic speed but rather the magnetosonic speed. Thus if the plasma beta is one or less, the phase velocity is much greater than the ion thermal velocity so the damping rate is substantially reduced. Some manipulations with Eq. (5) show that in this case, the growth rate is

$$\gamma_{Ms} \sim \frac{1}{2} \sqrt{\frac{\omega_{pi}}{\Omega} \frac{V_A}{V_s}} \; \omega_{po} \frac{V_{os}}{V_A} \tag{8}$$

where V_A is the magnetosonic velocity. Eq. (8) reduces to Eq. (7) in the limit of zero magnetic field.

Model for K-Line X-Ray Emission from Laser-Produced Aluminum Plasmas

By focusing a 0.25 ns, 2J pulse of neodymium laser radiation down to a spot-size of 50 microns, an average radiation intensity of 4×10^{14} W/cm^2 is obtained. A pulse of such intense radiation can break down solid target materials and generate hot, dense, non-LTE plasmas. For example, the x-ray spectrum emitted in the 5 to 8 Å region under the above conditions by a laser-produced aluminum plasma results from bound state transitions in hydrogen-and helium-like ions and indicates the plasma attained electron temperatures near or in excess of 1 keV.

Laser radiation energy is partitioned in many ways within the aluminum plasma and, in particular, some of it is converted into magnetohydrodynamic motion. However, as a prelude to a more complete fluid treatment, a non-hydrodynamic model can be set up, which describes the plasma's conversion of laser energy to keV x-rays. Such a model and its predictions, which surprisingly are in close agreement with experimental results, are described below.

In a non-uniform plasma, laser radiation energy is absorbed near the critical depth. For radiation with a 1.06μ wavelength, the critical electron density is 10^{21} cm^{-3}. At these densities, the electrons equilibrate rapidly among themselves (in 10^{-13} sec) and maintain a well-defined temperature, T_e, on 10 or 100 picosecond time scales. These considerations suggest the following hot-spot plasma model.

We imagine that, in the focal region of the laser beam, a uniform sphere of non-flowing aluminum plasma absorbs the neodymium laser energy. The sphere has volume ΔV and a constant total ion density N_o. N_o is determined by the condition that, in the state of total ionization, the electron density be equal to the critical density: $N_o = (10^{21}/13)\ cm^{-3}$. A pulse of laser radiation is absorbed into the plasma sphere by the electron gas. The electrons collide with ions within ΔV, heating and ionizing them, and due to a temperature gradient at the sphere's surface, electron kinetic energy is conducted from the sphere. Thus

$$\frac{d}{dt}\left(\frac{3}{2} N_e k T_e \Delta V\right) = \begin{pmatrix}\text{Energy absorbed from} \\ \text{the laser pulse}\end{pmatrix} + \begin{pmatrix}\text{Energy gained or lost} \\ \text{through collisions}\end{pmatrix} - \begin{pmatrix}\text{Energy lost by} \\ \text{conduction}\end{pmatrix} \qquad (9)$$

Conduction losses can be estimated by assuming a temperature gradient at the sphere surface of T_e/ℓ_e where

$$\ell_e = \begin{cases} 10^{-2}\ \text{cm} & t \le 10^{-10}\ \text{sec} \\ 10^{3}\sqrt{t} & t \ge 10^{-10}\ \text{sec} \end{cases}$$

The ion kinetic temperature changes can be computed from the equation

$$\frac{d}{dt}\left(\frac{3}{2} N_o k T_i \Delta V\right) = \frac{3 m_e}{m_i} \frac{N_e k(T_e - T_i)\Delta V}{\tau_e} \qquad (10)$$

where τ_e is the electron-electron $\pi/2$ collision time[51].

Energy transfer from the electrons to the internal degrees of freedom of the ions can be described by a set of rate equations, which are based on the energy level diagram of Figure 14 and the level transitions shown. The ion number densities N_μ for each state thus change in time according to the equations

$$\frac{dN_\mu}{dt} = \sum_\nu W_{\mu\nu} N_\nu \,, \quad \sum_\mu N_\mu = N_o \,. \qquad (11)$$

The collisional rates all depend on the electron number density, which is determined from the ionic charges Z_μ via $N_e = \sum_\mu Z_\mu N_\mu$, and on the electron temperature. Thus, Equations (9) through (11) represent a closed set of nonlinear equations of motion from which the emission of K-shell line radiation from the aluminum plasma can be computed.

A computer solution to Equations (9) through (11) was obtained for a case where 1.5 J of laser energy was deposited into the electron gas in 0.25 ns. The electrons and ions were initially taken to be in thermal equilibrium at a temperature of 0.1 keV. At this temperature, the electrons in ΔV store approximately 10^{-3} J of energy (taking $\Delta V = 5 \times 10^{-8}$ cm^3)* Thus, most of their dynamics of interest should occur after they are forced to absorb the additional 1.5 J of laser energy and should not be very sensitive to the initial condition chosen.

The computer integration of Equations (9) through (11) gave the following results. The electron temperature was driven to a maximum value of 2 keV in 0.3 ns, falling equally rapidly to 0.3 keV because of conduction losses. The ion temperature tracked the electron temperature with an 0.1 to 0.2 ns time delay, achieving a maximum value of only 1.5 keV. Most of the laser energy was conducted from the hot-spot and only a small portion (7×10^{-3} J) was radiated by Al XII and Al XIII excited states. The computed x-ray spectrum

---

*This corresponds to a sphere diameter of 46μ, which is roughly the focal-spot size.

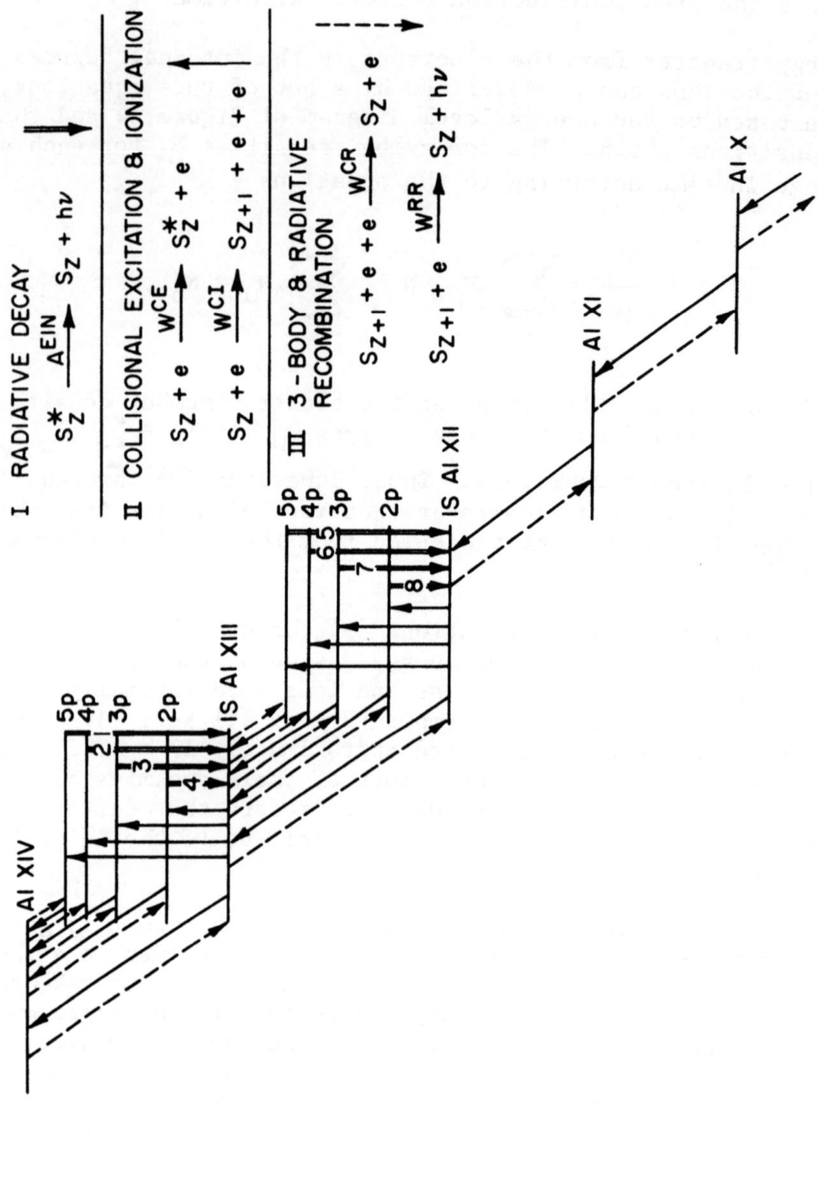

Figure 14. Energy Level Diagram for Aluminum

looked qualitatively and quantatively very much like the experimentally observed spectrum. The ratio of hydrogen K_α to helium K_α radiated energy had a computed value of 0.55 as compared with the experimentally measured value of 0.63. Experimentally it was found that 0.6% of 1.9 J of incident laser energy was converted to K-shell line radiation. In the computer solution to Equations (9) through (11), 1.5 J of laser energy was converted to 7×10^{-3} J of K-shell line radiation, a 0.5% conversion efficiency.

If it is assumed that the ion populations remain in coronal equilibrium as T_e changes, then Equation (11) can be replaced by the simpler set,

$$\sum_{\nu} W_{\mu\nu} N_{\nu} = 0 \ , \ \sum_{\mu} N_{\mu} = N_o \ . \tag{12}$$

Computer solutions to Equations (9), (10), and (12), the coronal model equations, were obtained for the same case as above; namely, 1.5 J of laser energy deposited in 0.25 ns into the electron gas. The coronal model predicted that only 3×10^{-3} J of the 1.5 J was radiated as K-shell line radiation, a 0.2% conversion efficiency. Moreover, the ratio of the He_α to H_α K-line radiation was 1.02. Finally, an estimate of the radiative recombination continuum radiation was made for both rate equation and coronal models. The rate equation model predicted that 2×10^{-3} J of continuum radiation was emitted while the coronal model predicted that only 1.1×10^{-3} J was emitted. Thus, the rate equation model predicted more line and continuum radiation output by a factor of 2 or more than the coronal model.

A 2-D Simulation of Laser-Produced Plasmas

A two-dimensional code has been developed for the processes involved in the interaction of a focussed laser pulse with a moderate atomic weight target. A system of fluid equations including self-generated magnetic fields[51,26,39] is solved in cylindrical geometry by a second-order algorithm using time-step split flux-corrected transport[52]. Representative results of numerical simulations are presented and the importance of the self-generated magnetic field is demonstrated.

Typically we consider nanosecond (or sub ns) Nd pulses with intensities > 10^{14} watts/cm^2 through the focal spot, which give rise to target plasma temperatures above 1 keV. Thermal energy spreads from the laser focal region via electron thermal conduction, fluid expansion, and line and continuum emission from highly charged ion states. Magnetic fields in the megagauss range are generated internally in the plasma, and tend to inhibit thermal conduction in the direction perpendicular to $\underline{B}$ in regions where $\Omega_e \tau_e > 1$. Radiation pressure also plays a role for intensities $> 10^{14}$ watts/cm^2, but is not included in the present code.

The fluid equations used are

$$\frac{\partial N}{\partial t} + \underline{\nabla} \cdot (N\underline{V}) = 0 \quad , \tag{13}$$

$$NM \frac{d\underline{V}}{dt} + \underline{\nabla} \cdot \left\{ P \underline{\underline{I}} + \underline{\underline{I}} \frac{B^2}{8\pi} - \frac{\underline{B}\,\underline{B}}{4\pi} \right\} = 0 \; , \tag{14}$$

$$\frac{\partial \epsilon}{\partial t} + \underline{\nabla}\cdot(\epsilon\underline{V}) = - P\underline{\nabla}\cdot\underline{V} + \underline{J}\cdot\underline{E}' + < \delta\underline{E}\cdot\delta J > + R_e + \underline{\nabla}\cdot(\underline{\underline{K}}_e\cdot\underline{\nabla}T_e), \tag{15}$$

$$(P, T_e) = \text{corona}\ (\epsilon, N) \; , \tag{16}$$

$$\frac{\partial \underline{B}}{\partial t} = \underline{\nabla} \times \left\{ \underline{V} \times \underline{B} - \frac{c^2}{4\pi} \underline{\underline{r}} \cdot (\underline{\nabla} \times \underline{B}) \right\} - \frac{ck}{eN} \underline{\nabla}N \times \underline{\nabla}T; \tag{17}$$

where

$$\underline{E}' = \underline{E} + \frac{1}{c} \underline{V} \times \underline{B} = - \frac{1}{eN} \underline{\nabla}(ZNkT_e) + \underline{\underline{r}} \cdot \underline{J} \; . \tag{18}$$

In these equations N is the total ion number density, $< \delta\underline{E} \cdot \delta\underline{J} >$ the laser energy deposition term, $P = NkT(1 + Z)$ the total pressure, ϵ

the thermal plus ionization energy density, and R_e the rate of change of ϵ due to radiation processes. The resistivity $\underline{\underline{r}}$, and electron thermal conductivity $\underline{\underline{K}}_e$, are taken from Braginskii[51] with the $\underline{B}$-dependence of $\underline{\underline{r}}$ and $\underline{\underline{K}}_e$ included.

Laser energy deposition by inverse bremsstrahlung is treated by an analytic path-integrated method which is energy conservative. Energy which reaches the critical density without being absorbed is deposited there. Emitted radiation is treated by a corona model but radiation transport is neglected. Numerical experiments have been performed with Mg and higher-Z targets. Here we report the results of work with Aluminum, which is representative.

We shall examine the effect of a laser pulse which is gaussian in space and time (50 micron half-width, 1 nsec pulse length) which deposits 20 joules in the target. The target is initially given by a solid region, plus an exponential density tail facing the laser, with a (small) uniform energy density, to approximate the state after a prepulse. The results we present here are insensitive to the initial tail parameters.

Figure 15 presents some important parameters versus distance along the incident laser beam. Ion number density, temperature, magnetic field strength, laser absorption power, and radiated power, are shown as a function of distance into the target. Profiles are for a cylinder of radius 40 μ about the axis of the incident beam at 1 nsec. The Nd laser pulse is assumed gaussian, 50 μ radial half-width, 20 joules total energy on target. Two persistent features are the crossed density and temperature ramps ($\underline{\nabla}N \times \underline{\nabla}T_e$ is a source for $\underline{B}$ in Eq. 17), and the separation of the critical density region, where the absorption peaks, from the region of maximum radiation emission. This separation explains why a plasma which may be quite hot (several keV) at the critical density emits radiation characteristic of a much lower temperature. This is because of the N^2 dependence of the emission which causes the cooler overdense plasma to emit most rapidly. A total conversion efficiency from laser energy to radiation emission of ~ 6% was obtained for the case in Fig. 15. Higher radiation efficiencies are expected for higher energy pulses.

The ripples in the laser absorption profile correspond to an acoustic wave propagating toward the laser (left), and indicate the presence of an absorption instability characterized by a critical wavelength ($\lambda > \lambda_c$ unstable),

$$\lambda_c \cong 2\pi \left(\frac{K_\perp kT}{3\ P_L} \right)^{\frac{1}{2}} \tag{19}$$

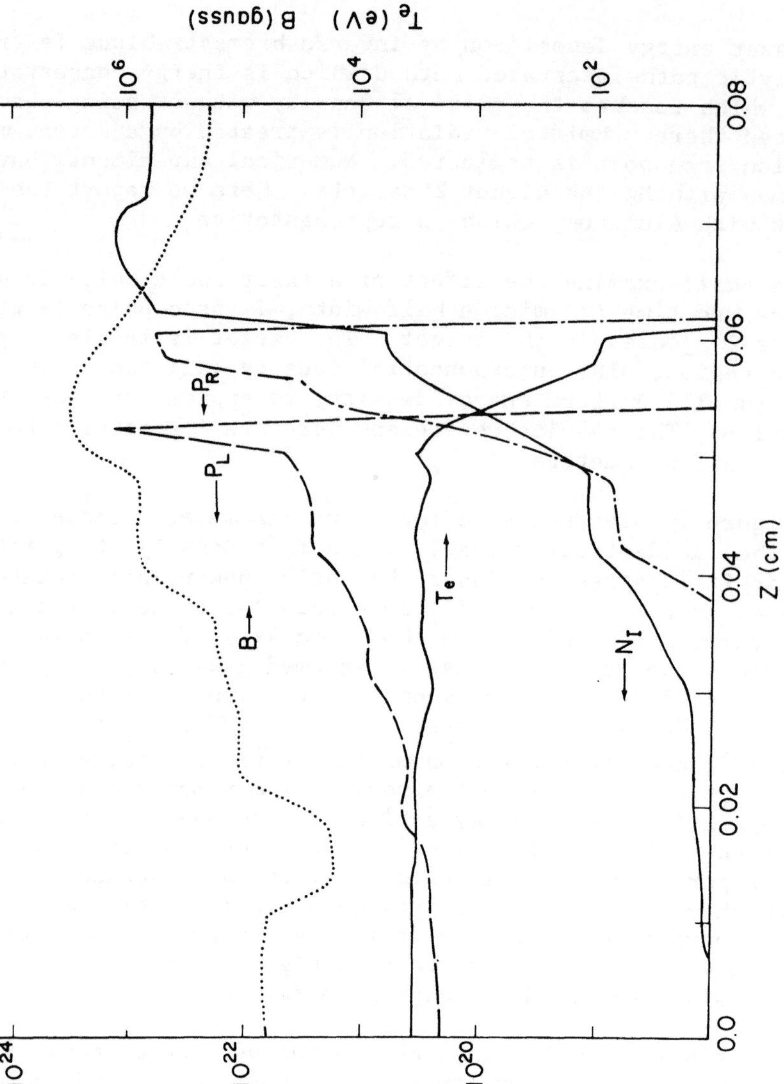

Figure 15. Axial Profiles of Parameters at a 40μm Radius

where P_L is the laser deposition power density and $K_\perp$ the thermal conductivity perpendicular to $\underline{B}$. It results from the density dependence of the inverse bremsstrahlung absorption, coupling energy into acoustic modes. It is approximately density independent. Further, it is observed when $\underline{B}$ is present, i.e., $K_\perp$ small since $\Omega_e \tau_e \gg 1$, and tends not to be observed if $\underline{B} = 0$.

The magnetic field is a maximum in the region surrounding the peak absorption and radiation layers. This reduces heat transport there and allows large temperature gradients to develop. In addition it alters the convection of plasma. It would also play a role in suprathermal particle generation and transport[42].

CONCLUSION

A variety of experimental and theoretical results have been presented. If there is a trend in these results and those of other laboratories, it is toward recognizing the complexity of laser-produced plasmas and their interactions with plasma and laser radiation. It is clear that a continuing, vigorous research program in laser-produced plasmas is needed to achieve the scientific and practical potential of these plasmas.

APPENDIX

Radiation Pressure and Plasma Dielectric Constant

The high frequency electromagnetic field oscillations in an inhomogeneous plasma produce a force on the electrons which results in a contribution to the electric field. This field has a solenoidal character which is associated with the form of the radiation pressure tensor. Magnetic fields are thus generated as a result of the radiation pressure force.

There are two ways to treat electromagnetic fields in a medium. In plasma phsyics, it is usually necessary to use the vacuum form and thus explicitly identify the current and charge contributions. This is the only safe approach. The other approach is to divide the charges and currents into free and constrainded and to treat the effects of constrained charges through constituitive relations. The high frequency electron oscillations or quiver induced by the electromagnetic field are of small amplitude below 10^{16} W/cm^2. Some of the effects of electron quiver can, in fact, be treated through a plasma dielectric constant and a useful expression for radiation pressure can be derived. However, a consistent treatment of radiation pressure effects in a plasma is not readily obtained by relying on a variable plasma dielectric constant.

Consider the electron velocity distribution to have a mean velocity which can be written as the sum of $\underline{V}$ which varies slowly in time and $\delta\underline{v}$ which undergoes high frequency oscillations (quiver) at the frequency of the electromagnetic field. If the electron velocities are non-relativistic than a unique $\delta\underline{v}$ can be defined for the districution since the force $(-e\delta\underline{E})$ is not velocity dependent. One can then write the electron velocity $\underline{v}$ as the sum of a slowly varying directed velocity $\underline{V}$, a random velocity $\underline{v}'$ and the rapidly oscillating velocity $\delta\underline{v}$.

We are interested in quantities which can be observed on the slow time scale, i.e. for times long compared to a laser light period but short compared to the laser pulse width. There are thus two averages needed: an average $\langle\rangle_v$ over the velocity distribution and a time average $\langle\rangle_t$ over the fast oscillations.

An expression is derived here for the averaged force on the electrons due to the high frequency fields. The fields $(\underline{E}, \underline{B})$ are assumed to have a low frequency part and a high frequency part $(\delta\underline{E}, \delta\underline{B})$. Maxwell's equations in the form

$$\nabla\cdot\underline{B} = 0, \qquad \nabla \times \underline{B} = \frac{1}{c}\frac{\partial E}{\partial t} + \frac{4\pi}{c}\underline{J}$$

$$\nabla\cdot\underline{E} = 4\pi\rho\,, \qquad \nabla \times \underline{E} = \frac{-1}{c}\frac{\partial B}{\partial t} \tag{A1}$$

can be used to derive the equation for conservation of momentum of the electromagnetic field.

$$\nabla\cdot\underline{\underline{P}}^{em} + \frac{\partial g}{\partial t} = -\left(\rho\underline{E} + \frac{1}{c}\underline{J} \times \underline{B}\right) \tag{A2}$$

where $\underline{g} = \dfrac{\underline{E} \times \underline{B}}{4\pi c}$ is momentum density and

$$\underline{\underline{P}}^{em} = \frac{1}{4\pi}\left[\frac{1}{2}\left(\underline{E}^2 + \underline{B}^2\right)\underline{\underline{I}} - (\underline{E}\,\underline{E} + \underline{B}\,\underline{B})\right] \tag{A3}$$

is momentum flux for the field.

Time averaging Eqs. (A2) and (A3) and subtracting the results from Eqs. (A2) and (A3) shows that these equations are also valid when each field is replaced by its high frequency part. In general, the high frequency fields could include plasma waves and thus not be entirely transverse. However, for this discussion, a quiescent plasma is assumed so that the waves are purely transverse and one can ignore the high frequency part δn of the density.

From the first velocity moment of the electron kinetic equation, one obtains the electron momentum conservation equation on the fast time scale:

$$\frac{\partial}{\partial t}\langle n\underline{v}\rangle_v + \nabla\cdot\langle n\underline{v}\underline{v}\rangle_v = -\frac{e}{m}n\langle \underline{E} + \frac{1}{c}\underline{v}\times\underline{B}\rangle_v + \frac{\underline{C}}{m} \tag{A4}$$

where

$$\underline{C} = ne\,(\underline{\underline{\rho}}\cdot\underline{J} + \underline{\underline{\alpha}}\cdot\nabla T) \tag{A5}$$

is the electron-ion collisional force, modeled by a resistivity $\underline{\underline{\rho}}$ and thermoelectric power $\underline{\underline{\alpha}}$.

Now, $\langle\langle n\underline{v}\rangle_v\rangle_t = n\underline{V}$ and $\langle\langle n\underline{v}\underline{v}\rangle_v\rangle_t = n\underline{V}\underline{V} + n\langle \underline{v}'\underline{v}'\rangle_v + n\langle\delta\underline{v}\;\delta\underline{v}\rangle_t$, so that by using conservation of electron number, the time average of the left hand side of Eq. (A4) can be written

$$n\frac{d\underline{V}}{dt} + \nabla\cdot(n\langle\underline{v}'\underline{v}'\rangle_v) + \nabla\cdot(n\langle\delta\underline{v}\delta\underline{v}\rangle_t) \tag{A6}$$

Expanding the time average of the right hand side of Eq. (A4) using both the low and high frequency forms of Eqs. (A2) and (A3), and combining with the expression (A6) for the left hand side gives the desired time averaged electron equation of motion written explicitly in terms of the slowly varying velocity and fields.

$$nm\frac{d\underline{V}}{dt} = -\nabla\cdot(\underline{\underline{P}}^e + \underline{\underline{P}}^r) - ne(\underline{E} + \frac{1}{c}\underline{V}\times\underline{B}) + \underline{C} - \frac{\partial \underline{g}}{\partial t} \tag{A7}$$

where $\underline{\underline{P}}^e = nm\langle \underline{v}'\underline{v}'\rangle_v$ is the electron pressure and where the radiation pressure, i.e. time-averaged pressure associated with the high frequency oscillations, is defined as

$$\underline{\underline{P}}^r = \frac{\underline{\underline{I}}}{8\pi}\langle(\delta\underline{E})^2 + (\delta\underline{B})^2\rangle_t - \frac{1}{4\pi}\langle \delta\underline{E}\,\delta\underline{E} + \delta\underline{B}\,\delta\underline{B}\rangle_t + mn\langle\delta\underline{v}\,\delta\underline{v}\rangle_t\ , \tag{A8}$$

and contains both field and particle contributions. Field momentum density is denoted by $\underline{g} = \langle \delta\underline{E}\times\delta\underline{B}\rangle_t/4\pi c$.

Since $\delta\underline{v} \cong -ie\,\delta\underline{E}/m\omega$, the particle term can be written as $(4\pi)^{-1}(1-\epsilon)\langle\delta\underline{E}\,\delta\underline{E}\rangle_t$ when $\epsilon = 1-(\omega_p/\omega)^2$ is the plasma dielectric constant in the under-dense region. Also, $\epsilon - n(d\epsilon/dn) = 1$, which means that $\underline{\underline{P}}^r$ can be written in the form

$$\underline{\underline{P}}^r = \frac{\underline{\underline{I}}}{8\pi}\langle(\delta\underline{E})^2 + (\delta\underline{B})^2\rangle_t - \frac{1}{4\pi}\langle\epsilon\delta\underline{E}\delta\underline{E} + \delta\underline{B}\delta\underline{B}\rangle_t \tag{A9}$$

given by Landau and Lifshitz[53] and Pitaevskii[54].

It is thus tempting to think that one has here a description where the high frequency electron oscillation effects can all be treated as electromagnetic through the plasma dielectric constant. This is not the case! The only way that a consistent treatment of electromagnetic momentum can be made is to introduce ϵ into Maxwell's equations, $\underline{\underline{P}}^{em}$ and $\underline{g}$ in such a way that Eq. (A2) is still formally valid with the source term $-(\rho\underline{E} + \frac{1}{c}\underline{J}\times\underline{B})$.

Using $\underline{D} = \epsilon\underline{E}$ in the inhomogeneous Maxwell's equations, using the expression in Eq. (A9) for $\underline{\underline{P}}^{em}$ and $\underline{g} = (1/4\pi c)\,\underline{D}\times\underline{B}$ gives

$$\nabla\cdot\underline{\underline{P}}^{em} + \frac{\partial\underline{g}}{\partial t} = -(\rho\underline{E} + \frac{1}{c}\underline{J}\times\underline{B}) + \frac{1}{8\pi}(1-\epsilon)\nabla\underline{E}^2 \tag{A10}$$

Equation (A10) is clearly in error as the electromagnetic momentum conservation equation. However, for the frequently considered case of a neutral ($\rho = 0$), unmagnetized ($\underline{B} = \underline{0}$) plasma, the truth is obscured since one may assume that the electromagnetic momentum source is $(8\pi)^{-1}$ $(1 - \epsilon)$ ∇ $\underline{E}^2$ rather than the correct value which vanishes. This would show that n^{-1} $\nabla \cdot \underline{\underline{P}}^r$ is approximately (since $\partial g/\partial t$ is small) the gradient of a scalar. The associated electric field would then be irrotational. In fact, for directed radiation, n^{-1} $\nabla \cdot \underline{\underline{P}}^r$ is solenoidal for any realistic spatial variation due to the anisotropy of $\underline{\underline{P}}^r$.

A correct procedure is clear from the treatment given here. An effective radiation pressure has been defined including a particle (quiver) contribution. The vacuum form of Maxwell's equations was used in deriving an electron equation of motion which includes this quiver contribution. Hence, a self-consistent treatment of particle and field variables can be made by using the time-averaged electron equation of motion, in the form of Eqs. A7, A9, along with the time-averaged, vacuum Maxwell equations (Eq. A1)[55].

Note that we have considered the effect of radiation pressure forces on the electrons rather than the plasma. The resulting electric field produces a motion of the electrons relative to the ions (electric current) whereas a radiation pressure force acting on a neutral plasma produces no such relative motion.

ACKNOWLEDGMENTS

The authors gratefully acknowledge the support of the U.S. Atomic Energy Commission and the Defense Nuclear Agency.

REFERENCES

1. M.A. Duguay and J.W. Hansen, "Measurement of the Nonlinear Index n_2 of Glass Using Picosecond Pulses", Proc. 2nd, ASTM Symp. Damage in Laser Materials (1971), NBS Spec. Publ. 341, pp 45-49.

2. J.M. McMahon,"Damage Measurements with Subnanosecond Pulses", Proc. 3rd ASTM Symp. Damage in Laser Materials (1972), NBS Special Bull. 372 pp 100-103.

3. E.S. Bliss, "Importance of Self Focusing in a Laser Amplifier with Large Beam Diameter", IEEE Jour. Quant. Elect. QE-8, 2, pp 273-276 (1972).

4. J.A. Fleck, in press at the present time.

5. The chamber design origniated at United Aircraft Corporation and was used at NRL by permission of Dr. Alan Haught.

6. G.H. McCall, F. Young, A.W. Ehler, J.F. Kephart, and R.P. Godwin, Phys. Rev. Letts. 30 (1973) 1116.

7. H.R. Griem, Plasma Spectroscopy, McGraw-Hill, New York (1964).

8. D.J. Nagel et.al., VI European Conference on Controlled Fusion and Plasma Physics, Moscow, Aug. 1973.

9. D.J. Nagel, C.M. Dozier and R.R. Whitlock, unpublished and J.F. Holzrichter, C.M. Dozier, J.M. McMahon, to be published.

10. W.E. Behring, R.J. Ugransky and U. Feldman, Appl. Optics, 12, 528 (1973).

11. D.J. Nagel et.al., to be published.

12. P.G. Burkhalter, D.J. Nagel and R.R. Whitlock, to be published.

13. B.M. Klein, D.J. Nagel and R.R. Whitlock, to be published.

14. B.M. Klein, to be published.

15. B.M. Klein et.al., VI European Conference on Controlled Fusion and Plasma Physics, Moscow, Aug. 1973.

16. U. Feldman et.al., to be published.

17. L. Cohen et.al., J. Opt. Soc. Am. 58, 843 (1968).

18. U. Feldman et.al., Astro. J. Letters 183, L43 (1973).

19. For a comparison of XUV sources see Gabriel, A.H., Nucl. Inst. and Methods 90, 157 (1970).

20. R.R. Whitlock, C.M. Dozier, D.J. Nagel, unpublished.

21. C.M. Dozier, unpublished.

22. D.J. Nagel, to be published.

23. D.J. Nagel and J.W. Criss, unpublished.

24. T.N. Lee, D.J. Nagel and R.H. Dixon, to be published.

25. D.J. Nagel, R. Pechacek and M. Gorbics, unpublished.

26. J.A. Stamper, K. Papadopoulos, R.N. Sudan, S.O. Dean and E.A. McLean and J.M. Dawson, Phys. Rev. Lett. 26, 1012 (1971).

27. F. Schwirzke and L.L. McKee, Proceedings of Fifth European Conference on Controlled Fusion and Plasma Physics, CEA, Grenoble, France, 1, 63 (1972).

28. R.S. Case, Jr., Bull. Am. Phys. Soc. 17, 1026 (1972).

29. D.V. Giovanielli and G.H. McCall, private communication.

30. J.A. Stamper, S.O. Dean and E.A. McLean, Laser Interaction and Related Plasma Phenomena, Vol. 2, (H. Schwarz and H. Hora, Eds.) Plenum, New York 1972, p. 281.

31. R.S. Bird, L.L. McKee, F. Schwirzke and A.W. Cooper, Phys. Rev. A, 7, 1328 (1973).

32. J.A. Stamper, NRL Report 7411, June 1972, p. 88.

33. Some of the VUV line spectra taken at NRL have a broadening which may have a zeeman contribution.

34. D.W. Forslund, J.M. Kindel, and E.L. Lindman, Phys. Rev., Letters 30, 739 (1973).

35. M.N. Rosenbluth and R.Z. Sagdeev, Comments on Modern Physics, Part E., 1, 129 (1972).

36. The results discussed here on backscatter are taken from a paper by E.A. McLean and J.M. McMahon which has been submitted for publication.

37. K. Eidmann and R. Sigel, Max-Planck-Institute für Plasma Physik, Garching; Report No. 2. p.p. IV/46, Dec. 1972.

38. M.M. Widner and T.P. Wright, Personal Communication.

39. J.B. Chase, J.L. LeBlanc and J.R. Wilson, to be published (approximately August 1973) in Physics of Fluids.

40. N. Winsor and D.A. Tidman, Communicated to Physical Review Letters.

41. Melvin M. Widner, to be published in Physics of Fluids.

42. D.A. Tidman and J.A. Stamper, Appl. Phys. Lett. 22, 498 (1973).

43. J. Nuckolls, L. Wood, A. Thiesen and G. Zimmerman, Nature 239, 133 (1972).

44. Reference 32, pgs. 2,3.

45. J.A. Stamper and D.A. Tidman, to be published in Physics of Fluids.

46. J.A. Stamper, NRL Memorandum Report 2492, August 1972.

47. Reference 32, pg. 6.

48. W.L. Kruer and J.M. Dawson, Phys. Fluids 15, 446 (1972).

49. Reference 32, pg. 6.

50. W. Manheimer and E. Ott, to be published.

51. S. I. Braginskii, "Reviews of Plasma Physics", ed. by Acad. M.A. Leontovich, 1, 205 (1965).

52. J.P. Boris, D.L. Book, J. Comp. Phys. 11, 38 (1973).

53. L.D. Landau and E.M. Lifshitz,"The Electrodynamics of Continuous Media", Addison-Wesley, Reading, Mass., 1960, p. 67.

54. L.P. Pitaevskii, Soviet Physics, J.E.T.P. 12, 1008 (1961).

55. The use of the vacuum form of Maxwell's equations also follows directly, on physical grounds. On the slow time scale, $\underline{D} = \underline{E} + 4\pi\langle\underline{P}\rangle_t$ where the polarization $\underline{P} = n(-e)\,\underline{\delta d}$ is due to the quiver excursion $\underline{\delta d} = e\,\underline{\delta E}/m\omega^2$. Thus, $\underline{D} = \underline{E} - (\omega_p/\omega)^2\langle\delta\underline{E}\rangle_t = \underline{E}$. The plasma dielectric constant is unity for the fields or Maxwell's equations (linear in the fields) on the slow (observable) time scale. The form $\epsilon = 1 - (\omega_p/\omega)^2$ is only valid on the fast (oscillatory) time scale. It occurs in the radiation pressure since this pressure is quadratic in the fields and velocities and these quantities are correlated on the fast time scale.

LASER-CTR DEVELOPMENTS IN AUSTRALIA*

J.L. Hughes

Laser Group
Department of Engineering Physics
The Australian National University
Canberra, Australia

ABSTRACT

Techniques which can be utilized for laser-CTR studies have been under development in Australia for many years. Particular reference is made to the properties of a single-beam laser-induced plasma system (LIP-1) and to the Nd-glass laser which supplies its excitation pulses.

INTRODUCTION

Although several research groups in Australia possess experience and techniques relevant to laser-CTR studies, the only direct effort at present is being undertaken with a single beam laser induced plasma system (LIP-1), powered by a rod/disc Nd-glass laser. This facility is located at the Australian National University in Canberra and is intended to establish techniques for the compression of matter to high densities. Laser-CTR orientated work is at present aimed at optimising LIP-1 to the equivalent of a one beam section of a multibeam breakeven system. Two topics of considerable importance in laser-CTR have been selected for initial study, namely, the suppression of "back-reflected" light from the target and the production of accurately reproducible tailored pulses of the form proposed by Nuckolls et al. [1]. Preliminary experiments [2][3] at gigawatts power levels indicate that LIP-1 should simplify many aspects of laser-CTR studies, particularly at high power levels and over an extended range of wavelengths where a better understanding

---

*Presented at the Third Workshop on "Laser Interaction and Related Plasma Phenomena" held at Rensselaer Polytechnic Institute, Troy, New York, August 13-17, 1973.

of the basic physics of laser-CTR and high density matter may be reached.

BACKGROUND

Modifications [2][3] to two pulse storage systems being developed for the study of non-linear optics in a vacuum [4][5][6][7][8][9] have resulted in basically simple but effective systems for laser-CTR studies. These modifications are shown in figure 1. The target excitation technique used in LIP-1 differs from those used in other laboratories. Here the target is excited via the rear surface. This method of target excitation has many distinct advantages over front surface excitation as will be discussed below.

NEODYMIUM-GLASS LASER

Basic System

The neodymium-glass laser system used to provide the excitation pulses for LIP-1 is being developed for studying non-linear optical effects in a vacuum and therefore differs in several important respects from Nd-glass lasers being developed specifically for laser-CTR studies. Among parameters indispensable to non-linear optical studies in a vacuum are short pulse durations and very high quality, diffraction limited laser beams. Trains of output pulses are required for such studies in order to increase the signal to noise ratio and hence operational requirements differ considerably from the single pulse operation in laser-CTR studies. The main Nd-glass system effort is being devoted to the production of the highest possible photon flux densities from a given pulse energy. For example, it should be possible to realise a photon flux density $\sim 10^{15}$ watts cm^{-2} over an area $\sim 10^{-6}$ cm^2 using only one joule per nanosecond, whilst one joule per picosecond under the same conditions should realise a photon flux density $\sim 10^{18}$ watts cm^{-2}. Unless these parameters can be realised in practice it is unlikely that the extremely high flux densities ($\sim 10^{21}$ watts cm^{-2}) necessary for a comprehensive study of non-linear optical effects in a vacuum can be achieved.

It is standard practice in Nd-glass laser systems to have an oscillator, pulse selector/trimmer followed by a chain of rod amplifiers interspersed with electro-optic, magneto-optic, dye and other switches, as well as an abundance of beam expanders.

The high power, Nd-glass laser system at the Australian National University is shown in figure 2. The components which

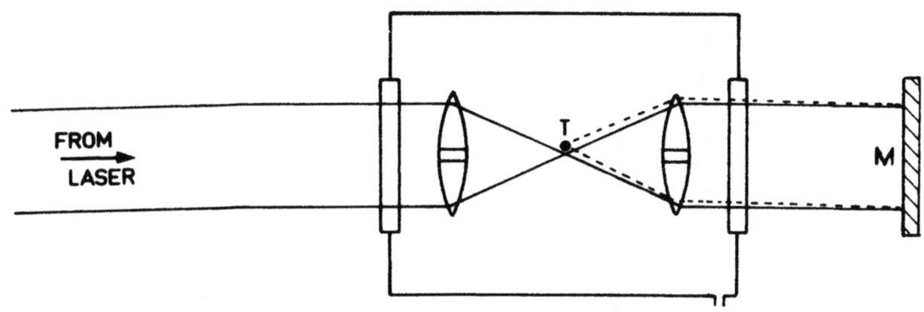

a

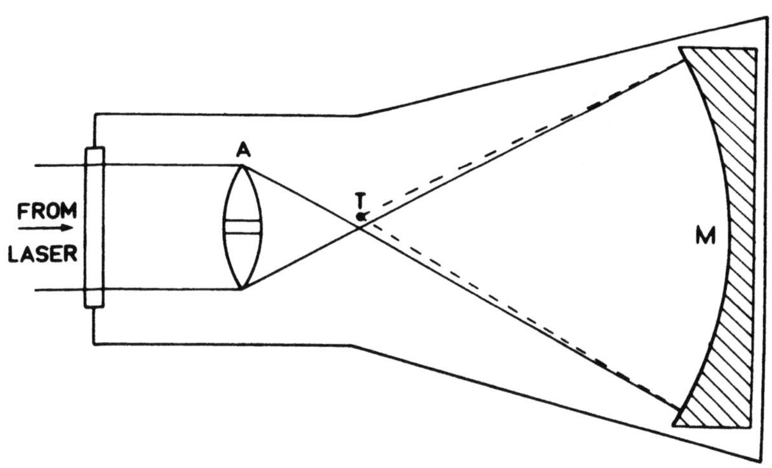

b

Figure 1.

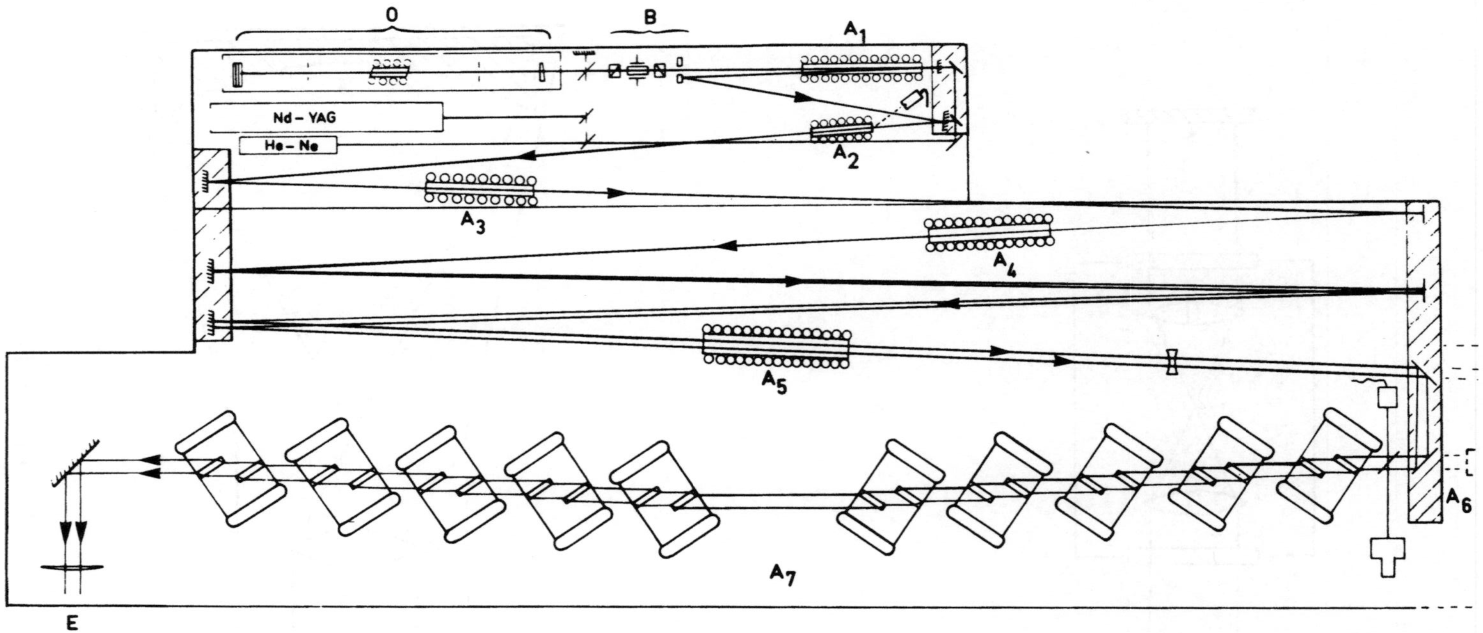

Figure 2.

Layout of the ANU high power Nd-glass laser system. O-mode-locked oscillator, B-DL11 Pockel's cell, A_1-A_5 rod amplifiers, A_7 disc amplifier, A_6 possible booster amplifier.

must remain rigid with respect to each other are fixed directly onto the polished surface of a 12 ton granite bench. The granite bench is mounted on a large concrete base which is sunk into, but isolated from, the laboratory floor. The relative movement between any component on the granite table and the laboratory floor is approximately 30 nanometers or 3×10^{-2} times the wavelength of the Nd laser. On the other hand the relative movement between the various components attached to the bench is expected to be much less, although this has not been measured. All the laser components which do vibrate, such as flash tube housings, are connected to a frame above the granite table and thereby isolated from the components fixed to the bench.

In the ANU Nd-glass laser, the maximum length of the granite table is 5 meters and its maximum width approximately 2 meters. The top of the granite bench provides for a minimum of 6 traverses from the beginning to the end of the rod amplifier chain, i.e. a path length of at least 30 meters, compared with a minimum oscillator to target path of 40 meters. This relatively compact laser layout allows for simpler control of environmental parameters, the temperature being maintained at $\pm 1^{\circ}$ C, the relative humidity at $<40\%$ and the dust level tolerable as a result of 1 mm excess pressure and other simple precautions within the laser room. Ultra clean room conditions which now appear to be essential for the reliable operation of very powerful laser systems have not been attempted to date.

The initial and output stages of the ANU laser can be seen in figure 3. In the foreground the output end of the 20 disc amplifier can be seen with two of the ten flashtube housings removed to show disc mountings. In the upper right hand corner the mode-locked oscillator can be seen with the flashtube housing tilted in the vertical plane to accommodate the Brewster ended rod.

Pulse Generator

To date all pulses have been generated in a Nd-glass mode-locked oscillator (15 cm x 0.6 cm Brewster ended ED^2 rod), which although far more temperamental than a Nd-Yag oscillator does provide the much shorter pulse durations required for the production of large flux densities.

The Nd-glass mode-locked oscillator has proved very reliable over extended periods of time, and without the use of a flowing dye cell, as long as certain basic procedures have been adhered to. In particular it is necessary to bring the rod to the same temperature as existed during the initial alignment of the oscillator. Although adequate care, or luck, in aligning the oscillator mirrors should make first shot operation possible, this has never been

Figure 3.

seriously attempted and a two or three shot pre-heat of the mode-locked oscillator rod is usually required each morning or prior to taking a series of shots after a long delay in the firing sequence. In an attempt to put the alignment of the mode-locked oscillator on a more automatic basis a T.V. alignment system was set up. Typical fringe patterns obtained from the interference of the input alignment beam and that reflected off the back reflector are shown in figure 4. Using a video recorder, it has been possible to compare the effects on mirror alignment of tapping various parts of the mode-locked oscillator. Interference from air conditioned air was also observed but cured by deflecting the air flow. With these precautions and the use of extremely rigid optical mounts, wedged mirrors and low threshold rods in the mode-locked oscillator, extremely reliable operation with nanosecond duration laser pulses has been achieved even without an aperture in the oscillator cavity. However, subnanosecond duration pulses require a 3 mm aperture for reliable operation.

The diffraction limited output beam profiles at meter intervals from the oscillator are shown in figure 5. It will be seen that the elliptical cross-section resulting from the Brewster ended rod rotates through 90° as the distance from the oscillator increases from the near to the far field. In normal operation a 3 mm aperture is inserted into the mode-locked oscillator cavity to ensure that the amplified laser beam is of circular cross-section. The individual pulse energy in the TEM_{00} output beam of the Nd-glass

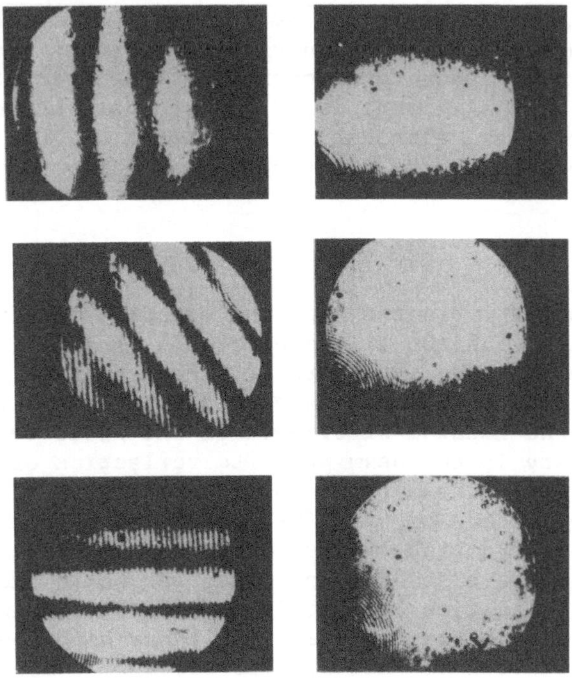

Figure 4.

Sequence of interference patterns obtained on a T.V. screen during the alignment of the mode-locked oscillator. (a) off horizontally (b) off horizontally and vertically (c) off vertically (d) two fingers off vertically (e) one finger off vertically (f) aligned.

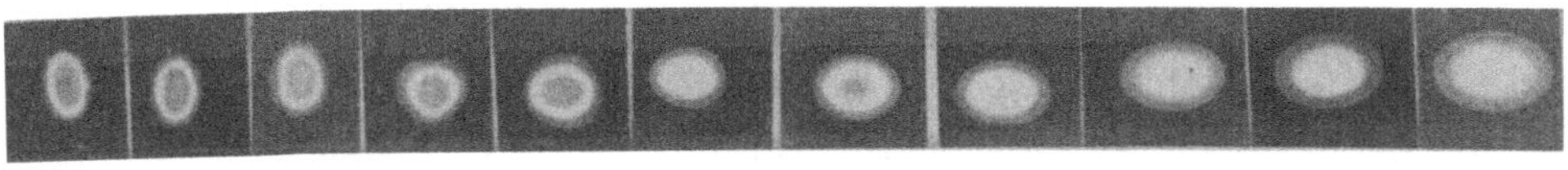

Figure 5.

Mode-locked oscillator beam profiles.

oscillator is approximately 10^{-3} joules and that of the whole pulse train approximately 40×10^{-3} joules. Increasing the pulse train energy to 70×10^{-3} leads to a considerable deterioration in the beam quality. The alignment laser used to date has been a 2.5 milliwatt He-Ne laser, although a c.w. Nd-Yag alignment laser will soon be incorporated in the position shown in figure 2.

Rod Amplifier Chain

A single pulse extracted from the output of the mode-locked oscillator with a CILAS DL 11 Pockel's cell is amplified in the rod amplifier chain as shown in figure 2. Each rod amplifier after the first should be positioned immediately after a given dielectric mirror so that the natural divergence of the laser beam can reduce the energy density in the beam prior to reflection off the next dielectric mirror. The mirror should be orientated at 45° when beam intensities become high. The process of maintaining the quality of the diffraction limited output beam of the mode-locked oscillator as it traverses the rod amplifier chain has presented considerable difficulties because it is easy to obtain a highly structured profile in the far field. It has been found that the best results are obtained when all "irrelevant" components are removed from the path of the laser beam and a series of apertures of slightly larger diameter than that of the beam inserted along the rod amplifier chain. The beam quality can also be detrimentally affected by utilizing too much of the cross-section of the amplifier rods. For this reason we have tended to use only the central portion of each rod in the ANU laser. In this way very high quality laser beams can be maintained in the rod amplifier chain during the amplification process. To date, the first three amplifiers are the only ones which have been used extensively. Under the conditions described above they provide gains of $\sim 10^3$ per pulse of $\sim 10^{-9}$ seconds durations in a very nearly diffraction limited beam. The gain can be increased if necessary by inclusion of the 2.5 cm diameter amplifier A_5. The peak power can also be increased by shortening the pulse duration. However, the problems encountered to date could best be tackled with nanosecond duration pulses of between 10^9 and 5×10^9 watts peak power. A trigger pulse for the disc amplifier of 1 to 5 joules and a few picoseconds duration will soon be available for non-linear optics studies while pulses of several tens of joules with durations between 10^{-10} and 10^{-9} seconds will be used for laser-CTR studies. It is believed that most of the pulse energy, whatever its duration, should be gained in an "exponential" amplifier [10] which is a divergent beam, [11][12] segmented amplifier capable of very high peak powers.

One of the unique features of the rod amplifier chain shown in figure 2 is the absence of most standard components normally associated with a rod amplifier chain. At the gains used to date, i.e. between 10^3 and 10^4, no self-oscillation of the chain has been detected. Experiments now in progress will determine the minimum separation and excitation levels required to operate the rod amplifier chain at higher gains. The effects of varying environmental parameters on beam quality have not been investigated. It now appears that adequate trigger pulses into the "exponential" amplifier can be obtained without the use of any switching device in the rod amplifier chain. Once inside the "exponential" amplifier the pulse will, under ideal conditions, have an axial gain of unity, the unique property of such an amplifier being the fact that its gain can be expressed as the ratio of output to input cross-sections only.

The present approach to the rod amplifier chain is to treat it merely as a convenient method of providing high quality trigger pulses for the "exponential" amplifier. In fact it may be necessary to get rid of rods altogether if a trend to sub-picosecond duration pulses materialises.

Disc Amplifier

The present disc amplifier system consisting of 20 identical, uncoated AO 5% doped discs, 25 mm thick and with a minor axis of 44 mm was initiated two years ago. Since then it has been used for the development of techniques [13] to harness part of the enormous energy storage capacity (500 MJ) of the ANU homopolar generator [14] [15][16]. Up to 10^5 joules per millisecond are now extracted from a compact inductive storage system driven by the homopolar generator. This energy has been used to excite the 20 disc amplifier via 40 linear flashtubes arranged in a series-parallel array. Preliminary measurements using uncoated AO discs give results comparable to those that have been reported using similar discs with condenser bank excitation. Detailed measurements will be published in due course.

The present 20 disc amplifier is now an integral part of the ANU laser system. It is however far from optimised for either the present experimental programme in laser-CTR or the non-linear optics programme. Instead, it is hoped that the developments now taking place will provide very large excitation energies in the multimegajoule range per beam, when such energies will be most needed for the excitation of a large "exponential" amplifier.

Of particular interest in the ANU disc amplifier is the "open" structure [17] approach where each segment would be treated as a separate section of the amplifier, as far as the excitation and mounting of the discs or slabs are concerned. In the ANU disc amplifier due to the relatively small disc size a compromise has been made by treating pairs of discs as separate entities.

Although Xenon flashtube excitation of the discs cannot be as efficient in the open structure system as it can be in the tightly packed zig-zag disc arrays of well known U.S. systems, by having the discs spread out the amplifier as a whole tends to lose those undesirable rod-like characteristics exhibited by the close packed disc structures.

The present disc amplifier system also acts as a polarizer in the non-linear optics experiments.

The "Exponential" Amplifier

The disc laser amplifier can be operated as the initial stage of an "exponential" amplifier, provided that the input beam diameter is small compared with the disc diameter and that beam divergence is increased to several milliradians near the input to the amplifier.

The exponential amplifier was proposed in order to amplify a sequence of pulses for non-linear optics investigations. Of particular importance is the fact that the axial gain cm^{-2} of the exponential amplifier under optimum operating conditions is unity. By installing pellicle polarizers between the segments of the amplifier it should be possible to increase the self-oscillation or pre-lasing threshold in off-axis directions through the amplifier. Under such conditions, the overall gain of an exponential amplifier will approach the ratio of output aperture area to input aperture area. Since this ratio can be very large, overall gain can also be very high. In existing disc systems the "exponential" characteristics (which can only be approximated in practice) will not be readily apparent until the diameters of succeeding discs are substantially increased.

Several problems have to be overcome before a trend from multiple to single chain disc amplifier systems becomes established. The first problem with large diameter discs or slabs arises from parasitic transverse oscillations which deplete the upper level prior to the passage of the laser pulse through the amplifier. There are two ways to reduce this difficulty. Firstly, progressively lower gain glass must be used as the diameter of successive disc segments increases and the "black glass" coating on the rim becomes less effective. Also, more discs or slabs will be required to maintain a given output pulse. Secondly two or more pulses must be used to excite the discs or slabs. The first pulse excites the discs to near threshold levels, and is then followed by a second pulse which brings the segments well above threshold in a period shorter than the time taken for the parasitic oscillations to significantly deplete the upper lasing level. Normally this would need the second pulse duration to be less than 100 microseconds in low gain media. Although this pulse duration is an order of magnitude shorter than the first pulse, it must be borne in mind that the efficiency above threshold would be much higher and it may even be possible to use arrays of photo-emitting diodes for the second pulse in the not too distant future. This would considerably increase the effectiveness of the exponential amplifier when using Nd-glass.

Under optimum conditions the initial sections of the exponential amplifier should consist of small, high gain, rim coated elliptical glass discs, spaced well apart though not necessarily equally spaced. The initial section of the exponential amplifier should be excited from a condenser bank. The later stages of the exponential amplifier should consist of low gain rim coated glass segments. Excitation on one side of the discs would be provided by a normal pulse of 10^{-3} seconds duration, whilst on the other side an array of short discharge flash lamps should be used. It is under these operating conditions that the ANU homopolar generator would be most effective. It can provide virtually unlimited energy for the threshold pulse on which would be superimposed the shorter duration pulse derived from a condenser bank. These techniques will be tested on our present disc system, which will be ideally suited for the purpose since it contains uncoated discs and is thus extremely sensitive to the parasitics. If the tests are successful it will be possible to construct a single beam Nd-glass laser system of very high power capabilities [10].

It may be appropriate here to reiterate the distinguishing characteristics of the exponential amplifier concept. Firstly, it is not limited to Nd-glass. Secondly, for the exponential amplifier to operate properly the segments must be well spread out. For this the present layout shown in figure 2 must be simplified even further, by eliminating the conventional rod amplifier stages altogether and leaving only the mirrors, which in such a case would have a substrate of Nd-glass and would themselves act as the amplifying medium. The resulting layout [18] is believed to be the simplest possible for a high power laser system.

LIP-1

A novel laser induced plasma system is shown in figure 6. This system, referred to as LIP-1, provides for single beam excitation of a target by means of an appropriately tailored pulse. It will be developed to be the equivalent of a single beam section of a multibeam breakeven system. LIP-1 consists of a pulse diagnostic section, a target chamber, a pulse tailoring system and a T.V. alignment system for positioning the tailored pulse on target. The initial form of LIP-1, shown in figure 6, has the target positioned approximately 100 microns above the common focus of two identical F1 lenses. The pulse tailoring system consists of three

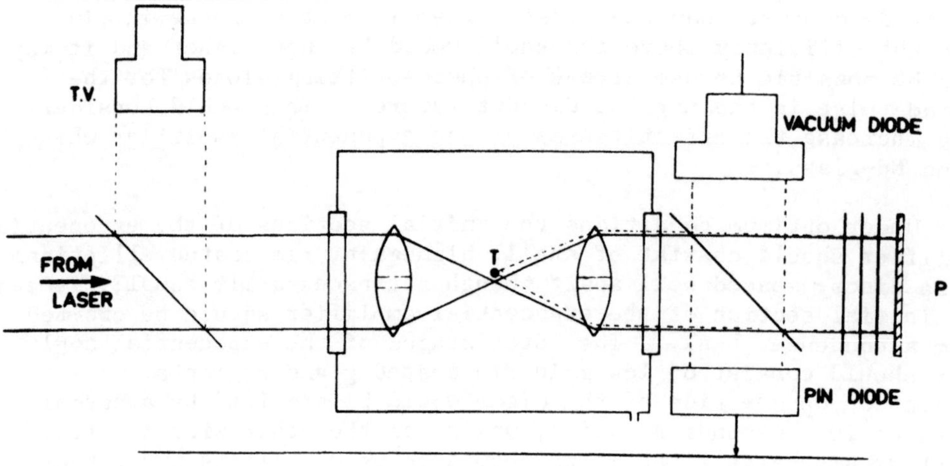

Figure 6.

optical flats and a dielectric mirror (or prism) set approximately 120 cms behind the target. A photograph of LIP-1 is shown in figure 7a, the pulse tailoring system is shown in figure 7b. Initial tests on LIP-1 have shown that basic conditions required to suppress any "back-reflected" light from the target can be realised in practice by trapping such light between the lens focus, into which the plasma expands, and the dielectric mirror. It is also possible to reproduce any pulse shape that can be used in laser-CTR studies by inserting appropriately positioned partial reflectors.

More sophisticated pulse tailoring systems, now under construction, will be enclosed within an evacuated chamber extending from the rear of the present chamber. The first of the new pulse tailoring systems shown in figure 8a will consist of nine appropriately coated pellicles which will produce a ten-pulse train of the required profile from the single, 10^{-10} seconds duration pulse entering the system. This pulse tailoring system will have the advantage of containing optical components which will be simple to move relative to each other when determining their optimum position in terms of various plasma parameters. The two other pulse tailoring systems are shown in figures 8b and 8c. In the latter systems the second lens of the system shown in figure 8a has been removed and the flat dielectric mirror has been replaced with a much larger 9 inch diameter concave mirror. In the system shown in figure 8b pulse tailoring will be carried out using appropriately antireflection coated concave plates. Unfortunately the systems

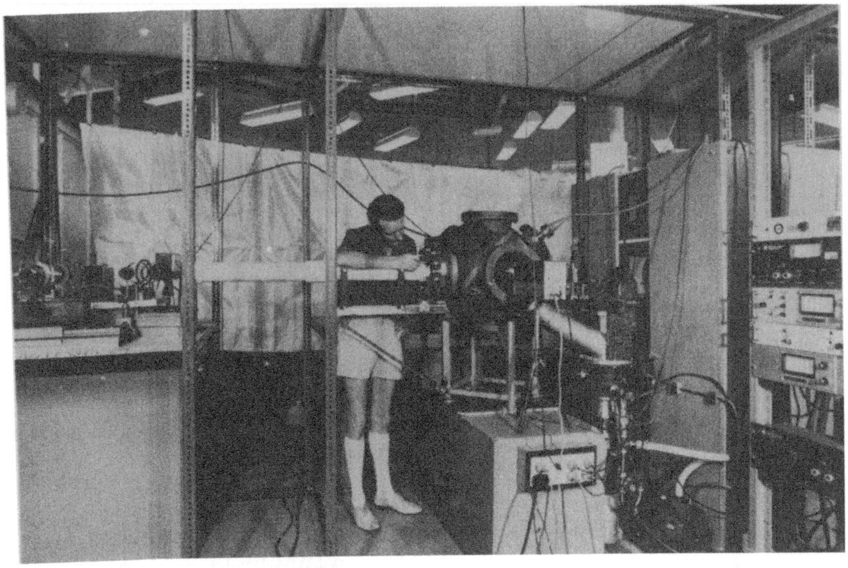

a. LIP-1 showing from left to right the pulse diagnostics area, the vacuum chamber and the pulse tailoring system behind the chamber.

b. The pulse tailoring system used for the initial experiments.

Figure 7.

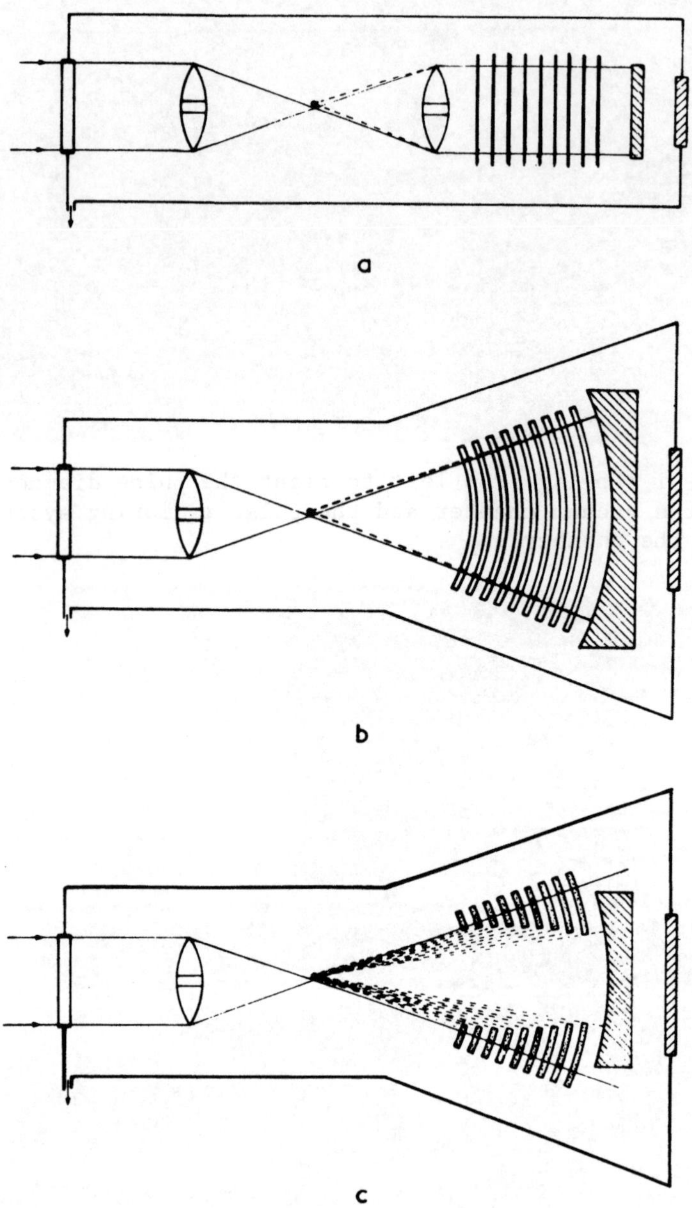

Figure 8.

Three types of pulse tailoring systems.

shown in figures 8a and 8b suffer from the disadvantages that they are only ~75% efficient in transforming the laser output pulse into an effective tailored pulse due to the unavoidable parasitic reflections. These two pulse forming systems also suffer from the fact that they involve transmission optics which limits their usefulness outside the visible region. These difficulties can be overcome in the pulse tailoring system shown in figure 8c which is a slight modification of the one shown in figure 8b. Here the concave discs of the second system are replaced with concave annuli.

The pulse tailoring system shown in figure 8c has considerable advantages because the tailoring process takes place purely by reflection. Since no transmission is involved this pulse tailoring system becomes very attractive for both CO_2 and vacuum U.V. laser pulses. Compared with the first two tailoring systems described it is also very efficient, converting the single laser output pulse into a usable tailored pulse with nearly 100% efficiency. However, the first nine pulses on target will probably have an annular energy distribution which may be disadvantage. The effectiveness of these pulse tailoring systems will be compared in terms of laser-CTR plasma parameters.

The first task in any laser induced plasma experiment is to accurately position the target so that it can be properly heated by the laser beam. In LIP-1 target alignment can be carried out simply but accurately, particularly if the alignment beam is of the same wavelength as the laser beam, i.e. a c.w. Nd-Yag laser should be used with a Nd-glass system. In all our experiments to date, LIP-1 has been aligned with a He-Ne laser, an undesirable method due to the unavoidable output wavelength difference of approximately 4,000 Å. However, the visible output of the He-Ne laser does allow a simple assessment of LIP-1 capabilities to be made. It is interesting to note that on looking down the optical path at the rear of the target chamber it is possible to see, through the second lens, the laser spot traversing the back of the 250 micron diameter target as the rear reflector is adjusted. This technique is now being extended for 1.06 microns operation and should allow pinpoint accuracy in positioning the beam on target. Initial target positioning is of considerable importance in LIP-1. Under no circumstances must the edge of the target penetrate the common focus region in the particular system used. If this happens, the target will be ionized by the incoming pulse, part of which could then be reflected back into the still activated, unprotected laser system. In all experiments performed to date tungsten targets ranging in diameter from 10 to 250 microns have been used. These targets have been positioned approximately 100 microns above the lens focus. To do this, the T.V. alignment system shown in figure 6 was used. In order to position the target accurately above or below the focus region, it is first of all necessary to position it in the centre of the focus region by adjusting the rear lens so that an approximately parallel He-Ne beam emerges from the rear of the chamber, with that portion of the beam which passes through the centre holes in both lenses being in the centre of the shadow produced by the thickness of the lenses. The target can then be easily located in the centre of the focus by observing the traverse on a T.V. monitor. The rear lens is then moved to a calculated position corresponding to a wavelength of 10,600 Å and the target is also moved, first horizontally and then vertically, to a position an estimated 100 microns above the 10,600 Å focus. The rear mirror is then moved a calculated amount and the laser is fired at low power, the plasma plume being observed on a T.V. monitor. Although the He-Ne alignment described above is not recommended for the final target alignment, remarkable accuracy has been achieved, as demonstrated by the vaporization holes in the target during a test run. Attempts have also been made to observe the target during alignment procedure using an optical fibre bundle, one end of which was positioned near the target. Tests to date have produced an image of the target which fills approximately one tenth of a T.V. monitor screen. The optical fibre monitor system which is being developed primarily for non-linear optics experiments, will also be used as described above for observing the position of the laser spot on the target.

The visible infra red T.V. camera used for target alignment as shown in figure 6, is fully protected from damage by the laser beam by a moveable beam splitter [19]. The laser system is deactivated whenever the beam splitter is in the upright position. This simple device has proved an invaluable asset in LIP-1. Prior to its installation protecting the T.V. camera had been an uncertain and time consuming process.

The initial experiments using LIP-1 have behaved according to expectations. However, many more tests are required, particularly at higher power levels, before the true contribution of LIP-1 to laser-CTR studies can be ascertained. One problem in particular now receiving attention is the relation between the optical delay time within the pulse storage system and the expansion period of the plasma. For example, to test the proposal of Nuckolls et al. [1] a tailored pulse of some 20 nanoseconds duration is required which would correspond to an optical delay line of 240 cms. If the plasma reflects during the last 100 picoseconds of the excitation period, then the plasma shutter must be effective for another 20 nanoseconds, a requirement which could present some problems. However, if the plasma does not reflect at these high flux densities, $\sim 10^{17}$ watts cm^{-2}, then the plasma shutter would have a much better chance of remaining effective for the required period. Obviously the first experiment that will be carried out in LIP-1 when the present modifications are completed is the determination of plasma reflections as a function of flux densities up to 10^{17} watts cm^{-2}.

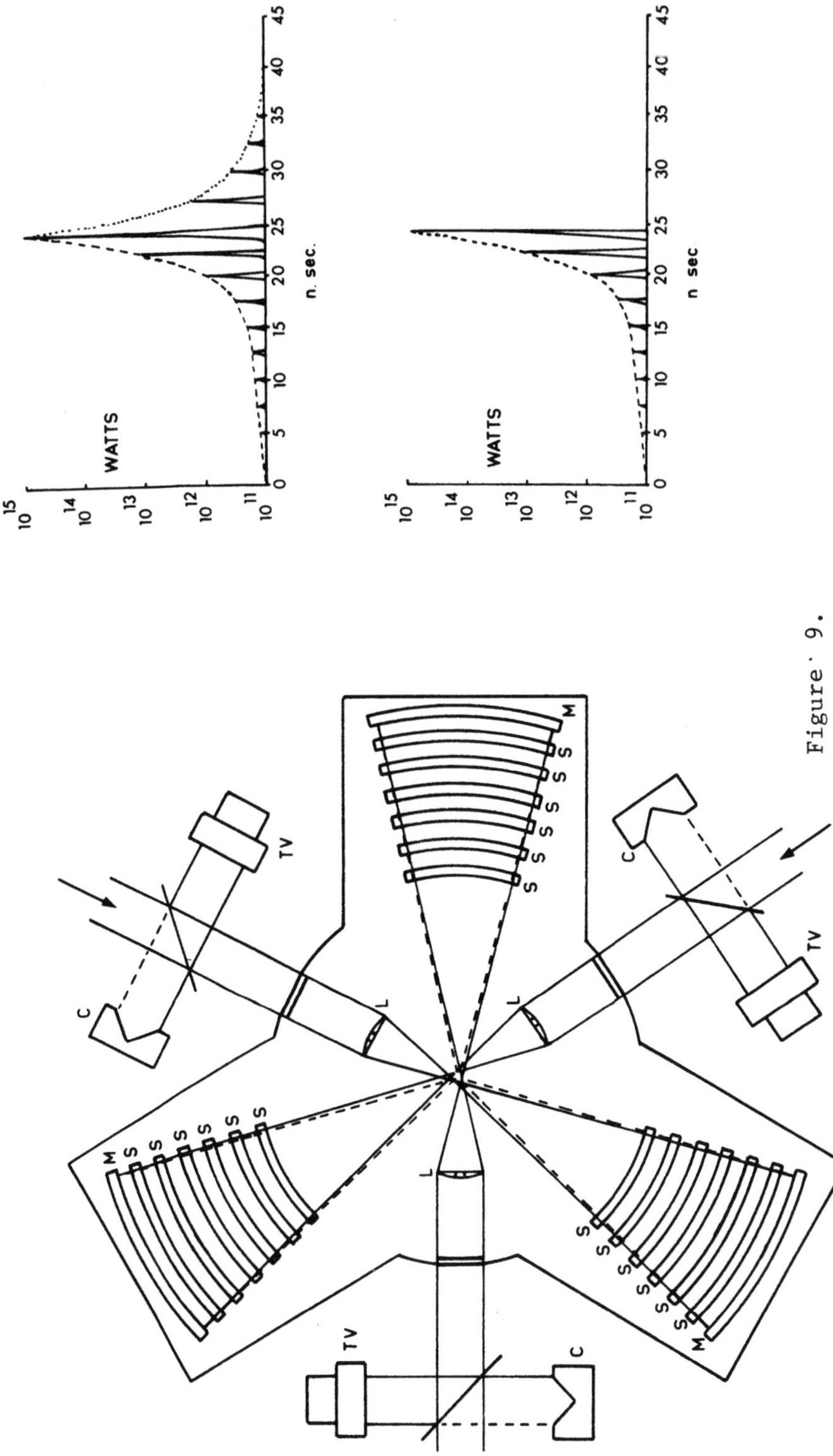

Figure 9.

A proposal for a three beam system for compressing targets in cylindrical co-ordinates. The two tailored pulses correspond to full and annular discs respectively.

To lower the breakeven threshold, it would be necessary to increase the number of laser beams onto the target. The simplest approach would take the form of an LIP-3 system of the type shown in figure 9, where the target is symmetrically excited in a cylindrical geometry.

In conclusion it should be stated that LIP-1 has behaved as anticipated to date and is capable of being used for many experiments in laser induced plasma physics. Whether or not it will be completely successful at the high power levels soon to become available from Nd-glass laser systems will depend on the reflective properties of plasmas at flux densities in the vicinity of 10^{17} watts cm^{-2}.

ACKNOWLEDGEMENTS

I would like to acknowledge contributions to the present work by colleagues at the Australian National University and the Australian Department of Supply as well as valuable discussions with members of overseas laboratories.

In particular I would like to acknowledge early contributions to the work by Mr W.N. Garwoli and Mr W.A.R. McFarlane and the subsequent contributions of Dr E.K. Inall, Mr R.G. Hawkins, Mr B. Smith and Dr M.J. Hollis.

I also wish to thank Mr A. Nagaiya, Mr V.S Pidde, Mr J.A. Wells and Mrs Tina Dixon for their help in preparing the manuscript.

The latest experiments are supported in part by a research grant from the A.I.N.S.E. Previous work relevant to the present paper has been supported by research funds made available by the British-Australian Joint Projects Board and the Australian National University.

REFERENCES

1. J. Nuckolls, L. Wood, A. Thiessen and G. Zimmerman, Nature, 239, 174, 1972.

2. J.L. Hughes and M.J. Hollis, Search, 4, 4, 117, 1973.

3. J.L. Hughes and M.J. Hollis, Optics Comm. 7, 3, 292, 1973.

4. H. Euler, Ann. Der Physik, 26, 398, 1936.

5. R. Karplus and M. Neumann, Phys.Rev. 83, 776, 1951.

6. J. McKenna and P.M. Platzmann, Phys.Rev. 129, 2354.

7. H.R. Reiss, Jrn. Maths, Phys. 3, 59, 1962.

8. A.I. Nikishov and V.I. Ritus, Soviet Phys. JETP, 25, 1135, 1967.

9. J.L. Hughes, Physics Letters B (in press).

10. J.L. Hughes, Applied Optics, 6, 1411, 1967.

11. J.M. Jeggo, Applied Optics, 9, 225, 1970.

12. Yu.V. Senatsky, Lebedev Institute, Research Report No. 87, 1971.

13. E.K. Inall, Jrn. Phys. E. Scientific Instruments, 5, 679, 1972.

14. J.W. Blamey, P.O. Carden, L.U. Hibbard, E.K. Inall, R.A. Marshall and M. Oliphant, Nature, 195, 113, 1962.

15. R.A. Marshall, Nature, 204, 1079, 1964, ibid, 221, 852, 1969.

16. G. Newstead, Science Jrn., 2, 3, 1967.

17. W.G. Garwoli and J.L. Hughes, Proceedings Aust. Inst. Phys. Seminar, the Flinders University, Adelaide, S. Aust., February 1968.

18. J.L. Hughes (patents pending).

19. L.J. Allinson, member of the Department of Engineering Physics Technical Staff 1972 (unpublished).

SOME REACTOR IMPLICATIONS OF LASER FUSION*

IAN SPALDING

UKAEA Culham Laboratory

Abingdon, Berks., U.K.

ABSTRACT

Some technological aspects of blanket design, operating frequency, and pellet production in a laser-compression reactor are discussed. Economic implications of the concept are then reviewed: assuming electrical pumping of the laser system, it is important to minimize the circulating power by using very efficient lasers ($n_e \sim 0.15 - 0.30$) and achieving a high burn up in the compressed core. The desirability of a high repetition rate ($f \sim 50\ H_z$), to allow an adequate margin in the capital cost of laser components, is demonstrated, and an estimate of typical costs for a 530 MW(e) deuterium-tritium fuelled system is presented. Economic operation of a pure DD system is shown to be much more difficult to achieve. Finally, the way in which these conclusions may be modified by alternative laser pumping techniques, and some physical factors which might affect published thermonuclear gain calculations, are summarized.

---

* Presented at the Third Workshop on "Laser Interaction and Related Plasma Phenomena" at Rensselaer Polytechnic Institute, Troy, New York, August 13-17, 1973.

1. INTRODUCTION

Several inertially-confined laser-ignited thermonuclear reactor appraisals have appeared recently[1-8], with conclusions which range from pessimism to optimism - reflecting the various[43] underlying assumptions. A major issue in such studies must be the magnitude of thermo-nuclear burn which might be achieved using laser pulses of convenient energy, at realistic powers and wave-lengths; it is therefore interesting that the relatively high energy gains predicted by the Livermore group, using fairly modest laser energies to produce and then ignite super-high density targets[5], have now been independently confirmed by at least three other groups[6,9,10]. Although these predictions may prove sensitive to many non-linear plasma effects which are difficult (and time-consuming) to model in computer codes, and which await detailed experimental confirmation, it can be useful to anticipate and explore the general characteristics of such laser-compression reactors (LCRs), for reasons similar to those already experienced in toroidal-confinement studies.

The most probable competitors for a LCR are the fast breeder fission reactor or a magnetically-confined fusion reactor. Since generating costs for the fast breeder reactor can already be assessed quite closely, these have been adopted as the basis for an economic criterion, and it has been assumed that fusion generating costs should not be greater than these anticipated fast breeder costs in order to provide a sufficient incentive for fusion to displace fission. This criterion may prove to be rather conservative since no allowance is made for any environmental advantage which may accrue to fusion.

2. TECHNOLOGICAL ASPECTS OF THE REACTOR

2.1. Reactor Vessel

The value of the electricity produced by a single thermo-nuclear pulse of some 10-1000 MJ is low, so that there are severe economic restraints on the cost of replacing damaged components in the vacuum vessel - and on the cost of pellet fabrication. It is therefore important to establish technological limits to the energies which might be contained without significant wear and tear to the LCR. (It may be noted that the blast damage to the reactor vessel should be considerably less than that caused by the equivalent energy release of a more massive chemical explosion[5]). The following three designs have been discussed in the open literature.

In the Blascon device[2,3] a free-standing vortex is formed by rotating a lithium pool about a vertical axis. The neutron flux and energy seen by the stationary structure will be low, because of the 2m thick lithium layer, with the advantages that radiation damage and build up of radioactive impurities normally expected in the structure are very much reduced. It has been estimated that if the mean density of the molten Li is reduced approximately 5% by the introduction of small gas bubbles, the primary shock to the pressure vessel would be attenuated by scattering and spalling at the bubble surfaces to such an extent that pulses of ~ 1500 MJ could be safely handled. However, access to the pellet is highly restricted, so that it seems extremely difficult to ensure the desired approximation to spherical symmetry of the incident laser beam. Moreover, the frequency of operation appears to be limited by the time required to re-establish the lithium vortex to approximately one pulse per ten seconds.

A slightly more conventional design, which is compatible with a laser input approximating to spherical symmetry, has been discussed in detail by the Los Alamos Group[4]. A thin ablative layer of lithium protects the inner wall of the reactor vessel, which is porous. Approximately 1.6 kg of lithium are vaporized per pulse and it is estimated that one second is required to achieve an adequate vacuum between the micro-explosions. Three further concentric walls provide:

(i) some structural restraint for the porous inner wall
(ii) the main pressure vessel wall, which is located deeply enough into the blanket to permit the high breeding ratios (~ 1.4) typical of Li coolant
(iii) an outer plenum, to collect the radial Li outflow from the inner blanket.

The overall plant efficiency, assuming an intermediate heat exchanger and a conventional steam cycle, is ~ 41%. Finally, the feasibility of a "dry wall" design, which should be more easy to pump rapidly, has been indicated by the Livermore Group[5].

Major parameters of the three systems are summarized in Table 1. The implications of these preliminary studies are encouraging, since they suggest that thermonuclear explosions of energy 10^9 J might be handled in specially designed reactor vessels and that pulses of ~ 50 MJ, which are more characteristic of a LCR, can be absorbed in relatively simple reactor vessels.

System	Blascon[2] (Oak Ridge)	Wetted Wall[4] (Los Alamos)	Dry Wall[5] (Livermore)
Thermonuclear energy, per pulse	1500 MJ	200 MJ	10-80 MJ
Repetition Rate of Laser	0.1 H_z	1 H_z	~ 100 H_z
Mean Power	150 MW(th)	200 MW(th)	~1000 - 8000? W(th)
Inner Wall	~ 2m Li (Vortex with 5% bubbles)	~ 2mm Li wetting 0.01m of austentitic stainless steel	10 chambers?
Li temp. before pulse	500-1000°C	400° C	-
LiVP (before any pulse)	10^{-2} - 50 Torr	~ 10^{-4} Torr	-
LiVP (before next pulse)	-	1 Torr	-
Inner Wall Diameter	~ 0.05 m	2m	6 m ?
O.D. of pressure vessel, or plenum	~ 4.3 m	~ 4.1 m	-
Li out-flow temp.	-	750° C	-

TABLE 1: PARAMETERS OF PROPOSED REACTOR VESSELS

2.2. Repetition Rate

Another important economic constraint on a LCR is likely to be the capital cost of a suitable laser, or of the capacitor store required for pumping it (assuming, for the present, electrical excitation). It is therefore desirable to operate the laser at as high a repetition rate as possible. When the thermonuclear energy released (E_f) is large this leads to impracticably high mean powers, as has already been noted[1]. With a thermonuclear energy release of order 10^8J, however, repetition rates (f) of 1 to 100 Hz may be quite practicable.

Four principal technological problems are likely to determine the optimum repetition rate:

(a) The lifetime of the laser and/or capacitor bank;

(b) The time required to evacuate the reactor vessel between explosions;

(c) The cost of the reactor vessel as a function of mean power;

(d) The cooling of the laser and associated optical components.

Pulse rates of order 50 Hz imply the need for reliable operation of components for approximately 4×10^{10} pulses, assuming a reactor lifetime (τ) of 25 years, which is an exacting requirement. The life of present experimental capacitor banks is typically 10^5 to 10^6 pulses. The reliable lifetime of most laser or electrical components can be economically increased, however, by under-rating them (cf. power-factor capacitors). As a specific example, present experience with energy storage capacitors at Culham suggests that their life scales as:

$$\tau f = D(V_o/V)^6 \tag{1}$$

where V_o is the maximum rated voltage, V is the peak operating voltage - of order V_o, and D is a constant characteristic of the capacitor design. Assuming a capital cost B£/J at the full rated voltage, it follows that the cost when derated to give the same life at a higher operating frequency is:

$$\begin{aligned} B' &= B(\tfrac{1}{2}\, CV_o^2)/(\tfrac{1}{2}\, CV^2) \\ &= B(\tau/D)^{\frac{1}{3}} f^{\frac{1}{3}} \end{aligned} \tag{2}$$

where C is the capacity. This cost increases much more slowly with repetition rate than the thermonuclear output, which is directly proportional to f, so that there is a distinct economic advantage in operating at higher frequencies[11]. It is concluded that the upper limit to the frequency of operation is unlikely to be determined by component lifetime considerations, provided that strong power-laws of the type given in Eq.(1) are valid. The operating frequency is therefore more likely to be determined by the requirement to evacuate the reactor vessel between explosions, so that the focussed laser beam can be efficiently transmitted to the target. If this evacuation time is estimated for "wet wall" designs to be of order one second[4], a high operating frequency implies a requirement for several reaction chambers within one reactor vessel.

The possible use of several reaction chambers complicates the definition of economically permissible laser costs. Although a variety of wavelength-dependent physical effects make it difficult to define an optimum choice of laser at the present time, it is already clear that the energizing laser must operate at very high peak and mean powers; it therefore seems highly probable that the final stages of amplification will utilize a fast-flow fluid lasing medium, and that distortion and damage of window materials, beam splitters and mirrors must be prevented by conservative optical design. The most obvious way of reducing the mean power loading on optical components is to design final amplifier stages which are physically integrated with optical ports and focussing elements on each reaction chamber. (This might prove particularly convenient if 'window-less' aerodynamic designs were employed, for example). If this approach is adopted one can envisage a single, central, capacitor bank and master oscillator which are used to pump and drive a multiplicity of amplifiers. The choice of the number of reaction chambers, and hence the operating frequency of the central capacitor bank, will be determined primarily by the capital cost of the efficiently pumped reaction chambers with their surrounding laser assemblies. In the subsequent discussion it will therefore be convenient to define the laser cost as the total capital cost of all the laser assemblies per pump joule of the central energy store.

For practical reasons, it seems desirable to restrict the number of reactor vessels to a low number. The operating frequency is arbitrarily set at 50 Hz in the following economic assessments, and it is noted that the optimization of the reactor vessel design to permit rapid repetition rates within one chamber may prove to be a problem of major technological importance.

2.3. Pellet Production

Solid hydrogen pellets having velocities of $\sim$ 10 m/sec and diameters of $\sim$ 250μm have been successfully used in laser-plasma experiments[12], and liquid hydrogen droplets as small as 15μm in diameter have been generated by ultrasonic techniques[13]. It seems probable that current developments will soon permit the ejection and selection of droplets of diameter 50-300μm, with high spatial reproducibility and at very high repetition rates[14]. Micron size pellets of iron have been accelerated electrostatically to velocities of 10^3m/sec[15] to 10^4m/sec[16]; it is estimated that hydrogen pellets of 1000μm diameter will require megavolt accelerators to achieve similar speeds[17].

The rudiments of a possible pellet technology therefore exist. Is there any problem of principle in using such techniques to inject DT pellets at cryogenic temperatures, or other types of target, into the hot environment of a working reactor?

Millar[18] has estimated the probable temperature history and evaporation rates of liquid and solid hydrogen pellets under conditions where radiation heating can be neglected; his model and working assumptions are detailed in Appendix 1. If we eject a droplet of (normal) hydrogen of radius a (cms) at an initial temperature T_o into a well-pumped cold environment, it cools from 20.4°K (BP at atmospheric pressure) to 14°K (FP) in a time

$$\tau_L \sim 5 \times 10^{-4} \text{ a sec} \tag{3}$$

and freezes in a further time

$$\tau_F \sim 3 \times 10^{-3} \text{ a sec} \tag{4}$$

The characteristic time for rapid thermal conduction towards the centre of a spherical pellet of diffusivity κ is given by $\tau_{COND} \sim a^2/40\ \kappa$, thus

$$\tau_{COND} \sim a^2 \tag{5}$$

for normal hydrogen at temperatures of interest. It follows that $\tau_F \geqslant \tau_{COND}$ for droplets having radii $\leqslant 3 \times 10^{-3}$ cm; however larger droplets will tend to cool and freeze preferentially at the surface, forming a slightly hollow shell. (The shape of the droplet is determined by surface tension and viscous forces within the time τ_F.) Small pellets have an approximately uniform temperature (T) as they continue to cool by sublimation, with the computed[18] time-history shown in Fig.1. After freezing, the fractional mass loss

$$G = \frac{\text{Time-integrated mass loss}}{\text{Initial mass at } 13.94^oK} \tag{6}$$

varies as indicated in Fig.1. The model therefore predicts that the mass lost by sublimation is less than that lost during the time

(τ_L) taken to cool to the freezing point ($\leqslant$ 7%), or the 13% lost during the freezing time (τ_F), vide Appendix 1. The numerical results embodied in equations 3-5 and Figure 1 are, of course, dependent on pellet composition, i.e. on the ratio of ortho to para hydrogen, deuterium or tritium and on the percentage of HD, DT etc. However, our qualitative conclusions are not strongly affected by the relative proportions of the various isotopes, since the vapour pressures giving strong sublimation cooling occur within a range of a few degrees about 10°K, i.e. below the freezing point, for both normal H_2 and D_2 and for HD. (The mean temperature and mass of the largest pellets of reactor interest will initially decrease more slowly with time than indicated in Fig.1, because of temperature gradients within the pellet.)

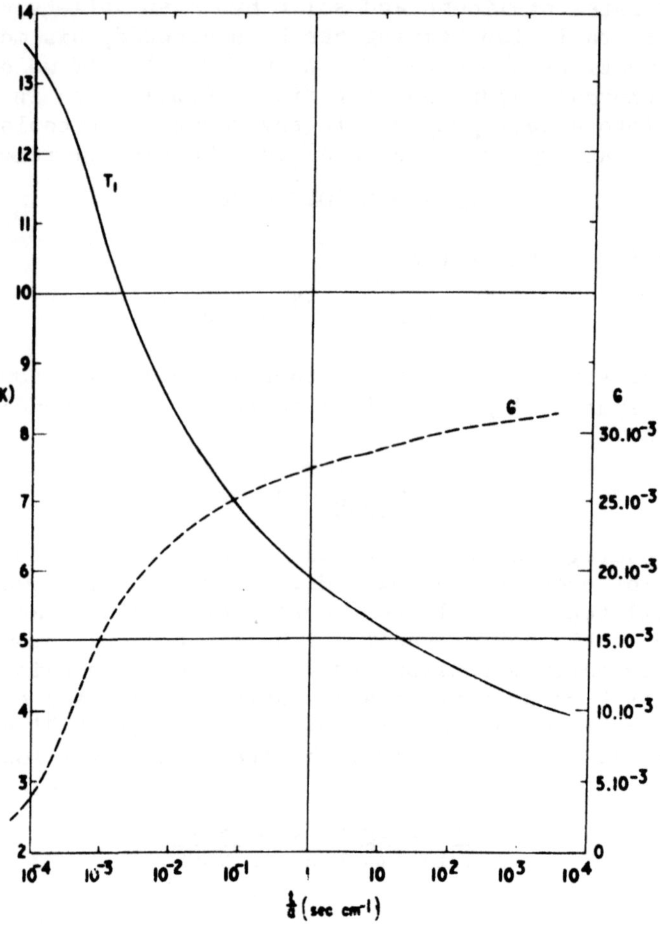

Fig.1. Temperature T and integrated fractional mass loss, G, as functions of t/a.

Metals which are thought to possess a promising combination of chemical, structural, and nuclear (radiation damage, after-heat etc.) characteristics for reactor vessel use include: stainless steels at bulk temperatures below 600°C, V and its alloys below 800°C, Nb and Mo and their alloys at ~ 1000°C.[19,20]. Thus, the inner surface of the LCR will have a temperature θ_R ranging from 400°C to an upper limit of approximately 1000°C, depending upon which of the reactor vessel concepts discussed in 2.1. is adopted. To order of magnitude, the thermal radiation intensity within the inner wall of the reactor vessel therefore approximates to black body radiation at ~ 1000°K. It would be difficult to shield the injected pellet from this ambient radiation in a working reactor. If the pellet is not shielded for the last metre or so of its path towards the focal zone of the laser, and if it absorbs a fraction β of the incident black-body radiation, it will approach a steady temperature during a transit time of some 10^{-1} - 10^{-2} seconds. The final, limiting, temperature (T_P) is determined by a balance between the sublimation cooling rate, heating by black body radiation, and heating by incident Li vapour, as indicated in equation 7:

$$\frac{1}{4}\frac{P_H}{kT_P}\bar{C}_H\, mL_s.4\pi a^2 = \beta.4\pi a^2.\sigma\theta_R^4 + \frac{1}{4}P_{Li}.\bar{C}_{Li}.4\pi a^2 \qquad (7)$$

where P_H is the equilibrium vapour pressure of hydrogen at the pellet temperature T_P

$\bar{C}_H$ is the mean velocity of the subliming molecule, of mass m.

L_s is the sublimation energy[21].

σ is Stephan's constant

P_{Li} is the vapour pressure of any Li within the vessel.

$\bar{C}_{Li}$ is the mean Li molecular velocity, which is assumed to exceed the pellet velocity, with a temperature of order θ_R.

T_P is seen to be independent of the radius. Neglecting, for the present, the effect of Li vapour, a maximum value for T_P is obtained by setting $\beta = 1$. This value is plotted in Fig.2, for normal hydrogen, and $P_{Li} = 0$; it is relatively insensitive to the wall temperature, or to β. At the limiting temperature the pellet evaporates at a rate given by:

$$- L_s\rho\, 4\pi a^2.\Delta a = 4\pi a^2(\beta\,\sigma\,\theta_R^4 + 0.25\, P_{Li}\,\bar{C}_{Li})\Delta t \qquad (8)$$

i.e.

$$\frac{da}{dt} = -\frac{\beta\,\sigma\,\theta_R^4 + 0.25\, P_{Li}\,\bar{C}_{Li}}{L_s\rho} \qquad (9)$$

where ρ is the density of the pellet. Thus, one half of the mass initially existing at the temperature T_P will sublime in a time

$$\tau_{\frac{1}{2}} \sim \frac{a}{5\beta}\left(\frac{L_s \rho}{\sigma \theta_R^{\ 4} + 0.25\ P_{Li}\ \bar{C}_{Li}}\right) \tag{10}$$

$\tau_{\frac{1}{2}}$ is proportional to the pellet radius and sensitive to θ_R. When $P_{Li} \leqslant 3 \times 10^{-2}$ Torr the term containing P_{Li} in equations 7-10 may be omitted, and a lower bound to $\tau_{\frac{1}{2}}$ may then be deduced for normal H_2 and D_2 by setting $\beta = 1$ in Figure 3. (β is, of course, a function of both pellet composition and θ_R and has a value less than unity.)

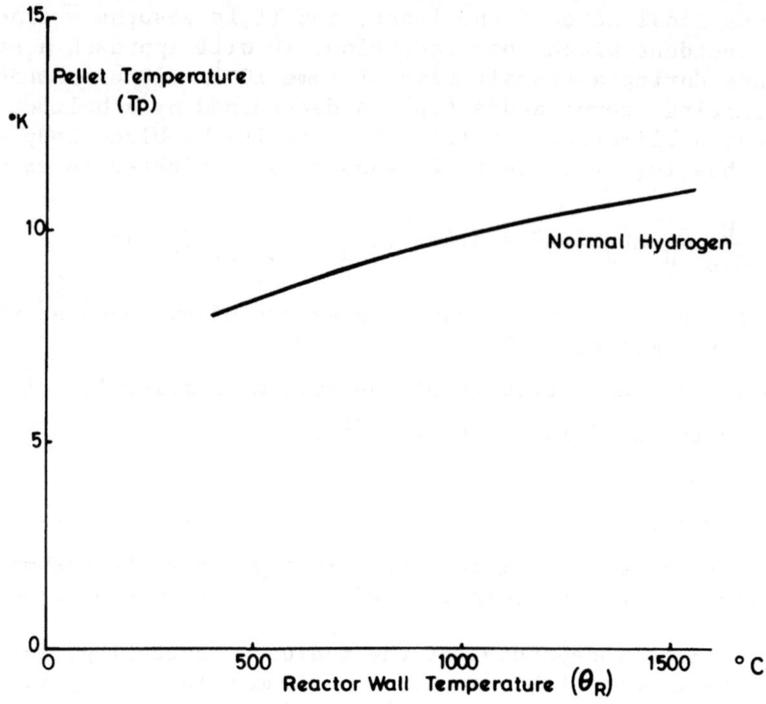

Fig.2. Limiting pellet temperature T_P, as a function of Wall Temperature $\theta_R (P_{Li} = 0)$.

Since the wetted-wall and Blascon systems can be designed to operate with $P_{Li} \leqslant 1$ Torr (cf Table 1), it follows from equations (3), (4), (7) & (10) that pellets of reactor interest ($a \sim 10^{-2} - 10^{-1}$ cm), having velocities between $10 - 10^3$m/sec, will arrive frozen at the centre of the reactor vessel. The mass loss due to sublimation (Fig.3) is then acceptable. (The complete evaporation of a pellet would produce a mean hydrogen density $\leqslant 5 \times 10^{13}$/cc in the reactor vessel. This density is too low to contribute heating terms of any significance in equations 7-10). Although a pellet tracking and laser interception technique has been successfully demonstrated[22], electrostatic acceleration and deflection techniques may produce sufficiently reproducible trajectories to ensure that pellets injected into the reactor vessel arrive within the focal zone of stationary laser beams.

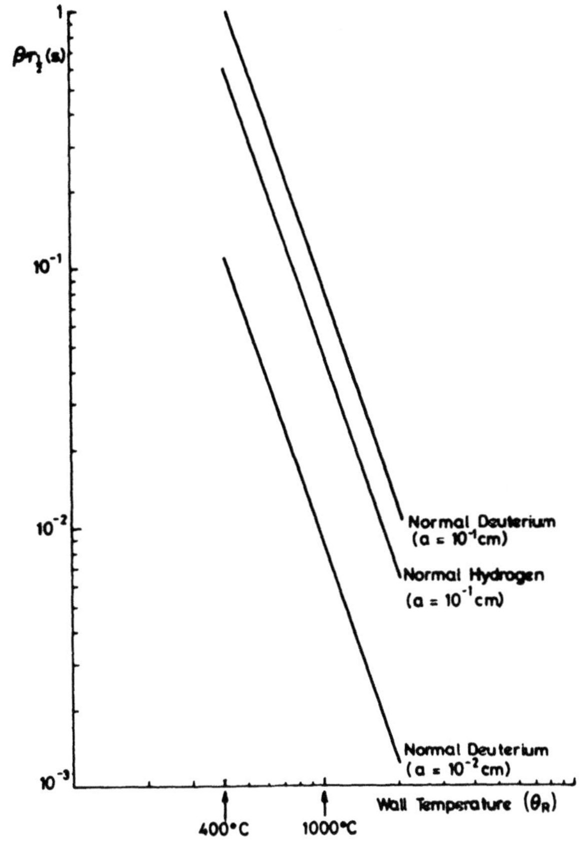

Fig.3. ($\beta\tau_{\frac{1}{2}}$) as a function of wall temperature θ_R, for pellet radii a = 10^{-1} and 10^{-2} cm and $P_{L,i}$ = 0. (The smallest possible sublimation time is obtained by setting β = 1)

3. ECONOMIC CONSTRAINTS

An energy flow diagram typical of a LCR operating with DT pellets, and electrical excitation of the laser, is shown in Fig.4. (The possibility of direct conversion of charged particle energy is neglected, although for some targets internal moderation[5] of the fusion neutrons may be sufficiently strong to justify additional capital investment required for direct conversion, with consequent higher plant efficiencies and lower thermal pollution.) Assuming a repetition rate of 50 Hz, but taking a thermonuclear gain A (ratio of fusion energy E_f to incident light energy E_r) intermediate between the low value ($A \sim 10$) discussed by Rosenbluth[7] and the higher values ($A \sim 80$) predicted in idealized codes[5,9,10,6] which postulate-inter alia - 100% absorption of laser energy, it has been shown[11] that costs could be comparable with alternative fission costs. The costing is summarized below.

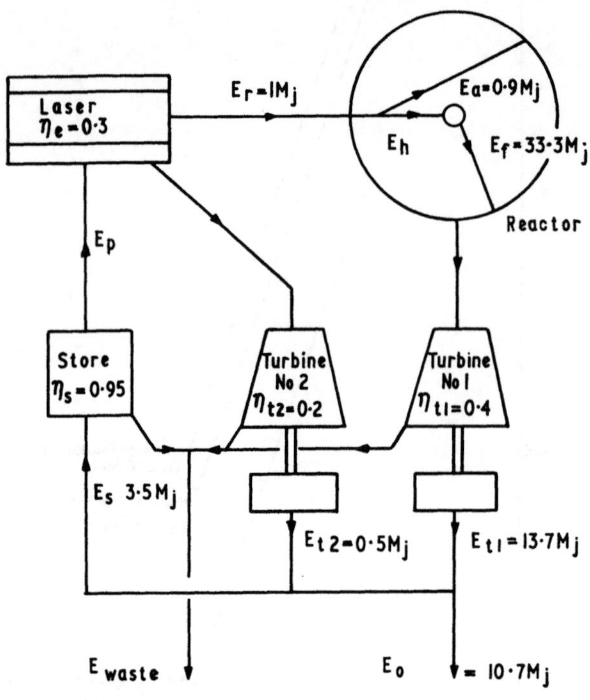

Fig.4. Typical energy flow diagram. At a repetition rate of 50 Hz the mean station output is 530 MW(e). If total laser costs amount to 8% of capital plant costs the laser could cost £3.3M, i.e. £3.3/laser joule or £1/pump joule.

3.1. Component Costs

The reference fast breeder (fission) reactor was a fully developed commercial 2 × 1250 MW(e) sodium-cooled reactor (CFR) operating with carbide fuel at a load factor of 65% and with a life of 25 years. It was assumed that the fuel inventory and replacement costs, and the refuelling penalty could be reduced by a factor four for the fusion system over the projected costs for the CFR. On this basis the allowable plant costs, excluding buildings and civil works, were £80/kW(e) at 1971 prices. It was argued that the reactor vessel design is simpler than that of the equivalent fission reactor, and the cost per kW(e) of the reaction chambers was therefore arbitrarily set at two thirds of that expected for the breeder vessel. Most of the reactor auxiliary costs will be similar in the two systems, with liquid sodium replaced by liquid lithium in the cooling system and the fission fuel handling equipment replaced by a tritium separation plant and fuel pellet manufacture and injection system.

The saving already outlined will be offset by the cost of a laser and its energy storage capacitors. The cost of capacitors will probably lie between the present cost of £0.1/joule for capacitors of limited life and £0.5/joule for long-life (power factor) capacitors. A figure of £1/joule was therefore tentatively assigned to cover capacitors having a life of the order of 4×10^{10} discharges, their charging equipment, and the laser. This may be compared with total laser costs for present one-off, limited-life, experimental neodymium or carbon dioxide facilities, which are of order £1 to £4 per pump joule.

3.2. Maximum Economic Level of Circulating Power

In calculating the total capital cost of the system the unit cost of each component is multiplied by the power which it handles. Since the most expensive components such as the reactor vessel, steam generators, and turbo-alternators all handle the total thermonuclear power, an important parameter is the ratio of the thermonuclear power to the useful output power. This internal power ratio R, is normalized by including the efficiency of the main turbo-alternator, so that:

$$\begin{aligned} R &= E_f \eta_{t1}/E_o \\ &= A.\left[A + 1 + \frac{(1-\eta_\ell)}{\eta_\ell} \cdot \frac{\eta_{t2}}{\eta_{t1}} - \frac{1}{\eta_s \eta_\ell \eta_{t1}}\right]^{-1} \\ &\approx E_{t1}/E_o \text{ when } A \gg 1 \end{aligned} \qquad (11)$$

Typical values of the reciprocal of this ratio are plotted in Fig.5 as a function of the thermonuclear gain for various values of the laser efficiency, assuming fixed values of the storage and thermal efficiencies. The minimum values of the thermonuclear gain required to satisfy the economic criterion are given by the intercepts with the lines of constant laser costs, assuming an operating frequency of 50 Hz.

The overall thermal efficiency of the system is simply

$$\eta = \eta_{tl}/R \qquad (12)$$

so that an excessively high value of the internal power ratio, R, not only leads to an economically unattractive system but to a low overall efficiency which could be a limiting factor in the siting and utilization of a LCR.

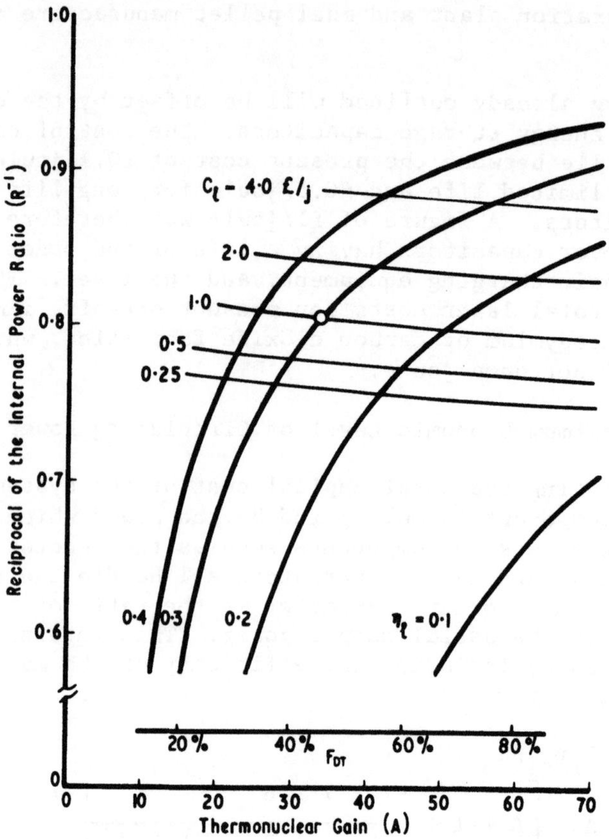

Fig.5. Variation of the reciprocal of the internal power ratio (R^{-1}) with thermonuclear gain (A), plotted for laser efficiencies (η_{ℓ}) of 0.1 to 0.4. The minimum value of the gain which would satisfy the economic criterion at an operating frequency of 50 Hz is given by the intercepts with the lines corresponding to laser costs (C_{ℓ}) of £0.25 to £4.0 per pump joule. The approximate core burn-up (F_{DT}) in a D-T reactor (assuming η_a = 0.1) is also indicated on the abscissa. The point indicated by a circle corresponds to the reactor parameters given in Fig.4 and Table II.

3.3. Burn-Up Required

In order to meet the economic criterion the thermonuclear gain in the reacting core must be high. This requirement can be expressed very approximately, but more vividly, in terms of the 'burn-up' required in the compressed core. Elementary computations of the ignition phase in the compressed core suggests that ignition with a 50% deuterium-tritium (D-T) fuel is obtained when the initial temperature approaches 5 keV, after which the ion temperature rises very rapidly to greater than 100 keV. Taking the mean energy from each reaction as 22.4 MeV, which includes the exothermic reactions of neutrons in lithium but neglects deuterium-deuterium (D-D) reactions, the energy gain is approximately:

$$A \approx 750\ F_{DT}\ \eta_a \tag{13}$$

where F_{DT} is the fraction of fuel ions consumed in the compressed core and η_a is the efficiency with which the core is heated by the laser, i.e.:

$$\eta_a = (E_f - E_a)/E_r$$

where E_a is the energy wasted by ablation etc. in the non-reacting outer corona. η_a is sensitive to a variety of physical factors, some of which are discussed in Section 4; with various simplifying assumptions η_a has been calculated to vary between 5% (6) and 10% (5,9). Taking $\eta_a = 0.1$, an approximate burn-up scale for D-T is indicated on Fig.4. Since the computed burn-up is sensitive to details of target size, shape and composition, the actual thermonuclear gain (A) should always be used to assess economic viability.

In a non-breeding reactor using only deuterium fuel the ignition temperature increases to about 50 keV. Even assuming direct conversion of both charged particle and neutron energies, the achievement of an economic energy gain appears much more difficult, since:

$$A \approx 75\ F_{DD}\ \eta_a \tag{14}$$

for D-D reactions in which the mean energy released for each primary reaction is 34.2 MeV. The ignition temperature could be significantly reduced by seeding the fuel with a small proportion of tritium[23], but at the expense of re-introducing the complexity of a tritium-breeding and separation cycle. It is also conceivable that the thermonuclear gain could be increased in pure D-D systems by triggering further reactions in unheated core material; although no computations have been published on this possibility the present discussion includes the economic consequences of enhancing the thermonuclear gain in this way.

The economic criterion of subsection 3.1 is illustrated in a slightly different manner in Fig.6 where the repetition rate (f) required to permit an economic plant cost of £80/kW(e) is plotted as a function of the thermonuclear gain for various values of the total laser costs, expressed in pounds sterling per pump joule for a laser system operating with $\eta_\ell = 0.3$, $\eta_{t1} = 0.4$, $\eta_{t2} = 0.2$, and $\eta_s = 0.95$. The tentative estimate of £1 per pump joule for laser costs and an assumed frequency of 50 Hz would permit economic operation with D-T fuel for thermonuclear gains greater than 33. A typical breakdown of costs for this system is given in Table II.

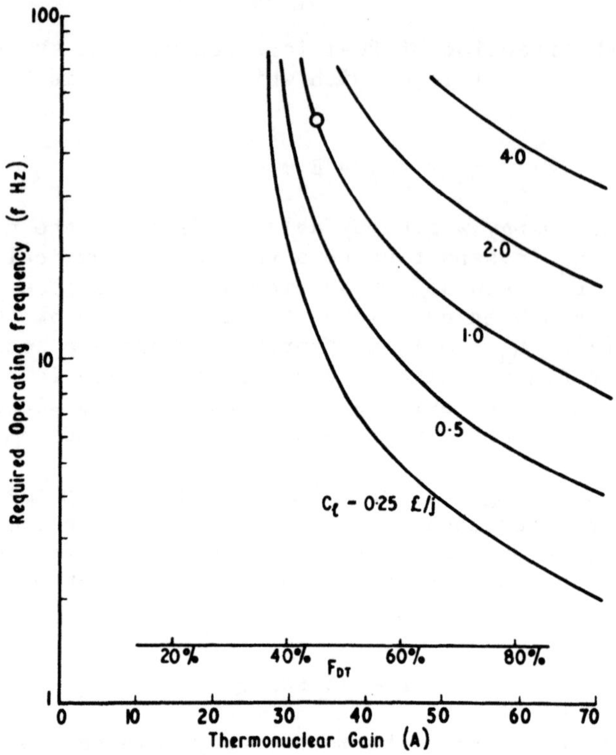

Fig.6. Required operating frequency (f) as a function of the thermonuclear gain (A) to meet the economic criterion with a laser efficiency (η_ℓ) of 0.3 and laser costs (C_ℓ) of £0.25 to £4.0 per pump joule. The approximate core burn-up (F_{DT}) in a D-T reactor (assuming $\eta_a = 0.1$) is also indicated on the abscissa. The point indicated by a circle corresponds to the reactor parameters given in Fig.4 and Table II.

Reactor vessel	11.8
Reactor auxiliaries	4.5
Liquid lithium	2.0
Control	4.5
Fuel processing	0.6
Steam generation	12.8
Turbines	27.8
Electrical plant	9.4
Capacitors and lasers	6.6
	80.0

Table II. Breakdown of Costs (£/kW(e)) for a D-T Reactor Operating with a Laser Efficiency $\eta_\ell = 0.3$ Assumptions: $\eta_{t1} = 0.4$; $\eta_{t2} = 0.2$; $\eta_s = 0.95$; $f = 50$ Hz; $A = 33.3$; $R = 1.25$

3.4. Economic Limitations

Two potential economic limitations can be identified from Figs 5 and 6:

(i) The necessity for a sufficiently high thermonuclear gain, equivalent to a relatively high burn-up. This requirement arises primarily because of the high capital cost of the steam generating plant and turbines, which makes it essential for the power handled by these components to be comparable with the output power; and

(ii) The need for lasers of adequate efficiency and reasonably low capital cost.

Provided possible technological problems associated with a reactor vessel operating at repetition rates of the order of 50 Hz can be overcome, an acceptable financial margin appears to exist for the (unspecified) laser capital costs; even if the thermonuclear gain does not reach values approaching 80 because of departures from spherical symmetry or for other physical reasons discussed in Section 4.

The criteria discussed above are approximately equivalent to the requirement:

$$A_{DT} > \frac{\{3 + 38(C_\ell/f)\}}{\eta_\ell \eta_{t1} \eta_s} \qquad (15)$$

where C_ℓ is the cost of the laser and its associated capacitors in £/pump joule, and f is the operating frequency of the reactor. This is more restrictive than the criterion for marginal operation given by Nuckolls et al[5].

4. SOME FACTORS AFFECTING THERMONUCLEAR GAIN

Several important physical effects which may adversely influence the thermonuclear gain A have been mentioned in the literature: in particular, the possible generation in low Z targets, at the longer laser wavelengths and at intensities of reactor interest, of supra-thermal electrons which might prematurely heat the compressed core(5,9,24). The value of $A \sim 30$ assumed in the reactor discussion of § 1-3 allows a factor ~ 2 for possible transmission and target reflection losses, because the computed values of $A \sim 80$(5,6,9,10) postulate 100% absorption of the incident laser light by inverse bremsstrahlung and plasma instabilities; the effects of non-classical absorption and stimulated scattering of incident light are therefore important and complex(25), since adequate account must be taken of strong plasma gradients, laser convergence and spectral band-width, the possibility of self-focussing and harmonic generation etc. These matters will be addressed in detail by other speakers at this meeting. However, two particular effects have not been extensively discussed, and may be of some practical interest:

(i) Stimulated Compton scattering and absorption(26-29) in the outer corona, at subcritical densities.

(ii) Relativistic over-dense propagation, in the presence of density gradients. (Neglecting attenuation, circularly polarized $10\mu m$ radiation should propagate(30) to densities of $\sim 7 \times 10^{20}/cc$, i.e. to densities a factor 70 above the critical density, assuming incident intensities as high as 10^{17} W/cm^2 and density scale lengths of ~ 1 mm; of course, strong attenuation due to relativistic(31), or other, instabilities would modify this conclusion.)

The gain could also be influenced by the state of polarization of the laser beam. For example, if the (coherent) beam is polarized the intensity distribution of the incident light does not have perfect spherical symmetry, for topological reasons(11); magnetic fields should then be generated by thermoelectric effects(32,33). Magnetic fields will additionally be generated through velocity anisotropy and radiation-pressure sources of assymmetry(32). Two-dimensional computations are then required to ensure that the fields are not strong enough to inhibit electron heat transport in the outer corona. (These problems can, of course, be avoided if unpolarised light, i.e. time averaging, is employed). The sensitivity of A to the heat transport mechanisms assumed in the computer codes has recently been investigated for a laser-driven compression at $\lambda = 10.6\mu m$(10); the effect of free-streaming limitations to electron thermal conduction(34-36) on a previously optimized compression was dramatic. (However, the beneficial effects of re-optimizing the

the laser pulse, and target, have not yet been fully investigated). It may be concluded that the choice of an optimum laser wavelength is a complicated one, which has not been discussed quantitatively in the open literature.

5. POTENTIAL ADVANTAGES OF THE LCR

Assuming these problems are solved, a laser compression reactor has several advantages over a magnetically confined fusion reactor:

(i) In common with all nuclear reactors, the structure will become radioactive and the fuel handing equipment will contain active materials, so that nuclear shielding and secondary containment vessels will restrict access to the reactor. The escape from the toroidal geometries preferred in magnetically confined fusion systems therefore provides a significant topological advantage for the spherical LCR geometry, since maintenance and replacement of sections of the structure are possible using simple modifications of present remote handling methods.

(ii) The abandonment of magnetic confinement not only removes the item of highest capital cost but also greatly simplifies the reactor design. The support structure which is required to withstand the large forces between sections of the magnet tends to dominate the reactor, and the need to cool the magnet and this massive structure to liquid helium temperatures will severely limit access for the blanket coolant pipes and other necessary auxiliaries.

(iii) A relatively modest vacuum is required, compared with other approaches to fusion.

(iv) Means of ignition, refuelling, and fusion-product removal which still have to be developed for most magnetically confined fusion reactor concepts, are automatically provided in this approach.

(v) The mean power output of the reactor can be readily controlled.

(vi) The various components can be developed and proved independently of the reactor itself. Thus, the lasers and fuel injection system can be developed separately, and scientific feasibility can be demonstrated on a single-shot basis in the equivalent of a very low power reactor.

Against these potential advantages must be balanced certain technological problems which are specific to laser fusion. These currently include the development of:

(a) Highly efficient, pulse-shaped lasers, working at appropriate wavelengths and operating at approximately 50 MW mean power;

(b) High repetition rate (50 Hz), long life (4×10^{10} pulse) pumping systems having costs of order £1 to £10 per laser joule;

(c) High repetition rate (5 to 50 Hz) reaction chambers compatible with high peak and mean power loadings, and capable of thermal cycling for more than 10^{10} pulses;

(d) Windows compatible with the nuclear environment, and capable of working with minimum distortion at high laser intensities, and at mean powers in the megawatt region;

(e) Means of positioning spherical pellets of controlled size, and of ensuring adequate symmetry of the incident laser flux.

At least four variants on the LCR discussed above can be conceived:

(I) LCR with nuclear excitation of the laser[37,38].

(II) LCR with nuclear preionization of the laser.

(III) LCR with chemical excitation of the laser.

(IV) LCR with fission, or fission-fusion, fuel[39,40,44]

Variants (I) and (IV) could significantly relax laser efficiency requirements; (I) and (III) would remove capacitor, and variant (II) preionizer, costs as economic constraints. If suitable chemicals and techniques for short-pulse use were discovered (III) might be operated at low laser efficiencies as a "fusion-augmented fossil" station. These four possibilities therefore allow some flexibility to the quantitative conclusions contained in the abstract.

ACKNOWLEDGEMENTS

This paper enlarges on previous joint work with R. Hancox[11]. The author also wishes to acknowledge fruitful discussions with W. Millar and useful comments by Dr R.J. Bickerton and Mr R. Carruthers.

REFERENCES

1. Report of a consultants meeting on fusion power and the environment, 15 - 17 Dec. 1971, IAEA, Vienna.
2. FRAAS, A.P., USAEC Rep. ORNL-TM-3231 (1971).
3. LUBIN, M.J., FRAAS, A.P., Sci. Amer. 224, 21 (1971).
4. BOOTH, L.A. et al., Los Alamos Report, LA-4858, 1 (Feb. 1972).
5. NUCKOLLS, J. et al., Nature (London) 239, 139 (1972) (and 7th Int. Quantum Electronics Conference, Montreal, May 1972).

6. BRUECKNER, K.A., KMS preprint KMSF-NP5, Apr. 1972, part of the compilation in Doublekay-7521, 27 Jul. 1972, Basel, Switzerland.
7. ROSENBLUTH, M., 5th European Conf. Controlled Fusion and Plasma Physics, Grenoble, Aug. 1972.
8. EDEN, M.J., SAUNDERS, P.A.H., Nucl. Fusion 11, 37 (1971).
9. CLARKE, J.S., FISHER, H.N. and MASON, R.J., Phys. Rev. Lett. 30, 89 (1973).
10. ASHBY, D.E.T.F. and CHRISTIANSEN, J.P., Sixth European Conf. on Controlled Fusion and Plasma Physics, Moscow (1973).
11. HANCOX, R. and SPALDING, I.J., Nucl Fusion 13, 385 (1973).
12. FRANCIS, G. et al., Phys. Lett. 25A, 486 (1967).
13. POWELL, B.A. and KING, C.D., J. Sci. Inst. (E) Series 2, Vol.1, 670 (1968).
14. WITKOWSKI, S., in "Proceedings of the Japan - U.S. Seminar on Laser Interaction with Matter", Sept. 1972, pp. 301. (Published by Japan Society for Promotion of Science, Tokyo).
15. SHELTON, H. et al., J. App. Phys. 31, 1243 (1960).
16. SMITH, D. and ADAMS, N.G., J. Phys. D: App. Physics 6, 700 (1973).
17. HENDRICKS, C.D., Charged Particle Research Laboratory Internal Report No. CPRL 71-4, Univ. of Illinois, Urbana, Illinois.
18. MILLAR, W., unpublished, Culham Memorandum CLM-M 74 (1968).
19. KULCINSKI, G.L., International Conference, World Energy Problems: Nuclear Solutions, pp. 240 (1972).
20. BLOW, S., J. Brit. Nucl. Energy Soc. 11, 371 (1972).
21. SCOTT, R.B., "Cryogenic Engineering", Van Nostrand, New York (1959); and "Technology and Uses of Liquid Hydrogen", Ed. Scott, Denton and Nicholls, Pergamon, London (1964).
22. WARD, B.A., Appl. Optics 11, 331 (1972).
23. NUCKOLLS, J. et al., Bull Am. Phys. Soc. Series II, Vol. 17, 6D2 pp. 1034 (1972) and UCRL 74116.
24. BOYER, K., in "Proceedings Japan - U.S. Seminar on Laser Interaction with Matter", Sept. 1972, Tokyo, pp. 69.
25. See, for example:
ROSENBLUTH, M.N. and SAGDEEV, R.Z., "Comments on Plasma Physics and Controlled Fusion" Vol. 1, pp. 129.
DAWSON, J.M. in Symposium on "Plasma Heating and Injection", pp. 155, Varenna, Italy. (International School of Plasma Physics, Sept. 1972).
MAKHANOV, V.G. and TSYTOVICH, V.N., Ibid.
NISHIKAWA, K., in "Proceedings Japan - U.S. Seminar on Laser Interaction with Matter", Sept. 1972, Tokyo.
26. ZE'DOVICH, Ya. B. et al., Zh. Eksp. Teor. Fiz. 62, 1392 (1972).
27. KRASYUK, I.K. et al., Zh. Eksp. Teor. Fiz. Red. 17, 130 (1973).
28. VINOGRADOV, A.V. et al., Zh. Eksp. Teor. Fiz. Red. 17, 271 (1973).
29. OFFENBERGER, A.A. and BURNETT, N.H., Phy. Lett. 42A, 527 (1973).
30. MAX, C. and PERKINS, F., Phys. Rev. Lett. 27, 1342 (1971).
31. MAX, C. and PERKINS, F., Phys. Rev. Lett. 29, 1731 (1972).

32. STAMPER, J.A., Naval Research Laboratory Report No. 7411 (1972).
33. TIDMAN, D.A. and STAMPER, J.A., App. Phys. Lett. 22, 498 (1973).
34. SALZMAN, H., Phys. Rev. Lett. 41A, 363 (1972).
35. BICKERTON, R.J., Nucl. Fusion 13, 457 (1973).
36. MORSE, R.L. and NIELSEN, C.W., Los Alamos Report LA-4896-MS (1972).
37. MILEY, H., "Laser Interaction and Related Phenomena", Volume 2, pp. 43 (1972).
38. THOM, K. and SCHNEIDER, R.T., AIAA Journal, 10, 400 (1972).
39. WINTERBERG, F., Nature 241, 449 (1973) and Z. Naturf. A28, 900 (1973).
40. ASKARJAN, G.A. et al., Zh. Eksp. Teor. Fiz. Red. 17, No. 10, (20 May 1973).
41. HONIG, R.E. and HOOK, H.O., RCA Rev. 21, 360 (1960).
42. WADD-TR-60-56 (Pt 1) Wright Air Development Division (Distributed as PS 171618).
43. KIDDER, R.E. "Proceedings Japan-US Seminar on Laser Interaction with Matter", Sept. 1972, Tokyo, pp. 331.
44. HOROSHKO, R.N. & HURWITZ, H., unpublished, University of Rochester Laboratory for Laser Energetics Report No. 19, (1973).

APPENDIX 1

Evaporation Rates for Liquid and Solid Hydrogen Spherical Pellets

If there is no radiation heating of the pellet, and no inward molecular flux, the number of molecules $/cm^2/sec$ evaporating from the pellet (φ), is given by

$$\varphi \leqslant \tfrac{1}{4}\, n\bar{c} \sim \tfrac{1}{4}\left(\frac{P_E(T)}{kT}\right)\left(\frac{8kT}{\pi m}\right)^{\frac{1}{2}} \qquad (A1)$$

where $P_E(T)$, the equilibrium vapour pressure at the pellet temperature T, is unaffected by the pellet curvature for radii $a \geqslant 10^{-6}$cm[18]. Assuming no supercooling and a uniform temperature within the pellet, the fractional mass lost by evaporation in a time Δt is

$$\frac{\Delta M}{M} = \frac{4\pi a^2 . \varphi . m . \Delta t}{(4\pi/3) a^3 \rho} \qquad (A2)$$

where ρ is the pellet density and m the molecular mass. This mass loss is related to the temperature drop (ΔT) in the pellet by the relation

$$\frac{\Delta M}{M} = \frac{L_F . \delta(T - T_{mp}) + c_s(T) . \Delta T}{L_v} \qquad (A3)$$

where L_v, L_F are the latent heats of vapourization and fusion, $c_s(T)$ is the specific heat and $\delta(T - T_{mp})$ is a delta function about the melting point T_{mp}.

It follows from (A3) that the mass lost during freezing is

$L_F/L_v \sim 12.9 - 15.6\%$ for normal $(n)H_2$, HD, or $n(D_2)$.

At other temperatures $\delta(T - T_{mp}) = 0$ and equations (A1 - A3) yield

$$\frac{1}{a}.\frac{dt}{dT} = \left(\frac{4\rho k}{3mL_v}\right) . \left(\frac{C_s(T).T}{\bar{c}.P_E(T)}\right) \qquad (A4)$$

Equation (A4) was numerically integrated[18], using tabulated values of $P_E(T)$[41,42] and $C_s(T)$[42], to deduce the variation of T with (t/a) reproduced in Fig. 1. Similarly, equation (A3) was numerically integrated to deduce the fractional mass lost after freezing:

$$G = \frac{1}{L_V} . \int_T^{T_{mp}} C_s(T)\,dT \qquad (A5)$$

The resulting variation of G with (t/a) is reproduced in Fig. 1; similar methods were used to estimate the time the drop remains liquid (τ_L) and the time taken to freeze (τ_F), cf. equations 3 and 4.

SUMMARY OF DISCUSSIONS AND COMMENTS

(V. Experiments for Nuclear Fusion)

One conclusion after several discussions was the essential breakthrough for the laser induced fusion can be expected when the compression and ignition of the pellets is verified. The decrease of reflectivity in laser radiation of interesting intensity as measured by several authors proved the nonexistence of invincible difficulties for this fusion concept. However, this had been concluded from instabilities.

A further aspect of laser produced fusion plasma was the revival of discussion on breeding fissionable material (P^{239} from U^{238} and U^{233} from Th^{232}) by the fast neutrons of the laser induced fusion reactions. The discussion on the paper by M. Lubin indicated that there are indeed new aspects of the problem to be examined before an economic way to such breeding can be realized.

The Institutes which reported definitely on fusion neutrons from laser produced plasmas are:

Lebedev-Institute (Group of Basov-Kryukov) (USSR)
Limeil Laboratories CEA (France)
Laser Energetics Laboratory, University of Rochester (USA)
Sandia Laboratories (USA)
Osaka University (Japan)
Max-Planck-Institute, Garching (Germany)
Lebedev-Institute (Group of Basov-Sklizkov) (USSR)
Lawrence Livermore Labs (USA)
Los Alamos Scientific Labs (USA)
Naval Research Labs (USA)
Lebedev Institute (Group of Prokhorov-Pashinin) (USSR)
KMS-Fusion Industries (USA)
Academy of Science, Peking (China)
Polish Academy of Science (Poland)

All the above with the exception of the Chinese group[1] presented essential results at this Laser Workshop (see these proceedings and the two previous volumes). The Soreq Nuclear Institute (Israel), the Australian State University and the Battelle Institute, Columbus, Ohio (USA) are very advanced in producing neutrons.

A. Feldman asked in the general discussion why quarter wavelength plates were not used for preventing the laser system from backscattered radiation. This method also proposed by C. G. Young[2] would have to overcome some difficulties of material damage which are still not resolved sufficiently for this purpose.

Besides the various methods for diagnostics discussed in the papers of this section, Bobin underlined the use of short time holography with green light or with shorter wavelengths.

Following the review of Spalding on laser-fusion reactors, a new design for the reactor "SATURN" developed by the KFA-Jülich group[3] was noted.

References

1. Laser Focus 9, August 1973, p. 4.

2. C. G. Young, Laser Focus 5, (21), 37 (1969).

3. F. H. Bohn, H. Conrads, J. Darvas and S. Förster, Some Aspects of Inertially Confined Fusion Reactors, Preprint KFA Jülich/ Germany, 1973.

VI

LASER APPLICATION TO HIGH ENERGY PHYSICS

PAIR PRODUCTION BY RELATIVISTIC ELECTRONS FROM AN INTENSE LASER FOCUS*†

J.W. Shearer, J. Garrison, J. Wong, and J.E. Swain

University of California, Lawrence Livermore Laboratory

Livermore, California 94550

ABSTRACT

A preliminary discussion is given of electron-positron pair production by means of electrons accelerated to relativistic velocities at the focus of a laser beam. First, the pair cross-section was numerically evaluated near its energy threshold. Then, two methods of relativistic electron production by focused laser light were considered: the coherent oscillation of electrons, and the acceleration of a few high velocity electrons by plasma waves excited by laser-driven instabilities. It was found that the first method would produce pairs for neodymium laser light which is close to the limits of achievable intensities (10^{19}-10^{20} W/cm^2). The second method, however, may be capable of producing pairs at lower intensities.

I. INTRODUCTION

In recent years the brightness of pulsed laser light sources has increased from 10^{12} W/cm^2/sr up to the range of 10^{17}-10^{20} W/cm^2/sr at the neodymium laser wavelength of 1.06 μ.[1-2] With a well-designed short focus lens, such laser pulses can be focused to corresponding intensities of 10^{17}-10^{20} W/cm^2. At these intensities electron-positron pair production by the strong

---

*presented at the Third Workshop on "Laser Interaction and Related Plasma Phenomena" held at Rensselaer Polytechnic Institute, Troy, New York, August 13-17, 1973.

electromagnetic field at the focus might be possible either by direct vacuum pair production or indirectly from relativistic electrons accelerated by these strong fields.

Two calculations of the probability of vacuum pair production by multiphoton absorption have been published[3,4]; both of them indicate that the production of a detectable number of pairs from this process would require many more orders of magnitude than contemporary lasers provide. A simple, order of magnitude, estimate can show this. For an appreciable pair production probability, the pair energy $2m_oc^2$ should be of the order of the electric field potential energy at one Compton wavelength $\lambda\!\!\!^{-}$. That is

$$eE_\ell \lambda\!\!\!^{-} \simeq 2m_oc^2 \ , \qquad (1)$$

where E_ℓ is the average electric field of the focused laser light. Therefore, the intensity I_ℓ of the laser light in vacuum can be written

$$I_\ell = c\varepsilon_o E^2 \simeq \frac{m_oc^3}{\pi r_o \lambda^2} \ , \qquad (2)$$

where r_o is the classical radius of the electron

$$r_o = e^2/4\pi\varepsilon_o m_o c^2 \ . \qquad (3)$$

Numerically, this is an intensity of 2×10^{30} W/cm^2, which is many orders of magnitude greater than available intensities, but is in rough agreement with values previously cited.[3,4]

However, it is well-known that focused laser pulses create hot plasmas in matter. It has been pointed out that there exists the "wholly real possibility" of observing pairs produced in such laser-plasma experiments by means of the excitation of high energy electrons.[5] When the electron kinetic energy E_v exceeds the pair production threshold $2m_oc^2$, then the fast electron can produce an electron-positron pair by scattering in the Coulomb potential of a nucleus as first calculated by Bhabha.[6] This is often called the "trident" process (see Figure 1). In this paper we shall discuss some mechanisms by which energetic electrons can be created in the laser-plasma focus, and we shall attempt to evaluate the number of electron-positron pairs which can be produced by the subsequent trident process.

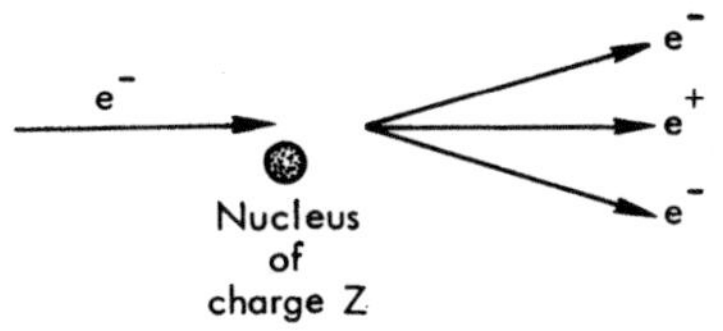

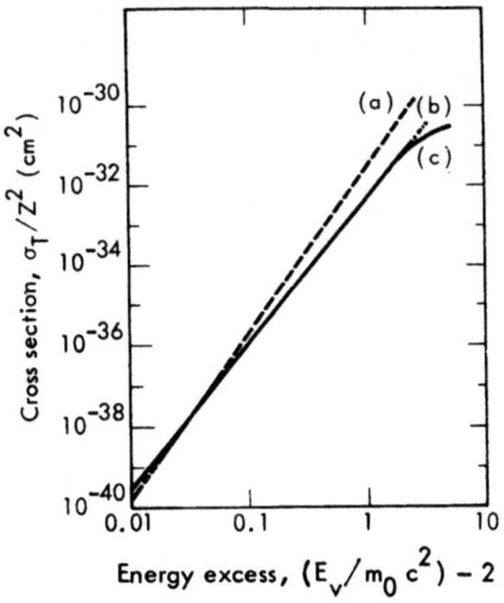

Figure 1. Schematic of "trident" process of pair creation.
Figure 2. Total cross section σ_T of trident process plotted versus the dimensionless energy excess above threshold, where E_V is the kinetic energy of the incident electron. Curve (a) is Bhabha's analytical calculation[6] for this range (see Eq. 4). Curve (b) is our approximate fit to the computer calculation. Curve (c) is our computer calculation based on Brodsky and Ting's method.[7]

II. TRIDENT CROSS-SECTION AT ELECTRON ENERGIES NEAR THE PAIR PRODUCTION THRESHOLD

We have calculated the cross-section σ_T for the trident pair production process by two different methods; the results are plotted in Fig. 2. In the first method we integrated Eq. (30) of Bhabha's paper,[6] and obtained

$$\sigma_T = \frac{(\alpha r_o Z)^2}{128}\left(\ell n \frac{1}{\gamma_B^2} - \frac{161}{60} + c_1 + c_2 + c_3\right)\left(\frac{E_v}{m_o c^2} - 2\right)^4 , \qquad (4)$$

where α is the fine structure constant, r_o is the classical electron radius, Z is the nuclear charge, and γ_B is defined by

$$\frac{1}{\gamma_B} \equiv 1 + \frac{E_v}{m_o c^2} . \qquad (5)$$

The symbols c_1, c_2, and c_3 in Eq. (4) are lengthy algebraic functions of γ_B which are given in Eq. (28) of Bhabha's paper. Numerical calculations of Eq. (4) are plotted in Fig. 2.

Bhabha's approach has the merit of yielding a convenient analytical expression for the cross-section; however he makes two approximations having attendant errors that are difficult to estimate. In the first place the initial electron is described by a classical straight line trajectory, and secondly the interference effects between the two electrons in the final state are neglected. Both approximations are avoided by evaluating the lowest order Feynman diagrams which are applicable to the trident process. Our numerical calculation uses a program for evaluating these diagrams which was written by S.J. Brodsky and C.C. Ting.[7] The program numerically evaluates the necessary products of the Dirac matrices and takes the trace to get the differential cross-section. This procedure is more accurate than the usual one which involves algebraic reduction of the matrix products. The differential cross-section is then numerically integrated to obtain the total cross-section as a function of incident electron energy. The results of these calculations are also plotted in Fig. 2.

It is seen that the second, more exact, calculation gives lower cross section values over most of the energy range of interest - near the trident production threshold. The reduction in cross-section is presumably due to inclusion of the interference between direct and exchange scattering, an effect which was omitted in Bhabha's calculation. For the subsequent portions of this paper we use the equation

$$\sigma_T \cong 9.6 \times 10^{-4} \, (\alpha r_o Z)^2 \left(\frac{E_v}{m_o c^2} - 2 \right)^{3.6} , \qquad (6)$$

which approximates the more exact cross-section curve. In Fig. 2 we see that the fit of the approximation is good up to kinetic energies of $\simeq 4 m_o c^2$, which is a sufficient range for our problem. The possibility of pair production by bremsstrahlung from the electrons, mentioned in a footnote of ref. 7, is not as likely as trident production in this energy range because it is a two-step process.

III. HIGH INTENSITY CIRCULARLY POLARIZED LIGHT

We consider first the coherent motion of plasma electrons in the electromagnetic wave of the laser light. Because the pair production threshold is twice the rest energy, this motion must be treated relativistically. This treatment is more difficult than the well-known nonrelativistic treatments.[8] We have not treated the linearly polarized wave, because it has been shown to be coupled to a longitudinal plasma wave.[9] However, the circularly polarized wave is a simple transverse wave at all intensities, and a convenient solution for this case has been obtained by Steiger and Woods.[10] We shall use these authors' results for the case which includes the relativistic mass change and the inverse Faraday effect, but we neglect energy losses caused by radiation by the electron.

Steiger and Woods[10] found that at high, circularly polarized, laser beam intensities where the coherent electron orbit velocity is relativistic, the electron kinetic energy E_v is a strong function of laser intensity but is almost independent of plasma density. We replotted this result (Fig. 3) and found that a useful approximation is given by the following simple power law

$$\frac{E_v}{m_o c^2} \cong \left(\frac{\lambda^2 I}{\lambda_o^2 I_o} \right)^n = \left(\frac{\lambda^2 I}{9\text{x}10^{18}} \right)^{0.657} , \qquad (7)$$

where the laser wavelength λ is measured in μ and the intensity I is in W/cm^2.

An equation that describes the rate dN_p/dt of creation of pairs by means of the trident process in a volume whose characteristic dimension is approximately ℓ times the wavelength λ is

$$\frac{dN_p}{dt} \text{ (pairs/sec)} = (\ell\lambda)^3 \; N_i N_e \sigma_T v_e , \qquad (8)$$

where N_i is the ion (nucleus) density and v_e is the velocity of the electron. Assuming that the thermal velocity can be neglected in comparison with the relativistic coherent velocity of gyration in the intense electromagnetic field, we have

$$v_e \equiv \beta c = \frac{c}{\gamma} (\gamma^2 - 1)^{1/2} , \qquad (9)$$

where the coefficient γ is the normalized total electron energy

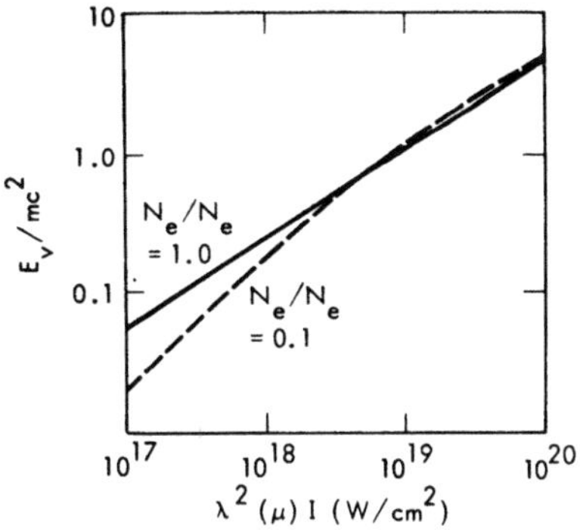

Figure 3. Electron kinetic energy E_v (in units of $m_o c^2$) plotted as a function of laser beam intensity (normalized to the wavelength λ) as obtained from the circularly polarized transverse wave solution of Steiger and Woods[10] at two different plasma density ratios.

$$\gamma = 1 + \frac{E_v}{m_o c^2} = 1 + \left(\frac{\lambda^2 I}{9\times 10^{18}}\right)^{0.657} , \qquad (10)$$

where we have substituted from Eq. (7).

Consider a plasma containing ions of charge $Z_i = N_e/N_i$, where Z_i is not necessarily the nuclear charge Z. Substitute Eqs. (6) and (9) into Eq. (8) to obtain the result

$$\frac{dN_p}{dt} = \frac{9.6\pi^2}{10^4}\left(\frac{c}{\lambda}\right)\ell^3\alpha^2\,\frac{Z^2}{Z_i}\left(\frac{N_e}{N_c}\right)^2 (\gamma-3)^{3.6}\,\frac{(\gamma^2-1)^{1/2}}{\gamma} , \qquad (11)$$

where the density N_c

$$N_c \equiv \varepsilon_o m\left(\frac{\omega_L}{e}\right)^2 = \frac{\pi}{r_o\lambda^2} \qquad (12)$$

is the nonrelativistic cutoff density parameter at which the plasma frequency ω_p equals the laser frequency ω_L. In convenient units, Eq. (11) becomes

$$\frac{dN_p}{dt}\,(\text{pairs/nsec}) = \frac{0.15}{\lambda(\mu)}\,\frac{Z^2}{Z_i}\,\ell^3\left(\frac{N_e}{N_c}\right)^2 (\gamma-3)^{3.6}\,\frac{(\gamma^2-1)^{1/2}}{\gamma} . \qquad (13)$$

Equations (7) and (13) are presented in graphical form in Figs. 4 and 5.

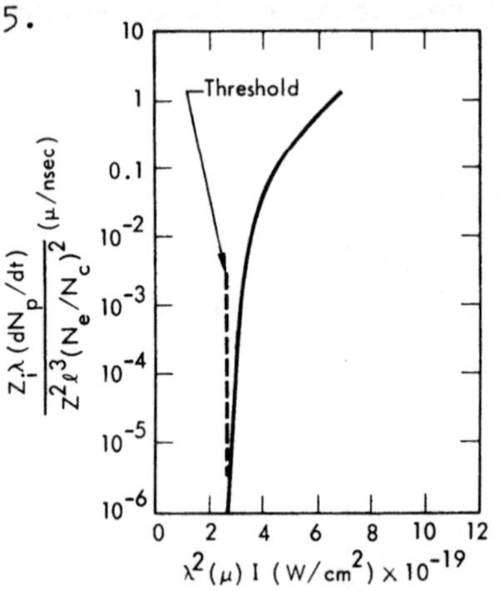

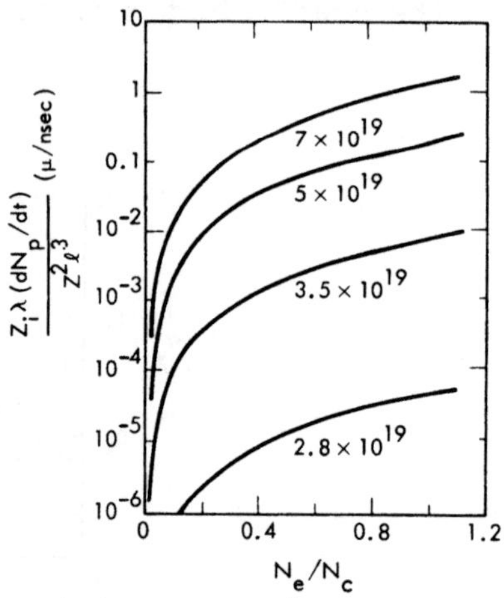

Figure 4. Density-normalized pair production rate plotted versus laser intensity, as computed from Eqs. (7) and (13).

Figure 5. Pair production rate plotted versus density ratio for several values of the laser intensity. The figures on the curves are $\lambda^2 I$ [$(\mu)^2$ (W/cm^2)]. The curves continue beyond cutoff ($(N_e/N_c)>1$) because of the possibility of relativistic beam penetration.[9,11]

The pair production rate threshold ($\gamma=3$) corresponds to a threshold laser beam intensity I_T of

$$I_T\left(\frac{W}{cm^2}\right)=\frac{2.6\times10^{19}}{(\lambda(\mu))^2} \quad . \tag{14}$$

The pair production rate then rises steeply to interesting values at somewhat higher intensities. These intensities (greater than 10^{19} W/cm^2 for neodymium lasers) are at the upper limit of current laser practice.[1-2]

Of course, this is an idealized single particle calculation which neglects unstable collective effects in the plasma. One should regard it mainly as an order of magnitude estimate that trident process pair production is hard to produce by means of the coherent electron motion alone.

When the electron motion is relativistic the laser beam can penetrate an overdense plasma,[9,10] because the plasma current is limited to the value $N_e ec$, instead of increasing with increasing beam intensity.[11] In such cases the pair production rate will be enhanced, provided the intensity is above the threshold.

IV. DISCUSSION OF PLASMA INSTABILITIES AT LOWER LIGHT INTENSITIES

Anomalous heating of plasmas by intense electromagnetic waves is now recognized as a significant effect both at radio wave frequencies and at laser frequencies.[12,13] The incident wave energy is coupled into plasma waves by plasma instabilities.[14] One of these is the ion-acoustic decay instability,[15] and the other is the oscillating two-stream instability.[16]

According to numerical computations of these processes,[13,17] the growth of the amplitude of the plasma waves is sufficiently rapid for the instability mechanism to saturate, causing the plasma to become turbulent. Many such calculations show a high velocity, high energy group of electrons, which is called a "supra-thermal electron tail" on the Maxwellian electron velocity distribution. One physical picture of this process is that some of the electrons are trapped by high phase velocity, high amplitude plasma waves.[13,17]

Although these numerical calculations were non-relativistic, the same qualitative arguments would be expected to hold for plasma waves whose phase velocity v_p is nearly the velocity of light ($v_p \to c$). Thus we must examine the plausibility of relativistic electron production by these plasma instabilities. If relativistic electrons ($E_V > 2m_oc^2$) are produced, the trident mechanism for pair production is possible.

The dispersion relation for longitudinal plasma waves can be written in the form[18]

$$v_p^2 = \frac{\omega_p^2}{K^2} + 3\,v_t^2 , \tag{15}$$

where K is the wave number, and v_t is the thermal velocity of the electrons. Let us examine the wave whose phase velocity v_p is equal to the velocity of light, because waves which trap relativistic particles will have velocities which closely approach this limiting solution. Then for $v_p = c$, we find

$$(K_c\lambda_D)^2 = \frac{v_t^2}{c^2 - 3v_t^2} , \tag{16}$$

where K_c is the corresponding wave number, and λ_D is the Debye length of one plasma ($\lambda_D\omega_p = v_t$). From Eq. (16) we find that waves which are potentially capable of trapping electrons in the relativistic energy range tend to have small values of K_c, corresponding to longer wavelength waves. As $v_t \to 0$, $K_c \to \omega_p/c$, as can be seen by eliminating λ_D from Eq. (16). Thus in the low temperature limit, the wavelength of the longitudinal wave of velocity c is equal to the vacuum wavelength of an electromagnetic wave whose frequency is equal to the plasma frequency.

Now we ask whether plasma waves of wave number K_c are likely to be excited in the plasma by laser-driven instabilities. Consider first the threshold conditions for the ion-acoustic decay instability. The most unstable wave number K_p can be written[19]

$$K_p^2 = \frac{\omega_L^2 - \omega_p^2}{3v_t^2} . \tag{17}$$

If we set $K_p = K_c$, and make the approximation $v_t \ll c$, we find from Eqs. (16) and (17) that

$$1 - \frac{\omega_p^2}{\omega_L^2} = 3\,\frac{\omega_p^2}{\omega_L^2}\,\frac{v_t^2}{c^2} . \tag{18}$$

Here Eq. (12) can be used to find the plasma density

$$\frac{\Delta N}{N_c} \simeq 3\,\frac{v_t^2}{c^2} , \tag{19}$$

where ΔN is the difference between the cutoff density N_c and the electron density N_e, and where $\Delta N << N_c$.

This result shows that electron plasma waves of velocity $v_p \simeq c$ can be excited by the parametric ion-acoustic instability at plasma densities close to the cutoff density. This is just the plasma density regime where the threshold intensity for this instability is low.[20] For example, if the electron temperature T_e is 1 keV, we find that $N_e = 0.994\ N_c$. Also, we find that in this case the wavelength of this longitudinal wave is approximately the same as the vacuum wavelength of the laser radiation. Because most focal spots used in practice are at least several wavelengths in diameter, several wavelengths can build up inside the focus.

At incident laser intensities high above threshold, the ion-acoustic decay instability will be excited at lower plasma densities where the most unstable mode will have a phase velocity less than the velocity of light. However, other modes will also be excited, particularly after saturation of the initial growth of the instability. Plasma simulation calculations[17] show that the wave number spectrum is enhanced in the low K regime after saturation. Thus, long wavelength plasma waves whose phase velocity is comparable to c should be excited over a wider range of densities in the plasma.

Another requirement for relativistic electron production is that the plasma wave amplitude be sufficiently high to provide the necessary acceleration electric field. In order to reach the threshold energy for trident pair production ($2m_oc^2$), we must have

$$2m_o c^2 \cong eE_p \frac{\lambda_p}{2} \pm \frac{1}{2} m_o v_{ep}^2 , \qquad (20)$$

where E_p is the average electric field of the plasma wave, λ_p is the wavelength, and v_{ep} is the velocity component of the individual electron in the direction of the electric field E_p. The number of electrons that will be accelerated thus depends not only on the wave intensity, but also on the shape of the velocity distribution on the "tail" of the distribution function. If the suprathermal "tail" is sufficiently large, an appreciable number of electrons can be accelerated to the threshold for trident pair production.

We conclude that it may be possible to obtain pair production from electrons accelerated in the turbulent plasma environment of anomalous absorption instabilities. However, we have not been able to estimate the probability of this process, or to make quantitative estimates of the production rate of the relativistic electrons. Such a capability awaits development of a relativistic theory or a relativistic plasma simulation numerical code which can be applied to the inhomogeneous plasmas which are produced within the small dimensions of the focal spot region of the focused laser light.

V. EXPERIMENTAL EVIDENCE FOR RELATIVISTIC ELECTRONS

In the previous section we have shown the plausibility of relativistic electron production by the plasma instabilities associated with the anomalous absorption of laser radiation. Experimental indications of such electrons have recently been seen at this laboratory.

These experiments were done with our "long path" neodymium-glass disc laser system.[21,22] A 6 nsec double pulse was used whose intensity peaks were separated by 3 nsec; the pulse shape was as shown in ref. 21, not ref. 22. The total output energy in each pulse was 80 ± 10 joules. One difference from the older work was that the light path through the system made only five passes through the discs rather than nine passes. Optical spectrometer measurements of this new pulse indicated a narrower overall spectral width (≃ 60 Å) than the previous pulse (≃ 100 Å). A detailed account of the substructure of the output pulses from the long path laser is available.[23]

Although the narrowing of the overall spectral width seemed like a minor change, an unusually penetrating hard component of the x-rays was seen when this new pulse was incident on our standard polyethylene target (approximate composition $(CH_2)_n$). The detector was a plastic fluor (15 cm diam. x 20 cm long) originally intended for neutron measurements with $(CD_2)_n$ targets; its front face was located 22 cm from the target. The absorption curve obtained from these measurements is plotted in Fig. 6.

The absorption coefficient μ of the hard component in Fig. 6 is approximately $\mu = 0.05\ \mathrm{cm}^2/\mathrm{g}$, corresponding to x rays in the 1-10 MeV range (independent of the absorbers used).[24] Such extremely hard x rays were not seen in such abundance in earlier experiments;[22] the reasons for the difference are not known. Whatever the reasons, it appears that in at least one set of experimental conditions, hard x rays were seen which are best interpreted as bremsstrahlung created by relativistic electrons of energy greater than $2m_oc^2$ (∿ 1 MeV).

However, we are unable to estimate the absolute number of relativistic electrons produced, because we do not know where the x rays were produced. One MeV electrons have a range in cold material of 0.4 g/cm^2 or more.[24] The range in hot plasma would be higher, but even the cold material range is already much greater than the dimensions of the laser focal spot. Thus,

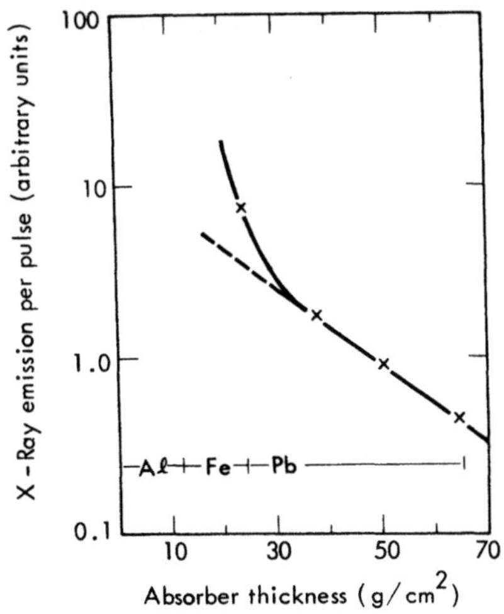

Figure 6. X-ray absorption curve for laser-target experiment described in text. The ordinate scale is based upon the energy absorbed by the detector. The abscissa represents absorber thickness; the aluminum and iron were never removed.

relativistic electrons created at the target would be expected to escape into the vacuum chamber, where they would travel to the walls, creating bremsstrahlung x-rays at many locations. Until further experimental work is done to isolate and measure these effects, it is not possible to estimate the absolute number of electrons produced.

These conditions also cast doubt on whether pair production takes place at the focus, since in some directions the distance from the focus that the relativistic electron travels is much greater than the focal spot dimensions.

VI. COMPARISON OF PAIR PRODUCTION AND BREMSSTRAHLUNG BY FAST ELECTRONS

For experimental purposes it is of interest to compare the average energy loss of fast electrons by pair production to the average energy loss by bremsstrahlung x-radiation. To do this, consider the probability dP that pairs are produced in distance dx

$$dP = \sigma_T \, N_n \, dx \ , \tag{21}$$

where σ_T is the trident cross-section, and N_n is the density of nuclei per unit volume. Near the threshold we can assume that approximately all of the kinetic energy E_v is lost when the trident process occurs, so that the averaged energy loss (over many electrons) can be written

$$\left(\frac{dE}{dx}\right)_{pairs} \cong -E_v \frac{dP}{dx} = -9.6\text{x}10^{-4}\ \alpha^2 r_o^2 Z^2 N_n E_v \left(\frac{E_v}{m_o c^2} - 2\right)^{3.6} , \quad (22)$$

where we have substituted from Eq. (6).

It is well-known that a similar equation exists for the radiative energy loss due to bremsstrahlung[25]

$$\left(\frac{dE}{dx}\right)_{rad} = -4\ \alpha r_o^2 Z^2 N_n E_v \ln\ (183/Z^{1/3})\ . \quad (23)$$

This expression is the average energy loss at all x-ray frequencies. For comparison with the experiment described in the previous section, however, we want to know the average radiation loss by emission of hard x rays ($E \gtrsim 2m_oc^2$). Because the x-ray spectrum is almost constant, this can be written approximately as

$$\left(\frac{dE}{dx}\right)_{\text{Hard x}} \simeq -21\ \alpha r_o^2 Z^2 N_n\ (E_v - 2m_o c^2)\ , \quad (24)$$

where we have put Z=1 in the slowly varying logarithmic term.

The approximate ratio of average pair production energy loss to average hard x-ray emission is then found from Eqs. (22) and (24)

$$\frac{(dE)_{pairs}}{(dE)_{\text{Hard x}}} \simeq 4.6\text{ x }10^{-5}\ \alpha\left(\frac{E_v}{m_o c^2}\right)\left(\frac{E_v}{m_o c^2} - 2\right)^{2.6} . \quad (25)$$

This result can be used to make an estimate of the possibility of pair production in an experiment where bremsstrahlung x rays have been produced. In our earlier experiment,[22] 10^{-9} joules of x-ray energy was emitted by the source as x rays of 100 keV or greater. One would expect much less energy to have been emitted as x rays of $2m_oc^2$ or greater. For electrons near threshold $[(E_v/m_oc^2 - 2) << 1]$, Eq. (25) predicts that the ratio of pair energy to hard x-ray energy would be much less than 10^{-6}. So, one concludes that the average pair production energy loss in this example was much less than 10^{-15} joules. However, the threshold energy $2m_oc^2$ for production of a single pair is of the order of 10^{-13} joules, which is still greater than this extreme upper limit. Thus, we conclude that no pairs were produced in our early experiment,[22] and that one should not expect to see pair production in similar experiments unless orders of magnitude greater x-ray bremsstrahlung intensities are detected.

In the most recent experiment, described above, in which more hard x rays were detected, we cannot tell whether pairs were produced, because we could not estimate the absolute x-ray intensity for that experiment.

VII. SUMMARY

We have examined various mechanisms for electron-positron pair production by intense, focused laser light pulses. Vacuum pair production was estimated to be unobservable, in agreement with previous authors. The "trident" process of pair production by high energy electrons was then considered, and the cross-section was calculated. The remaining question is how the high energy (kinetic energy > $2m_oc^2$) electrons can be created at the laser focus.

In one case, that of the coherent "quivering velocity" of an electron in a circularly polarized beam, we were able to obtain a result for the threshold of pair production which was at the extreme upper end of contemporary feasible focused intensities.

In another case, that of the parametric ion-acoustic instability which is excited in laser-produced plasmas, we have given qualitative arguments for the plausibility of production of at least a few relativistic electrons of the longitudinal plasma waves. On the basis of available experimental information, it seems unlikely that electron-positron pairs have been produced at experimental intensities of 10^{12}-10^{15} W/cm^2. When new experiments are done in the intensity range 10^{16}-10^{18} W/cm^2, however, a few pairs may possibly be produced. Thus, an experimental search for positron-electron pairs need not wait for a focused laser intensity as high as 10^{19}-10^{20} W/cm^2, as previously estimated.[5]

VIII. ACKNOWLEDGMENTS

We would like to acknowledge interesting discussions of these problems with Ray E. Kidder, Stephen Bodner, and James Eddleman. We are indebted to Arno Steiger and C.H. Woods for discussions and access to some of their calculations. We wish to thank Stanley J. Brodsky for sending us his calculation program.

REFERENCES

1. W.F. Hagen, J. Appl. Phys. 40, 511 (1969).

2. P.G. Kryukov and V.S. Letokhov, Usp. Fiz. Nauk 99, 169 (1969) [English translation: Sov. Phys.-Usp. 12, 641 (1970)].

3. F.V. Bunkin and I.I. Tugov, Dokl. Akad. Nauk SSR 187, 541 (1969) [English translation: Sov. Phys.-Dokl. 14, 678 (1970)].

4. E. Brezin and C. Itzykson, Phys. Rev. D 2, 1191 (1970).

5. F.V. Bunkin and A.E. Kazakov, Dokl. Akad. Nauk SSSR 193, 1274 (1970) [English translation: Sov. Phys.-Dokl. 15, 758 (1971)].

6. H.J. Bhabha, Proc. Roy. Soc. 152, 559 (1935).

7. S.J. Brodsky and S.C.C. Ting, Phys. Rev. 145, 1018 (1966).

8. V.L. Ginzburg, The Propagation of Electromagnetic Waves in Plasmas, J.B. Sykes and R.J. Taylor, Translators (Pergamon Press, New York, 1964).

9. P. Kaw and J. Dawson, Phys. of Fluids 13, 472 (1970).

10. A.D. Steiger and C.H. Woods, Phys. Rev. A 5, 1467 (1972).

11. C. Max and F. Perkins, Phys. Rev. Lett. 27, 1342 (1971).

12. P. Kaw, J. Dawson, W. Kruer, C. Oberman, and E. Valeo, Soviet Journal of Quantum Electronics 1, 205 (1971).

13. W.L. Kruer and J.M. Dawson, Phys. Fluids 15, 446 (1972).

14. K. Nishikawa, J. Phys. Soc. of Japan 24, 1152 (1968).

15. D.F. DuBois and M.V. Goldman, Phys. Rev. Lett. 14, 544 (1965).

16. P.K. Kaw and J.M. Dawson, Phys. Fluids 12, 2586 (1969).

17. J.S. DeGroot and J.I. Katz, Physics of Fluids 16, 401 (1973).

18. L. Spitzer, Jr., Physics of Fully Ionized Gases 2nd Ed., (Interscience Publishers, New York, 1962), p. 56.

19. J.W. Shearer and J.J. Duderstadt, Nuclear Fusion 13, No. 3, 401 (1973).

20. C. Yamanaka, T. Yamanaka, H. Kang, K. Yoshida, M. Waki, and T. Shimamura, Phys. Letters 38A, 495 (1972).

21. S.W. Mead, R.E. Kidder, J.E. Swain, F. Rainer, and J. Petruzzi, Applied Optics 11, 345 (1972).

22. J.W. Shearer, S.W. Mead, J. Petruzzi, F. Rainer, J.E. Swain, and C.E. Violet, Phys. Rev. A 6, 764 (1972).

23. L.W. Coleman, J.E. Swain, F. Rainer, and R.A. Saroyan, Pulse Characterization for the Long Path Laser at LLL, Rept. UCRL-74626 (1973), unpublished.

† Work performed under the auspices of the U.S. Atomic Energy Commission.

RELATIVISTIC OSCILLATIONS OF ELECTRONS AND PROTONS IN LASER FIELDS AND COLLECTIVE EFFECTS STIMULATING PAIR PRODUCTION*

Heinrich Hora**
Rensselaer Polytech. Institute Graduate Center
Hartford, Conn. 06120

ABSTRACT

The oscillation energy of electrons and protons in laser fields of high intensity resulting in relativistic motion is calculated, and used to discuss the thresholds and efficiencies of pair production with the well known formulae of the quantum electrodynamics. The relativistic values of the cut-off density and of the collision frequency, and the relativistic optical constants are derived to find the strong increase of the effective laser field strengths in plasmas due to collective effects near the cut-off density. If plasma densities below that of the solid state should be used, only the CO_2 laser could produce proton-antiproton pairs at intensities of 10^{18}W/cm^2. Neodymium glass lasers should need densities of nearly 100 times of the solid state. Electron-positions-pairs should be produced by neodymium glass laser pulses of 4×10^{16}W/cm^2 of 0.1 nsec. duration. The efficient production of antihydrogen for triggering micro-explosions of solid DT or pure deuterium for energy production is discussed.

---

*Presented at the Third Workshop "Laser Interaction and Related Plasma Phenomena", Rensselaer Polytech. Inst., Troy, New York, Aug. 13-17, 1973.
**Visiting Faculty Member, Laboratory for Laser Energetics, University of Rochester, Roch., N.Y. 14627

1) INTRODUCTION

The production of electron-position pairs in the focus of laser radiation of extremely high intensities has been treated by Prokhorov e.a.[1)2)] and Basov and Krokhin[3)]. Recent work was known by Hora[4)], Shearer e.a.[5)] and Lubin[6)]. The preceding work of Prokhorov and Bunkin[7)] started from the threshold of laser radiation for wave lengths λ of 1μ (neodymium glass $\lambda = 1.06\mu$ or iodine laser[7)] of intensities $I \approx 10^{26} W/cm^2$ to produce electron-position pairs in vacuum due to the polarization of the vacuum[8)]. This intensity may reached, because light of any higher intensity in vacuum will be absorbed by such a pair production and will be attenuated.

Besides of this pair production in vacuum, which will not be considered here, pairs are produced by charged particles located in the focus of a laser beam. There is to distiguish between the interaction of a single charged particle (electron or proton) located in the focus, interacting with the vacuum field to produce pairs[1)2)] and the case of many particles in the focus where collisions occur. The thresholds for pair production in the case of collisions decrease considerably[1)2)]. It is the merit of Prokhorov and Bunkin to have shown these conditions in principle, where the total energy of the particles was used, including the oscillation energy and the energy of translation. The translation energy depends on initial conditions, generation of standing waves and the time dependence of the laser pulse, which have to be defined separately. This paper uses the oscillation energy only as an exact solution of the equation of motion. Following the case without retardation of the laser field[4)], it will be demonstrated here that with retardation the same value of oscillation energy results. With these exact oscillation energies, the threshold will be derived.

Starting from the nonlinear increase of the actual laser field strength near the cut-off density, conditions are discussed, where the threshold of electron pair production within underdense plasma can come up to proton pair production due to the dielectric effects. We discuss here the change of the cut-off density due to the relativistic change of the electron mass.

Finally the consequences of anti-proton generation will be considered under very crude and speculative assumptions to estimate the possibility of a transfer of laser energy into anti-hydrogen in an efficient way.

2) RELATIVISTIC OSCILLATION OF PARTICLES

To evaluate the oscillation energy of charged particles in high intensity laser fields, we start from the equation of motion. The relativistic motion of a single particle (electron or ion) of the mass m_o and charge e (which sign is to be determined by an unimportant phase) is given for the electric and magnetic field $\underline{E}$ and $\underline{H}$ of the actual amplitudes E_o and H_o respectively of a wave propagating towards the +x direction with alinear polarization of $\underline{E}$ parallel to the y direction, by the transversal motion (parallel to $\underline{E}$)

$$\frac{d}{dt}\frac{m_o v_y}{[1-(v_x^2+v_y^2)/c^2]^{1/2}} = e\, E_o \cos(-kx+\omega t) \tag{1a}$$

and by the longitudinal motion (parallel to $\underline{E} \times \underline{B}$)

$$\frac{d}{dt}\frac{m_o v_x}{[1-(v_x^2+v_y^2)/c^2]^{1/2}} = \frac{e}{c} v_y H_o \cos(-kx+\omega t+\phi) \tag{1b}$$

ϕ is the phase between $\underline{E}$ and $\underline{H}$. A motion parallel to $\underline{H}$ (in the z direction) will not occur. If the electromagnetic wave is incident in a plasma with a complex refractive index n and if spatial changes of n(x) are small enough to fulfill the WBK condition, the following substitution is possible, using the Maxwell equations

$$E_o = \frac{E_v}{|\tilde{n}|^{1/2}} \quad ; \quad H_o = E_v|\tilde{n}|^{1/2} \tag{2}$$

where E_v is the corresponding amplitude of the incident light wave in vacuum before entering the plasma.

Using the substitution

$$Z = [1-(v_x^2+v_y^2)/c^2]^{1/2} = [1-\underline{v}^2/c^2]^{1/2}, \tag{3}$$

Eq. (1) can be integrated

$$v_y = \frac{eE_v Z}{m|\tilde{n}|\omega} \sin(-kx+\omega t) \tag{4}$$

Using Eqs. (3) and (4), from Eq. (1b) follows

$$\frac{d}{dt}\,[c^2-Z^2(c^2+e^2E_v{}^2\sin^2(-kx+\omega t)/(m_o{}^2|\tilde{n}|\omega^2))^{1/2}]Z =$$

$$Z\,\frac{e^2E_v{}^2}{m_o{}^2c\omega}\left[\frac{1}{2}\sin 2(-kx+\omega t)\cos\phi-\sin^2(-k+\omega t)\sin\phi\right] \qquad (5)$$

Obviously, the expression under the radical cannot become negative, therefore Z can vary only between

$$1 \geq Z^2 \geq Z^{*2} > 0 \qquad (6)$$

where

$$Z^{*2}\left(1+\frac{e^2E_v{}^2}{c^2m_o{}^2|\tilde{n}|\omega^2}\right) = 1 \qquad (7)$$

The minimum value of Z is

$$Z^* = [1-v_{max}{}^2/c^2]^{1/2} = 1/[1+\frac{e^2E_v{}^2}{m_o{}^2\omega^2c^2|\tilde{n}|}]^{1/2}. \qquad (8)$$

This can be used to evaluate the maximum value of the kinetic energy ε_{kin} of the oscillation (which can be identified by the oscillation energy ε_{osc} of the electron) without solving the equation of motion (Eqs. (1a) and (1b)) directly.

$$\varepsilon_{kin} = m_oc^2\left(\frac{1}{[1-v^2/c^2]^{1/2}}-1\right) = m_oc^2$$

$$\times\left(\left(1+\frac{e^2E_v{}^2}{m_o{}^2\omega^2c^2|\tilde{n}|}\right)^{1/2}-1\right) \qquad (9)$$

For $E_v \ll E_v{}^r$ we reproduce the well-known non-relativistic value of the oscillation energy:

$$\varepsilon_{kin} = \frac{1}{2}\;\frac{e^2E_v{}^2}{m_o\omega^2|\tilde{n}|}\quad,\quad (E_v \leq E_v{}^r) \qquad (10)$$

where $E_v{}^r$ is the relativistic limit

$$\frac{E_v^{r2}}{|\tilde{n}|} = m_o c^2 \frac{m_o \omega^2}{e^2} \quad . \tag{11}$$

This limit for neodymium glass lasers with a wave length $\lambda = 2\pi c/\omega = 1.06\mu m$ (Nd) and for CO_2 lasers ($\lambda = 10.6\mu m$) in the case of electrons (e) and protons (p) is:

$$\frac{E_v^r}{\sqrt{|\tilde{n}|}} = \begin{cases} 3 \times 10^{10}\ \mathrm{V/cm}\ (e)\ (Nd) \\ 5.5 \times 10^{13}\ \mathrm{V/cm}\ (p)\ (Nd) \\ 3 \times 10^{9}\ \mathrm{V/cm}\ (e)\ (CO_2) \\ 5.5 \times 10^{12}\ \mathrm{V/cm}\ (p)\ (CO_2)\ . \end{cases} \tag{12a}$$

The corresponding laser intensities $I_v = I_v^r$ in vacuum are

$$\frac{I_v^r}{|\tilde{n}|} = \begin{cases} 1.1 \times 10^{18}\ \mathrm{W/cm^2}\ (e)\ (Nd) \\ 3.7 \times 10^{24}\ \mathrm{W/cm^2}\ (p)\ (Nd) \\ 1.1 \times 10^{16}\ \mathrm{W/cm^2}\ (e)\ (CO_2) \\ 3.7 \times 10^{22}\ \mathrm{W/cm^2}\ (p)\ (CO_2), \end{cases} \tag{12b}$$

where the refractive index $\tilde{n}$ is to be used at the actual location $\underline{r}_o$ within the plasma, neglecting absorption between the vacuum and $\underline{r}_o$.

For the pure relativistic case at very high laser intensities I and amplitudes $E_v > E_v^r$ we find from Eq. (7)

$$\varepsilon_{kin} = \frac{ecE_v}{\omega\sqrt{|\tilde{n}|}} \approx \sqrt{I} \quad . \quad (E_v >> E_v^r) \tag{13}$$

This result demonstrates that there is no dependence of the oscillation energy of relativistic particles on their rest mass m_o, and the increase is proportional to the square root of the intensity I only.

3) CONDITIONS FOR PAIR PRODUCTION

Besides the concept of interaction of a high intensity laser field with the vacuum for producing electron proton pairs by vacuum polarization[1)8)] which has a threshold near an intensity of 10^{26} W/cm^2 for neodymium glass laser, another mechanism was discussed by Erber[10)] and Mitter[11)] where a free electron is oscillating in the laser field without any collision. The pair production is given by a transition rate

$$\exp(-8/e\gamma) \ ,$$

where[2)]

$$\gamma = \frac{\varepsilon_{kin}}{m_o c^2} \frac{E_o}{E_{cr}} \quad \text{and} \quad E_{cr} = m_o^2 c^3/(e\hbar) \ . \qquad (14)$$

Using $E_o = E_v$ (putting $|\tilde{n}| = 1$) and ε_{kin} from Eq. (13) we arrive at

$$\gamma = \frac{e^2\hbar}{\omega m_o^3 c^2} E_v \ . \qquad (15)$$

With a value γ=0.1, resulting in an exponential function before Eq.(14) of 2.6×10^{-12} we find field strengths E_v^v and intensities I_v^v:

$$E_v^v = \begin{cases} 4.4 \times 10^{13} \text{ V/cm (e) (Nd)} \\ 2.7 \times 10^{23} \text{ V/cm (p) (Nd)} \\ 4.4 \times 10^{12} \text{ V/cm (e) } (CO_2) \\ 2.7 \times 10^{22} \text{ V/cm (p) } (CO_2) \end{cases} \qquad (16)$$

$$I_v^v = \begin{cases} 2.7 \times 10^{24} \text{ W/cm}^2 \text{ (e) (Nd)} \\ 1.03 \times 10^{44} \text{ W/cm}^2 \text{ (p) (Nd)} \\ 2.7 \times 10^{22} \text{ W/cm}^2 \text{ (e) } (CO_2) \\ 1.03 \times 10^{42} \text{ W/cm}^2 \text{ (p) } (CO_2) \end{cases} \qquad (17)$$

which are not very much different from the result with the uncorrected value[2)] of ε_{kin} and which are higher than the laser intensities considered below. Intensities exceeding the absolute threshold of pair production in vacuum 10^{26} W/cm^2 for Nd glass and of 10^{24} W/cm^2 are only of academic nature.

The other mechanism, the pair production due to collisions in a plasma results at lower thresholds by the simple condition only of $\varepsilon_{kin} \gg m_o c^2$ or $I_v > I_v^r$ or $E_v > E_v^r$ of Eqs. (12). The order of magnitude for the cross section of pair production[2] is given by

$$\sigma = \frac{e^8}{\pi \hbar^2 m_o^2 c^6} \ell n^3 \left(\frac{\varepsilon_{kin}}{m_o c^2} \right) . \tag{18}$$

Where use was made of Bhabha's formulae[5]. A correction of this cross section by a more refined treatment was given by Shearer et al.[5] where for circular polarized radiation the laser field amplitudes have to be multiplied by a factor 2 to 3 to reach the conditions of this treatment. The number N_p of pairs produced in a plasma volume V during a time τ and a density n_e of electrons is

$$N_p = \frac{e^8 n_e^2}{\pi \hbar^2 m_e^2 c^5} V\tau \; \ell n^3 [\varepsilon_{kin}/(m_o c^2)] , \tag{19}$$

or by using Eq. (13)

$$N_p = \frac{e^8 n_e^2}{\pi \hbar^2 m_e^2 c^5} V\tau \; \ell n^3 [eE_v/(\omega \sqrt{|\tilde{n}|} m_o c)] . \tag{20}$$

Because of the logarithmic function in Eq. (20), the difference of our result to that with the uncorrected value[2] of ε_{kin} is indeed not very strong, at least it is not a question of many orders of magnitude.

Because the following considerations are also directed to questions of pair production of protons besides of that of electrons, the use of Eq. (20) is then a very rough approximation, because the interaction of the protons with other particles and quanta is neglected. The mass m_e in the factor before V appearing in Eq. (20) was derived from electrodynamic quantities only, given by the fine structure constant, therefore at this special point the electron mass m_e will remain. The condition for proton pair production is that of intensities I_v and fields E_v exceeding the relativistic thresholds of Eq. (12) for the protons.

4) COLLECTIVE EFFECTS

The result of pair production in plasmas due to collisions, Eq. (20), has a remarkable rate, if the particle energy exceeds the energy of the rest mass by a factor of 3 or more. For electrons, the needed laser intensities I_v and field strength E_v are then indeed not too far from the present state of technology if thin plasmas with n=1 are presumed. But, if the pair production of protons is desired, the results of Eq.(12) indicate intensities exceeding strongly the present aspects of laser technology. In this case we have a chance only by assuming collective effects to cause very small values of $|\tilde{n}|$ to increase the vacuum values E_v or I_v very considerably. These small values of $|\tilde{n}|$ [9,12] are expectable at such electron densities n_e of the plasma that the laser frequency ω is equal to the plasma frequency ω_p

$$\omega_p^2 = \frac{4\pi e^2 n_e}{m_e} [1-v^2/c^2]^{1/2} \tag{21}$$

where the relativistic correction of the rest mass $m_o=m_e$ for the electron was used, as introduced by Kidder[13]. The dielectric effects will be determined by the electrons only. Following our results of Eqs. (9) and (13) for $E_v \gg E_v^r$, we find

$$\omega_p^2 = \frac{4\pi e^2 n_e}{m_e} \frac{m_e c^2}{\varepsilon_{kin}} = 4\pi e n_e \omega \sqrt{|\tilde{n}|}\, c\, /E_v \tag{22}$$

From this result follows a relativistic change of the cut-off density of the electrons n_{eco}^{rel}, if it can be presumed that the frequency ω of the incident laser radiation will not be changed

$$n_{eco}^{rel} = \frac{\omega^2 m_e}{4\pi e^2} \frac{\varepsilon_{kin}}{m_o c^2} = n_{eco} \frac{\varepsilon_{kin}}{m_o c^2} \tag{23}$$

where n_{eco} is the nonrelativistic cut-off density of $10^{21}cm^{-3}$ for Nd glass lasers and of $10^{19}cm^{-3}$ for CO_2 lasers.

From the result of Eq. (23) one derives immediately the preferred wave length of the CO_2 lasers: if proton pair production would be desired, $\varepsilon_{kin}/m_o c^2$ is 1836.

Therefore, only CO_2 laser cut-off densities can be used, being then an increased of the cut-off density of up to some $10^{22}cm^{-3}$, if densities still in the order of magnitude of the solid state density ($6x10^{22}$ cm^{-3}) should be applied. Proton pair production with neodymium glass lasers using the dielectric effects near the cut-off density would require a compression of the plasma up to hundred times the solid state density.

The collision frequency ν of the electrons in a plasma is changing at relativistic conditions from the usual value derived for high laser intensities[14)15)]. At non-relativistic conditions ($I<<I^r$; $E<E^r$), the collision frequency is

$$\nu = \frac{\pi^{1/2}\omega_p{}^2 m_e{}^{1/2} e^2 \ell n\ \Lambda}{16x2^{1/2}(kT+\varepsilon_{kin})^{3/2}} \tag{24}$$

This value was derived from the dc-resistivity of the plasmas and its applicability to the high frequency conditions was justified[16)] because absorption constants resulted well comparable with that of the quantum mechanical theory of inverse bremsstrahlung for densities below cut-off ($n_e<<n_{eco}$). There still exists a difference by a factor $3\pi^{3/2}/8$ against the derivation of the collision frequency ν from the model of Dawson and Oberman[17)] based on the collisionless Vlasov equation and using the ion correlation to describe the resistivity, if the factor 3 of a former misprint is neglected[17)]. The justification of these values is given, in principle at least, by the comparability with the values derived quantum mechanically ($n_e<<n_{eco}$) and with that of Eq. (24) derived from the dc-resistivity for $n_e> n_{eco}$[18)]. The suppression of Spitzer's low frequency corrections of the electron-electron collisions was shown before[4)], especially under the view of the coherent oscillation of the electrons at high intensities, where electron-electron collisions are to be neglected.

The plasma temperature T in Eq. (24) can be neglected against the oscillation energy ε_{kin} of the electrons in the laser field. Using the relations of Eqs. (22) and (24), we find

$$\nu = \frac{\pi^{1/2}e^2\omega^2\ell n\ \Lambda}{2^{9/2}\ c\ \varepsilon_{kin}} \quad \text{for } \varepsilon_{kin} \geq m_o c^2 \tag{25}$$

these relativistic collision frequencies ν are remarkably small. As an example, for ε_{kin} = 1 MeV, the value of ν is in the order of 10^5 Hz for CO_2 lasers and of 10^7 Hz for Nd glass lasers. Therefore the plasma absorbs less energy at relativistic conditions and becomes more transparent. A similar result of increasing transparency was derived for the relativistic instability[18].

The Coulomb logarithm $\ln \Lambda$ used in Eq. (24) and (25) is to be corrected relativistically where use was made from the value[4])

$$\Lambda = \frac{c}{2e^2\omega(\pi n_e)^{1/2}} \frac{E_v}{\sqrt{|\tilde{n}|}} \tag{26}$$

where n_e is to be always the cut-off value, if $|\tilde{n}|$ is to have a strong difference from unity. Using Eq.(22) we find

$$\Lambda = \begin{cases} 1.2\text{x}10^9 \left(\dfrac{\varepsilon_{kin}}{m_e c^2}\right) \text{for Nd glass lasers} \\ 1.2\text{x}10^7 \left(\dfrac{\varepsilon_{kin}}{m_e c^2}\right) \text{for } CO_2 \text{ lasers} \end{cases} \tag{27}$$

Now we can follow the aim to evaluate the minimum absolute value of the refractive index

$$|\tilde{n}| = [\mathrm{Re}^2(\tilde{n}) + \mathrm{Im}^2(\tilde{n})]^{1/2} \tag{28}$$

which was shown before[19]) and remains also in the relativistic case to be

$$|\tilde{n}|_{min} = \left(\frac{\nu}{\omega}\right)^{1/2} \quad \text{at } \omega_p^2 = \omega^2 + \nu^2 \quad (\nu \ll \omega) \tag{29}$$

From Eq. (25) and from Eq. (24) for $kT \ll \varepsilon_{kin}$, we arrive at $\omega_p^2 = \omega^2$

$$|\tilde{n}|_{min} = \begin{cases} \dfrac{\pi^{1/4}\omega^{1/2} e\ell n\,\Lambda}{8\sqrt{2}\; c^{1/2}\varepsilon_{kin}^{1/2}} & \text{for } \varepsilon_{kin} \geq m_e c^2 \\ \dfrac{\pi^{1/4}\omega^{1/2} m_o^{1/4} e\ell n\Lambda}{8\sqrt{2}\;\; \varepsilon_{kin}^{1/2}} & \text{for } \varepsilon_{kin} \leq m_e c^2 \end{cases} \qquad (30)$$

The resulting transparency of the plasma at $\varepsilon_{kin} > m_o c^2$, avoiding losses of laser energy by collisions, is also favorable to the dielectric increase of the effective field E due to the denominator $\sqrt{|n|}$ in the WBK approximation. As an example, we find for ε_{kin}=1MeV for CO_2 lasers a value $|\tilde{n}|_{min}$ = 1.2x10^{-5}, and for neodymium glass lasers $|\tilde{n}|_{min}$ = 4.1 x 10^{-5}.

The result of Eq. (30) shows the possibility of electron-position pair production at laser intensities in the vacuum much less than the relativistic thresholds, Eq. (12b), if the increased effective field strength at the cut-off density is exceeding the relativistic thresholds. Such mechanism is indeed a highly complicated resonance process, so get a first approximation only by the following discussion, where we start from a non-relativistic ε_{kin} at their plasmas using Eq. (30) to reach an effective electric field of 10 times that of the relativistic threshold for CO_2 and (Nd glass) laser respectively to reach e-e-production

$$\frac{Ev}{1/2(4.1)\text{x}10^{-5}} \left[\frac{\varepsilon_{kin}}{2m_o c^2}\right]^{+3/2} = 10\ E_v^r \qquad (31)$$

Taking into account the non-relativistic relation

$$E_v = E^r(\varepsilon_{kin}/m_o c^2)^{1/2} \qquad (32)$$

$$\varepsilon_{kin} = 1.8(3.7) \times 10^{-2} m_o c^2 \qquad (33)$$

Using Eq. (12b), we find then necessary laser intensities for e-e-production of

$$I = \begin{cases} 2 \times 10^{14}\ \text{W/cm}^2 & \\ 4 \times 10^{16}\ \text{W/cm}^2 & \end{cases} \text{for} \begin{matrix} CO_2 \\ \text{Nd glass} \end{matrix} \text{ lasers} \qquad (34)$$

which is indeed reachable by the present best laser intensities. It is to take into account that this calculation is in the first step of iteraction only, and due to the stimulated resonance mechanism of decreasing $|\tilde{n}|$, it may be even less than given in Eq. (34). The only condition is to reach a WBK-like or similar penetration of the laser radiation down to the relativistic cut-off density. This may be possible at times long enough for shrinking the electromagnetic energy density due to the decreasing $|\tilde{n}| \simeq 10^{-2}$, e.g., if the plasma has a depth of 100μ until reaching the cut-off density, laser pulses of 100 psec. or more are necessary.

For an efficient proton pair production, the field E_v should be in the relativistic range, therefore

$$E_v \geq E_v^r \left(\frac{m_p c^2}{m_e c^2}\right) \qquad (m_p = \text{proton rest mass}) \tag{35}$$

with the threshold E_v^r for electrons (12a). Taking into account a relativistic relation instead of Eq.(31)

$$\frac{E_v}{1.2(4.1)\times 10^5} \left(\frac{\varepsilon_{kin}}{2m_o c^2}\right)^{1/2} = 500\ E^r\ \frac{m_p}{m_e} \tag{35a}$$

for an effective ion pair production, we result at

$$\varepsilon_{kin} = 5.2(13) m_e c^2 \tag{35b}$$

which corresponds to laser intensities in vacuum of

$$I_v^2 = \begin{cases} 1.3 \times 10^{17}\ \text{W/cm}^2 \text{ for } CO_2 \text{ lasers} \\ 2.5 \times 10^{19}\ \text{W/cm}^2 \text{ for Nd glass lasers} \end{cases} . \tag{36}$$

There are upper bounds for the efficient proton pair production if the cut-off densities are given and if the laser field had been built up by shrinking according to that of $|\tilde{n}|$.

Finally another case of increase of the effective field E by the decreasing denominator $|\tilde{n}|$ near cut-off is mentioned, causing nonlinear net forces of electrodynamic kind in laser produced plasmas, as derived by the author generally and on the basis of the WKB approximation[9]. A further case of an exact evaluation avoiding the WKB-approximation was given by Lindl and Kaw[10], while the predominance of the nonlinear force over the thermal forces was shown by Steinhauder and Ahlstrom[21] at high temperatures for all intensities. Their evaluation was used to prove the predominance for laser intensities exceeding 10^{14} W/cm^2 (Nd) and 10^{12} W/cm^2 (CO_2) generally[15]. Also numerical evaluations demonstrated the predominance in a WBK-case for 10^{15} W/cm^2 (Nd) immediately[19] and a non-WBK-case including the reflection of light by Mulser and Green[19] showed the larger nonlinear force for the global (net) acceleration over the thermokinetic force even for intensities of $5\mathrm{x}10^{14}$ W/cm^2 (Nd). It does not cancel these results when for a dynamic calculation with laser pulse incident on a plasma at selected conditions of long laser pulses it was demonstrated that the forces in the shocked plasma below the absorption region can determine the expansion to a greater degree[22] than the mechanisms within the interaction region of the laser light with the plasma. The influence of the nonlinear force also on the whole dynamical process was demonstrated by Shearer, Kidder, and Zink[23][24]. The nonlinear forces, caused by changes of $\tilde{n}$ for cases of infrared lasers and microwaves, were treated by Chen[25] and Klima[26] and were included in the calculations of laser compression of fusion plasmas by Nuckolls, et al.[27].

5) CONCLUSIONS ON EFFICIENT PROTON PAIR PRODUCTION

From Eqs. (36) and (22) we conclude the possibility of producing proton pairs by firing a CO_2 laser beam of an intensity of 1.1×10^{18} W/cm^2 onto a hydrogen target of solid state density. Indeed the mechanisms involved furtherly, are very little known with respect to the lack of detailed knowledge of the instabilities in relativistic conditions and of the processes of multiple radiation emission of electrons[6] and of following processes of reabsorption and perhaps meson generation, etc.[11]. Taking into account this leak of knowledge, we can estimate very roughly only the size of an apparatus for most optimistic conditions, resulting in an effective conversion of laser energy into proton pairs.

Assuming the logarithmic function in Eq. (20) has a value 25, we arrive at

$$N_p = 10^{24} V\tau = 10^{24} Fl\tau \quad , \tag{37}$$

if F is the cross section of the laser bundle and l is the length of the reaction zone. We compare the total energy E_p of the pairs ($m_p \sim$ proton rest mass, $E_p = 2m_p c^2 N_p$) with that of the incident CO_2 laser energy E_L,

$$E_L = F\tau I_v \tag{38}$$

where

$$I = 1.1 \times 10^{18} \text{ W/cm}^2 .$$

We arrive at

$$\frac{E_p}{E_L} = \frac{2m_p c^2 N_p}{F\tau I_v^*} = 2.8 \times 10^{-4} \ell \tag{39}$$

Neglecting the details of absorption, Eq. (39) results in a first order length l of the reaction zone of $l=l_o$, if a conversion efficiency $E_p/E_L = 1$ should be reached, of

$$l_o = 36 \text{ meter} \quad . \tag{40}$$

The use of a self-focussed filament of a diffraction limited beam will be preferable because of stabilizing quasi-stationary conditions of the reaction. A diffraction-limited beam from a CO_2 laser ($F=10^{-5}cm^2$) as described in Eq. (38) would result, using Eq. (36), in

$$E_L = 1.1 \times 10^{+12} \tau \quad \text{Joules}, \tag{41}$$

which corresponds to a feasible laser power of 2×10^{12} Watts. In our gedanken experiment we have used a value $|\tilde{n}| = 10^{-5}$ which causes an effective velocity of light of 3×10^3 m/sec. in the reaction zone, resulting in a minimum time τ of operation of 6×10^{-3} seconds. This comes to a laser pulse energy of 6.6×10^9 joules.

The conditions of a 36 meter long reaction zone and laser pulses of 6.6 joules energy and of 6 nsec. length are not too fantastic conditions for an efficient production of antiprotons, taking into account that the energy to be stored within one second in the magnetic fields of a future tokamak device for nuclear fusion is of the same order of magnitude.

The further procedure of a nearly lossless extraction of the antiprotons and positrons from the plasma by strong electric fields, their recombination, condensation and levitation within vacua of a few tenths of a degree absolute would then be a further large technological problem. But the aim could result in an alternative fusion reactor, if laser-compression for inertial confinement should not succeed. Pellets of antihydrogen of 10^{-8}g falling into a solid shell of deuterium-tritium or of 10^{-4}g falling into solid deuterium should trigger a breakeven fusion reaction, if the trigger energies of 1.6M Joule and 16 GJoule are used respectively for inertial confinement without compression[28].

REFERENCES

1. F.V. Bunkin, A.M. Prokhorov, Polarization Matiere et Rayounement, Vol. Jubilaire en l'Honneur d' Alfred Kastler, Paris (1969).
2. F.V. Bunkin, A.E. Kazakov, Sov. Phys. Doklady 15, 758 (1971).
3. N.G. Basov and O.N. Krokhin (Seminar at the Lebedev Inst., Moscow, 1969).
4. H. Hora, Nature, Phys. Sc. 243, No. 124, 34 (1973); Opto-Electronics 5 (November 1973).
5. J.W. Shearer, J. Garrison, and J. Wong, Lawrence Livermore Labs. UCRL-74423, Feb. 1973, Phys. Rev. D (to be published).
6. M.J. Lubin (Private Communication, June 1973).
7. K. Hohla, and K. Kompa, Appl. Phys. Letters 22, 77 (1973).
8. W. Heisenberg, Z. Physik 90, 209 (1934).
9. H. Hora, Phys. Fluids 12, 182 (1969).
10. Th. Erber, Rev. Mod. Phys. 38, 626 (1966).
11. H. Mitter (Seminar at the Max-Planck-Inst. of Physik, Munich, June 1973).
12. H. Hora, D. Pfirsch, A. Schlüter, Z. Naturforsch 22a, 278 (1967).
13. R. Kidder, Varenna Summer School (July 1969), UCRL-Preprint 71775 (1969).
14. S. Rand, Phys. Rev. 136B, 231 (1964).
15. H. Hora, Opto-Electronics 2, 201 (1970).
16. H. Hora, IPP Garching, Rept. 6/27 (1964); H. Hora, and H. Wilhelm, Nucl. Fusion 10, 111 (1970).
17. T.W. Johnston and J.M. Dawson, Phys. Fluids 16, 722 (1973).
18. P. Kaw and J. Dawson, Phys. Fluids 13, 472 (1970).

19. H. Hora, in Laser Interaction and Related Plasma Phenomena, H. Schwarz and H. Hora eds., (Plenum, New York, 1972), Vol. 2, p. 341.
20. J. Lindl and P. Kaw, Phys. Fluids 14, 371 (1971).
21. L.C. Steinhauer, and H.G. Ahlstrom, Phys. Fluids 13, 1103 (1970).
22. B. Green and P. Mulser, in Laser Interaction and Related Plasma Phenomena, H. Schwarz and H. Hora eds., (Plenum, New York, 1972), Vol. 2, p. 381.
23. J.W. Shearer, R.E. Kidder, J.W. Zink, Bull. Am. Phys. Soc. 15, 1483 (1970).
24. R. Kinsinger, Conf. on Numerical Treatment of Plasma Instabilities, Los Alamos, March 1973.
25. F.F. Chen, Comments on Plasma Physics 1, 81 (1972).
26. R. Klima, Plasma Physics 12, 123 (1970).
27. J. Nuckols, L. Wood, A. Thiessen, G. Zimmerman, Paper presented at the VII Intern. Quantum Electronics Conf., Montreal (May 1972); Phys. Today, (August, 1973).
28. H. Hora and D. Pfirsch, in Laser Interaction and Related Plasma Phenomena, H. Schwarz and H. Hora eds., (Plenum, New York, 1972), Vol. 2, p. 515.

QUANTUM THEORY OF INVERSE BREMSSTRAHLUNG ABSORPTION AND PAIR PRODUCTION*

John F. Seely
Department of Physics and Astronomy
The University of Tennessee
Knoxville, Tennessee 37916

I. INTRODUCTION

The inverse Bremsstrahlung process is believed to play an important role in the heating of plasma electrons by laser radiation. In this process, an electron absorbs energy from the laser beam during a collision with a nucleus. From a classical viewpoint, the electron oscillates in the electric field of the laser beam. During a collision with a nucleus, the electron is knocked out of phase with the electric field, and the oscillatory energy of the electron is converted to random thermal energy. From a quantum viewpoint, the electron can gain energy only in units of $\hbar\omega$, where ω is the frequency of the laser radiation, and it is not clear that the classical theory is valid. In this paper, we discuss the quantum theory of inverse Bremsstrahlung absorption, and compare the quantum results to the classical results.

In the inverse Bremsstrahlung process, a positive-energy electron absorbs energy from the laser beam during a collision with a nucleus. But according to the Dirac theory of the electron, both positive-energy and negative-energy electrons may exist. When a negative-energy electron makes the transition to a positive-energy state by absorbing energy from the laser beam during a collision with a nucleus, this is observed in the laboratory as the production of an electron-positron pair. This pair production process is discussed in Section IV.

---

*Presented at the Third Workshop on "Laser Interaction and Related Plasma Phenomena" held at Rensselaer Polytechnic Institute, Troy, New York, August 13-17, 1973.

II. CLASSICAL THEORY OF INVERSE BREMSSTRAHLUNG

In this section, we discuss the classical theory of inverse Bremsstrahlung absorption. Let us begin by considering the equation of motion for an electron in an oscillatory electric field

$$m\frac{d\vec{v}}{dt} = e\vec{E}_0 e^{i\omega t} - \frac{1}{2}m\nu_{eff}\vec{v}$$

where ν_{eff} is an effective collision frequency. After solving the equation of motion for velocity, the time-average kinetic energy of the electron is found to be

$$\varepsilon_0 = e^2E_0^2 \ / \ 2m\omega^2 \tag{1}$$

where we have assumed that the collision frequency ν_{eff} is much less than the electric field frequency ω. This is a valid assumption for optical laser frequencies, where $\omega \sim 10^{15}$ Hz. Taking the product of velocity and force, the average rate of energy absorption by the electron from the electric field is

$$\frac{d\langle\varepsilon\rangle}{dt} = \frac{e^2E_0^2}{2m\omega^2}\,\nu_{eff} \ . \tag{2}$$

Thus the oscillation energy of the electron $e^2E_0^2/2m\omega^2$ is converted to random thermal energy during a collision with a nucleus. If no such collision occurs, then the electron continues to oscillate in phase with the electric field and gains no thermal energy.

In order to find an expression for the effective collision frequency, we must use an appropriate kinetic equation for the electrons. Dawson and Oberman[1] have used the Vlasov set of equations to investigate the collisional absorption of high-frequency electromagnetic radiation. Their results are valid for the weak-field case $\varepsilon_0 \ll \varepsilon_t$, where ε_t is the thermal energy of the electron

$$\varepsilon_t = \frac{3}{2}\,kT \ . \tag{3}$$

For laser radiation of wavelength 1μ and intensity 10^{17}W/cm^2, the oscillation energy found from Eq. (1) is 20 KeV. Thus in an intense laser beam, the oscillation energy of the electron may be larger than the thermal energy. Silin[2] has used a kinetic equation[3] to find effective collision frequencies for the two cases of $\varepsilon_0 \ll \varepsilon_t$ and $\varepsilon_0 \gg \varepsilon_t$. If the electromagnetic radiation is circularly polarized, the effective collision frequencies are[2]

$$\nu_{eff} = \frac{4\sqrt{2\pi}\ Z^2 e^4 N_e \ln\Lambda}{3m^{1/2}(kT)^{3/2}} \tag{4}$$

for the weak-field case and[4]

$$\nu_{eff} = 8\pi Z^2 em\omega^3 N_e \ln\Lambda / E_0^3 \tag{5}$$

for the strong-field case, where N_e is the electron density and $\ln\Lambda$ is the Coulomb logarithm. Equations (4) and (5) agree with an analysis by Bethe[6] in which the collision frequency is found using the Rutherford scattering cross section.

The rate of energy absorption is found by substituting ν_{eff} into Eq. (2). The weak-field absorption rate contains the well-known $T^{-3/2}$ dependence on electron temperature. The strong-field absorption rate is proportional to $I^{-1/2}$, where I is the laser intensity. Thus the inverse Bremsstrahlung absorption rate decreases as the laser intensity is increased. This is because the velocity of an electron oscillating in an intense laser beam is high, and the frequency of collisions between a high-velocity electron and a nucleus is small.

III. QUANTUM THEORY OF INVERSE BREMSSTRAHLUNG

According to quantum theory, an electron can gain energy from a laser beam only in units of $\hbar\omega$, and it is not clear that the classical theory of inverse Bremsstrahlung absorption discussed in Section II is valid. In this section, we discuss the quantum theory of inverse Bremsstrahlung absorption.

An electromagnetic field may be treated classically if a large number of photons are in the same energy state. The electromagnetic field of a monochromatic laser beam satisfies this requirement. We treat the laser field as a uniform circularly polarized plane wave propagating in the z direction. The vector potential is

$$\vec{A}(t) = (cE_0/\omega)(\vec{e}_x \cos\omega t + \vec{e}_y \sin\omega t). \tag{6}$$

The electrons are described by the solution

$$\psi_{\vec{p}}(\vec{x}, t) = \exp\left\{\frac{i}{\hbar}\vec{p}\cdot\vec{x} - \frac{i}{2m\hbar}\int^t |\vec{p} - \frac{e}{c}\vec{A}(t')|^2 dt'\right\} \tag{7}$$

to the Schrödinger equation for an electron in the plane wave (6). The Coulomb field of a fixed scattering nucleus is treated as a perturbation.

THE TRANSITION PROBABILITY

The inverse Bremsstrahlung process is shown schematically in Fig. 1. An electron with momentum $\vec{p}_1$ is scattered by a Coulomb potential $V(\vec{x})$ and acquires momentum $\vec{p}_2$. The probability for transitions between states $\vec{p}_1$ and $\vec{p}_2$ is found using first order perturbation theory in a manner similar to previous authors.[7,8,9] However, we use second-quantized theory to describe the electrons rather than the first-quantized theory used previously.

The total Hamiltonian of an electron in the fields of the plane wave (6) and the Coulomb potential $V(\vec{x})$ is

$$H = H_0 + H_I$$

where

$$H_0 = \int d^3x \psi^\dagger(\vec{x},t) \frac{1}{2m} \left|\frac{\hbar}{i}\vec{\nabla} - \frac{e}{c}\vec{A}(t)\right|^2 \psi(\vec{x},t) \cdot \tag{8}$$

$$H_I = \int d^3x \psi^\dagger(\vec{x},t)\, V(\vec{x})\psi(\vec{x},t). \tag{9}$$

Here H_0 is the unperturbed Hamiltonian, and H_I is the interaction Hamiltonian. Using second-quantized theory,[10] we write

$$\psi(\vec{x},t) = \sum_{\vec{p}} b_{\vec{p}}(t)\psi_{\vec{p}}(\vec{x},t) \tag{10}$$

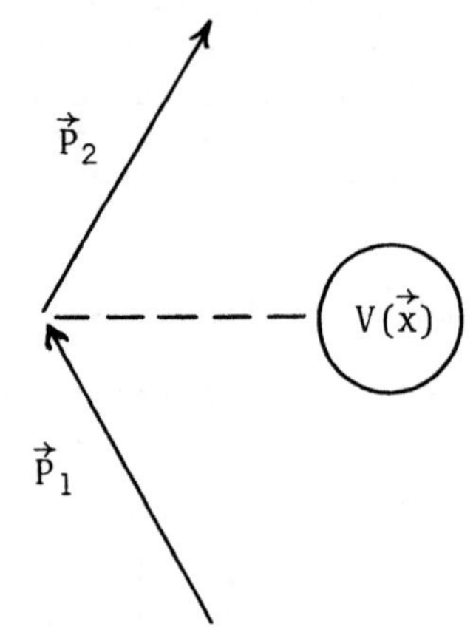

Fig. 1. The inverse Bremsstrahlung process.

where $\psi_{\vec{p}}(\vec{x},t)$ is given by Eq. (7). Substituting Eq. (10) into Eqs. (8) and (9), we find

$$H_0 = \sum_{\vec{p}} \frac{1}{2m} \left| \vec{p} - \frac{e}{c}\vec{A}(t) \right|^2 b_{\vec{p}}^{\dagger} b_{\vec{p}}$$

$$H_I = \sum_{\vec{p}_1} \sum_{\vec{p}_2} \langle \vec{p}_2 | V(\vec{x}) | \vec{p}_1 \rangle b_{\vec{p}_2}^{\dagger} b_{\vec{p}_1} \tag{11}$$

where

$$\langle \vec{p}_2 | V(\vec{x}) | \vec{p}_1 \rangle = \int d^3x \psi_{\vec{p}_2}^{*}(\vec{x},t) V(\vec{x}) \psi_{\vec{p}_1}(\vec{x},t). \tag{12}$$

The operators $b_{\vec{p}}$ and $b_{\vec{p}}^{\dagger}$ are the Fermion destruction and creation operators. Their matrix elements[10] are $[N_e(\vec{p})]^{1/2}$ and $[1-N_e(\vec{p})]^{1/2}$ respectively, where $N_e(\vec{p})$ is the number of electrons with momentum $\vec{p}$. From Eq. (11), we see that the interaction Hamiltonian induces transitions in which an electron with momentum $\vec{p}_1$ is destroyed and an electron with momentum $\vec{p}_2$ is created. Let there initially be $N_e(\vec{p}_1)$ electrons with momentum $\vec{p}_1$ and $N_e(\vec{p}_2)$ with $\vec{p}_2$. The transition matrix element between the initial state $|i\rangle$ and the final state $|f\rangle$ is

$$\langle f|H_I|i\rangle = [N_e(\vec{p}_1)]^{1/2}[1-N_e(\vec{p}_2)]^{1/2}\langle \vec{p}_2|V(\vec{x})|\vec{p}_1\rangle. \tag{13}$$

From first order perturbation theory, the transition probability amplitude is

$$a(i\to f) = -\frac{i}{\hbar}\int_{-T/2}^{+T/2} \langle f|H_I|i\rangle dt . \tag{14}$$

The integrals in Eq. (12) have been evaluated by several authors.[7,8,9] When Eq. (13) is substituted into Eq. (14) and $a(i\to f)$ is squared, the transition probability per unit time is found to be

$$\frac{|a(i\to f)|^2}{T} = \sum_{n=-\infty}^{\infty} T(n, \vec{p}_1 \to p_2)$$

where

$$T(n, \vec{p}_1 \to \vec{p}_2) = (2\pi/\hbar) N_e(\vec{p}_1)[1 - N_e(\vec{p}_2)]|\bar{V}(\Delta\vec{p}/\hbar)|^2$$

$$\times J_n^2(\lambda/\hbar\omega)\,\delta(\Delta E - n\hbar\omega) \tag{15}$$

$$\bar{V}(\vec{\Delta}p/\hbar) = -4\pi Z e^2 \hbar^2/|\vec{\Delta}p|^2$$

$$\Delta E = (p_2^2 - p_1^2)/2m$$

$$\lambda = e\, E_0 \Delta p_{\perp}/m\omega \ . \tag{16}$$

Here $\vec{\Delta}p = \vec{p}_2 - \vec{p}_1$, the subscript $\perp$ refers to the direction perpendicular to the z axis, $\bar{V}(\Delta\vec{p}/\hbar)$ is the Fourier transform of $V(\vec{x})$, and J_n is the Bessel function of order n. The parameter λ is a function of the intensity and frequency of the laser radiation, and ΔE is the change in kinetic energy of the electron. From the δ function of Eq. (15), we see that $T(n, \vec{p}_1 \to \vec{p}_2)$ is the transition probability per unit time for the transition in which an electron with momentum $\vec{p}_1$ is destroyed, an electron with momentum $\vec{p}_2$ is created, and $|n|$ laser photons are absorbed (n>0) or emitted (n<0). We note here that $T(n, \vec{p}_1 \to \vec{p}_2) = T(-n, \vec{p}_2 \to \vec{p}_1)$.

THE KINETIC EQUATION

The change in $N_e(\vec{p}_2)$, the number of electrons with momentum $\vec{p}_2$, may be written schematically as[11]

$$\frac{\partial N_e(\vec{p}_2)}{\partial t} = \sum_{n=1}^{\infty} \sum_{\vec{p}_1}$$

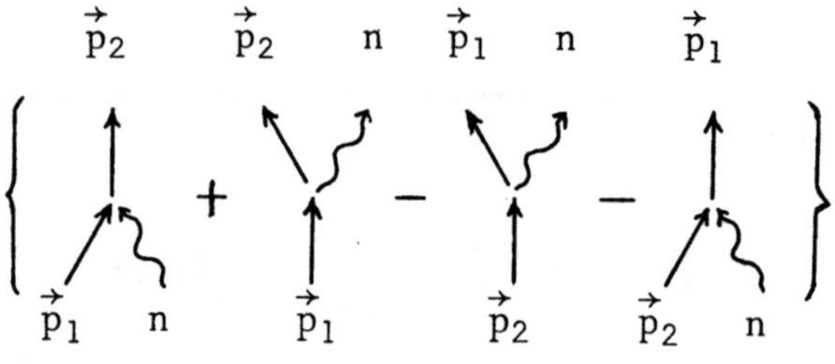

(17)

The processes in which an electron with momentum $\vec{p}_2$ is destroyed are subtracted from the processes in which an electron with momentum $\vec{p}_2$ is created. This difference gives the net increase in $N_e(\vec{p}_2)$. The schematic equation may be converted to a mathematical equation by replacing the diagrams in Eq. (17) with the transition probabilities per unit time for the processes given by Eq. (15). This mathematical equation is

$$\frac{\partial N_e(\vec{p}_2)}{\partial t} = \sum_{n=1}^{\infty} \sum_{\vec{p}_1} \{T(n, \vec{p}_1\to\vec{p}_2)+T(-n, \vec{p}_1\to\vec{p}_2) - T(-n, \vec{p}_2\to\vec{p}_1) - T(n, \vec{p}_2\to\vec{p}_1)\} \tag{18}$$

We now take the classical limit of Eq. (18). We assume that the electrons are nondegenerate so that $N_e(\vec{p})<<1$. We let the volume of the box in which the system is normalized become infinite so that the sum over $\vec{p}_1$ becomes an integral over $\vec{V}_1$. A Maxwell distribution is assumed for the electrons. The classical limit of Eq. (18) is

$$\frac{\partial f_e(\vec{V}_2)}{\partial t} = \frac{4Z^2e^4N_e}{m}\left(\frac{m}{2\pi kT}\right)^{3/2}\int d^3V_1|\vec{V}_2 - \vec{V}_1|^{-4} \times(e^{-mv_1^2/2kT}-e^{-mv_2^2/2kT})\sum_{\substack{n=-\infty\\ n\neq 0}}^{\infty} J_n^2(\lambda/\hbar\omega)\ (\Delta E - n\hbar\omega) \tag{19}$$

where $f_e(\vec{V})$ is the electron distribution function and N_e is the electron density. Equation (19) is the kinetic equation for the electrons.

EFFECTIVE COLLISION FREQUENCY

We use the kinetic equation (19) to find effective collision frequencies for the two cases of weak and strong laser field. These effective collision frequencies are then compared to the classical expressions (4) and (5).

For the weak-field case, $\lambda<<\hbar\omega$ and we use an approximation for the Bessel function in Eq. (19) which is valid for small argument. The kinetic equation for electrons in a weak laser field is

$$\frac{\partial f_e(\vec{V}_2)}{\partial t} = \frac{4Z^2e^4N_e}{m}\left(\frac{m}{2\pi kT}\right)^{3/2} e^{-mV_2^2/kT}\int d^3V_1|\vec{V}_2-\vec{V}_1|^{-4} \times(\lambda/2\hbar\omega)^2[(e^{\hbar\omega/kT}-1)\delta(\Delta E-\hbar\omega)+(e^{-\hbar\omega/kT}-1)\delta(\Delta E+\hbar\omega)]. \tag{20}$$

The first δ function corresponds to the absorption of a photon, and the second to the emission of a photon. Thus only single-photon processes are significant for the weak-field case.

We let $\hbar\to 0$ in Eq. (20) and evaluate the integral over $\vec{V}_1$.

The result is then substituted into

$$\frac{d\langle\varepsilon\rangle}{dt} = \int d^3V_2 \frac{mV_2^2}{2} \frac{\partial f_e(\vec{V}_2)}{\partial t} \tag{21}$$

which gives the change in the average kinetic energy of the electrons. Comparing this expression to Eq. (2), the effective collision frequency is found to be[9]

$$\nu_{eff} = \frac{4\sqrt{2\pi}\; Z^2 e^4 N_e \ln\Lambda}{3m^{1/2}(kT)^{3/2}} \; . \tag{22}$$

This result agrees exactly with the classical effective collision frequency given by Eq. (4). The absorption coefficient may be found by substituting Eq. (22) into

$$K = I^{-1} N_i \frac{d\langle\varepsilon\rangle}{dt} = 4\pi e^2 N_i \nu_{eff}/m\omega^2 V_g$$

where $V_g = c(1 - \omega_p^2/\omega^2)^{1/2}$ is the group velocity of the laser radiation and ω_p is theplasma frequency. The result agrees exactly with the absorption coefficient found by Dawson and Oberman.[1]

For the strong-field case, $\lambda \gg \hbar\omega$ and we use an approximation for the Bessel function in Eq. (19) which is valid for large argument. The kinetic equation for electrons in a strong laser field is

$$\frac{\partial f_e(\vec{V}_2)}{\partial t} = \frac{2Z^2 e^4 N_e}{m} \left(\frac{m}{2\pi kT}\right)^{3/2} e^{-mV_2^2/2kT} \int d^3V_1 |\vec{V}_2 - \vec{V}_1|^{-4}$$

$$\times \left[(e^{\lambda/kT} - 1)\delta(\Delta E - \lambda) + (e^{-\lambda/kT} - 1)\delta(\Delta E + \lambda)\right]. \tag{23}$$

The first δ function corresponds to the absorption of $s \equiv \lambda/\hbar\omega$ photons and the second to the emission of s photons where $s \gg 1$. Thus multiphoton absorption and emission processes are dominant over single-photon processes for the strong-field case. From Eq. (15), we see that $T(n, \vec{p}_1 \to \vec{p}_2) = 0$ for $\Delta p_\perp = 0$. Most photons are absorbed or emitted by electrons whose paths are nearly perpendicular to the direction of propagation of the laser radiation. Using $(\Delta p_\perp)^2 \cong 2m\hbar\omega$ and Eq. (16), we find that

$$s \cong 2e^2 E_o^2/m\hbar\omega^3 \; . \tag{24}$$

The number of photons absorbed or emitted is the same order of magnitude as the ratio of the classical electron kinetic energy (1) to the photon energy. Equation (24) implies that multiphoton processes are dominant for neodymium laser intensities greater than $10^{12}W/cm^2$.

We now assume that the electron temperature is sufficiently low so that $\lambda >> kT$. When Eq. (23) is substituted into Eq. (21) and the result is compared to Eq. (2), the effective collision frequency for the strong-field case is found to be[9]

$$\nu_{eff} = 8\pi^2 Z^2 em\omega^3 N_e / E_0^3 \ . \tag{25}$$

This result is to be compared to the classical effective collision frequency given by Eq. (5). We note that the classical expression (5) contains the Coulomb logarithm which does not occur in the quantum mechanical calculation. The Coulomb logarithm normally enters because it is necessary to cut off a divergent integral. In the calculation leading to Eq. (25), the integrals are convergent so that no cutoff is necessary. Equation (25) agrees qualitatively with the work of Osborn[12] and Rand,[13] but differs considerably from the work of Hughes and Nicholson-Florence[14] who consider only single-photon processes.

Other effects may be included in the above analysis of inverse Bremsstrahlung absorption by simply using the appropriate wave function to describe the electrons. For example, the effect of a uniform magnetic field may be included by using the solution[5] to the Schrödinger equation for an electron in the plane field (6) and a uniform magnetic field. Relativistic effects are taken into account by using the Volkov[15] solution to the Dirac equation for an electron in a plane electromagnetic wave. If a uniform magnetic field parallel to the laser beam is present, then the Redmond[16] solution to the Dirac equation should be used.

IV. PAIR PRODUCTION

According to the Dirac theory of the electron, both positive-energy and negative-energy electrons may exist. In the normal state of the vacuum, all negative-energy states are filled and all positive-energy states are unfilled. Thus the vacuum is actually a "sea" of negative-energy electrons. Only changes from the normal state of the vacuum are observable. If a negative-energy electron (carrying charge $e<0$) is raised to a positive-energy state, a "hole" in the negative-energy states is left behind. Since the charge of the sea of negative-energy states has been decreased by e, the hole appears to have charge $+|e|$. The positive-energy electron and the hole in negative-energy states are observed as an electron-positron pair. Thus the transition of a negative-energy electron to a positive-energy state is interpreted as the production of an electron-positron pair.

A plasma may be viewed as being composed of positive-energy electrons and ions in a sea of negative-energy electrons. When a laser beam passes through the plasma, the positive-energy and negative-energy electrons oscillate in the electromagnetic field of the laser beam. When perturbed by scattering from a nucleus, a negative-energy electron may make the transition to a positive-energy state by absorbing energy from the laser beam. We derive an expression for the cross section of this pair production process using procedures similar to those used in Section III.

The pair production process is shown schematically in Fig. 1. Particle 1 is a negative-energy electron and particle 2 is a positive-energy electron. The potential of the scattering nucleus is treated as a perturbation. Using first order perturbation theory, the probability amplitude for the transition is

$$a(1 \rightarrow 2) = -\frac{i}{\hbar}\int_{-T/2}^{T/2} dt \int d^3x \psi_2^\dagger V(\vec{x}) \psi_1 \ . \tag{26}$$

where ψ_1 is the wave function of the negative-energy electron and ψ_2 is the wave function of the positive-energy electron.

For an electron in a strong electromagnetic field, it may be shown[5] that the interaction between the field and the electron charge is much stronger than the interaction between the field and the electron spin. Thus an electron in an intense laser beam behaves like a spinless particle and may be described by the Klein-Gordon equation. The wave functions ψ_1 and ψ_2 in Eq. (26) are solutions[17] to the Klein-Gordon equation for an electron in a circularly polarized electromagnetic wave of strength E_0 and frequency ω. We note that Yakovlev[18] has used the more complicated Volkov[15] solutions to the Dirac equation to investigate pair production induced by scattering from a nucleus.

When Eq. (26) is squared, the transition probability per unit time is found to be

$$\frac{|a(1\rightarrow 2)|^2}{T} = \sum_{n=-\infty}^{\infty} S(n,\ 1\rightarrow 2)$$

where

$$S(n,\ 1\rightarrow 2) = (2\pi/\hbar)\ M_1^2 M_2^2 |\bar{V}(\Delta\vec{p}\,'/\hbar - n\vec{k})|^2$$

$$\times\ J_n^2(\lambda'/\hbar\omega)\,\delta(\Delta E' - n\hbar\omega) \tag{27}$$

$$\Delta E' = R_2 + R_1 - (e^2E_0^2/2\omega)(1/k\cdot p_2 - 1/k\cdot p_1) \tag{28}$$

$$\vec{\Delta p}' = \vec{p}_2 - \vec{p}_1 - (e^2E_0^2\vec{k}/2\omega^2)(1/k\cdot p_2 - 1/k\cdot p_1) \tag{29}$$

$$\lambda' = eE_0(\alpha_x^2 + \alpha_y^2)^{1/2}$$

$$\alpha_i = e_i\cdot p_1/k\cdot p_1 - e_i\cdot p_2/k\cdot p_2, \quad i = x \text{ or } y$$

$$R_i = (\vec{p}_i\cdot\vec{p}_i c^2 + m^2c^4)^{1/2}, \quad i = 1 \text{ or } 2\ .$$

Here $p_2 = (\vec{p}_2, iR_2/c)$ and $p_1 = (\vec{p}_1, -iR_1/c)$ are the energy-momentum four-vectors of the positive-energy and negative-energy electrons, respectively, and M_2 and M_1 are wave function normalization factors. The propagation four-vector of the electromagnetic field of the laser beam is $k = (\vec{k}, i\omega/c)$ where $\vec{k} = \omega\vec{e}_z/c$ and

$$k\cdot p_2 = \vec{k}\cdot\vec{p}_2 - \omega R_2/c^2.$$

Equation (27) is the probability for the transition of an electron from a negative-energy state to a positive-energy state. The difference in energy between the two states is $\Delta E'$. Equation (28) implies that the presence of the electromagnetic field of the laser beam changes the energy of an electron by the amount $-e^2E_0^2/2\omega k\cdot p$. In the non-relativistic limit, $k\cdot p$ is equal to $-m\omega$ for a positive-energy electron and $+m\omega$ for a negative-energy electron. The energy of a nonrelativistic positive-energy electron is increased by an amount ε_0 given by Eq. (1), and the energy of a negative-energy electron is decreased by the same amount. Thus the energy gap between the positive-energy and negative-energy states, which is $2mc^2$ in the absence of the electromagnetic field, is widened when the field is turned on.

To find the cross section for pair production, we sum the transition probability (27) over n (from 1 to ∞) and over $\vec{p}_1$ and $\vec{p}_2$. In the limit of large normalization volume the sums over $\vec{p}_1$ and $\vec{p}_2$ become integrals over $\vec{V}_1$ and $\vec{V}_2$. We assume the strong-field case $\lambda' \gg \hbar\omega$ and use an approximation for the Bessel function in Eq. (27) which is valid for large argument. The cross section for pair production is[5]

$$\sigma = 8r_0^2m^8c^3\omega E_0^{-2}(2\pi\hbar)^{-2}\int d^3V_1\int d^3V_2\ M_1^2\ M_2^2$$

$$\times\ |\vec{\Delta p}' - \Delta E'\vec{k}/\omega|^{-4}\ \delta(\Delta E' - \lambda') \tag{30}$$

where $r_0 = Ze^2/mc^2$.

Although Eq. (30) is not an explicit expression for the cross section, some general conclusions may be drawn. It is not necessary that the electrons be accelerated to relativistic velocities for pair production to occur. This is in contrast to the trident[19,20,21] process in which the scattering of a relativistic particle by a nucleus induces pair production. The cross section (30) is expected to decrease as the electron velocity increases, just as the inverse Bremsstrahlung absorption rate decreases for high electron velocity. Since $S(n, 1\to 2) = 0$ when $\lambda' = 0$, most electron-positron pairs are emitted nearly perpendicular to the laser beam.

RELATIVISTIC OSCILLATION OF ELECTRONS

Let us now consider the quantum theory for the relativistic oscillation of electrons in a laser beam. The electrons may be described by the Volkov[15] solution to the Dirac equation or, if the laser beam is very intense, by the corresponding solution to the Klein-Gordon equation. In either case, the time-average kinetic four-momentum (or quasimomentum[22]) of the electron is

$$q_\mu = p_\mu - re^2E_0^2k_\mu/2\omega^2k\cdot p$$

where $r = 1/2$ if the laser radiation is plane polarized and $r = 1$ if it is circularly polarized (cf. Eqs. (28) and (29)). Similar to the definition of the rest mass $p\cdot p = -m^2c^2$, we may define the "effective" mass of an electron in an electromagnetic field by $q\cdot q = -m_*^2c^2$, where

$$m_* = m(1 + re^2E_0^2/m^2c^2\omega^2)^{1/2}.$$

This is the well-known mass shift[23] which occurs in both the classical and quantum treatments of an electron in an electromagnetic field. The energy m_*c^2 may be compared to Hora's[21] results. When a uniform magnetic field parallel to the plane-polarized laser beam is present, the effective mass of an electron with spin S in Landau level N = 0, 1, 2,... is[24]

$$m_* = m[1 + \hbar\omega_c(2N + 1 - S)/mc^2 + e^2E_0^2(k\cdot p)^2/2m^2c^2\omega^2((k\cdot p)^2 - m^2\omega_c^2)]^{1/2}$$

which reduces to

$$m_* = m[1 + \hbar\omega_c(N + 1/2 - S/2)/mc^2+e^2E_0^2/4m^2c^2(\omega^2-\omega_c^2)]$$

in the nonrelativistic limit. Thus a resonance occurs when the laser frequency ω is equal to the electron cyclotron frequency ω_c.

ACKNOWLEDGEMENTS

This work was supported by the U.S. Atomic Energy Commission under Contract AT-(40-1)-2598.

REFERENCES

1. J. M. Dawson and C. Oberman, Phys. Fluids 5, 517 (1962); T. W. Johnston and J. M. Dawson, Phys. Fluids 16, 722 (1973).
2. V. P. Silin, Zh. Eksp. Teor. Fiz. 47, 2254 (1964) [Sov. Phys.-JETP 20, 1510 (1965)].
3. V. P. Silin, Zh. Eksp. Teor. Fiz. 38, 1771 (1960) [Sov. Phys.-JETP 11, 1277 (1960)].
4. Equation (5) is derived from Eqs. (5.1) and (5.5) of Ref. 2 with $\delta_1 = \pi/2$ and $\delta_2 = \pi$; see Ref. 5.
5. J. F. Seely, Dissertation, University of Tennessee (1973).
6. H. A. Bethe, Los Alamos Scientific Laboratory Informal Report No. LA-5031-MS (1972).
7. F. V. Bunkin and M. V. Fedorov, Zh. Eksp. Teor. Fiz. 49, 1215 (1965) [Sov. Phys.-JETP 22, 844 (1966)].
8. Helmut Brehme, Phys. Ref. C3, 837 (1971).
9. J. F. Seely and E. G. Harris, Phys. Rev. A7, 1064 (1973).
10. E. G. Harris, A Pedestrian Approach to Quantum Field Theory (John Wiley and Sons, New York, 1972).
11. E. G. Harris in Advances in Plasma Physics (John Wiley and Sons, New York, 1969), Vol. III, p. 157.
12. R. K. Osborn, Phys. Rev. A5, 1660 (1972).
13. S. Rand, Phys. Rev. 136, B231 (1964).
14. T. P. Hughes and M. B. Nicholson-Florence, J. Phys. A1, 588 (1968).
15. D. M. Volkov, Z. Physik 94, 250 (1935).
16. P. J. Redmond, J. Math. Phys. 6, 1163 (1965).
17. L. S. Brown and T. W. Kibble, Phys. Rev. 133, A705 (1964).
18. V. P. Yakovlev, Zh. Eksp. Teor. Fiz. 49, 318 (1965) [Sov. Phys.-JETP 22, 223 (1966)].
19. F. V. Bunkin and A. E. Kazakov, Dokl. Akad. Nauk SSSR 193, 1274 (1970) [Sov. Phys.-Dokl. 15, 758 (1971)].
20. J. W. Shearer, J. Garrison, J. Wong, and J. E. Swain, Phys. Rev. A (to be published).
21. H. Hora, Nature Phys. Sc. 243, 34 (1973).
22. Ya. B. Zel'Dovich, Zh. Eksp. Teor. Fiz. 51, 1492 (1966) [Sov. Phys.-JETP 24, 1006 (1967)].
23. J. H. Eberly in Progress in Optics (John Wiley and Sons, New York, 1969), vol. VII, p. 359.
24. V. P. Oleinik, Zh. Eksp. Teor. Fiz. 61, 27 (1971) [Sov. Phys.-JETP 34, 14 (1972)].

COLLECTIVE PHOTON EFFECTS: INTENSITY DEPENDENT REFRACTIVE INDEX OF INTENSE LASER BEAMS?*

J.L. Hughes

Laser Group
Department of Engineering Physics
The Australian National University
Canberra, Australia

ABSTRACT

Although the Euler-Heisenberg theory of the scattering of light by light in a vacuum must be considered correct when only a few optical photons are involved, it is by no means certain that this theory remains valid under experimental conditions when the number of photons becomes very large. In this paper the possibility of the existence of collective photon effects is considered, with particular reference to their ability to produce an intensity-dependent refractive index in an optical centre-of-mass interaction region. Experimental preparations being made to search for such an effect are described.

INTRODUCTION

In 1936, Euler and Heisenberg [1][2] considered the scattering of light by light in a vacuum and proposed that the scattering cross-section σ was given by a relationship of the form

$$\sigma \propto \left(\frac{h\nu}{m_e c^2}\right)^6 \qquad (1)$$

provided that the photon energy $h\nu \ll m_e c^2$, where m_e is the rest mass of the electron. Over the past thirty-six years the scattering of light by light in a vacuum has been reconsidered by several other authors [3][4]. All arrived at the same conclusion as Euler and Heisenberg, namely, that the minute size of the scattering cross-

* "Presented at the Third Workshop on 'Laser Interaction and Related Plasma Phenomena' held at Rensselaer Polytechnic Institute, Troy, New York, August 13-17, 1973"

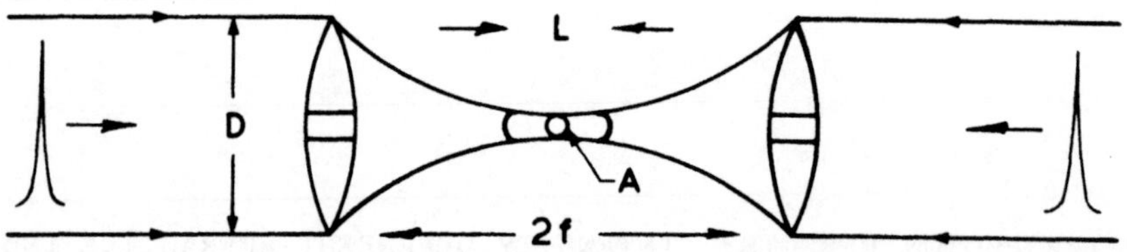

Figure 1. Optical centre-of-mass region formed by two lenses.

section for optical photons precludes any possibility of observing the effect in practice.

However, when we consider the experimental situation in detail it is by no means certain that it corresponds to the theoretical situation envisaged by Euler and Heisenberg. In order to investigate the scattering of light by light in a vacuum it is necessary to form an optical centre-of-mass region of the type shown in figure 1, where two equal but oppositely directed laser pulses are superimposed in the common focus region of two identical lenses. Under such experimental conditions, the photon energy density within the optical centre-of-mass region can for short periods of time become very large, containing $> 10^{20}$ photons for several pico seconds. It is therefore uncertain whether the single particle Euler-Heisenberg theory can still apply under such conditions, because these conditions satisfy $2nh\nu \sim m_e c^2$, rather than $h\nu \ll m_e c^2$, where n is the number of laser photons from each pulse within the optical centre-of-mass region at a given time. If the Euler-Heisenberg theory is not valid under the experimental conditions envisaged then entirely new non-linear optical effects may be operating within an optical centre-of-mass region. Such interactions could well arise from collective photon effects. In this paper we shall consider the practical consequences of the existence of collective photon effects without delving too deeply into their origin.

THRESHOLDS FOR DETECTING COLLECTIVE PHOTON EFFECTS

The upper limit for the detection of collective photon effects would correspond to the threshold for the detection of the scattering of light by light in a vacuum given by the Euler-Heisenberg theory.

The vacuum count rate for the scattering of light by light in the experimental arrangement shown in figure 1 is given according to Euler and Heisenberg by

$$N(\text{VAC}) = \frac{n^2 L\sigma}{AC\tau} \quad \text{counts per pulse} \qquad (2)$$

where n is the number of laser photons per pulse, L the length of the cylindrical centre-of-mass interaction region of cross-sectional area A, σ the Euler-Heisenberg scattering cross-section, c the velocity of light in a vacuum and τ the laser pulse duration. Taking the diameter of the optical centre-of-mass region as $d = f\Theta$, where f is the focal length of the lenses, Θ the divergence of the diffraction limited laser beam, and the F number of the lenses is f/D, with D the diameter of the laser beam, then the cross-sectional area A is given by $F^2\lambda^2$. Taking $L/c\tau \sim 1$ and N(VAC) = 1 and substituting these parameters in relation (2) we obtain,

$$n^2 \sim \frac{A}{\sigma} = \frac{F^2\lambda^2}{\sigma} \qquad (3)$$

Since $L/c\tau \sim 1$, $F \sim 6$ and $\sigma \sim 10^{-64}\text{cm}^2$ for a laser wavelength λ of 1.06 microns, we obtain the number of laser photons within the optical centre-of-mass region as $n \sim 10^{29}$ photons per pulse. This corresponds to a flux density of 10^{28} watts cm^{-2}.

This level of laser flux density is well beyond the practical capabilities of Nd-glass lasers and would imply that such lasers could never provide a means for producing the scattering of light by light in a vacuum.

COLLECTIVE PHOTON EFFECTS

Collective photon effects could arise within intense photon distributions such as would occur in the optical centre-of-mass regions of the type shown in figure 1. The origin of such effects need not concern us here but the result could lead to a drastic reduction in the threshold for observing the scattering of light by light. The drastically reduced threshold would be due to collective photon effects having created a field, akin to an external field, in which the scattering process could take place. According to McKenna and Platzmann [4], the scattering of light by light in an external field is enhanced by a factor expressed as

$$N = N(\text{VAC}) \left(\frac{\text{Energy stored in the field}}{\text{energy of one photon}}\right)$$

$$= \frac{N(\text{VAC})E^2}{8\pi h\nu} \text{ counts per pulse} \qquad (4)$$

where E is the strength of the external field. In practice it is difficult to produce large fields within such a small volume as that characteristic of an optical centre-of-mass region $\sim 10^{-8}\ \text{cm}^3$. For this reason it is not possible to improve the scattering count rate by more than 10^4 using macroscopic fields available in the laboratory. On the other hand a photon field would be effective only within the optical centre-of-mass region and it is possible that a significant enhancement of the scattering of light by light in the presence of such a field would occur in practice.

Taking the collective photon effects to imply that each photon senses the presence of a field, whatever its nature, due to the 2n-1 other photons within the optical centre-of-mass region, we obtain from relation (4),

$$\frac{E^2}{8\pi h\nu} = \frac{K(2n-1)h\nu}{h\nu} \tag{5}$$

where K is the measure of the contribution of each photon to the collective photon effect.

MOST FAVOURABLE THRESHOLD FOR DETECTING COLLECTIVE PHOTON EFFECTS

The presence of collective photon effects could be detected via the enhanced scattering of light by light. Under the most favourable conditions, all the photons within the optical centre-of-mass region would contribute fully to the photon field, i.e. K = 1 in relation (5) which then gives,

$$N = N(VAC)2n \tag{6}$$

or

$$N = \frac{2n^3\sigma}{F^2\lambda^2} \quad \text{counts per pulse} \tag{7}$$

Taking N = 1, F = 6, $\lambda = 10^{-4}$ cms and $\sigma = 10^{-64}$ cm^2 we obtain $n^3 \sim 5 \times 10^{57}$ or $n \sim 2 \times 10^{19}$ photons per laser pulse. This corresponds to a flux density of 4×10^{18} watts cm^{-2}, a level well within the present capability of Nd-glass lasers.

PRACTICAL CONSIDERATIONS

It is unlikely that K~1 above leading to enhancements $\sim 10^{19}$ compared with classical field enhancement of only 10^4. However, quantum field enhancements $\sim 10^{10}$ would not be unreasonable which would result in K $\sim 10^{-9}$ in the above example, implying that each of the 2n photons contributed $\sim 10^{-9}$ of their energy to the collective photon effects. Taking K $\sim 10^{-9}$, and bearing in mind that collective photon effects are here considered to be directly proportional to the total number of photons within a given optical centre-of-mass region, a practical threshold for detecting the presence of collective photon effects would be given by relation (7) as $n^3 \sim 5 \times 10^{66}$ or $n \sim 1.5 \times 10^{22}$ photons per laser pulse. This photon flux would correspond to 4×10^{21} watts cm^{-2} and the required laser parameters would be, E ~4,000 joules, $\tau \sim 10^{-12}$ seconds and the peak power output of the diffraction limited beam 10^{16} watts. These parameters are considered to be within the capabilities of Nd-glass lasers [5]. However, considerable development of present Nd-glass laser systems will be required before the beam quality can be achieved at the very high power levels involved.

INTENSITY DEPENDENT REFRACTIVE INDEX OF INTENSE LASER BEAMS

The foregoing argument indicates that present understanding of the physics of photons may be far from complete. The onset of collective photon effects within the optical centre-of-mass region would imply that the optical photons traversing this region travel more slowly than in a vacuum. Initially this process would either lead to an increase in the number of photons within the optical centre-of-mass region or to a decrease in the volume of such a region depending on whether $L \gtrless c\tau$, both effects leading to an increase in the photon number density of the region. Such an effect would eventually lead to the automatic collapse of the optical centre-of-mass region as a whole, and hence to the realization of extreme photon energy densities in the laboratory. Intensity dependent refractive index effects leading to self-focusing in material media have been considered by several authors [6][7][8].

Determining the behaviour of the collapsing optical centre-of-mass system would be of fundamental scientific importance. For example, does the photonenergy density increase until very high order, as well as the lowest order, particle pairs can be produced, or does some photon-photon scattering process prevent the collapsing optical centre-of-mass region from achieving the thresholds where such pairs can be produced? These are but a few of the exciting possibilities, entirely new approaches to the study of particle physics, that would be open if evidence for an intensity dependent refractive index of intense laser beams could be obtained experimentally.

EXPERIMENTAL TECHNIQUES

Experimental investigations of the properties of optical photons require an optical centre-of-mass region in order to satisfy energy-momentum requirements. The two techniques being developed to realise optical centre-of-mass regions with circularly polarized light are shown in figure 2. The system [9] shown in figure 2a is now operational and is the first apparatus devised specifically to test the Euler-Heisenberg theory for the scattering of light by light in a vacuum. A photograph of the apparatus is shown in figure 3. To form the optical centre-of-mass region, the first pulse in a modelocked pulse train is stored in the optical delay line extending from the common focus of the two lenses to the dielectric coated mirror M (figure 2). The pulse returns to the common focus of the two lenses at the instant that the second pulse in the train enters this region for the first time. By overlapping the two pulses in the focus, the required optical centre-of-mass region is formed. A typical modelocked pulse train used to form the optical centre-of-mass region is shown in figure 4. A much simpler version of the apparatus is shown in figure 2b. This system is now under construction utilizing low F number optics dictated by laser-CTR studies. Most of the initial work to determine the effect of residual air molecules on the signal to noise ratio can be carried out with low F number optics. Once sufficient operating experience has been obtained it is hoped that a much larger system with F6 optics ($L \sim c\tau$) can be installed. It should be noted that the use of circularly polarized light simplifies the problem of preventing the returning pulses in a perfectly aligned system from damaging the laser system because they are strongly attenuated on their return journey through the disk amplifier. If quarter wave plates can be produced with larger diameters than the three inch diameters now available then it will be possible to use this desirable polarization to realise optical centre-of-mass regions characterised by photon flux densities well in excess of 10^{18} watts cm^{-2} in the apparatus described above. If such components cannot be manufactured, then another approach [10] using plane polarized laser beams must be used at very high power levels.

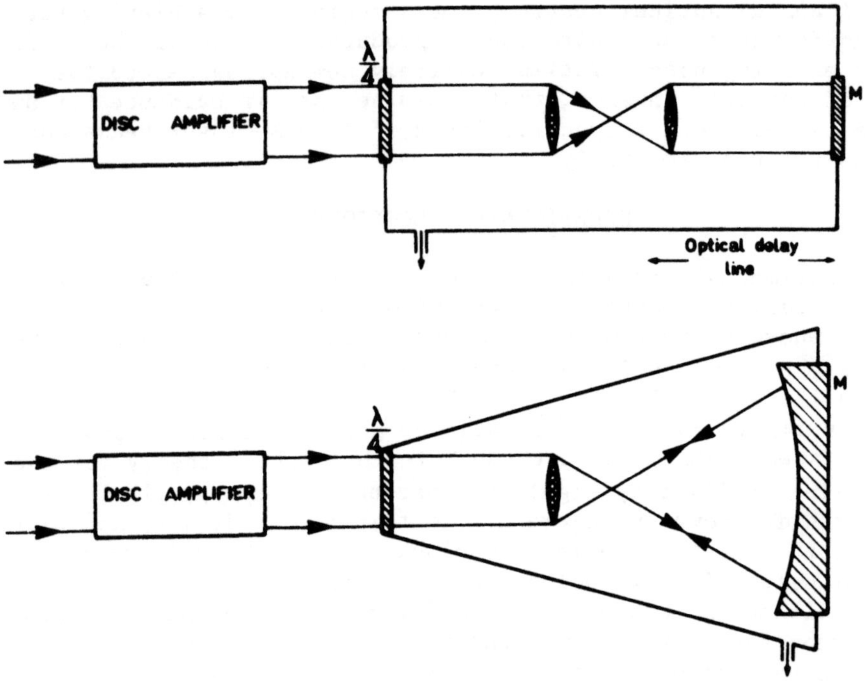

Figure 2. Two types of pulse storage, optical centre-of-mass systems.

Figure 3.
One of the optical centre-of-mass systems mounted on a carriage which can be extracted from the vacuum chamber. The detectors will be mounted on the holder above the two lenses, one of which is hidden by its mounting block. The 2" diameter quarter wave plate is mounted on the end plate assembly.

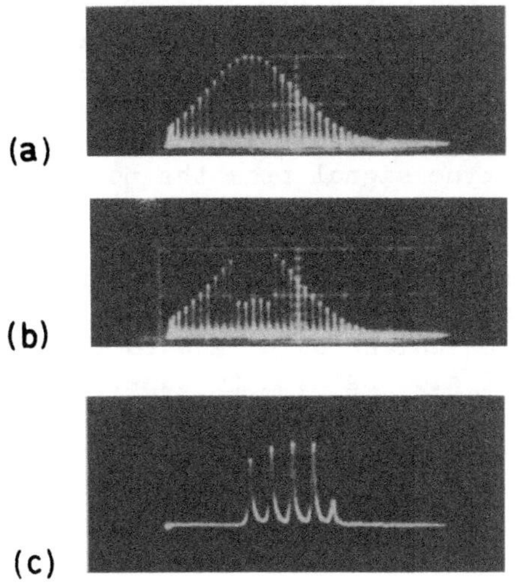

Figure 4.
Sequence for producing the required pulse train. (a) 50n secs per division (b) 50n secs per division (c) 20n seconds per division.

The initial experimental effort involves the determination of noise due to the presence of residual air molecules both inside and in the vicinity of the optical centre-of-mass region. Although these residual particles, which number between 1-1,000 prior to the arrival of the first pulse, are considered a nuisance as far as the study of photon physics in a vacuum is concerned, they could be used to realise the experimental conditions required for the study of relativistic particles in the presence of intense electromagnetic fields in the manner discussed by Hora [11] in this conference. In fact this field of research considerably enhances the scope of the present apparatus. The layout of the apparatus and diagnostic techniques under development are shown in figure 5. To detect any intensity dependent refractive index it is necessary to compare the profiles of the pulses before and after they pass through the optical centre-of-mass region so that any change in pulse profile can be correlated with any change in photon scattering both at the fundamental and possibly the second harmonic laser wavelength. Pulse profile changes are being sought using two different techniques, two photon absorption and a high speed streak camera. Both pulse profiles can be compared during the same exposure under identical conditions. Considerable difficulties remain, however, before picosecond resolutions become routine.

Techniques being developed to detect enhanced photon-photon scattering at both fundamental and second harmonic wavelengths are shown in figure 6. The aim is to confirm that two equal signals can be detected which are emitted simultaneously on either side of the optical centre-of-mass region. To do this, truncated optical fibre bundles will be used so that the signals picked up on one side of optical centre-of-mass region can be attenuated and delayed by precisely known amounts, a process which should resolve the true signal from the noise for different operating modes of the laser system.

Charged particle detector plates are incorporated into the system for two reasons. They will be used to detect multiphoton ionized air molecules which could lead to the generation of optical noise, and they will also be used to superimpose the picosecond duration pulses within the common focus region in conjunction with the optical fibre detectors. To do this the air pressure in the chamber will be raised from $\sim 10^{-9}$ torr to 10^{-3} torr so that multiphoton ionization will be more pronounced. Superimposing two pulses to form an optical centre-of-mass region is a fairly simple process with the present $\sim 10^{-9}$ seconds duration pulses which extend for ~ 30 cms. With pulses of picoseconds duration, however, this process becomes difficult to achieve. It is hoped that the proposed technique may also provide a means of determining the mean duration of the picosecond pulses by traversing one of the modelocked oscillator mirrors by the appropriate amount.

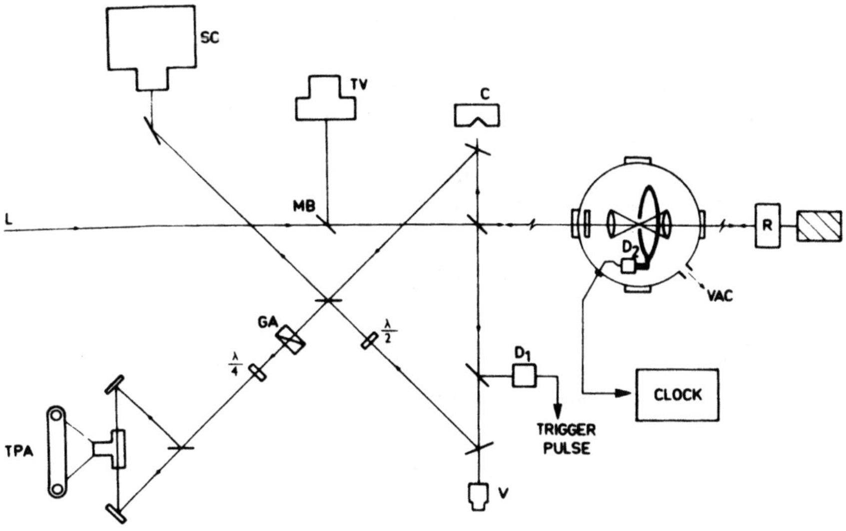

Figure 5.
Layout of the apparatus. L-laser output, MB-movable beam splitter SC-Streak camera, TV-infrared-visible, C-calorimeter, D_2 - optical detector, R-dielectric mirror, GA-Glan air prism, TPA-two photon absorption.

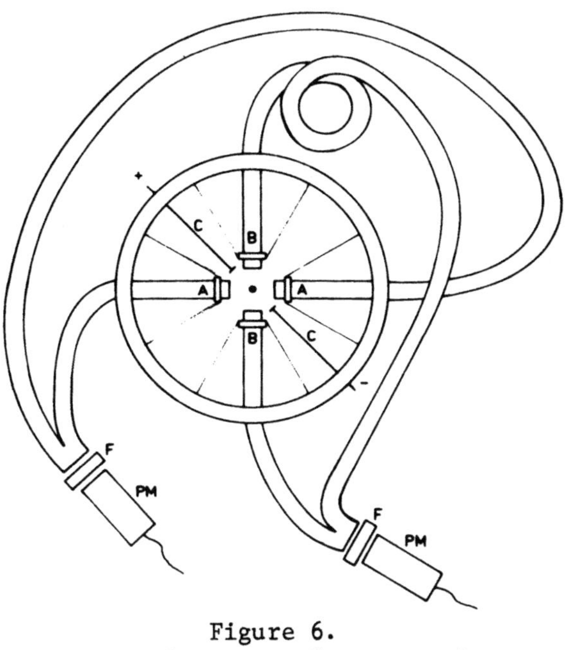

Figure 6.
Plane through the optical centre-of-mass region perpendicular to the direction of the laser beams showing the arrangement of optical fibre and charged particle detectors.

SUMMARY

Attention has been drawn to the possible existence of an intensity dependent refractive index associated with intense laser beams within an optical centre-of-mass region. It has been estimated that the intensity dependent refractive index could be observed at laser beam intensities $\sim 10^{21}$ watts cm^{-2}.

ACKNOWLEDGEMENTS

I would like to thank Dr. Alma Werner for background discussions and Mr. P. Donohue for help with operating the laser and optical centre-of-mass apparatus.

REFERENCES

1. H. Euler, Ann. Der. Physik, 26, 398, 1936.
2. W. Heisenberg and H. Euler, Z. Physik 98, 714, 1936.
3. R. Karplus and M. Neuman, Phys. Rev., 83,776, 1951.
4. J. McKenna and P.M. Platzman, Phys. Rev., 129, 2354, 1963.
5. J. L. Hughes, Applied Optics, 6, 1411, 1967.
6. G. A. Askar'gan, Soviet Phys.JETP 15, 1088, 1962.
7. V. I. Talanov, Radio Phys.7, 254, 1964.
8. R. Y. Chiao, E. Garmaire and C. H. Townes, Phys. Rev. Letters 13, 479, 1964.
9. J. L. Hughes "A pulse storage system for studying the properties of laser photons". Q.E. Jm. Vol. QE-8, No.6, 536, 1972.
10. J. L. Hughes, Optics communications, Vol. 3, No. 6, 374, 1971.
11. H. Hora, "Relativistic oscillations of electrons and protons in laser fields and collective effects stimulating pair production". Presented at this conference.

FEASIBILITY OF X-RAY LASERS†‡

Benjamin Lax

Francis Bitter National Magnet Laboratory* M.I.T.

Cambridge, Massachusetts 02139

A. H. Guenther

Air Force Weapons Laboratory

Kirtland AFB, New Mexico

D.R. Cohn and W. Halverson

Francis Bitter National Magnet Laboratory,* M.I.T.

Cambridge, Massachusetts 02139

ABSTRACT

The basic requirements for an x-ray laser are discussed. Consideration is given to producing soft x-ray laser action by focussing high power radiation from an Nd: glass laser on solid targets. Requirements for x-ray laser gain, pump power and the volume of the laser medium are derived. The feasibility of soft x-ray lasers is compared to that of hard x-ray lasers. Proposed methods to create inversion and current experimental efforts are described.

† Presented at the Third Workshop on "Laser Interaction and Related Plasma Phenomena" held at Rensselaer Polytechnic Institute, Troy, New York, August 13-17, 1973.

‡ This work is supported by the U.S. Air Force Office of Scientific Research.

* Supported by the National Science Foundation.

INTRODUCTION

The development of visible and near infrared lasers in the early 1960's generated speculation about the possibility of extending coherent radiation into the x-ray region of the electromagnetic spectrum, which is usually defined as the wavelength range from 1 Å to 100 Å. It became quickly apparent, however, that no energy source existed which could deposit enough pumping power per unit volume into a medium which might be a laser in the x-ray region. In 1966 the invention of the mode-locked laser offered the possibility that a mode-locked oscillator combined with a train of amplifiers might provide an adequate pump source. During the last few years laser technology has evolved to the point where we can now give serious consideration to using high power lasers for exciting an x-ray laser. A number of other proposals have been made to make an x-ray laser using pumping sources such as x-rays, electron beams, and ion beams.[1, 2, 3] In general, these proposals suffer from the fact that the excitation source cannot impart high enough power density to the x-ray laser medium.[4] Even if inversion were attained, the density of inverted atoms would be relatively small and the gain of the x-ray laser medium would be very low.[5] As will be shown, however, the physical constraints involved in producing an x-ray laser dictate that the system be characterized by a high gain.

Several experiments have been carried out which have a direct bearing on the x-ray laser problem. Very strong x-ray and extreme-ultra-violet line emission has been observed from targets irradiated by high-power lasers. Furthermore, some evidence has been found of anomalous processes in the laser-heated plasmas which may produce non-equilibrium populations of excited atomic states--a requirement for the establishment of an inversion. In experiments at the Culham Laboratory,[6, 7] a ruby laser in the 100 megawatt range was used to irradiate carbon, iron, nickel, and other materials. These experiments gave clear evidence of the generation of a number of strong, discrete emission lines in the x-ray and ultra-violet region from multiply-ionized atoms in the laser-produced target plasmas. Similar experiments were performed by Mallozzi at Battelle.[8] He used a Nd^{3+} laser with 100 J energy and 1-2 nsec pulses, corresponding to a power level of about 10^{11} W. The observed x-ray emission from high Z targets indicated a non-thermal electron energy distribution. On the basis of this Mallozzi suggested the possibility of obtaining stimulated emission under such high level of excitation. Similar work was also carried out at Livermore, Limeil, and the Lebedev Institute. High intensities of x-ray line emission again suggested an anomalous electron energy distribution in the target plasmas. Although these studies indicate that the plasma electron distribution function may be strongly non-thermal and has a considerable

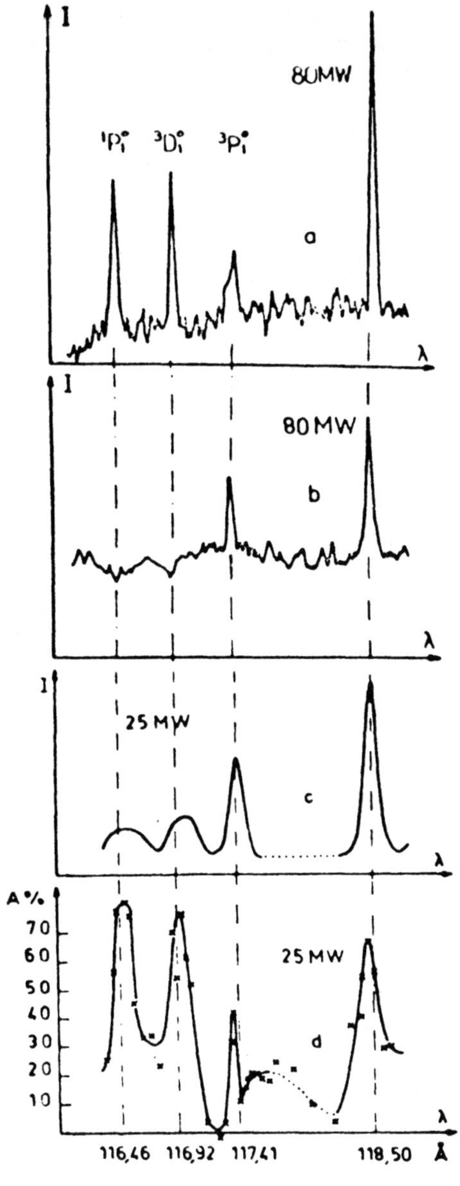

Fig. 1

Spectra from aluminum laser-plasma.
(a) Emission spectrum from hot plasma zone 1.25 mm from target surface;
(b) 0.4 mm from target, 80 MW laser power.
(c) and (d) emission and absorption spectrum from hot zone, 25 MW laser power. From Jaegle, et al. (9).

high energy component, the lack of definitive experimental measurements or theoretical predictions of the form of the electron distribution prevents calculation of inversion rates of excited atomic states in the plasma.

The most definitive evidence of a non-equilibrium population of excited atomic states in a laser-produced plasma was obtained by Jaegle, et al.[9] at the University of Paris in Orsay. These investigators irradiated an aluminum target with a Q-switched laser and observed multiple ionization as expected. They compared the emission spectrum from the hot plasma region with that of the cooler plasma. Radiation from the cool zone was characteristic of a plasma in Boltzmann equilibrium. However, in the hot zone the 117.4 Å line, corresponding to a $2p^6\,{}^1S_0 \rightarrow 2p^5\,4d\,{}^3P_1$ transition, exhibited an anomalously high intensity, as shown in Fig. 1. The absorption of the 117.4 Å line in the plasma also was much less than that of neighboring lines. Peacock et al. have also recently observed anomalous line intensities under similar conditions.[10] These experiments suggest that an inversion might be obtained in the plasma formed by intense radiation on a solid target.

BASIC REQURIEMENTS FOR AN X-RAY LASER

There are three basic requirements for an x-ray laser.

1. There must be a source of very high power density to pump the laser and produce inversion. Presently, only the mode-locked laser-amplifier system appears to fulfill this requirement.

2. A mechanism must be present for creating a population inversion in the medium. This aspect of the x-ray laser is the least understood theoretically. Nevertheless, it has been established experimentally that a non-equilibrium situation exists within the hot plasma created near the surface of targets irradiated by high power lasers. This is a necessary but not sufficient condition to create an inversion. There are several suggestions for providing an inversion mechanism. Slutz et al. at Livermore[11] argue that it may be possible to take advantage of the fact that, under proper conditions, the hydrodynamic expansion time of a laser-produced plasma can be short relative to the radiative recombination time. In this case, it may be possible for three body processes to fill the upper bound levels faster than two body radiative processes fill the lower levels, thereby creating an inversion of the higher states relative to the lower ones. Thus, a 3p state could be inverted relative to the 2s state. The Livermore calculations indicate this is not a likely process for inverting the 2p relative to the 1s state.

A second possible mechanism for inversion which should be considered is that of charge exchange. It may be possible to irradiate a compound in which one of the atoms becomes preferentially excited and gives off its electrons to the other atomic species to produce inversion in a manner somewhat analogous to that of the He-Ne laser.

A third possible means of obtaining inversion is to use laser radiation to produce x-rays which would then selectively depopulate certain levels by inner-shell photo-ionization.[12] An important question to be considered in this approach is the total efficiency of the two-stage process of converting the incident laser energy into stored energy in the inverted atoms. If this efficiency is low, it may be difficult to satisfy the high gain requirements described in the next section.

A fourth possibility, suggested by Bristow et al, at the University of Rochester, would be to take advantage of the difference in the excitation rates of electrons to higher levels from the ground state of a multiply ionized atom.[13] If the excitation rate to the higher of the two excited levels is faster than that to the lower level it may be possible to attain an inversion between these two levels. However, the strong dependence of the excitation rates in question upon electron energy may prevent the attainment of the desired inversion.

In summary, it is extremely difficult to predict under what physical circumstances inversion will be attained. Detailed and careful experimentation will be required to provide an answer.

3. The third requirement for an x-ray laser is that its gain be very high. The x-ray laser action would occur on a very short time scale--at most for the duration of the mode-locked laser pulse, $\tau_L \approx 10^{-11}$ sec. This time is too short for light to travel around a Bragg crystal resonant cavity or a soft x-ray metallic structure at grazing incidence. Hence, an x-ray laser must be a super-radiant source in which a spontaneously emitted photon at one end of the lasing medium triggers a cascade of stimulated transitions which amplify the initial disturbance. For such a source to exhibit laser-like behavior, the product of the gain per unit length, g, and the active distance, z, must satisfy the inequality,

$$gz > c_1 \qquad (1)$$

where c_1 should be the order of 10. Hence, for a laser line to exhibit line-narrowing and other attributes of coherence, large gain is required. It is also necessary that

$$g > \ell \tag{2}$$

where ℓ represents the rate of photon loss. In the cases of interest to us, Eq. (2) will be satisfied if the high-gain requirement of Eq. (1) is satisfied. An additional demand on the gain is imposed in some cases by the fact that the population inversion may be maintained only for the characteristic lifetime of the laser transition (i.e., the laser is self-terminating). In this case the length of the active laser medium is given by

$$z_{max} = c\tau \tag{3}$$

where τ is the time for a transition between the two laser levels and c is the speed of light. τ may be equal to either the Auger time or the spontaneous decay time. The requirement on gain imposed by self-termination is then given by Eq. (1) and Eq. (3) as

$$gc \geq \frac{c_1}{\tau} . \tag{4}$$

The problem of self-termination may be obviated by use of a traveling wave excitation where the laser media is inverted in phase with the movement of the propagating super-radiant x-ray pulse. Traveling wave excitation might be achieved by orienting the pump laser radiation such that it impinges at grazing angle relative to the surface of the target. In this case a heating front would propagate along the surface of the target and net medium gain would develop transiently behind this front.

SOFT X-RAY LASER

We now consider the physical considerations involved in providing a soft x-ray laser.[4] As noted above the question of whether inversion can be obtained is an open one. In this section we assume that inversion can be achieved and consider the physical processes involved in the attainment of sufficiently high gain.

The particular configuration which we will discuss is one in which radiation from a Nd^{3+}: glass laser amplifier with a short (10^{-11} sec) pulse is focussed onto the surface of a low Z solid target, as shown in Fig. 2. Two suitable targets for ~30-60 Å x-ray are carbon, where we are interested in a L→K transition, and aluminum (M→L transition). The pumping laser beam would be perpendicular to the target face or be at an angle in the x-y plane.

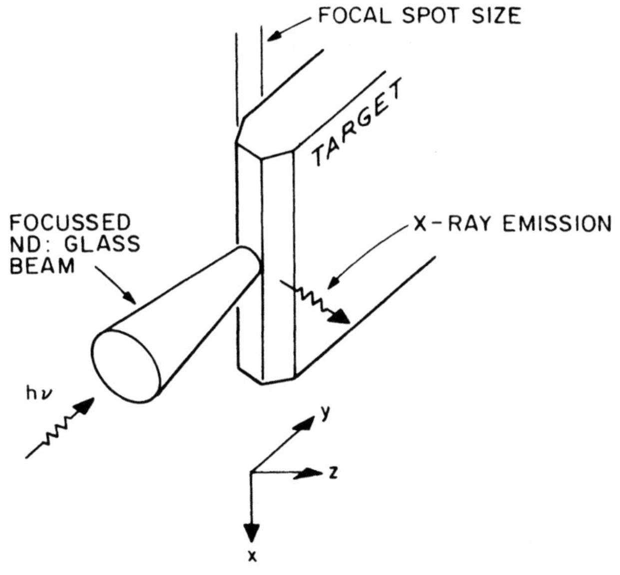

Fig. 2

Schematic diagram of target design and orientation.

The x-ray lasing action is expected to take plase in the hot plasma formed on the plane surface of the target and the coherent x-rays would emerge in fan-shaped beams centered on the $\pm$ z axis. Along the slab (x axis) the x-rays would be re-absorbed by the material which is not excited. The smallest practical dimension which has been achieved in the focal region of a mode-locked Nd^{3+}: glass laser is on the order of 30 μm. Hence, the maximum achievable energy density would be on a region of area $A \approx 10^{-5}$ cm^2. The surface of the target will be fully ionized in a very short time and the laser radiation will penetrate the plasma a distance on the order of the skin depth or about 100 Å. We will assume that thermal conduction and other energy losses will increase the depth of the region of intense heating to about 1000 Å. The pumping laser pulse should have a very short rise time and be free of a lower power precursor in order to heat rapidly the dense plasma.

The minimum pumping power required to create an inversion is given by

$$P \approx \frac{\Delta N \Delta V \hbar\omega}{\epsilon \tau} \tag{5}$$

where ω is the frequency of the x-ray transition in question, ΔN is the net population inversion density in a volume ΔV, ϵ is the efficiency of converting pump energy to energy stored in the inverted atoms, and τ is the characteristic relaxation time for the upper level. The efficiency factor ϵ represents the effect of plasma energy losses, pump beam reflection,[14] the effective ionization and thermal energy of the plasma, and the ratio of inverted atoms to those absorbing pump power without being inverted. If we assume that 10% of the atoms are inverted and that 1% of the incident laser energy is used to excite the upper laser level, then ϵ will be on the order of 0.001. The relaxation time τ will be equal to the spontaneous lifetime, τ_s, except when the time for relaxation by an Auger process, τ_a, is shorter than τ_s. If we take $\tau = \tau_a = 10^{-14}$ sec, $\epsilon = 0.001$., $\hbar\omega = 300$ eV, $\Delta V = 10^{-10}$ cm^3, and $\Delta N = 10^{22}$ cm^{-3} then $P \approx 5 \times 10^{12}$ W. This power level is available today with mode-locked lasers. The power of x-ray laser radiation would be about 5×10^9 W. If $\tau = \tau_s = 3 \times 10^{-12}$ sec, then $P \approx 2 \times 10^{10}$ W, and the power of the x-ray laser radiation would be about 2×10^7 W.

Let us consider a diamond or aluminum target irradiated by 10^{12} W of radiation from a mode-locked laser. Assuming classical absorption, the electrons in the conduction band of the target can gain energies on the order of 1 keV in less than a psec. This energy is sufficient to ionize the core electrons.

The gain of the excited medium is given by the expression

$$g \approx \frac{\lambda^2}{8\pi} \frac{\Delta N}{\Delta \nu \tau_s} \tag{6}$$

where $\Delta\nu$ is the linewidth of the transition in question, and λ the wavelength. For the high density plasmas under consideration here $\Delta\nu$ will be determined primarily by Stark broadening. As a rough approximation we assume that the Stark broadening can be described by the expression for strong Stark broadening for quasi-static hydrogen lines. In this case $\Delta\nu(\text{sec}^{-1}) \approx 0.25\, n^{2/3}$ where n is the electron density in cm^{-3}. If we take $\Delta N \approx 10^{22}\ \text{cm}^{-3}$ and $n = 10^{23}\ \text{cm}^{-3}$ we find that

$$g \approx 4 \times 10^4\ \text{cm}^{-1}$$

and that (7)

$$gz \approx 100$$

This result shows that Eq. (1) is satisfied. An evaluation of photon loss processes such as Compton scattering and photo-ionization has indicated that Eq. (2), i.e., $g > \ell$, is also satisfied. If some mechanism existed such that the inversion lasted longer than the characteristic relaxation time, τ, the additional requirement $gc > c_1/\tau$ would not have to be satisfied. However, it may be the case that inversion will only last for a time τ and that a traveling wave type excitation is not possible. In this situation we must satisfy the additional requirement on gain given by Eq. (4). If we assume that $\tau = \tau_s$, $c_1 = 10$, and $\tau \approx 3 \times 10^{-12}$ sec the gain required by Eq. (4) is

$$g \gtrsim 10^2\ \text{cm}^{-1} \tag{8}$$

This requirement is indeed satisfied by the gain of 4×10^4 given in Eq. (7). However, if $\tau = \tau_a = 10^{-14}$ sec then we require $g \gtrsim 3 \times 10^4$ and Eq. (4) is only marginally satisfied by this value for the gain.

SCALING RELATIONS: HARD VS. SOFT X-RAY LASER

It is possible to develop some scaling relations based upon the requirements that were previously discussed. These scaling relations indicate that substantially more power is required to produce a hard x-ray laser than is necessary for a soft x-ray laser. We begin with a determination of the minimum volume required for sufficient gain and then use this relation in conjunction with Eq. (5) to calculate the minimum power required for x-ray laser action.

Let us assume that the volume of excited atoms is in the form of a rectangular solid with dimensions x, y and z and that the pump laser radiation is impinging along the y direction. We desire that lasing occur preferentially in the z direction and thus require that $z \gg y$ and x. We will assume that z and x are determined primarily by the size of the laser focal spot and that they are related by $c_2 x = z$ where c_2 is a constant. The y dimension will be determined by the effective absorption length of the pump laser radiation. In order to satisfy the requirement $gz > c_1$ the minimum volume ΔV that is necessary is given by

$$\Delta V = xyz = \frac{yz^2}{c_2} = \frac{yc_1^2}{c_2 g^2} \tag{9}$$

Now since $\tau_s \propto \lambda^2$, we can write the gain as

$$g = \frac{\lambda^2}{8\pi} \frac{\Delta N}{\Delta\nu\tau_s} = \frac{c_3 \Delta N}{\Delta\nu} \tag{10}$$

where c_3 is a constant. If we take $\Delta\nu$ to be equal to $c_4 n^{2/3}$ then Eq. (10) becomes

$$g = \frac{c_3 \Delta N}{c_4 n^{2/3}} \tag{11}$$

Combining Eqs. (9) and (11) gives the result

$$\Delta V = \frac{yc_1^2 c_4^2 n^{4/3}}{c_2 c_3^2 \Delta N^2} \tag{12}$$

Now, if we let $\Delta N = fn$, where f represents the ratio of inversion density to electron density,

$$\Delta V = \frac{c_1^2 c_4^2 y}{c_2 c_3^2 f^2 n^{2/3}} \tag{13}$$

The minimum power given by Eq. (5) is then

$$P = \frac{c_1^2 c_4^2 y \hbar\omega n^{1/3}}{c_2 c_3^2 \epsilon f \tau} \tag{14}$$

From Eq. (14) it can be seen that if $\tau = \tau_s \propto \lambda^2$ then $P \propto \omega^3$. Thus, the minimum power required for a 10 keV x-ray laser is at least 10^3 times greater than that required for a 1 keV laser Our previous calculations showed that if $\tau = \tau_s$, approximately 2×10^{10} W would be required for a soft x-ray laser in the 300 eV range; we deduce that about 10^{15} W would then be required for a hard x-ray laser in the 10 keV range. This amount of power is beyond the limits of present technology.

A requirement on the plasma density can be derived by considering the implications of a finite time of inversion. Combing Eqs. (4) and (6) we have the result

$$\frac{\lambda^2 \Delta N \tau}{8\pi \Delta\nu \tau_s} > \frac{c_1}{c} \tag{15}$$

Writing $\Delta N = fn$ and $\Delta\nu = c_4 n^{2/3}$ Eq. (15) becomes

$$n > \left(\frac{8\pi c_4 \tau_s}{\lambda^2 fc\tau} \right)^3 \tag{16}$$

Or, if $\tau = \tau_s$

$$n > \left(\frac{8\pi c_4}{\lambda^2 fc} \right)^3 \tag{17}$$

Thus, we see that a short duration of inversion time implies that n must be greater than some critical value. Furthermore, since n varies as λ^{-6}, a decrease in wavelength in going from a soft to a hard x-ray laser greatly increases the minimum value for n. Using reasonable numbers it can be shown on the basis of these considerations that a self terminating x-ray laser in the 10 keV range does not look feasible. However, if the density can be increased by implosion or compression, it may be possible to decrease the minimum wavelength for self-terminating laser action.

EXPERIMENTAL STUDIES

Kepros et al.[15] at the University of Utah claimed to have obtained x-ray lasing action by focussing approximately 10^9 W of radiation from an Nd: glass laser onto a $CuSO_4$ solution. It was claimed that the x-rays were copper-K radiation around 1.5 Å; however, the requirements on minimum power given by Eq. (14)

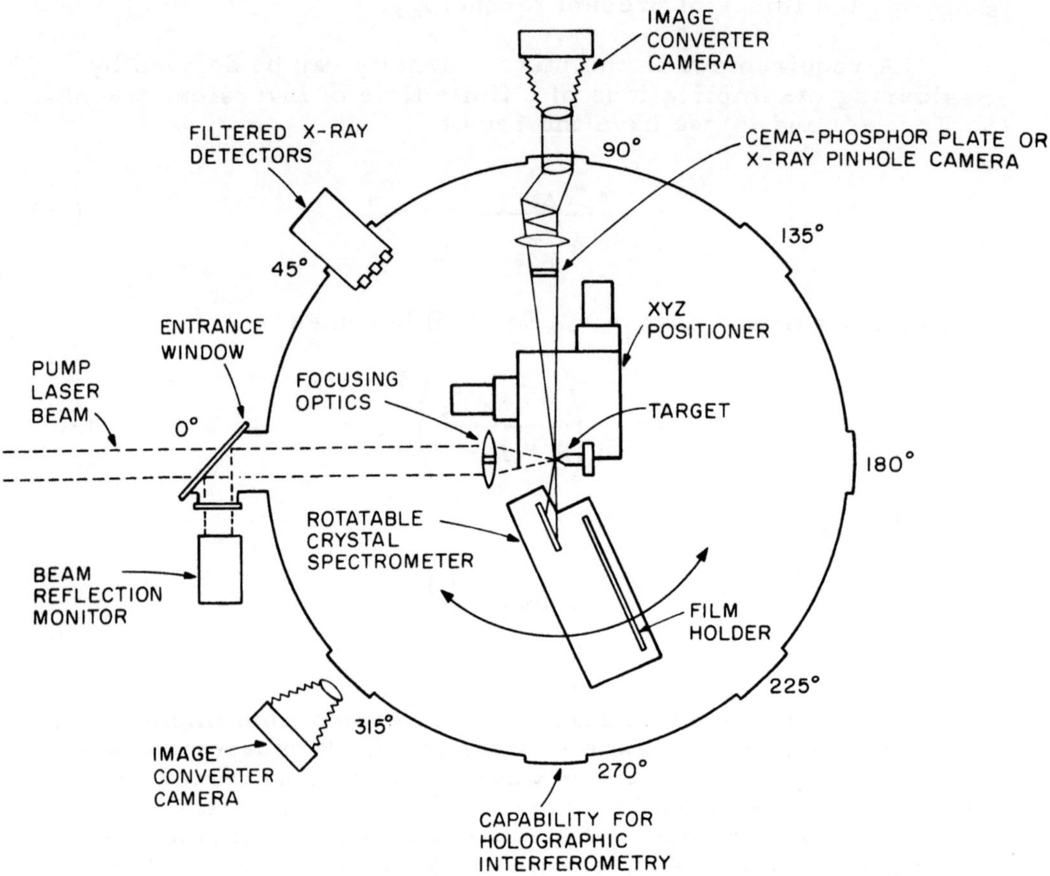

Fig. 3

Schematic drawing of soft x-ray laser experimental chamber.

indicate that for the dilute solution of $CuSO_4$ used, a minimum pump power of approximately 10^{18} - 10^{19} W would be necessary to achieve laser action. This amount of power is about nine decades greater than was used. Furthermore, this requirement of 10^{18} - 10^{19} W is sufficient only in the case where it can be assumed that the inversion lasted for longer than $\tau = \tau_s \approx 10^{-15}$ sec. It is more likely that a "quasi-cw" inversion could not be maintained, in which case the requirement on density given by Eq. (17) would also have to be fulfilled. This requirement could not possibly be satisfied in a dilute $CuSO_4$ solution. Recently the experimental validity of the Utah results has been questioned by Boster at Livermore.[16] He has claimed that the spots on photographic film that were attributed to x-rays may well have resulted during the handling of the x-ray film packs.

Two experimental programs to demonstrate the feasibility of attaining soft x-ray laser action are presently under way. At the University of Rochester an attempt will be made to achieve lasing action at 128 Å from oxygen VII. A collaborative program between the Air Force Weapons Laboratory and the M.I.T. Francis Bitter National Magnet Laboratory is being initiated to study the feasibility of creating x-ray laser action in the 30 to 60 Å range from carbon and aluminum targets.

Figure 3 shows a schematic diagram of the experimental arrangement which is currently being assembled for the AFWL-MIT experiment. The output from an acousto-mode-locked Nd:YAG oscillator will be used to generate ~20 psec pulses. After pulse selection and sharpening the low energy pulse will be used as an input for a four-pass YAG amplifier prior to amplification by a CGE VD-640 laser. This system will produce output powers in the 10^{12} - 10^{13} W range. The 1.06 μm laser radiation will impinge upon specially prepared targets in the form of small knife edge slabs as shown in Fig. 2. Both carbon and aluminum targets will be used in the initial experiments; later work may involve the use of boron, silicon, and composite targets.

The detection of x-radiation and verification of laser action will be undertaken by the coordination and comparison of a number of x-ray detection techniques.

(1) We will look, first of all, for directional effects. If laser action occurs, the x-radiation will be the shape of a fan whose broad side is in the plane of the excited layer and narrow in the direction perpendicular to both the pump laser beam and the target axis. An angular concentration of x-ray energy into a beam will be detected by a channel electron multiplier array (CEMA). The small size of each element in the array will permit the determination to high resolution of the angular intensity distribution from the target.

(2) The second effect which we shall study is the wavelength narrowing of the x-ray transition at the onset of laser action. The line narrowing will be observed with a bent crystal spectrometer in conjunction with a phosphor backed electron multiplier array or photographic recording film. As shown in Fig. 3, the bent crystal spectrometer will be rotatable around a point centered at the target irradiation site to study the x-ray line width and intensity as a function of angle. The wavelength resolution associated with crystal diffraction techniques in the 40 Å range is about 1% and is probably sufficient to enable an unambiguous determination of the presence of lasing action within the target medium. Intense continuum x-radiation from the target may limit the resolution of this technique.

(3) A third experiment to demonstrate the presence of laser action will be to vary the output of the Nd: glass laser system above and below the threshold for laser action and to look for nonlinear changes in the output of the x-ray line radiation. Here we may be limited by the dynamic range of input laser power for which laser action is observed.

CONCLUSIONS

The detailed physics of producing x-ray lasing action is not well understood. However, basic considerations such as those discussed in Secs. 2-4 can be used to estimate both minimum power requirements and requirements on the density of inverted atoms. These considerations indicate that it may be possible to demonstrate soft x-ray laser action using a presently available high power mode-locked laser as a pump. It appears to be significantly more difficult to develop a hard x-ray laser; a hard x-ray laser is probably not possible with present technology. The continued development of high power lasers could eventually change this situation. The least understood aspect of creating an x-ray laser is that of obtaining inversion. However, experimental results showing non-equilibrium atomic electron populations give hope that inversion may indeed be obtainable. The experimental difficulty in obtaining unambiguous evidence of x-ray laser action requires very refined diagnostics; great care must be taken in planning and interpreting experiments.

If x-ray lasers were developed they could make very important contributions to a variety of activities.[17] An x-ray laser would be an important tool for the metalurgist in the study of structural defects and failure in materials. High speed photography of material response and behavior using x-ray laser radiation would greatly increase our knowledge of basic material properties. In addition, x-ray lasers would make possible studies of the interaction of x-rays with materials at very high incident intensity levels. The attainment of x-ray lasing action may make possible the development of high resolution phase contrast spectroscopy in the x-ray region. It may also be possible to use an x-ray laser to obtain holographic images. Three dimensional holography in the x-ray region could greatly add to knowledge of the structure of large biological molecules. X-ray lasers might also play an important role in medicine in the treatment of tumors. For physicists an x-ray laser would provide a powerful new tool for solid state spectroscopy. Finally, x-ray lasers could be used for measurements of plasma density needed to assess the feasibility of obtaining controlled thermonuclear fusion by the compression of DT pellets by intense laser radiation. Therefore, in spite of the theoretical and experimental difficulty of demonstrating the feasibility of x-ray lasers, the potential applications justify serious efforts to achieve the goal of producing x-ray laser action.

REFERENCES

1. M.A. Duguay and P.M. Rentzepis, Appl. Phys. Lett. 10, 350 (1967).
2. R. McCorkle, Phys. Rev. Lett. 29, 982 (1972).
3. M. Scully, private communication, 1973.
4. B. Lax and A.H. Guenther, Appl. Phys. Lett. 21, 361 (1972).
5. Recent developments in the focussing of relativistic electron beams and observations of extremely fast braking of the beams in targets suggest that a high gain e-beam pumped x-ray laser might eventually be possible.
6. B.C. Fawcett, A.H. Gabriel, F.E. Irons, N.J. Peacock and P.A.H. Saunders, Proc. Phys. Soc. (London) 88, 1051 (1966).
7. B.C. Boland, F.E. Irons and R.W.P. McWhirter, J. Phys. 13, 1180 (1968).
8. P.J. Mallozzi, H.M. Epstein, R.G. Jung, D.C. Applebaum, B. P. Fairand and W.J. Gallagher, Battelle Research Report, January 1972.
9. P. Jaegle, A. Carillon, P. Dhez, G. Jamelot, A. Sureau and M. Cukier, Phys. Lett. A36, 167 (1971).

10. N.J. Peacock, private communications, 1973.

11. S. Slutz, G. Zimmerman, W. Lokke, G. Chapline and L. Wood, talk at USNRL X-Ray Laser Symposium, May 1973.

12. M.H. Key, private communication, 1973.

13. T.C. Bristow, M.J. Lubin, J.M. Forsyth, E.B. Goldman and J.M. Soyres, Optics Commun. 5, 317 (1972).

14. It has been reported recently by G.V. Sklizkov and others that the plasma reflectivity decreases quite dramatically at high laser intensity.

15. J. Kepros, E. Eyring and F. Cagle, Proc. Nat. Acad. Sci. 69, 1744 (1972).

16. T. Boster, Appl. Opt. 12, 433 (1973).

17. A. Schawlow, CLEA Conference, Washington, D.C. May 1973.

IS THERE A HIGH FREQUENCY LIMIT TO LASER ACTION?*+

George C. Baldwin

Division of Nuclear Engineering
Rensselaer Polytechnic Institute
Troy, New York

The usefulness of radiation of any frequency is greatly enhanced by even a modest improvement in its degree of coherence. In the very short wavelength end of the spectrum, coherent sources have yet to be developed; if and when they are, they should enable new processes and new kinds of measurement, just as they have in the optical, infrared and microwave regions. Thus, the development of new coherent sources, and, in particular, of lasers for generating radiation of shorter wavelengths is desirable.

The evolution of stimulated emission devices, now collectively known as "lasers," has been marked by a not-always steady, but nevertheless sustained progression toward ever-shorter wavelengths or higher frequencies. It is generally recognized that higher frequency entails greater difficulty, and this difficulty can be traced to the frequency dependence of Einstein's A and B coefficients. However, aside from indicating further increase in difficulty as one goes to still higher frequencies, there is nothing in the third-power dependence of this ratio which points to any ultimate limit to the frequency at which stimulated emission can be employed to generate radiation. In fact, astrophysicists tell us that stimulated emission plays an important role in stellar interiors even at photon energies much higher than those which have so far been generated by means of lasers.

The first member of the family, the ammonia beam maser (1), which appeared in 1955, operated at a frequency of 23,800 megahertz, or a wavelength of 1.26 cm. Soon other devices for generating mi-

---

*Presented at the Third Workshop on "Laser Interaction and Related Plasma Phenomena" held at Rensselaer Polytechnic Institute, Troy, New York, August 13-17, 1973.

+Supported by the U.S. Army Research Office-Durham and the National Aeronautics and Space Administration.

crowave radiation by stimulated emission were constructed, and the acronym "maser," characterized such devices.

In 1958, Schawlow and Townes (2) showed that stimulated emission devices should be possible for infrared and optical frequencies, and, within two years (3) the first optical maser was operated at 0.7 micrometer. Soon a great variety of infrared and optical devices employing stimulated emission were developed, and the new acronym, "laser," now seems to have usurped the role of designating the family surname.

Schawlow and Townes had also briefly considered the possibility of stimulation devices for still higher frequencies, but, concluding that this was not feasible, stated that, "Unless some radically new approach is found, they cannot be pushed to wavelengths much shorter than those in the ultraviolet....."

That prediction proved ultraconservative:

Within the decade since publication of the Schawlow-Townes paper, ion lasers were operating at wavelengths as short as 3000 Å (4).

In 1970, the IBM and NRL groups (5,6) developed lasers, based on the traveling wave excitation method of Shipman (7), which might be regarded as a "radically new approach." These have operated in the Werner and Lyman bands of molecular hydrogen down to 1100 Å.

Recent announcements of lasing in the xenon molecule system at 1700 Å encourage hopes that many other laser systems will in time be devised for the far ultraviolet.

Lax and Guenther (9) have discussed the feasibility of x-ray lasers. Demonstration of stimulated emission, if not of actual lasing in the soft x-ray region, appears likely (10).

Finally, proposals which appeared in the early 1960's for gamma ray (11) lasers are being revived, beginning with a paper by R. V. Khokhlov (12) at the International Quantum Electronics Conference in 1972. The number of outstanding Soviet laser experts who expressed the conviction that gamma ray lasers are feasible was an outstanding feature of the recent Vavilov Conference in Novosibirsk. They were not deterred by a detailed recital by this author (13) of the very great difficulties which it would involve.

The "limit" postulated by Schawlow and Townes (2) was based on technological difficulty rather than on a fundamental phenomenon. As is always the case with such limits, it has been breached. It is appropriate therefore to reopen the question, but in a form less likely to invite refutation of the answer; i.e., to ask whether any fundamental limit does in fact exist to the generation of radiation by means of the laser principle.

Should the answer prove affirmative, that, too, may eventually be proved wrong. We hope so, since our purpose is to pose a challenge rather than to discourage scientific ingenuity.

However, before a meaningful answer can be given to the question, one must define what is meant by the term "laser action," since there are so many kinds of lasers. In all external aspects; for example, physical appearance, microwave masers and ultraviolet lasers have absolutely nothing in common; their one common feature is an abstract physical principle by which they operate; namely, the coherent addition of quanta to a limited number of modes of the radiation field through the mechanism of stimulated emission.

Since we have posed a general question, its answer must not depend on the specific details of any special type of laser system or upon a temporary deficiency of technology. It is possibility, rather than feasibility or practicality, which we seek to establish.

Expecting challenge or refinement, I offer the definition of the generic term "--ASER" as follows:

> a device, by means of which energy, initially in incoherent form, is temporarily stored in quantum states of a material medium, and subsequently transformed into a limited number of modes of electromagnetic radiation, by means of stimulated transitions between those states.

This definition, by specifying an incoherent input, excludes optical frequency conversion devices which use lasers as their pumps. However, it will turn out that the high frequency limits of either type of device are comparable.

The pump is to provide energy at sufficient intensity that the density of excited states is adequate to assure net gain by stimulated emission. There are, of course, practical limitations to the intensity of every known kind of pumping source. However, "practical" limitations often prove to be functions of time; i.e., they are technological rather than absolute. Progress in the development of lasers has been largely due, not only to improvements in existing pumps, but also to the development of entirely new techniques of pumping, including electron beams, traveling wave arc discharges, laser heating of plasmas, etc.

One paper at this conference considers the use of nuclear radiation for pumping optical lasers (14). Nuclear reactions can be expected to join the list of pumping techniques if lasers are to be developed for the hard x-ray and gamma-ray regions.

Thus, the variety of methods and apparatus for pumping lasers is so great that no general assertions are safe as to ability to pump. Admittedly, power requirements become fantastically high for wavelengths shorter than the soft x-rays discussed by Lax and Guenther (9), if those wavelengths are to be generated by atomic transitions. Before jumping to conclusions, however, it is well to look for other types of transition, since the properties required of the pump are dependent on those of the system to be pumped.

Resuming the task of definition, we describe the laser medium as

> a system of many identical members, each characterized by a pair of combining energy levels separated by an energy interval proportional to the frequency to be amplified.

Present-day lasers (15) employ a great diversity of such systems: vibrational, rotational and electronic states of molecules, electronic states of atoms, ions and radicals, impurity ions in crystals and glasses, chelates, spin systems, etc.

Notably absent from the list of operating laser systems are deeply lying inner shell states of atoms or ions; excited states of compound or isomeric nuclei; nuclei on the point of fission; particle beams; positronium; matter-antimatter pairs. Before asking whether any particular one, all of these, or one we have overlooked can be used as a laser medium, let us try to establish a limit to the laser principle which does not depend on properties of individual systems which are not essential to the stimulation process.

Our point in suggesting these additions to the list of possible systems is simply that several kinds of quantized systems do exist which have transition energies far higher than those used in existing lasers, and that means are known for exciting them, at least in small density, giving hope that appropriate pumping procedures may eventually be devised, if, when they have been pumped, such systems can be made to lase.

The important properties of the quantum system are:

transition energy $E_2 - E_1 = h\nu$

lifetimes of the respective states T_1, T_2

radiation width of the upper level γ_r

natural width of the line γ

statistical weight ratio of the states g_1/g_2

stimulation cross section $\sigma_s = (\lambda^2/2\pi)(\gamma_r/\gamma)$

removal cross section σ_i

The total population of laser-active states in the material

medium is apportioned among excited and lower states in a manner determined by the pump mechanism and intensity and on the history of the system. We designate the respective concentrations of excited and lower states and of the various inert species present by N_2, N_1, and N_i $(i > 2)$, respectively.

Corresponding to these concentrations and cross sections, the laser medium possesses interaction coefficients

$$\mu = \sum_i N_i \sigma_i$$

for photon removal and

$$\varkappa = (N_2 - \frac{g_1}{g_2} N_1)\ \sigma_s(f\gamma/\Gamma)$$

for stimulation. The new factors f and Γ will be explained shortly.

Turning now to the radiation field, the laser may amplify radiation in either traveling-wave or standing-wave modes, depending on its construction. The more efficient standing-wave type is preferred when reflecting surfaces can be employed. Unless it proves possible to use Bragg reflections or the anomalous transmission of Bragg-directed radiation in crystals (see below), it is unlikely that reflectors can be provided. It will be necessary to depend on the geometry of the active medium to limit the number of modes into which stimulated emission can develop appreciable intensity, and to use a traveling wave system. The condition for growth of standing wave amplitude is more stringent than that for traveling waves, which is simply that the linear gain coefficient for stimulated emission $\varkappa$ exceed the linear removal coefficient μ . The standing wave case involves additional losses at reflecting surfaces. In optical lasers, these losses are more important than those due to imperfect transparency of the medium. At higher frequencies, that is no longer the case.

The pumping power requirement may be expressed as

$$P \geq N_2 h\nu/T_2 \geq N_2 h\nu\gamma_r .$$

The threshold pump power is that which can establish an upper-state population N_2 which permits the gain-in-excess-of-loss inequality to be satisfied. One might jump to conclusions that, if no limit is to be placed on pump power, it will always be possible to satisfy the gain condition. That is not the case.

Before showing this, however, one should take note of new aspects which appear as one passes beyond the soft x-ray region.

The first is the availability of states of relatively long lifetime. Lifetimes of atomic inner-shell vacancy states are indeed short, and the corresponding pump power required to saturate the populations in material of average density is fantastically great. However, this is not the case with the long-lived nuclear states known as "isomers" which are metastable in consequence of high nuclear spin.

Figure 1 compares several kinds of system, with lifetimes ranging from 10^{-18} to 10^{9} seconds. The shortest are associated with electric dipole transitions, the next group, with electric quadripole and magnetic dipole, etc. The energy-lifetime relationships one might at first expect are smeared out somewhat by alternative processes of de-excitation: Auger effect in atomic levels, internal conversion and beta emission in nuclear isomers. These competing processes tend to reduce the stimulation cross section through the factor (γ_r/γ) because they increase the natural width. The long lifetimes of the isomers of course imply small radiation widths γ_r; how that affects the stimulation cross section depends on the ratio of the total line width Γ to the natural line width γ. This point will be considered below. For the present, the point is that long lifetime permits accumulation of excitation over extended time intervals, so that high transition energy is not inevitably associated with high pump power. However, the cost of escaping from that problem is that such lasers will be single-pulse devices, and that one must use nuclear systems.

Another new aspect is recoil from the high momentum associated with each stimulated emission event (Figure 2). As one goes to higher quantum energies, the energy of recoil, which must be furnished at the expense of the transition energy, eventually exceeds the width of the natural line. The center of the emission line is displaced to lower, and that of the absorption line to higher energy. Marcuse (16) pointed out that this might eliminate the need for population inversion. Baldwin, Neissel and Tonks (17) pointed out that recoil also furnishes a mechanism for separating the excited state at the instant of its formation, if the recoil energy is sufficient to rupture a chemical bond.

Unfortunately, the gamma ray line is not merely shifted, but broadened. There is, however, an unshifted component if the emitting atom or nucleus is bound to a crystal lattice, which absorbs the recoil momentum as a rigid body (Figure 3). This is called the Mossbauer effect (18); it is observed for many gamma ray emitting nuclei in the shortest-lifetime group of Figure 2. It should also be seen for x-ray emission, but to my knowledge no one has reported it. The effect is manifested by the resonant absorption of gamma radiation from a solid source by ground state nuclei in a second solid. Were the latter in excited states, one would observe stimulated emission.

The fraction of all emission events which are not recoil-shifted

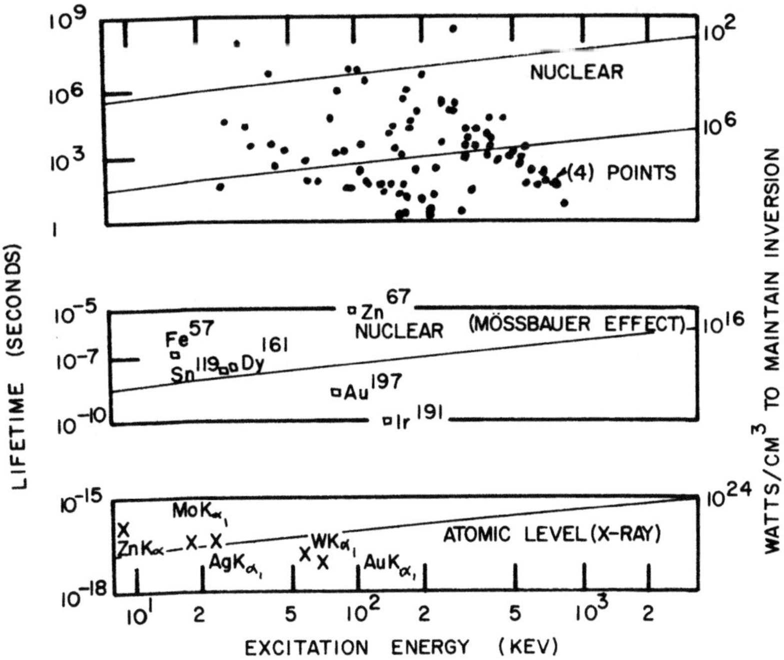

Fig. 1. Half-lives and transition energies of isomeric pairs; the left-hand scale for sloping lines is the spontaneously emitted power at saturation.

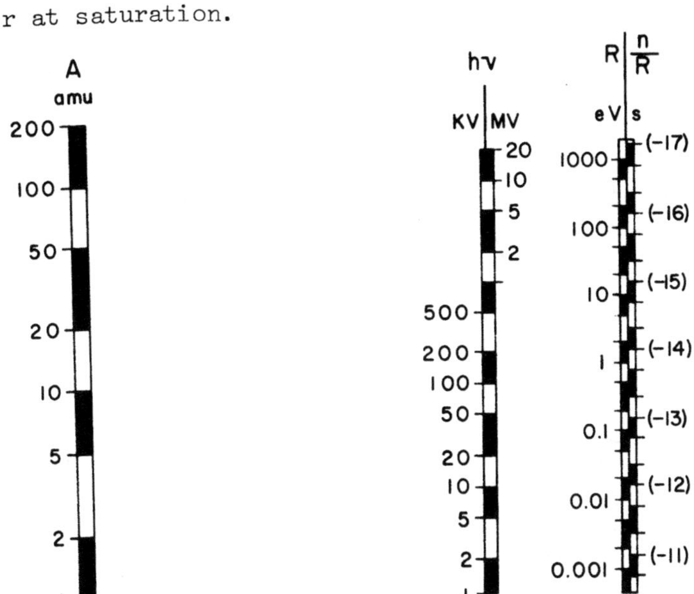

Fig. 2. Nomographic relationship of recoil shift of the energy hν of a photon emitted by a free atom of mass A.

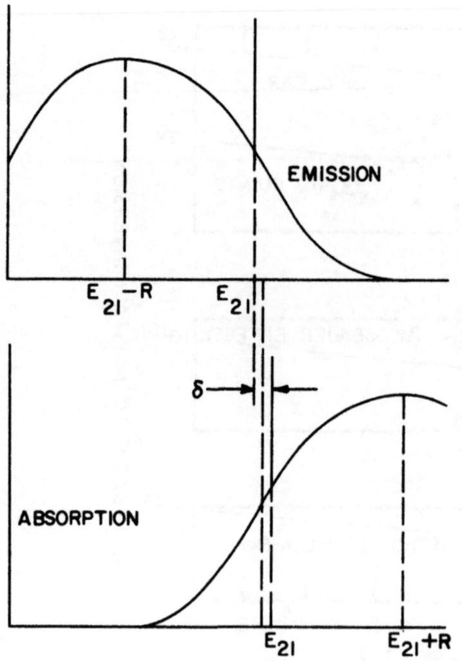

Fig. 3. The thermally broadened "Doppler" and the recoilless "Mossbauer" lines for photon emission and absorption, respectively, slightly displaced by inhomogeneous line perturbation δ.

Fig. 4. Origin of the Bragg-Bormann effect in a perfect lattice, showing the two possible normal modes.

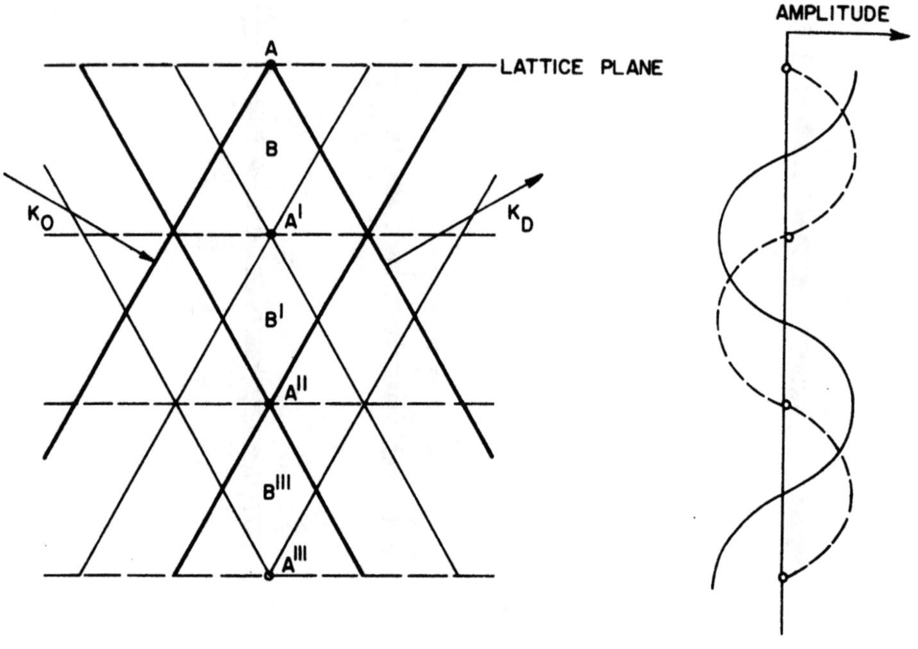

is called the Mossbauer fraction, f (19); it is practically equal to the Debye-Waller factor of x-ray diffraction theory (20). Although the Mossbauer line is not thermally broadened, but has the natural width, it is perturbed by many effects (see below) which produce an inhomogeneous line width Γ appreciably greater than the natural width γ. However, resonance absorption is a well-known phenomenon in many nuclear isomers.

Another new phenomenon appears when the wavelength shortens to values comparable to the interatomic spacing. Bragg diffraction then becomes possible; moreover, the interference of direct and scattered waves propagating at the Bragg angles (Figure 4) establishes pairs of electromagnetic field modes, one, with nodes at the lattice points, the other, with antinodes at the lattice points. The former is transmitted with relatively little absorption in perfect crystals at low temperature (21). This phenomenon, "anomalous transmission," has been considered as a possible basis for a waveguide or resonator (17), but, as the weakly absorbed mode will also be less capable of stimulating emission, it is unlikely to be the basis of a laser active medium.

Finally, and most important to the subject of this paper, is the appearance of new mechanisms for photon removal, which decrease the transparency of the laser medium to the radiation which is to be stimulated within it.

In the soft x-ray region, the photoelectric effect (22) is the dominant mechanism of absorption. An atom absorbs the photon to form an ion pair. The recombination radiation is of lower frequency and incoherent with the primary radiation. The cross section for photoionization varies as the 5-th power of the atomic number and as the 7/2-power of frequency; thus, at increasing energies, transparency improves, and, as is well known, x-rays become more penetrating.

As one nears the relativistic region, Compton scattering by free or loosely bound electrons becomes the dominant removal mechanism. Its effect is greater than one might infer from tables of energy absorption, since the secondary, scattered photon is of lower frequency and incoherent with the incident radiation. Every Compton scattering event removes a photon from the traveling wave. The cross section per atom for Compton scattering is proportional to the number of electrons which are relatively unbound, approaching Z at high photon energy, where it varies nearly inversely with energy. Thus, it too, decreases with increasing frequency of radiation.

However, this trend is neither fast enough nor does it persist indefinitely. Above the threshold of 1.02 MeV, positron-electron pair production in the Coulomb field of each nucleus and of each electron (22) increases the removal cross section, contributing an amount proportional to $Z^2 + Z$ and to a logarithmically-rising function of the photon energy above threshold. In the extreme relativistic limit, the logarithmic factor approaches a constant value, but never decreases.

The basic gain condition for traveling waves in an extended laser medium is

$$N^* \sigma_s > \Sigma N_i \sigma_i$$

with N* representing the effective density of excited states. If we accept the unlikely suggestion of Marcuse (16) that recoil makes inversion unnecessary, or the assurances of radiochemists that some isomeric nuclei can be isolated in a condition of nearly complete inversion, and if we admit only negligible amounts of inert substances, then, in the extreme limit

$$N^* \leq N_2$$

$$N_i \rightarrow 0 \qquad i \neq 2$$

and the gain condition reduces to a comparison of stimulation and removal cross sections.

The stimulation cross section contains two factors, $\lambda^2/2\pi$ and γ_r/Γ. The second depends on the specific system and on its environment. Its maximum value is unity.

There remains, finally, the comparison of $\lambda^2/2\pi$ with the total removal cross section.

Figure 5 shows this comparison for photon energies between 10 keV and 10 MeV in H, Na, Fe, Mo and Pb.

Those substances of highest atomic number exhibit the rapidly decreasing absorption characteristic of the photoeffect at low energies, and Mo and Pb have absorption edges in the range of the graph. Between 100 and 300 keV, however, the decreasing trend is reversed, because Compton scattering cross section decreases with energy less rapidly than the resonance cross section. The nearly constant slopes, within the range of the graph, for H and Na, are determined by the ratio of these two cross sections.

In the heavier elements, the contribution of pair production becomes appreciable below 10 MeV.

Nuclear interactions contribute somewhat to the total removal cross section but only to a small extent in comparison with pair production, and, as soon as the giant resonance energy is exceeded,

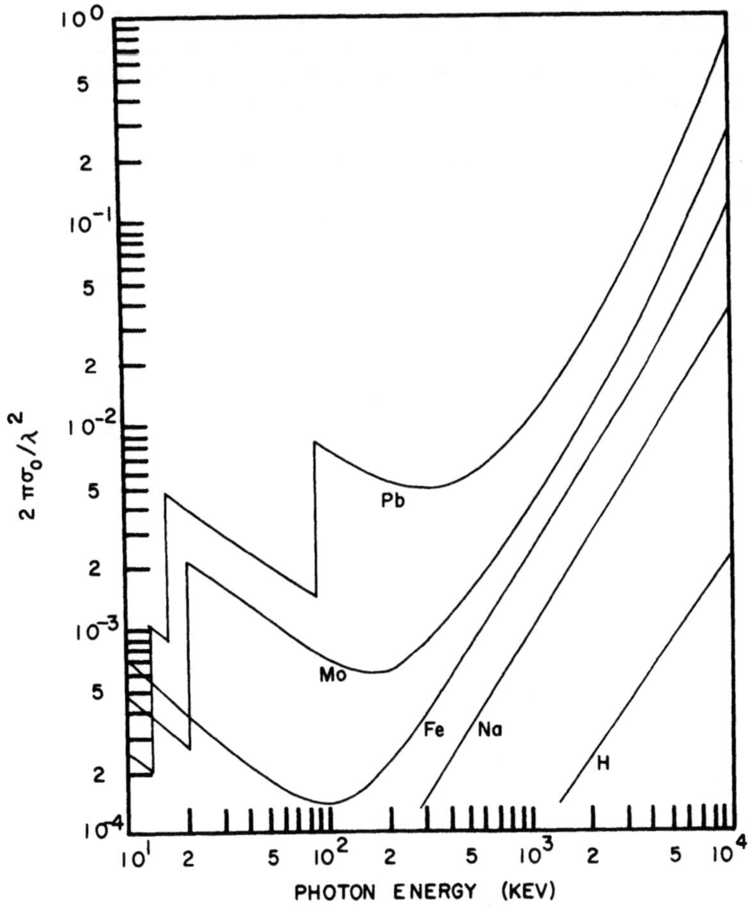

Fig. 5. Ratio of the total non-nuclear removal cross section of several substances to $\lambda^2/2\pi$ in the 10 keV to 10 MeV energy range.

Fig. 6. Energy level diagram illustrating the production of nuclear isomeric states by neutron capture in a parent isotope and their decay to a daughter isotope.

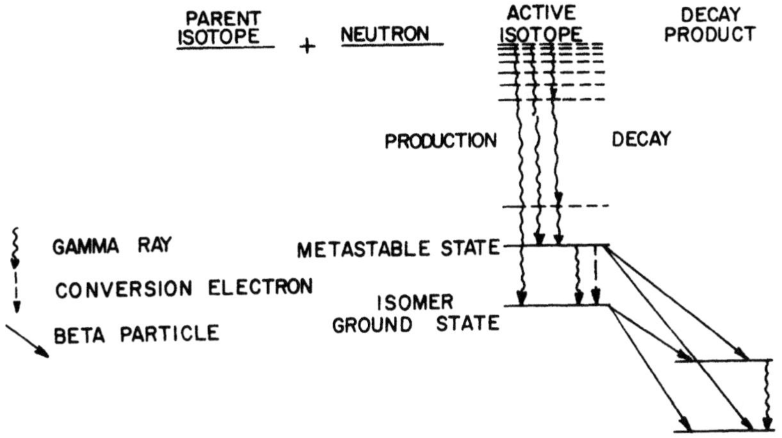

the nuclear interactions decrease in cross section. However, an important feature of pair production is that, unlike the nuclear interaction, there is no energy for which it begins to decrease. This feature of the process has been amply verified experimentally.

Thus, we must conclude that an absolute upper bound exists to the photon energies for which laser action can occur, and that depending on the laser medium, it lies in the range between 1 MeV and 100 MeV. At higher energies, a wave cannot grow as rapidly as it is depleted. This same restriction applies also to frequency conversion devices employing nonlinear interactions.

A practical limit to laser action will be found to lie well below this absolute ideal limit, of course. The practical man will at once wish to know, "How near this limit can we expect to approach?"

Even were all the special assumptions we have made justified, only a limited intensity could be generated before two-photon and higher-order processes of scattering and pair production would become appreciable. Similar considerations limit the maximum intensity in a semiconductor laser (23).

Two other, complementary effects, keep the practicable range well below the MeV region.

We have noted that short-lived states involve pumping powers which are so high that the integrity of the laser medium cannot be maintained.

Long-lived states, on the other hand, involve the difficulty associated with the factor $f\gamma_r/\Gamma$. Nevertheless, most studies of the gamma ray laser problem to date have dealt with long-lived nuclear isomers, formed by neutron capture, which can be radiochemically isolated.

Figure 6 illustrates the formation of nuclear isomers. Each parent nucleus which captures a thermal neutron becomes a compound nucleus with about 8 MeV of excitation, the neutron binding energy. Most of that energy is removed in times much less than 10^{-12} s by so-called capture gamma radiation. Recoil from the capture radiation is sufficient to rupture chemical bonds, so that the reaction products can be isolated. In favorable cases, inverted populations result, either during bombardment or subsequently as a result of beta decay of the lower level.

Were it possible to obtain all nuclei of the chemically concentrated isomer at rest in identical environments, the stimulation cross section would be (25)

$$\sigma_s = \frac{\lambda^2}{2\pi} \frac{\beta}{1+\alpha} \frac{T_1}{T_1+T_2} f$$

and many cases are known for which this exceeds the cross section for removal by a comfortable margin.

Unfortunately, there are a number of inhomogeneous perturbations of the Mossbauer line in real solids, which reduce the resonance cross section (19), so much so that initial optimism in the early sixties rapidly gave way to complete discouragement. These effects include:

1) spin-spin interactions among the nuclei;
2) gravitational red shift between nuclei at different gravitational potentials;
3) second-order doppler shifts, due either to unequal temperatures or to different lattice vibrational spectra in different regions of the medium;
4) chemical concentration shifts due to randomly distributed products of radioactive decay;
5) localized heating and lattice defects caused by beta decay.

Most of the recent proposals to revive the gamma ray laser have been addressed to the resolution of these difficulties.

Khokhlov (8,26) has proposed a method for eliminating the spin-spin broadening.

Letokhov (27) and Rivlin (28) have pointed out that the Mossbauer effect will be profoundly modified and can be enhanced by intense optical frequency fields.

Goldanskii and Kagan (29) have pointed out that the middle group of isomeric states, with lifetimes in the range of microseconds to perhaps tenths of seconds, has not been adequately studied, and have argued that lasing may prove possible for such isomers.

Letokhov (30) has proposed an extremely fast photochemical method for isolating nuclear isomers of even very short lifetime.

However, it is not within the scope of this paper to evaluate specific proposals of this nature. Our purpose has been to define the region within which laser action is possible in principle. Any part of this region presents a major challenge, but one which offers great rewards for success.

References

1. J. P. Gordon, H. J. Zeiger and C. H. Townes, Phys. Rev. 99, 1264 (1955).
2. A. L. Schawlow and C. H. Townes, Phys. Rec. 117, 1940 (1958).
3. T. H. Maiman, Nature 187, 493, 1960.
4. Handbook of Lasers, Chemical Rubber Co. Press, Cleveland, Ohio, 1971, Chapter 7.
5. R. Hodgson, Phys. Rev. Letters 25, 494, 1970.
6. R. W. Waynant, Proc. VII IQEC, Montreal, Paper D-4, May 1972.
7. J. D. Shipman, Jr., App. Phys. Letter 10, 3, 1967.
8. Physics Today, 26, 17, 1973.
9. B. Lax and A. H. Guenther, App. Phys. Letters 21, 361, 1972.
10. P. Malozzi, III Vavilov Conference on Nonlinear Optics, Novosibirsk, USSR, June 1973; A. Jaegle, ibid.
11. L. Rivlin, USSR Patent Application, January, 1961; W. Vali and V. Vali, Proc. IEEE 51, 182, 1963; G. C. Baldwin, J. P. Neissel and L. Tonks, ibid, 51, 1247, 1963; G. C. Baldwin, J. P. Neissel and L. Tonks, Trans. Am. Nuc. Soc. 6, 176, 1963.
12. R. V. Khokhlov, Proc. VII IQEC, Montreal, Paper A4.
13. G. C. Baldwin, III Vavilov Conference On Nonlinear Optics, Novosibirsk, USSR June 1973.
14. R. T. Schneider, this workshop.
15. Handbook of Lasers, Chem, Rubber Co. Press, Cleveland, Ohio 1971.
16. D. Marcuse, Proc. IEEE 51, 849, 1963.
17. G. C. Baldwin, J. P. Neissel and L. Tonks, loc. cit.
18. R. L. Mossbauer, Zeitschr. f. Phys., 151, 124, 1958.
19. H. Frauenfelder "The Mossbauer Effect", W. A. Benjamin, New York, 1962.
20. R. W. James, "The Optical Principles of the Diffraction of X-Rays", Bell, New York, 1962.
21. B. Batterman and H. Cole, Rev. Mod. Phys. 36, 861, 1964.
22. W. Heitler, "The Quantum Theory of Radiation" 3rd ed., Oxford, 1957.
23. A. M. Prokhorov, Science 149, 830 (1965).
24. G. C. Baldwin, J. P. Neissel and L. Tonks, loc. cit.
25. W. Vali and V. Vali, loc. cit.
26. R. V. Khokhlov, III Vavilov Conference on Nonlinear Optics, Novosibirsk, USSR June 1973.
27. V. S. Letokhov, ibid.
28. L. Rivlin, ibid.
29. V. Goldanskii, private communication; V. Goldanskii and Y. Kagan, Uspekhi Phys. Sci. 110, 445, 1973.

SUMMARY OF DISCUSSIONS AND COMMENTS

(VI. Laser Application to High Energy Physics)

In the discussion of this section it was pointed out that during this conference for the first time the application of laser physics to high energy physics and the physics of elementary particles was seriously considered. The first report on a probable measurement of electron-pair production by a laser beam focused into a plasma was presented (J. W. Shearer et al.). Though the lecturer hesitated to confirm the final statement of pair production, the measurement of X-rays from 1 to 10 MeV energy (behind 6 cm lead) could be considered a result. But as Shearer mentioned in the discussion, the measurements have to be extensively repeated. There exist some difficulties at present due to the complete reconstruction of the laser system. On the question of how to measure the generation of positrons, J. W. Shearer answered: "I have not done that calculation carefully enough to tell you how much counting statistics you would need. It's not easy to think of doing that kind of experiment unless you have a longer pulse laser because everything comes out all at once. It seems to me that a more fruitful experimental approach would be to use magnetic focusing to get the positrons away from the source. Then shield out most of the bremsstrahlung in order to insulate them in some other point of the apparatus which is well shielded and look for coincidence in that."

The nonlinear swelling of the actual laser intensities in the plasmas near cut-off density due to dielectric mechanisms and consequently the strong decrease of the thresholds for electron or proton pair production comprised the paper by H. Hora. Brueckner asked how such a field could be built up. The answer was that a transparent, but very special view can be given by the WKB approximation as used in the presentation and as is also well known in the similar case of the nonlinear forces. The longitudinal shrinking of the group velocity, the dilation of the effective wavelength and the simultaneous increase of the actual laser field strength need some time. For example, the interaction length in a plasma irradiated by a neodymium glass laser is at least 1 mm if the wavelength becomes $\eta^{-1} = 10^3$ times its vacuum value. This is done to get a 10^3 higher value for the oscillation energy of the electrons over its vacuum value and to decrease the threshold intensities by the same magnitude. The interaction time is then, at the very least, the time for light traveling or 10^{-3} times the speed of light along 1 mm, 3 nsec. This WKB-model is a measure of first approximation only. There exist numerical results which may lead to even better conditions (shorter times, higher swelling).

F. Winterberg indicated on Hora's paper that there may exist an analogy to the Clausius-Mosotti relation of dielectrics for the vacuum polarization or the conditions in a plasma.

H. Hora mentioned that 10^{-7} grams of anti-protons are produced in Serpukho every year under the aspect of the highest possible explosive or energy storage in anti-matter[1]. R. Kidder asked Hora what advantage proton pair production by lasers will have compared to the anti-proton production by accelerators. F. Winterberg specified the difference in density. Hora interpreted the equation of the pair production in vacuum, showing a square law of density in such a way that an accelerator would produce GeV-protons of a density $n_p \simeq 10^{10}$ cm^{-3} or less. If fired onto a target of solid state density n, the value $n_p n$ is decreased by 10^{12} in contrast to the laser causing the particles of solid state density to reach the necessary energies. In any beam of accelerated particles the particles diverge by electrostatic forces while a laser field interacts with the charges within a very dense space charge neutral plasma.

The problems in application of X-ray lasers for making single molecules visible by X-ray holography was also discussed. In connection with this any X-ray quantum being changed in phase and direction by a molecule has a certain probability of destroying the molecule before it can be used to produce the desired information. The impossibility of X-ray diagnostics in single molecules or in amorphous materials is well known[2]. The X-ray diagnostics of crystals is based on the symmetric repetition of events only where a certain number of defects (30%) can be taken into account.

In the discussion of K. A. Brueckner on the paper by B. Lax it was confirmed that the essential idea is the expansion and cooling of the plasma and then body recombination filling the upper levels. This is according to a scheme proposed by L. Wood.

References

1. Yu. D. Prokoshkin, Umschau (Frankfurt) 73, 153 (1973).

2. W. Hoppe, R. Mössbauer, Acta Cryst. A26, 414 (1970)

VII

APPENDIX AND INDEXES

LIST OF CONTRIBUTORS* AND ATTENDEES

Charles A. Andrade
Watervliet Arsenal
Watervliet, New York 12189

David S. Bailey
University of California
Lawrence Livermore Lab.
P. O. Box 808
Livermore, California 94550

Hector A. Baldis
INRS-Energie
University du Quebec
Varennes, C.P. 1020
Quebec, Canada

*George C. Baldwin
Nuclear Science Dept.
Rensselaer Polytechnic Institute
Troy, New York 12181

C. E. Bell
U.S. Naval Ordinances Lab.
White Oak
Silver Spring, Maryland 20910

Gerald Ben-David
Israel Atomic Energy
Plasma Physics Lab.
Sorey Research Center
Yavne, Israel

Robert E. Beverly, III
Battelle Mem. Inst.
505 King Avenue
Columbus, Ohio 43201

D. D. Bhawalkar
Scientific Officer 'E'
Bhabha Atomic Research Center
Electronics Division
Trombay, Bombay 400095 India

*Jean Louis Bobin
C. E. Limeil-BP 27
94 Villeneuve
St. Georges, France

A. F. Brayman, III
Student
Rensselaer Polytechnic Institute
Troy, New York 12181

Henry E. Breed
Physics Dept.
Rensselaer Polytechnic Institute
Troy, New York 12181

Terrence S. Brown
Atomic Energy Canada Ltd.
Chalk River Nuc. Lab.
Chalk River, Ontario, Canada

*K. A. Brueckner
KMS Fusion, Inc.
3941 Res. Park Drive
P. O. Box 1567
Ann Arbor, Michigan 48106

S. Burnett
Atomic Energy Commission
Washington, D.C.

*Wayne G. Burwell
Mgr., Kinetics & Gas Laser
United Aircraft Res. Lab.
400 Main Street
E. Hartford, Connecticut 06108

W. J. L. Buyers
Atomic Energy of Canada
Chalk River, Ontario, Canada

*Gregory H. Canavan
Chief, Physics Gp
AFWL/DYT
Kirtland Air Force Base
New Mexico 87117

Carl T. Case
Assistant Prof. Physics
Air Force Inst. of Tech.
AFIT/ENP
Wright-Patterson AFB
Ohio 45324

*Francis F. Chen
Boelter Hall 7731, UCLA
405 Hilgard Avenue
Los Angeles, California 90024

*Augustine Y. Cheung
Institute for Fluid Dyn. &
Applied Math.
Univ. of Maryland
College Park, Maryland 20742

Marvin G. Clark, Jr.
Director, Special Programs
Rensselaer Polytechnic Institute
Troy, New York 12181

Ralph S. Cooper
Assoc. L-Division Leader
Los Alamos Scientific Lab.
P. O. Box 1663
Los Alamos, New Mexico 87544

J. F. Cuderman
Sandia Laboratories
P. O. Box 5800
Albuquerque, New Mexico 87115

Michael I. Current
197 10th Street
Troy, New York 12181

David Dingee
Chief, Plasma Physics Sec.
Battelle Mem. Inst.
505 Ring Avenue
Columbus, Ohio 43201

*D. F. DuBois
Visiting Professor
University of Colorado
821 Gamow Tower
Dept. of Astrogeophysics
Boulder, Colorado 80302

Donald W. Drawdaugh
(Advisory Scientist Westinghouse
Research Lab)
5032 Highland Avenue
Bethel Park, Pennsylvania 15102

Walter Eppenstein
Physics Dept. 1C24 SC
Rensselaer Polytechnic Institute
Troy, New York 12181

Kent Estabrook
Physicist
Lawrence Radiation Lab.
Box 808
Livermore, California 94550

Albert Feldman
13822 Flint Rock Road
Rockville, Maryland 20853

Roland H. Fisher
Science Appl. Inc.
P. O. Box 2351
La Jolla, California 92037

Jerry H. M. Fu
Senior Scientist, EG & G
P. O. Box 809
Los Alamos, New Mexico 87544

Sergio Garriba
Associate Professor
Polytechnic Di Milano
CESNEF
Via G. Ponzio 34/3
Milano, Italy 20133

J. S. Gieger
Senior Research Officer
Atomic Energy of Canada
Chalk River Nuclear Lab.
Chalk River, Ontario, Canada

Damon Giovanielli
Staff Member
Los Alamos Scientific Lab.
P. O. Box 1663
Los Alamos, New Mexico 87544

*Alexander J. Glass
Head, Basic Studies Sec.
Lawrence Livermore Lab. L-215
Livermore, California 94550

*Robert P. Godwin
Los Alamos Scientific Lab.
P. O. Box 1663
Los Alamos, New Mexico 87544

Robert R. Goforth
Research Associate
University of Maryland
College Park, Maryland 20742

*Arthur H. Guenther
Scientific Director
AF Weapons Lab/DYD
Kirtland AFB
Albuquerque, New Mexico 87117

Mark P. Hacker
Rm. 4, 170 Albany Street
Cambridge, Massachusetts 02139

G. M. Halpern
University of Rochester
Laboratory for Laser Energetics
Rochester, New York 14627

Todd A. Halpern
Physicist, U.S. Dept. of Defense
DIADT - IC
Washington, D.C. 20301

Paul Harteck
Chemistry Dept.
Rensselaer Polytechnic Institute
Troy, New York 12181

E. C. Hathaway
Dept. of Physics
Kansas State University
Manhattan, Kansas 66506

Robert L. Hickok
Electronic Engineering Dept.
Rensselaer Polytechnic Institute
Troy, New York 12181

*Heinrich Hora
Max-Planck-Institut fur
Plasmaphysik
Garching, Germany
Rensselaer Polytechnic Institute
Hartford Graduate Center
275 Windsor Street
Hartford, Connecticut 06120

*John L. Hughes
Dept. of Physical Sciences
The Australian National U.
Box 4
Canberra, Australia ACT2601

*S. Kaliski, Prof dr. hab.
01-489 Warszawa, Poland

*Ray E. Kidder
Workshop Advisor
Lawrence Livermore Lab.
P. O. Box 808
Livermore, California 94550

Lawrence A. Klein
New York University
Dept. of Elec. Engr. &
Computer Science
University Heights
Bronx, New York 10453

*W. L. Kruer
University of California
Lawrence Livermore Lab.
Livermore, California 94550

Craig M. Kullberg
(RPI Student)
638 Page Street
Mankato, Minnesota 56001

B. Kuyel
2536 Overlook Road, Apt. 404
Cleveland Heights, Ohio 44106

*Benjamin Lax
Director, Francis Bitter Nat'l Lab, MIT
Albany Street
Cambridge, Massachusetts 02139

*M. J. Lubin
Mech. & Aerospace Science Dept.
University of Rochester
Rochester, New York 14627

Roy C. McCord
Research Associate
RPI of Connecticut
275 Windsor Street
Hartford, Connecticut 06120

Olin K. McDaniel, III
1327 Pinckney Avenue
Florence, South Carolina 29501

R. W. MacPherson
DREV/CROV Defense Re. Board
Box 880
Courcelette, Quebec, Canada

Edward C. Malarkey
6904 40th Avenue
Hyattsville, Maryland 20782

Robert Malone
Graduate Research Assistant
Cornell University, Physics Dept.
Ithaca, New York 14850

*P. L. Mascheroni
Research Scientist Assoc.
Statistical Mechanics
University of Texas
Physics Dept.
Austin, Texas 78712

B. P. Mathur
Inst. of Gas Tech.
3424 S. State St., IIT Campus
Chicago, Illinois 60616

Farres P. Mattar
Polytechnic Inst. of Brooklyn
333 Jay Street
Brooklyn, New York 11201

D. H. Maylotte
General Electric CR & D
1 River Road
Schenectady, New York 12345

Heinrich A. Medicus
Physics Dept.
Rensselaer Polytechnic Institute
Troy, New York 12181

Stephen A. Metz
Metallurgist
U.S. Naval Research Lab.
Overlook Avenue S.W.
Washington, D.C. 20375

*R. L. Morse
University of California
Los Alamos Scientific Lab.
P. O. Box 1663
Los Alamos, New Mexico 87544

*John Nuckolls
Assoc. Div. Lab., "A" Div.
Lawrence Livermore Lab.
P. O. Box 808
Livermore, California 94550

Kanwal J. Parbhakar
I.N.R.S.-Energie
C.P. 1020
Varennes, Quebec, Canada

*G. J. Pert
Dept. of Applied Physics
The University of Hull
Hull, HU6 7RX, England

J. F. Reintjes
Research Physicist
Naval Research Lab.
Washington, D.C.

*M. C. Richardson
National Research Council
of Canada
Ontario, Canada

David Salzman
Soreq Nuclear Research Center
Yavneh, Israel

*Richard T. Schneider
University of Florida
Dept. Nucl. Eng. Science
Gainesville, Florida 32601

*Helmut Schwarz
Rensselaer Polytechnic Institute
275 Windsor Street
Hartford, Connecticut 06120

*F. Schwirtzke
Naval Postgrad. School
Monterey, California 93940

*John F. Seely
University of Tennessee
Box 8646
Knoxville, Tennessee 37916

*J. W. Shearer
Lawrence Livermore Lab.
University of California
P. O. Box 808
Livermore, California 94550

*Richard Sigel
Max-Planck-Institute fur
Plasmaphysik
D-8046
Garching, West Germany

*Ian J. Spalding
United Kingdom Atomic Energy
Authority
Abingdon, Berkshire, England

*John Stamper
Research Physicist
Naval Research Lab.
Washington, D.C. 20375

*Edward Teller
Lawrence Livermore Lab.
University of California
P. O. Box 808
Livermore, California 94550

J. J. Thomson
Physicist
Lawrence Radiation Lab.
Box 808
Livermore, California 94550

Alan L. Throop
MIT, 20B-124
Cambridge, Massachusetts 02139

David B. van Hulsteyn
Dept. Elec. Eng. ENS 124
University of Texas
Austin, Texas 78712

Craig T. Walters
505 King Street
Columbus, Ohio 43210

Irvin M. Winer
International Laser System
3404 N. Orange Blossom Trail
Orlando, Florida 32804

*F. Winterberg
University of Nevada
Desert Research Institute
Reno, Nevada 89507

*E. A. Witalis
Research Institute of the
Swedish National Defense
S-10450 Stockholm, Sweden

*Chiyoe Yamanaka
Osaka University
Suite Osaka 565
Japan

Zeev Zinamon
The Weizmann Institute
Dept. of Physics
Rehouvot, Israel

---

1. H. J. Schwarz
2. H. Hora
3. G. C. Baldwin
4. J. S. Geiger
5. B. Kuyel
6. F. Winterberg
7. R. H. Fisher
8. C. M. Kullberg
9. E. Teller
10. M. P. Hacker
11. A. L. Throop
12. K. Estabrook
13. J. Stamper
14. J. F. Seely
15. J. H. M. Fu
16. M. Clark
17. J. L. Bobin
18. C. Yamanaka
19. T. A. Halpern
20. A. Y. Cheung
21. H. A. Baldis
22. S. Garriba
23. A. Feldman
24. J. J. Thomson
25. D. F. DuBois
26. M. C. Richardson
27. P. L. Mascheroni
28. R. McCord
29. C. T. Walters
30. T. S. Brown
31. D. Dingee
32. O. K. McDaniel, III
33. G. H. Canavan
34. R. P. Godwin
35. A. F. Brayman, III
36. C. E. Bell
37. D. S. Bailey
38. M. J. Lubin
39. R. L. Hickok
40. L. A. Klein
41. R. R. Goforth
42. J. Halpern
43. A. J. Glass
44. A. H. Guenther
45. D. Salzman
46. J. F. Cuderman
47. C. A. Andrade
48. D. W. Drawbaugh
49. E. A. Witalis
50. R. T. Schneider
51. Z. Zinamon
52. R. Malone
53. D. Giovanielli
54. R. S. Cooper
55. K. A. Brueckner
56. R. Sigel
57. E. C. Malarkey
58. I. J. Spalding
59. K. J. Parbhakar
60. F. P. Mattar
61. B. P. Mathur
62. J. F. Reintjes
63. C. T. Case
64. C. E. Hathaway
65. J. L. Hughes
66. M. I. Current
67. J. W. Shearer
68. G. J. Pert
69. D. B. van Hulsteyn
70. S. Kaliski
71. G. Ben-David
72. S. A. Metz
73. S. Burnett
74. R. E. Kidder
75. J. Albritten
76. R. E. Beverly
77. D. H. Maylotte

Those present at the workshop but not shown on photograph:

1. D. D. Bhawalkar
2. H. E. Breed
3. W. G. Burwell
4. W. J. L. Buyers
5. F. Chen
6. W. Eppenstein
7. P. Harteck
8. W. L. Kruer
9. B. Lax
10. R. M. Lichtenstein
11. R. W. MacPherson
12. H. A. Medicus
13. R. L. Morse
14. J. Nuckolls
15. F. Schwirtzke
16. I. M. Winer

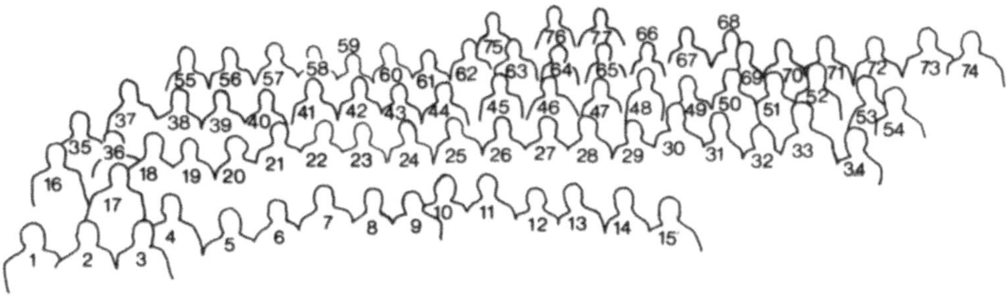
1 2 3 4 5 6 7 8 9 10 11 12 13 14 15
16 17 18 19 20 21 22 23 24 25 26 27 28 29 30 31 32 33 34
35 36 37 38 39 40 41 42 43 44 45 46 47 48 49 50 51 52 53 54
55 56 57 58 59 60 61 62 63 64 65 66 67 68 69 70 71 72 73 74
75 76 77

AUTHOR INDEX

PAGES 1 TO 396 WILL BE FOUND IN VOLUME 3A AND PAGES 397 TO 899 IN VOLUME 3B.

Numbers in parentheses following the text page numbers are reference numbers, and are included to assist in locating a reference at the end of each contribution when the author's name is not cited at the point of reference in the text.

A

I

L

M

N

O

Q

R

SUBJECT INDEX

PAGES 1 TO 396 WILL BE FOUND IN VOLUME 3A AND PAGES 397 TO 899 IN VOLUME 3B.

Volume ionization 103

GPSR Compliance

The European Union's (EU) General Product Safety Regulation (GPSR) is a set of rules that requires consumer products to be safe and our obligations to ensure this.

If you have any concerns about our products, you can contact us on ProductSafety@springernature.com

In case Publisher is established outside the EU, the EU authorized representative is:

Springer Nature Customer Service Center GmbH
Europaplatz 3
69115 Heidelberg, Germany

Batch number: 10684222

Printed by Printforce, the Netherlands